# Nanopesticides

Leonardo F. Fraceto
Vera Lucia S. S. de Castro
Renato Grillo • Daiana Ávila
Halley Caixeta Oliveira • Renata Lima
Editors

# Nanopesticides

## From Research and Development to Mechanisms of Action and Sustainable Use in Agriculture

Springer

*Editors*
Leonardo F. Fraceto
Institute of Science and Technology
of Sorocaba
São Paulo State University
Sorocaba, São Paulo, Brazil

Renato Grillo
Department of Physics and Chemistry
São Paulo State University
Ilha Solteira, São Paulo, Brazil

Halley Caixeta Oliveira
Department of Animal and Plant Biology
State University of Londrina
Londrina, Paraná, Brazil

Vera Lucia S. S. de Castro
Embrapa Environment
Jaguariuna, São Paulo, Brazil

Daiana Ávila
Laboratory of Biochemistry and Toxicology
in Caenorhabditis Elegans
Graduation Program in Biochemistry
Federal University of Pampa
Campus Uruguaiana
Uruguaiana, Brazil

Renata Lima
Labiton
Sorocaba University
Sorocaba, Brazil

ISBN 978-3-030-44872-1 ISBN 978-3-030-44873-8 (eBook)
https://doi.org/10.1007/978-3-030-44873-8



This Springer imprint is published by the registered company Springer Nature Switzerland AG
The registered company address is: Gewerbestrasse 11, 6330 Cham, Switzerland

*This book is dedicated to those who made remarkable scientific discoveries and imagined their applications that ameliorate our lives and promote environmental sustainability*

# Foreword

One of the growing interest areas of nanotechnology applications is the development of novel formulations of pesticides that are more efficient, targeted, and release controlled. Such features may not only bring about cost savings but may also lower the overall use of pesticides and thus contribute towards reducing the negative impacts on human health and the environment. However, despite the novelty of the approach, only a few research groups have so far ventured into this area of R&D due to the current major gaps in knowledge. This book will, therefore, be useful in enhancing the understanding of fundamental concepts, underlying mechanisms, and state-of-the-art regarding nanopesticides.

The book is comprised of 11 chapters that are written by leading experts in their respective fields. It takes an account of the cutting-edge of the technology, gives pointers to the future directions of R&D, and identifies the areas where further research is needed. In describing the new technology, the authors have taken a balanced view by highlighting both the likely benefits and the potential risks of formulating pesticides at the nano-scale. In particular, the nano-scale formulation of natural pesticidal compounds, together with biodegradable polymers, can open up numerous new possibilities for the development of safer products for the control of a variety of pests of agricultural and public health importance. The Experts' perspectives on the novelty, the future prospects and trends of the technology, and regulatory aspects will be equally informative for the developers, the regulators, and the users of nanopesticides.

In this context, Chap. 1 has looked into the use of natural degradable polyhydroxyalkanoates (PHAs) for the development of embedded extended-release formulations of herbicides and fungicides. It also provides example applications in laboratory soil ecosystems in wheat plant communities infected with *Fusarium sp.* and weeds.

Chapter 2 provides another interesting example of the development of smart formulations based on biodegradable and eco-friendly nano-chitosan and their application in fungal disease control.

The use of RNA interference (RNAi) is an interesting field of biotechnology that has also been studied for potential applications in pest control. However, such applications generally suffer from limitations in the delivery of dsRNAs to diffused and dispersed pest populations in the field. This is where the use of nano-carriers has been studied as a delivery vehicle for RNAi-based pesticides for the control of agricultural pests. Chapter 3 presents an overview of the literature on this topic, and discusses biosafety considerations in relation to the use of formulations based on nano-carrier containing RNAi.

Chapter 4 is focused on discussing the interaction of nanopesticides with plants. Such an understanding is fundamentally important to drive effective, safe, and sustainable application of nanopesticides in agriculture. Whilst the discussion centres around the conceptual basis, it also provides different examples of the uptake, mode of action, and effects of nanopesticides in the context of physiological and metabolic responses of plants exposed to nanopesticides. It also demonstrates prediction models that can provide a hint on the likely response of the plants to a specific type of nanopesticide.

Chapter 5 discusses the methods that can be used to study the behaviour and fate of nanopesticides in aquatic and terrestrial environments. It discusses the advantages and disadvantages of individual methods and highlights the important considerations that are needed due to the nano-scale characteristics of nanopesticides when using conventional environmental risk assessment methods.

Chapter 6 continues the theme of Chap. 4 to further discuss the interactions of nano-enabled agrochemicals with soil microbiome that plays a vital role in maintaining the soil quality as well as plant nutrition and health. Using examples of formulations based on nano forms of copper and silver, as well as nanocarriers of conventional pesticides, the chapter discusses the current state of knowledge in regard to the potential prospects and implications of nanopesticides for the soil microbiota.

Chapter 7 discusses bioactivity and environmental impacts of nano-formulated insecticides. Whilst the comprehensive overview presented in this chapter includes nano-formulation of synthetic pesticides, a particular focus is also on formulations of natural insecticidal compounds, the use of which can be expected to further minimize the environmental impacts. A similar theme is discussed in Chap. 8 in relation to the environmental toxicity of nanopesticides against non-target organisms. The comprehensive state-of-the-art overview concerns environmental safety of nanopesticides against non-target model organisms (microbes, plants, worms, insects, algae, daphnids, and fish). It also touches upon the various methods for characterization for the study of interactions of nanopesticides with biological and environmental systems, the use of nano-informatics, safety-by-design, environmental risk analysis and management for responsible development and regulation of nanopesticides.

Chapter 9 provides an overview of the aspects relating to environmental safety and regulation. Using a case study of nano-atrazine, the chapter discusses the current limited knowledge in relation to the behaviour and fate, and the potential adverse environmental impacts of nanopesticide formulations. It not only takes a note of the new advancements but also highlights the main challenges in regard to risk analysis of nanopesticides. This theme is further elaborated in Chap. 10 that discusses risk assessment of nanofertilizers and nanopesticides. The review shows that environmental and human health impacts of the nano-agrochemicals are of general concern. It highlights the scarcity of the relevant toxicological data to allow adequate risk assessment. The impact of the such knowledge gaps is considered a barrier to the development of regulatory policies, and, as a consequence, an obstacle to new marketable products.

Finally, Chap. 11 provides a market analysis of nanopesticides at different stages from R&D to the market. The market scenario depicts a continuous investment in the technology and innovation to develop more effective products, in a framework of mergers, ventures, and partnerships to accelerate the development and launch of the products. The analysis indicates that the development of nano-encapsulated pesticide formulations is currently an open field that can enable the development of new original materials and formulations. The overview identifies the current status and trends in the market, and provides a summary of the forthcoming technologies. It discusses the key challenges in the scale-up, and identifies encapsulation of microorganisms and dsRNA as new and emerging disruptive technologies.

In summary, the book provides an up-to-date account of the cutting-edge technology for the development of nanopesticides, its pros and cons, and potential applications in agriculture. It provides a balanced view of the innovations in this field in consideration of both benefits and risks. The book is highly commended to all those involved in R&D and safety/regulation of pesticides in an academic, research, industrial, or regulatory setting.

University of Chester
Chester, UK

Qasim Chaudhry

# Preface

Nanomaterials have been contributing to agricultural science and technology in various phases of production and commercialization. Especially, nanopesticides can improve crop yields and are believed to reduce harmful effects on the environment. Their benefits may include better stability, permeability, and dispersion of the active ingredient, improved targeting to pest species, higher efficacy, decreased application doses, prevention of premature degradation, and increased environmental safety. Despite their promising use, it is necessary to study their accumulation in the environment and their impact on non-target organisms and consequently on biodiversity and human health. Nowadays, there is a lack of worldwide data on nanopesticide efficacy compared to conventional ones and on their environmental effects. Considering these facts, we discuss in this book some recent features of nanopesticide development, application, and toxicity evaluation. The book is organized into 11 chapters. Chapters 1–3 describe the use of different carriers for the controlled release of active ingredients aiming at agriculture applications. Chapters 4–6 describe some methods used to understand the fate and behaviour of nanopesticides in plants, soil, and water. Chapters 3 and 6 discuss their potential toxicity and impacts on the environment. Chapters 7 and 8 showed the potential toxicity of nanopesticides and their impacts on environment. Chapters 9 and 10 provide an overview of environmental safety aspects and regulatory issues regarding nanopesticides. Finally, Chapter 11 discusses the commercial aspects of nanopesticides in crop production.

In this context, with this book, we intended to contribute to a broader perspective of nanopesticide characteristics and risk assessment, regulation, application, and marketing.

Sorocaba, Brazil — Leonardo F. Fraceto
Jaguariuna, Brazil — Vera Lucia S. S. de Castro
Ilha Solteira, Brazil — Renato Grillo
Uruguaiana, Brazil — Daiana Ávila
Londrina, Brazil — Halley Caixeta Oliveira
Sorocaba, Brazil — Renata Lima

# Acknowledgements

The editors would like to heartily acknowledge all authors for the valuable sharing of their findings, knowledge, and opinions in this book. Also, the editors would like to thank Springer for the opportunity to publish this book. In addition, we would like to thank Brazilian fellow agencies: São Paulo Research Foundation (FAPESP #2017/21004-5), CAPES, and CNPq (#306583/2017-8). We are also grateful to Prof. Qasim Chaudhry for his kind words in the foreword.

# Contents

# Contributors

**Alok Adholeya** TERI-Deakin Nanobiotechnology Centre, The Energy and Resources Institute (TERI), Gurugram, Haryana, India

**Danielle Araújo Agarrayua** Laboratory of Biochemistry and Toxicology in Caenorhabditis elegans, Graduation Program in Biochemistry, Federal University of Pampa Campus Uruguaiana, Uruguaiana, Brazil

**Gilberto J. Arruda** Plant Resources Study Group, Center for Natural Resources Study, Mato Grosso do Sul State University, Dourados, Brazil

**Daiana Ávila** Laboratory of Biochemistry and Toxicology in Caenorhabditis Elegans, Graduation Program in Biochemistry, Federal University of Pampa Campus Uruguaiana, Uruguaiana, Brazil

**Deepa Bhagat** ICAR-National Bureau of Agricultural Insect Resources (NBAIR), Bangalore, India

**Leandro S. Bortolozzo** Brazilian Nanotechnology National Laboratory (LNNano), Brazilian Center for Research in Energy and Materials (CNPEM), Campinas, São Paulo, Brazil

School of Technology, University of Campinas (UNICAMP), Limeira, São Paulo, Brazil

**Anderson R. L. Caires** Optics and Photonics Group, Institute of Physics, Federal University of Mato Grosso do Sul, Campo Grande, Brazil

School of Life Science, University of Essex, Colchester, UK

**Poliana Cardoso-Gustavson** Oxiteno S.A., Research and Development Crop Solutions, São Paulo, Brazil

**Olivier Christiaens** Department of Plants and Crops, Faculty of Bioscience Engineering, Ghent University, Ghent, Belgium

**Francine Côa** Brazilian Nanotechnology National Laboratory (LNNano), Brazilian Center for Research in Energy and Materials (CNPEM), Campinas, São Paulo, Brazil

Center for Nuclear Energy in Agriculture (CENA), University of São Paulo (USP), Piracicaba, São Paulo, Brazil

**Ian Colbeck** School of Life Science, University of Essex, Colchester, UK

**Rubens S. Costa** Brazilian Nanotechnology National Laboratory (LNNano), Brazilian Center for Research in Energy and Materials (CNPEM), Campinas, São Paulo, Brazil

Interdisciplinary Laboratory of Advanced Materials, Federal University of Piauí (UFPI), Teresina, Piauí, Brazil

Federal Institute of Education, Science and Technology of Piauí (IFPI), Oeiras, Piauí, Brazil

**Gabriela H. Da Silva** Brazilian Nanotechnology National Laboratory (LNNano), Brazilian Center for Research in Energy and Materials (CNPEM), Campinas, São Paulo, Brazil

**Felícia Pereira de Albuquerque** Institute of Science and Technology of Sorocaba, São Paulo State University (UNESP), Sorocaba, Brazil

**Vera Lucia S. S. de Castro** Embrapa Environment, Jaguariúna, Brazil

**Fabrício S. Delite** Brazilian Nanotechnology National Laboratory (LNNano), Brazilian Center for Research in Energy and Materials (CNPEM), Campinas, São Paulo, Brazil

**Aline M. Z. de Medeiros** Brazilian Nanotechnology National Laboratory (LNNano), Brazilian Center for Research in Energy and Materials (CNPEM), Campinas, São Paulo, Brazil

Center for Nuclear Energy in Agriculture (CENA), University of São Paulo (USP), Piracicaba, São Paulo, Brazil

**Flávia Suelen de Oliveira Pereira** Laboratory of Biochemistry and Toxicology in Caenorhabditis elegans, Graduation Program in Biochemistry, Federal University of Pampa Campus Uruguaiana, Uruguaiana, Brazil

**Khaidem Aruna Devi** Nano Research Facility Laboratory, Department of Molecular Biology and Biotechnology, Rajasthan College of Agriculture, Maharana Pratap University of Agriculture and Technology, Udaipur, Rajasthan, India

**Peter Fedor** Department of Environmental Ecology, Faculty of Natural Sciences, Comenius University, Bratislava, Slovakia

**Leonardo F. Fraceto** Institute of Science and Technology of Sorocaba, São Paulo State University (UNESP), Sorocaba, Brazil

**Renato Grillo** Department of Physics and Chemistry, São Paulo State University, Ilha Solteira, Brazil

**Josef Jampílek** Department of Analytical Chemistry, Faculty of Natural Sciences, Comenius University, Bratislava, Slovakia

**Marta Simão Kfouri** Oxiteno S.A., Research and Development Crop Solutions, São Paulo, Brazil

**Latif U. Khan** Brazilian Nanotechnology National Laboratory (LNNano), Brazilian Center for Research in Energy and Materials (CNPEM), Campinas, São Paulo, Brazil

**E. G. Kiselev** Siberian Federal University, Krasnoyarsk, Russia

Institute of Biophysics SB RAS, Federal Research Center "Krasnoyarsk Science Center SB RAS", Krasnoyarsk, Russia

**Katarína Kráľová** Institute of Chemistry, Faculty of Natural Sciences, Comenius University, Bratislava, Slovakia

**Ashok Kumar** Nano Research Facility Laboratory, Department of Molecular Biology and Biotechnology, Rajasthan College of Agriculture, Maharana Pratap University of Agriculture and Technology, Udaipur, Rajasthan, India

**Rocío López-Cabeza** RECETOX, Faculty of Science, Masaryk University, Brno, Czech Republic

**Ronald Maldonado-Rodriguez** The Fluoromatics Lab. 10, Geneva, Switzerland

**Diego Stéfani T. Martinez** Brazilian Nanotechnology National Laboratory (LNNano), Brazilian Center for Research in Energy and Materials (CNPEM), Campinas, São Paulo, Brazil

Center of Nuclear Energy in Agriculture (CENA), University of São Paulo (USP), Piracicaba, São Paulo, Brazil

School of Technology, University of Campinas (UNICAMP), Limeira, São Paulo, Brazil

**Carlos H. Z. Martins** Laboratory of Solid State Chemistry (LQES), Institute of Chemistry, University of Campinas (UNICAMP), Campinas, São Paulo, Brazil

**Bruna Mendonça** Oxiteno S.A., Research and Development Crop Solutions, São Paulo, Brazil

**N. G. Menzyanova** Siberian Federal University, Krasnoyarsk, Russia

**Halley Caixeta Oliveira** Department of Animal and Plant Biology, State University of Londrina, Londrina, Brazil

**Ajay Pal** Department of Biochemistry, College of Basic Sciences and Humanities, Chaudhary Charan Singh Haryana Agricultural University, Hisar, Haryana, India

**Marko Petek** Department of Biotechnology and Systems Biology, National Institute of Biology, Ljubljana, Slovenia

**Romana Petry** Brazilian Nanotechnology National Laboratory (LNNano), Brazilian Center for Research in Energy and Materials (CNPEM), Campinas, São Paulo, Brazil

Center for Natural and Human Sciences, Federal University of ABC (UFABC), Santo André, São Paulo, Brazil

**Rafael Pires-Oliveira** Oxiteno S.A., Research and Development Crop Solutions, São Paulo, Brazil

**Barbara Plancot** Aix-Marseille University, Laboratory of Microbial Ecology of the Rhizosphere, Bioscience and Biotechnologies Institute of Aix-Marseille, St Paul lez Durance, France

**Montcharles S. Pontes** Plant Resources Study Group, Center for Natural Resources Study, Mato Grosso do Sul State University, Dourados, Brazil

**Damyanti Prajapati** Nano Research Facility Laboratory, Department of Molecular Biology and Biotechnology, Rajasthan College of Agriculture, Maharana Pratap University of Agriculture and Technology, Udaipur, Rajasthan, India

**Ana Cristina Preisler** Department of Animal and Plant Biology, State University of Londrina, Londrina, Brazil

**S. V. Prudnikova** Siberian Federal University, Krasnoyarsk, Russia

**Caroline Brandão Quines** Laboratory of Biochemistry and Toxicology in Caenorhabditis elegans, Graduation Program in Biochemistry, Federal University of Pampa Campus Uruguaiana, Uruguaiana, Brazil

**Carla M. S. Sabino** Brazilian Nanotechnology National Laboratory (LNNano), Brazilian Center for Research in Energy and Materials (CNPEM), Campinas, São Paulo, Brazil

Laboratory of Advanced Functional Materials (LaMFA), Department of Physics, Federal University of Ceará (UFC), Fortaleza, Ceará, Brazil

**Vinod Saharan** Nano Research Facility Laboratory, Department of Molecular Biology and Biotechnology, Rajasthan College of Agriculture, Maharana Pratap University of Agriculture and Technology, Udaipur, Rajasthan, India

**Catherine Santaella** Aix-Marseille University, Laboratory of Microbial Ecology of the Rhizosphere, Bioscience and Biotechnologies Institute of Aix-Marseille, St Paul lez Durance, France

**Etenaldo F. Santiago** Plant Resources Study Group, Center for Natural Resources Study, Mato Grosso do Sul State University, Dourados, Brazil

**E. I. Shishatskaaya** Siberian Federal University, Krasnoyarsk, Russia

Institute of Biophysics SB RAS, Federal Research Center "Krasnoyarsk Science Center SB RAS", Krasnoyarsk, Russia

**Braj Raj Singh** TERI-Deakin Nanobiotechnology Centre, The Energy and Resources Institute (TERI), Gurugram, Haryana, India

**Guy Smagghe** Department of Plants and Crops, Faculty of Bioscience Engineering, Ghent University, Ghent, Belgium

**Clauvis Nji Tizi Taning** Department of Plants and Crops, Faculty of Bioscience Engineering, Ghent University, Ghent, Belgium

**S. Thomas** International and Interuniversity Centre for Nano Science and Nano Technology, Mahatma Gandhi University, Kottayam, Kerala, India

**T. G. Volova** Siberian Federal University, Krasnoyarsk, Russia

Institute of Biophysics SB RAS, Federal Research Center "Krasnoyarsk Science Center SB RAS", Krasnoyarsk, Russia

**N. O. Zhila** Siberian Federal University, Krasnoyarsk, Russia

Institute of Biophysics SB RAS, Federal Research Center "Krasnoyarsk Science Center SB RAS", Krasnoyarsk, Russia

# Natural Degradable Polyhydroxyalkanoates as the Basis for Creation of Prolonged and Targeted Pesticides to Protect Cultivated Plants from Weeds and Pathogens

**T. G. Volova, S. V. Prudnikova, N. O. Zhila, N. G. Menzyanova, E. G. Kiselev, E. I. Shishatskaaya, and S. Thomas**

**Abstract** The present chapter is a synthesis of the authors' data on the design and use of extended-release formulations of herbicides and fungicides embedded in a degradable matrix of polyhydroxyalkanoates (PHA). The structure and physicochemical properties of the experimental formulations and the kinetics of their degradation in soil and pesticide release have been reported. The favorable effects of the application of the experimental pesticide formulations in laboratory soil ecosystems in wheat plant communities infected with *Fusarium* plant pathogen and weeds have been described.

**Keywords** Slow-release formulations · Poly-3-hydroxybutyrate · Antifungal activity · Herbicidal effect · Controlled release · Physiological effects

## 1 Introduction

Increased accumulation of toxic and unrecyclable waste products caused by uncontrolled use of chemicals is one of the main global environmental problems. A way to meet this challenge is to expand the use of tools and methods of biotechnology, which may help to protect beneficial biota and enhance productivity in agriculture

T. G. Volova (✉) · N. O. Zhila · E. G. Kiselev · E. I. Shishatskaaya
Siberian Federal University, Krasnoyarsk, Russia

Institute of Biophysics SB RAS, Federal Research Center "Krasnoyarsk Science Center SB RAS", Krasnoyarsk, Russia

S. V. Prudnikova · N. G. Menzyanova
Siberian Federal University, Krasnoyarsk, Russia

S. Thomas
International and Interuniversity Centre for Nano Science and Nano Technology, Mahatma Gandhi University, Kottayam, Kerala, India

 
L. F. Fraceto et al. (eds.), *Nanopesticides*,
https://doi.org/10.1007/978-3-030-44873-8_1

as well as reduce toxic impacts of chemicals on agroecosystems and the whole biosphere (Gavrilescu et al. 2015). Intensive farming involves the use of enormous amounts of various chemicals to control weeds, pests, and pathogens of crops. However, most of these substances are accumulated in biological objects, contaminate soil and water environments, harm living organisms, and upset the balance in natural ecosystems (Carvalho 2017).

One of the new directions aimed at reducing the risk of uncontrolled spread and accumulation of pesticides in the environment is the development and use of pesticides with a controlled release of the active substance embedded in a biodegradable matrix or covered by biodegradable coating. Nanotechnology is currently an important tool for increasing agricultural productivity. Nanotechnology-based systems perform an active compound sustained release, keeping between minimal concentration and maximum safe concentration. The nanotechnology-based systems reduce the amount of active compound required for biological response, also reducing environmental contamination risks, energy consumption, and labor costs (Oliveira et al. 2018, 2019).

An important component of the creation of such formulations is the availability of suitable materials with the following properties (Yusoff et al. 2016; Sarkar et al. 2018):

- ability to fit into the environment and global biosphere cycles, i.e. degradability;
- safety for living organisms and the environment;
- prolonged (weeks and months) presence in the environment and controlled degradation, during which non-toxic products are formed;
- chemical compatibility with fertilizers and pesticides;
- processability by generally accessible methods, which are also compatible with technologies for the production of fertilizers and pesticides.

Encapsulation of pesticides is a relatively new approach, although the first papers were published in the 1990s (Greene et al. 1992). Interest in such research is increasing every year. The authors of those studies noted the following advantages of using pesticide controlled-delivery systems:

- prolonged action due to continuous release of pesticides at a level sufficient for effective function over a long period;
- fewer treatments due to prolonged action after a single application;
- shorter time needed to apply such pesticides;
- lower contamination of the environment;
- longer activity of pesticides unstable in the aqueous medium;
- conversion of the liquid pesticide into the solid formulation, which simplifies shipping and decreases flammability of the formulation;
- lower toxicity to biota due to the reduction of pesticide mobility in soil and, therefore, lower accumulation in the food chain.

The key ingredient for the construction of slow-release formulations is the availability of the appropriate biodegradable carrier. Thus, it is important to find and investigate materials with the necessary properties. The materials extensively studied as matrices for embedding agrochemicals are synthetic nondegradable polymers (polystyrene, polyacrylamide, polyethylene acrylate, polyamide, polyurethane, polycyanoacrylate) (Sarkar et al. 2018). One of the new research areas is the use of new pesticides formulations with controlled and targeted release of pesticide encapsulated in biodegradable polymer matrix (Grillo et al. 2014; Roy et al. 2014). In the environment, the polymer matrix undergoes degradation by soil microorganisms and there is gradual pesticide release into the environment (Ong and Sudesh 2016). The use of such products will help to reduce the amount of pesticides used and ensure the controlled delivery of pesticides during the whole growing season of the plant, preventing sharp releases into the environment that occur when plants are treated with free pesticides. These formulations can only be constructed if materials with the following properties are readily available.

Achievements in science and technology determine a wider use of products synthesized in biotechnological processes. Production of environmentally friendly materials possessing new useful properties remains one of the main problems today. The diversity of polymers with widely varying stereo-configuration and molecular weight and the feasibility of producing various composites with different materials create the basis for obtaining a wide range of new materials with valuable properties. Recently, there has been growing interest in biopolymers (polymers of biological origin). There are two major kinds of biopolymers: polymers synthesized by biological systems (microorganisms) and chemically synthesized polymers based on biological feedstocks (amino acids, sugars, fats) (Chanprateep 2010).

## 2 Polyhydroxyalkanoates as a Basis for Pesticide Deposition

Among the biodegradable polymers that have already been developed or are being developed now for various applications, including medical ones, there are aliphatic polyesters, polyurethanes, polyamides, polylactides, polyglycolactides, silicon, polyethylene terephthalate, etc. These polymers are promising materials for fabricating biomedical devices, controlled drug delivery systems, degradable packaging for food and drinks, products for agriculture and public utilities (Lobo et al. 2011; Heng et al. 2017; Keskin et al. 2017).

Today, polyesters of monocarbon acids, polylactides (PLA) and polyglycolides (PGA), are the most widely used biodegradable polymers. The second most popular type of biodegradable polymers is polyhydroxyalkanoates (PHAs)—polymers of hydroxy-derived alkanoic acids. PHAs have lots of attractive properties including biodegradability and biocompatibility that make them promising materials for various applications, including biomedical ones (Sudesh and Hideki 2010; Volova et al. 2013, 2017b; Singh et al. 2012). PHAs have significant advantages in comparison with other biomaterials (Chen 2010):

- the high biocompatibility of PHAs, polyhydroxybutyrate in particular, is accounted for by the fact that the monomers constituting this polymer—hydroxybutyric acid—are natural metabolites of body cells and tissues;
- PHAs undergo true biological degradation, which occurs via the cellular and the humoral pathways; the resulting monomers of hydroxybutyric acid do not cause abrupt acidification of tissues and, therefore, do not give rise to any pronounced inflammatory reaction;
- PHA bioresorption rates are much lower than those of polylactides and polyglycolides; PHA-based implants can function in vivo for 2–3 years, depending on their form and implantation site; moreover, PHA degradation can be controlled;
- PHAs are produced by direct fermentation; no multistage technology is needed (monomer synthesis, polymerization, addition of plasticizers and modifying components);
- PHAs can be synthesized on such feedstocks as sugars, organic acids, alcohols, mixtures of $CO_2$ and $H_2$, products of plant biomass hydrolysis, industrial wastes of sugar and palm oil production, hydrogen-containing products of processing of brown coals and hydrolysis lignin;
- PHAs constitute a family of polymers of various chemical structures, consisting of monomers containing 4–12 and more carbon units, including high-crystallinity thermoplastic materials and rubber-like elastomers;
- PHA properties (crystallinity, mechanical strength, temperature characteristics, and biodegradation rates) can be controlled by varying the composition of the culture medium and tailoring the chemical structure of the polymer;
- PHAs can be processed from various phase states (powder, solution, gel, melt) using conventional techniques.

PHAs are very promising polymers as, being thermoplastic, like polypropylene and polyethylene, they also have antioxidant and optical properties as well as piezoelectricity. PHAs are highly biocompatible and can be biodegraded in biological media. In addition to poly(3-hydroxybutyrate) (P(3HB)), there are various PHA copolymers, which, depending on their monomeric composition, have different basic properties (degree of crystallinity, melting point, plasticity, mechanical strength, biodegradation rate, etc.). The properties of PHA polymers provide wide prospects for applications in various fields (public and agriculture, medicine and pharmacology, electronics, etc.).

PHAs are used to manufacture agricultural devices. These are films for greenhouses, packages for fertilizers and vegetables, pots, nets, ropes, etc. A new and environmentally important PHA application may be delivery of pesticides and fertilizers. Researchers of the Siberian Federal University and the Institute of Biophysics SB RAS were the first to prove that PHA can be used as a degradable base providing controlled release of fungicides and herbicides during the growing season of plants (Volova et al. 2008); pre-emergence formulations have been developed. That provided a basis for the new important use of PHAs—construction of slow-release formulations, in which chemicals for crop protection would be embedded in the matrix of these degradable polymers.

The ability of polyhydroxyalkanoates to break down in biological media is one of their most valuable properties. PHAs are degraded to water and $CO_2$ or to methane and water under aerobic and anaerobic conditions, respectively. Biodegradation of PHAs in the environment is carried out by extracellular depolymerases of microorganisms. Depolymerases are characterized by different molecular organization and specificity to substrate.

The analysis of the available literature shows that rather few authors reported on integrated studies of various aspects of PHA degradation, which is a very complex process. Most of the studies were performed in the laboratory, and they mainly dealt with the mechanism of interaction between the PHA supramolecular structure and PHA-depolymerizing enzymes, the structure and molecular organization of various depolymerases and microorganisms secreting extracellular PHA depolymerases.

An important question is the pattern of polymer breakdown in the natural environment. Extensive pioneering research on PHA biodegradation behavior in natural soil ecosystems was performed at the Siberian Federal University and Institute of Biophysics SB RAS.

We studied the kinetics and laws of the degradation of PHA in natural ecosystems in various regions and received answers to key questions of the PHA biodegradation process:

- which microorganisms are the most effective PHA degraders;
- how do the PHA properties change during degradation;
- how do environmental conditions (temperature, salinity, oxygen availability, pH, etc.) affect this process;
- how the process of PHA degradation will be affected by weather and climate of different regions.

PHA degradation influenced the total counts of microorganisms and composition of soil microflora. The microbial community formed on the polymer surface and the soil microbial community were different in the composition and percentages of the species. By employing the clear zone technique, we, for the first time, showed that each of the PHA types studied had specific degraders. PHA degradation behavior was studied in different environments: Siberian soils under broadleaved and coniferous trees, tropical soils (in the environs of Hanoi and Nha Trang), seawater (the South China Sea), a brackish lake (Lake Shira), and freshwater recreational water bodies in Siberia. Those studies showed that degradation occurred at different rates depending on the polymer composition, shape of the specimen (film or 3D construct), climate and weather conditions, and microbial community composition. The time over which the polymer loses 50% of its mass may vary between 68.5 and 270 days in Siberian soils, between 16 and 380 days in tropical soils of Vietnam, between 73 and 324 days in the brackish lake (Shira), between 127 and 220 days in the seawater of the South China Sea, and between 17 and 65.9 days in freshwater lakes (Prudnikova and Volova 2012). (Reprinted by permission from Springer Nature Customer Service Center GmbH: Springer Nature, Microbial Ecology, Microbial degradation of polyhydroxyalkanoates with different chemical compositions and their biodegradability, Volova TG, Prudnikova SV, Vinogradova ON, Syrvacheva DA, Shishatskaya EI, 2017).

The studies of PHA degradation in different soils showed that the following conditions affect the degradation of PHA: polymer composition, its geometry and the technique used to process it, weather conditions, the type of the ecosystem and its microbial component in particular, as the factor determining the mechanism of PHA biodegradation: preferential attack of the amorphous regions of the polymer or equal degradation of both crystalline and amorphous phases. PHA degrading microorganisms that dominate microbial populations in some soil ecosystems have been isolated and identified.

The data on the degradation of PHA under natural conditions are very important and they form the basis for the use of these polymers as a basis (matrix) for the deposit of pesticides in order to create long-term and targeted plant protection products.

## 3 Experimental Formulations of the Fungicide Tebuconazole and Their Efficacy

Fungicides are necessary for modern high-performance crop farming to protect crops from pests and diseases. The total crop loss in the world from pests is about 35%, and it is even higher in developing countries (48%). Approximately 1/3 of these losses are due to plant diseases caused by bacteria, fungi, and viruses, which reduce the quality of products and cause poisoning of animals and people. Mycotoxins, which are produced by some disease agents, pose a serious danger. One of the most common diseases of crops is fusarium infection, caused by a soil pathogenic fungus belonging to the genus *Fusarium*. The use of fungicides leads to a fusarium infection decrease and reduction of mycotoxin content in grain.

Triazoles are the largest group of fungicides that can be used for treating plants in the early stages of the disease development or for preventive treatment. One of the widely used triazoles nowadays is tebuconazole (TEB). TEB is a broad-spectrum systemic fungicide against crop diseases (fusarium infection, rust, rots, powdery mildew, and others), some diseases of grapes, soya bean, rapeseed, sunflower, and vegetables. TEB inhibits the process of ergosterol biosynthesis in the cell membranes of plant pathogenic fungi, resulting in the disruption of cell membranes, causing the death of pathogen. Studies addressing the use of PHA as a matrix for embedding pesticides are few. The use of the P(3HB/3 HV) copolymer for production of microspheres loaded with the ametrine and atrazine herbicides was shown by Lobo et al. 2011; Grillo et al. 2011. Suave et al. (2010) reported encapsulation of the malathion insecticide in microspheres from P(3HB) blended with polycaprolactone. There is no information in the available literature on the use of PHA as a matrix for embedding fungicides. Commercial formulations of TEB, represented by suspensions or emulsions used for spraying plants, are used widely. The fungicide is released from these formulations too quickly, which affects its effectiveness, and the fungicide has to be applied again. Thus, in order to increase the effectiveness of TEB and reduce its harmful effects on the environment, new formulations with controlled release of TEB are needed.

To construct environmentally friendly forms of TEB, biodegradable polymer P(3HB) was used as a matrix. The procedure for creating slow-release formulations of TEB in the form of films and pellets was described by Volova et al. (2016a). P3HB/TEB formulations were studied using X-ray structure analysis, differential scanning calorimetry, and Fourier transform infrared spectroscopy. Another study described TEB release from P3HB/TEB formulations into sterile distilled water and soil (Volova et al. 2017a). The fungicidal effect of P3HB/TEB formulations against the plant pathogen *Fusarium verticillioides* (formerly *Fusarium moniliforme*) was compared with that of Raxil Ultra (commercial formulation) (Volova et al. 2017a). In the first 2–4 weeks after the application, there was a noticeable fungicidal effect of the P3HB/TEB formulations, and it lasted for 8 weeks. In addition, no significant impact of experimental formulations on the soil aboriginal microflora was revealed. TEB release was found to depend on the TEB loading and the geometry of the formulation constructed, and TEB release in the soil occurred gradually, as P(3HB) was degraded.

Particular attention was paid to the study of potential for designing embedded target-delivery formulations of polymeric fungicides by nanotechnology-based systems (Shershneva et al. 2019).

The surface morphology of the P(3HB)/TEB microparticles was studied using scanning electron microscopy (SEM) (Fig. 1a, b). SEM analysis showed the presence of large undissolved crystals of TEB on the surface of microparticles. That was probably caused by the high concentration of TEB, which did not dissolve completely because of the presence of high-molecular-weight chains of P(3HB) in the solution. With the TEB increase in microparticles from 10 to 50%, the amount of TEB crystals on the surface of microparticles increased too. Apparently, the increase in encapsulation efficiency resulted from the high adsorption of TEB crystals on the surface of microparticles with the initial 50% TEB concentration in the solution (Table 1).

Moreover, a direct relationship between the TEB loading and the average diameter of microparticles was noted: with TEB loading increased from 10 to 50%, the average diameter of microparticles increased from 41.3 to 71.7 μm (Table 1). By contrast, as the TEB loading was increased, the yield of microparticles, regardless of the polymer initial mass, decreased. As for zeta potential, no effect of the TEB loading on the zeta potential was detected (Table 1).

Evaluation of the size distributions of microparticles showed that, as a percentage, particles with a diameter of about 50 μm prevailed over all concentrations of TEB loading. The proportion of the smallest particles with a diameter of 25 μm

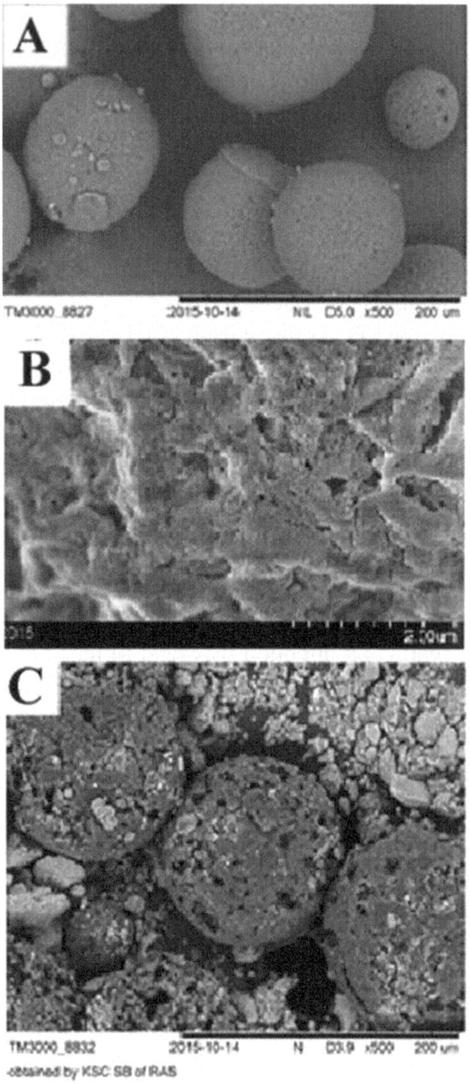


**Fig. 1** SEM images of P(3HB)/TEB microparticles before (**a**, **b**) and after (**c**) exposure to the soil; bars—200 μm (**a**, **c**) and 2 μm (**b**)

**Table 1** Characteristics of P(3HB)/TEB microparticles with different amounts of TEB

| Sample | Encapsulation efficiency (%) | Yield of particles (%) | Average diameter (μm) | Zeta potential (mV) |
|---|---|---|---|---|
| P(3HB)/TEB-10 | 59 | 70.9 | 41.3 | −35.7 ± 2.0 |
| P(3HB)/TEB-25 | 65 | 63.0 | 63.2 | −32.6 ± 0.9 |
| P(3HB)/TEB-50 | 86 | 58.5 | 71.7 | −35.3 ± 2.1 |

increased while the TEB load was reduced to 10%. Conversely, with an increase in TEB loading to 50%, the proportion of large microparticles with a diameter of 125 μm and more increased significantly. Thus, the average diameter of the microparticles increased with the load of TEB from 41.3 to 71.7 μm. The emulsion technique makes it possible to obtain nanoscale microparticles that can penetrate plant tissue and are suitable for post-harvest processing and protection from damage to the aerial parts of plants (Ding et al. 2011; Qian et al. 2013). Larger microparticles obtained in our research can be used for pre-sowing treatment of seeds or pre-emergence introduction of fungicides into the soil together with the seeds.

TEB release from P(3HB)/TEB microparticles into sterile distilled water and soil was studied. TEB release from microparticles in distilled water during 60 days is indicated in Fig. 2. TEB release from microparticles with the TEB loading of 25 and 50% was similar. A possible reason for this may be low water solubility of TEB. This is probably associated with the low water solubility of TEB, and therefore, when the concentration of TEB in water reached its highest possible level, the rate of TEB release from the microparticles with 25 and 50% of TEB loading slowed down. TEB crystals were found on the surface of the 50% loaded microparticles at the end of the experiment, suggesting partial TEB release from microparticles, and thus, prolonged release of TEB was achieved. By the end of the experiment, TEB release from microparticles with 10, 25, and 50% of TEB was 43, 38, and 25%, respectively. Thus, the reason for slow TEB release from microparticles is apparently low water solubility of TEB. These results suggest that release of the fungicide can be regulated by changing TEB content in microparticles.

The exposure of TEB-loaded microparticles in soil microcosms led to the degradation of the polymer matrix of microparticles and a more intensive TEB release into the soil compared with the release to water. Obvious changes in the morphology of microparticles after 21 days of exposure can be seen on SEM images: partial destruction, the appearance of surface erosion, hollows, and cavities (Fig. 1c). After

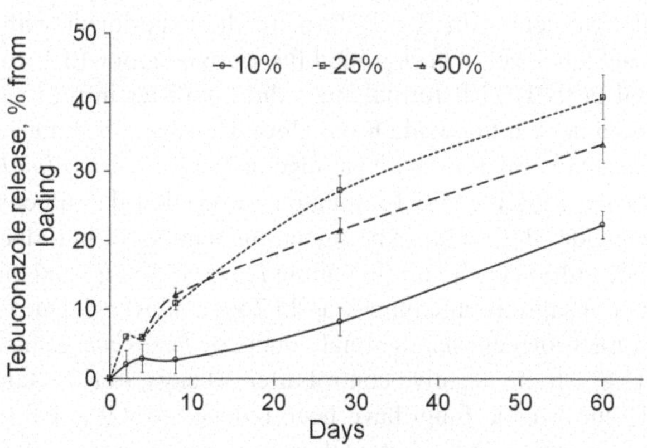


**Fig. 2** Release kinetics of TEB from P(3HB)/TEB microparticles with 10, 25, and 50% of TEB loadings

35 days, the microparticles degraded by 80% and looked like small fragments of irregular shape with through holes and tunnels.

Antifungal activity of P(3HB)/TEB formulations with the TEB loading of 25% was studied in experiments with pathogenic fungi *Fusarium verticillioides*. This species of *Fusarium* genus is dangerous for people, because it not only damages the grain yield, but also produces mycotoxin (fumonisin), causes mycoses in immunocompromised people and has oncogenic potential (Voss et al. 2002).

The experiment was performed in vitro by growing *Fusarium verticillioides* on malt-extract agar in Petri dishes. As a positive control, 200 μL of commercial fungicide Raxil Ultra (Bayer AG, Germany) containing 120 mg $L^{-1}$ of TEB was added to an agar-well. This dose was consistent with the load of TEB in the formulations. The experiment showed that the growing zone of the *F. verticillioides* decreased by two to three times under the influence of commercial fungicide and experimental TEB formulations. No significant differences were observed between the diameters of colonies in the positive control group and in the group of P(3HB)/TEB microparticles. Thus, the antifungal activity of P(3HB)/microparticles is comparable with the antifungal activity of commercial TEB, and it follows that experimental formulations of embedded TEB hold promise for constructing long-term formulations of TEB.

The efficacy of P(3HB)/TEB formulations was investigated in rhizosphere soil of wheat plants infected by plant pathogen *F. verticillioides* (Volova et al. 2018). TEB was embedded in degradable microbial polymer, P(3HB), designed as microgranules and films. Germination test of wheat seeds on the nutrient medium showed the presence of phytopathogenic fungi *Fusarium*, *Bipolaris,* and *Alternaria*. The total contamination of wheat seeds reached 9.5%, and 5.6% of which were *Fusarium* species. Thus, internal seed infection leads to the development of seedling disease in the early stages, inhibits the growth of plants, and reduces their productivity.

The developed experimental formulations of P(3HB)/TEB were placed into the soil simultaneously with the sowing of wheat seeds, and their fungicidal activity was compared with the effect of traditional used preparations: pre-sowing treatment of seed or soil treatment with Raxil Ultra. In the experiment with the initially infected seeds and low level of background fusarium infection ($3.1 \times 10^3$ CFU $g^{-1}$), the experimental P(3HB)/TEB formulations did not differ in root pathogens suppression from commercial fungicide Raxil Ultra. However, in simulated conditions of high infectious load of the soil with pathogenic fungi *F. verticillioides*, the fungicidal activity of the P(3HB)/TEB formulations exceeded the effectiveness of the commercial fungicide. Before the experiment, the number of introduced *Fusarium* fungi reached one million per/g soil (including *F. verticillioides* and minor species), while the number of saprotrophic fungi was $25.2 \times 10^3$ CFU $g^{-1}$. Due to competitive relationships in microbiocenosis, the total counts of *Fusarium* genus decreased to $21.2 \times 10^3$ CFU $g^{-1}$ in the negative control after 30 days. For the same reason the total counts of saprotrophic fungi have been reduced to $4.9 \times 10^3$ CFU $g^{-1}$. The counts of saprotrophic and phytopathogenic fungi were $9.2 \times 10^3$ and

$8.4 \times 10^3$ CFU $g^{-1}$, respectively, when Raxil Ultra was used. Therefore, fungicidal activity of P(3HB)/TEB formulations in soil with a high concentration of *F. verticillioides* was higher than when using commercial fungicide.

The infection of seeds and plants in contaminated soil by plant pathogens cause significant damage of roots. Nevertheless, even in case with naturally infected soil, *Fusarium* infection was also found in the first 10 days in all groups of plants, including the groups with TEB treated soil. This happened due to the fact that the seeds were infected with phytopathogenic microscopic fungi, and the infection had already appeared at an early stage of the seedlings. Then, the infection of plant roots not treated with fungicide increased. From 10 to 30 days, the number of plants infected with root rot increased (from 17 to 30% of the total number of tested plants). It was shown that infection caused by fungi of the genus *Fusarium* made its main contribution to the etiology of root rot (50–80% of all infections).

So, TEB is an effective fungicide used to protect different cereal crops. However, triazole fungicides, including TEB, are phytotoxic. Fungicides of triazole group suppress biosynthesis of ergosterol in cell membranes of pathogens and cause their death. Thus, crops infected by *Fusarium* and treated with triazole fungicides are affected by two negative factors: phytopathogens and pesticides. To identify the mechanism of the damaging effect of these factors, culture of *Triticum aestivum* infected with phytopathogens (*Alternaria*, *Fusarium*) and treated with triazole fungicides (tebuconazole) was used. The morphology of root apexes with population of border cells and the composition of exometabolites (proline, carbonylated proteins, and malonic dialdehyde) were analyzed (Shishatskaya et al. 2018). Proline (an integral indicator of the activity of antioxidant root systems), carbonylated proteins (CP), and malonic dialdehyde (MDA) are the indicators of the level of oxidative modification of proteins and activity of membrane lipids peroxidation.

At Day 10, the contents of MDA, CP, and proline in roots of the control wheat plants (group 1, without TEB application) did not differ significantly from their contents in plants roots of groups 2 (the treatment with Raxil Ultra applied to the soil) and 3 (the treatment with seeds pretreated with Raxil Ultra). At Day 20, the amount of MDA and proline in roots of group 1 increased considerably (by a factor of 8.5 and by a factor of 19) compared to Day 10, while CP decreased slightly (by a factor of 1.8). At Day 30, proline content in the roots of group 1 decreased dramatically, while MDA and CP contents did not change significantly. In group 2, contents of MDA, CP, and proline in the roots did not differ significantly from the control, suggesting that phytotoxic effects of TEB were softened as soil contamination with phytopathogens decreased. However, in group 3, contents of proline, MDA, and CP in the roots were higher than in group 1 by a factor of 2.2, 2.0, and 1.7, respectively. That was indicative of the activation of phytotoxic stress and free radical processes, as the effect of TEB used to pretreat the seeds before sowing must have been exhausted by Day 30. This study showed that the effect of TEB on redox homeostasis in wheat roots varied depending on the growth stage of plant and was considerably different in ecosystems with plants and soil infected by *Fusarium*

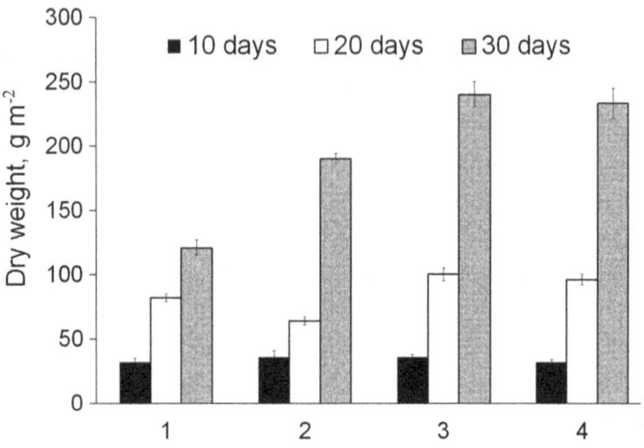


**Fig. 3** The effect of TEB delivery mode on the increase in wheat aboveground biomass: *1*—negative control, *2*—positive control (Raxil applied to soil), *3*—P(3HB)/TEB microparticles (10% TEB), *4*—P(3HB)/TEB microparticles (25% TEB)

phytopathogens. At Day 20 of plant growth, during the tillering stage, TEB produced the strongest phytotoxic effect on wheat plants.

The results of evaluating the productivity of wheat communities in experiment with high degree of soil infection and root damage caused by rot are shown in Fig. 3. At Day 10, the aboveground biomass of wheat plants was comparable in the negative and positive control groups and in the treatment groups (P(3HB)/TEB microparticles). At Day 30, in the group with Raxil, the aboveground biomass reached 190 g $m^{-2}$, while in the treatment groups it was higher (230–240 g $m^{-2}$).

The fungicidal activity of the experimental slow-release formulations of TEB embedded in the matrix of degradable P(3HB) against fusarium infection of wheat was comparable to that of TEB in commercial formulation Raxil in early stages (Day 10). In the later stages, P(3HB)/TEB formulations more effectively suppressed the development of *Fusarium* in soil and inhibited the growth of plant root rot.

## 4 Experimental Formulations of Herbicides and Evaluation of Their Efficacy

[Weeds cause great damage to agriculture, and herbicides constitute the most extensively used group of pesticides (40–50%), their commercial varieties accounting for about 40% of all commercial pesticides. Weed control using herbicides is one of the major components of modern efficient agriculture. However, herbicides, as well as other pesticides, persist in the soil, posing a hazard to human health, leading to the emergence of herbicide-resistant weed species, threatening the stability of agroecosystems and leaving the ground almost permanently barren. Much research effort has been recently focused on constructing new formulations and investigating their behavior in the environment. The main purpose of such studies is to produce less toxic and more selective pesticides and reduce the rate of pesticide application.] (Reprinted by permission from Springer Nature Customer Service Center GmbH: Springer Nature, Environmental Science and Pollution Research,

Poly(3-hydroxybutyrate)/metribuzin formulations: characterization, controlled release properties, herbicidal activity, and effect on soil microorganisms, Volova T, Zhila N, Kiselev E, Prudnikova S, Vinogradova O, Nikolaeva E, Shumilova A, Shershneva A, Shishatskaya E, 2016).

Triazines are commonly used broad-spectrum selective herbicides, which do not persist for a very long time in soil. Metribuzin (MET) is a pre-emergence and post-emergence herbicide which is used to treat different crops and has high biological activity in various climate zones (Fedtke 1981). [MET has been used by many researchers as a herbicide for constructing slow-release formulations based on various synthetic and natural materials: polyvinylchloride, carboxymethyl cellulose (Kumar et al. 2010a), acrylamide (Sahoo et al. 2014), methacrylic acid combined with ethylene glycol and dimethacrylate (Zhang et al. 2009), sepiolite (Maqueda et al. 2008), alginate (Flores-Céspedes et al. 2013), phosphatidylcholine (Undabeytia et al. 2011), kraft lignin (Chowdhury 2014), lignin/polyethylene glycol blends (Fernández-Pérez et al. 2011, 2015), chitin, cellulose, starch (Fernández-Pérez et al. 2010; Rehab et al. 2002), bentonite, activated carbon (McCormick 1985), etc. Thus, by varying the shape of the carrier, the technique employed to construct it, and the material used, one can influence MET release kinetics and design-controlled delivery systems for this herbicide.] (Reprinted by permission from Springer Nature Customer Service Center GmbH: Springer Nature, Environmental Science and Pollution Research, Poly(3-hydroxybutyrate)/metribuzin formulations: characterization, controlled release properties, herbicidal activity, and effect on soil microorganisms, Volova T, Zhila N, Kiselev E, Prudnikova S, Vinogradova O, Nikolaeva E, Shumilova A, Shershneva A, Shishatskaya E, 2016).

Degradable polymers of various origins are being tested as materials for constructing pesticide carriers. A review of current literature shows that polymers based on derivatives of carbonic acids have attracted the attention of many researchers. Special attention is given to polyhydroxyalkanoates (PHAs)—microbial polymers having many useful properties. [Production of polyhydroxyalkanoates (PHAs) is a rapidly developing branch of the industry of degradable bioplastics, and they are regarded as candidates to eventually replace synthetic polymers (Chen 2010; Ienczak et al. 2013; Kaur and Roy 2015; Volova et al. 2013).] (Reprinted by permission from Springer Nature Customer Service Center GmbH: Springer Nature, Environmental Science and Pollution Research, Poly(3-hydroxybutyrate)/metribuzin formulations: characterization, controlled release properties, herbicidal activity, and effect on soil microorganisms, Volova T, Zhila N, Kiselev E, Prudnikova S, Vinogradova O, Nikolaeva E, Shumilova A, Shershneva A, Shishatskaya E, 2016).

The purpose of this study was to investigate the herbicidal activity of MET embedded in the polymer matrix based on degradable poly(3-hydroxybutyrate) [P(3HB)] by exposing in laboratory soil ecosystems with higher plants. For the first time construction and investigation of slow-release MET formulations of different geometries with metribuzin embedded in the P(3HB) were described in Volova et al. (2016b). The P(3HB)/MET mixtures (powders, solutions, and emulsions) were used to construct MET-loaded pellets, films, granules, and microparticles and tested. Using X-ray, DSC, and FTIR methods the absence of chemical bonds between the components of MET and P(3HB) has been shown. [The kinetics of polymer degradation, MET release, and accumulation in soil were studied in laboratory soil microecosystems with higher plants. The study showed that MET release can be controlled by using different techniques of constructing formulations and by varying MET loading. The herbicidal activities of P(3HB)/MET formulations and commercial formulation Sencor Ultra were tested on the *Agrostis stolonifera* and *Setaria macrocheata* plants. All P(3HB)/MET formulations had pronounced herbicidal activity, which varied depending on MET loading and the stage of the experiment (Volova et al. 2016c).] (Reprinted by permission from Springer Nature Customer Service Center GmbH: Springer Nature, Environmental Science and Pollution Research, Poly(3-hydroxybutyrate)/metribuzin formulations: characterization, controlled release properties, herbicidal activity, and effect on soil microorganisms, Volova T, Zhila N, Kiselev E, Prudnikova S, Vinogradova O, Nikolaeva E, Shumilova A, Shershneva A, Shishatskaya E, 2016). Moreover, the herbicidal activity of P(3HB)/MET microgranules and films was tested against weeds such as *Chenopodium album* and *Melilotus albus* in the presence of wheat (*Triticum aestivum*, cv. Altaiskaya 70) (Zhila et al. 2017). The experimental P(3HB)/MET formulations showed pronounced herbicidal activity against these weeds. The effectiveness of the experimental formulations in inhibiting the growth of *Chenopodium album* and *Melilotus albus* was comparable to and, sometimes, higher than that of the Sencor Ultra (commercial formulation).

Using emulsion technique, P(3HB)/MET microparticles, with the 10 and 25% of MET loadings, were prepared. The best conditions for preparing P(3HB)/MET microparticles are as follows: the concentration of P(3HB) and PVA (30 kDa) was 1%, agitation speed was 750 rpm. The average size of microparticles P(3HB)/MET with the 10 and 25% of MET loadings was comparable—54 μm (Table 2). The SEM analysis showed that the microparticles, regardless of their size, had a wrinkled surface.

[The value of the ξ-potential, which is an important parameter of particles characterizing their stability in solutions, was −26.2 and − 33.2 mV for the microparticles with the 10 and 25% MET loadings, respectively. The yield of the particles from emulsions with different MET loadings was rather high, more than 60%, but

**Table 2** Characteristics of the P(3HB)/*MET* microparticles with the 10 and 25% of MET loadings

| MET loadings | EE[a] (%) | *Y*[b] (%) | The average size (μm) | ξ-potential (mV) |
|---|---|---|---|---|
| *P(3HB)/MET microparticles* | | | | |
| 10% | 21 | 76.5 | 54.0 | −30.8 ± 2.3 |
| 25% | 18 | 71.6 | 54.4 | −26.2 ± 2.9 |

[a]EE—is the efficiency of MET encapsulation in the microparticles. EE was calculated using the following formula: EE = $(M_{enc}/M_{init}) \times 100\%$, where $M_{enc}$ is the mass of MET encapsulated in P(3HB) (mg) and $M_{init}$ is the initial mass of MET (mg)

[b]*Y*—the microparticles yield (percent of the P(3HB) mass used to construct microparticles). *Y* was calculated using the following formula: $Y = (M_m/M_p) \times 100\%$, where $M_m$ is the mass of the P(3HB)/*MET* microparticles, mg, and $M_p$ is the mass of P(3HB) and MET used for microparticles preparing, mg

MET encapsulation efficiency was low, 18–21%.] (Reprinted by permission from Springer Nature Customer Service Center GmbH: Springer Nature, Environmental Science and Pollution Research, Poly(3-hydroxybutyrate)/metribuzin formulations: characterization, controlled release properties, herbicidal activity, and effect on soil microorganisms, Volova T, Zhila N, Kiselev E, Prudnikova S, Vinogradova O, Nikolaeva E, Shumilova A, Shershneva A, Shishatskaya E, 2016).

MET release from the P(3HB)/MET microparticles with 25% of MET loading in sterile distilled water was studied. By the end of the experiment (49 days), about 95% of MET embedded in the polymer matrix were released from the microparticles (25% of MET loading). As P(3HB) does not dissolve and does not hydrolyze in water, MET was passively released from the polymer matrix as well, diffusing through the pores. The MET release rate from microparticles in the first 3 days was 7.7 mg $d^{-1}$ reduced to 1.5 mg $d^{-1}$ in the next 11 days. The lowest MET release rate (0.2–0.27 mg $d^{-1}$) was at the end of the experiment.

For describing metribuzin release from microparticles, the Korsmeyer-Peppas model was used:

$$M_t / M_\infty = Kt^n$$

$M_t$ is the MET amount released at time $t$, $M_\infty$ is the MET amount released over a very long time (it generally corresponds to the initial MET amount). $K$ is a kinetic constant and $n$ is the diffusional exponent.

Exponent $n$ was 0.405, which suggests MET diffusion from polymer matrix according to Fick's law. The value of $K$ was 0.081 $h^{-1}$. The ζ-potential and morphology of the microparticles incubated in water did not change. Moreover, no significant changes in physicochemical properties were detected (crystallinity degree and temperature parameters).

[Kinetics of MET release from P(3HB)/MET microparticles and degradation of P(3HB) were studied in laboratory soil microecosystems with higher plants. All microparticles, irrespective of the amount of metribuzin loading, were almost completely degraded after 30–40 days of incubation in soil (Fig. 4); the average degradation rates of the microparticles with the 10 and 25% MET loadings were 0.15 and 0.17 mg $d^{-1}$, respectively. As the polymer matrix was degraded, molecular weight of the polymer decreased, while its polydispersity and degree of crystallinity increased, suggesting preferential disintegration of the amorphous phases of the polymer.

The dynamics of degradation of the polymer matrix, which determines MET release, influenced herbicide accumulation in soil (Fig. 4). The MET concentrations released from microparticles were comparable with metribuzin concentration in soil from Sencor Ultra and were measured after 20–30 days of incubation of the formulations loaded at 25 and 10% MET. Concentrations reached about 4.8–6.8 and 1.5–2.4 μg $g^{-1}$ soil, respectively. Thus, the 100% release of MET was observed from the microparticles, which were completely degraded during the experiment. The relationship between herbicide release rate and the level of loading was shown in a previous study (Prudnikova et al. 2013).

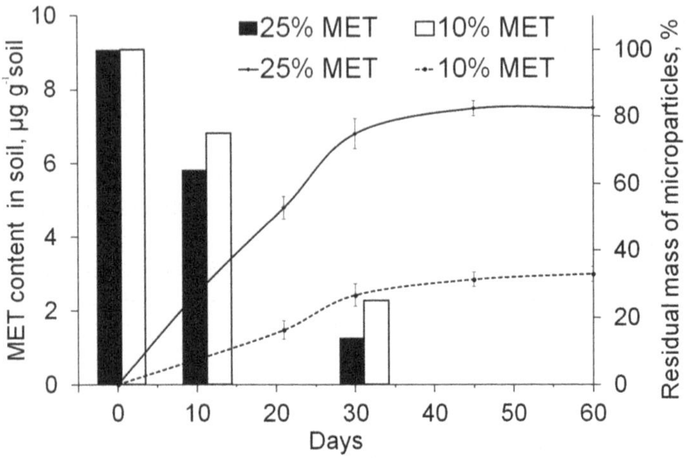


**Fig. 4** Degradation dynamics of P(3HB)/*MET* microparticles with 10 and 25% of MET loadings in soil (histograms) and MET release (curves) from them into the soil in laboratory conditions

Constant *K* and exponent *n*, characterizing kinetics of metribuzin release from the P(3HB)/MET microparticles were obtained by using the Korsmeyer-Peppas model. Metribuzin release from microparticles was characterized by the anomalous case-II transport. The values of the diffusional exponent (*n*) at 10 and 25% loadings were 0.98 and 0.91, respectively. Constant *K*, which contains diffusion coefficient and structural and geometric data on the formulations, was 0.0013 and 0.0024 $h^{-1}$ at 10 and 25% loadings, respectively.] (Reprinted by permission from Springer Nature Customer Service Center GmbH: Springer Nature, Environmental Science and Pollution Research, Poly(3-hydroxybutyrate)/metribuzin formulations: characterization, controlled release properties, herbicidal activity, and effect on soil microorganisms, Volova T, Zhila N, Kiselev E, Prudnikova S, Vinogradova O, Nikolaeva E, Shumilova A, Shershneva A, Shishatskaya E, 2016). Parameter $t_{50}$ characterizes the time when MET is released with the highest rate. The values of the $t_{50}$ at 10 and 25% of MET loadings were 21 days.

[The weeds *Agrostis stolonifera* and *Setaria macrocheata* were used to study the herbicidal activity of the P(3HB)/MET microparticles. P(3HB)/MET microparticles had comparable effects on the plants (Fig. 5). [In the previous study, we also showed that formulations of the herbicide Zellek Super shaped as microgranules and films successfully suppressed the growth of *Agrostis stolonifera* (Prudnikova et al. 2013). Moreover, the effectiveness of MET embedded in carboxy methyl cellulose–kaolinite composite (CMC-KAO) against weeds growing in wheat crops was shown in the field experiment by Kumar et al. (2010a, b).

The herbicidal effect of the experimental P(3HB)/MET microparticles on the plants was stronger than the effect achieved in the positive control (Sencor Ultra). The analysis of the parameters of MET effect on the plant density and the weight of fresh green biomass showed that P(3HB)/MET microparticles exhibited herbicidal

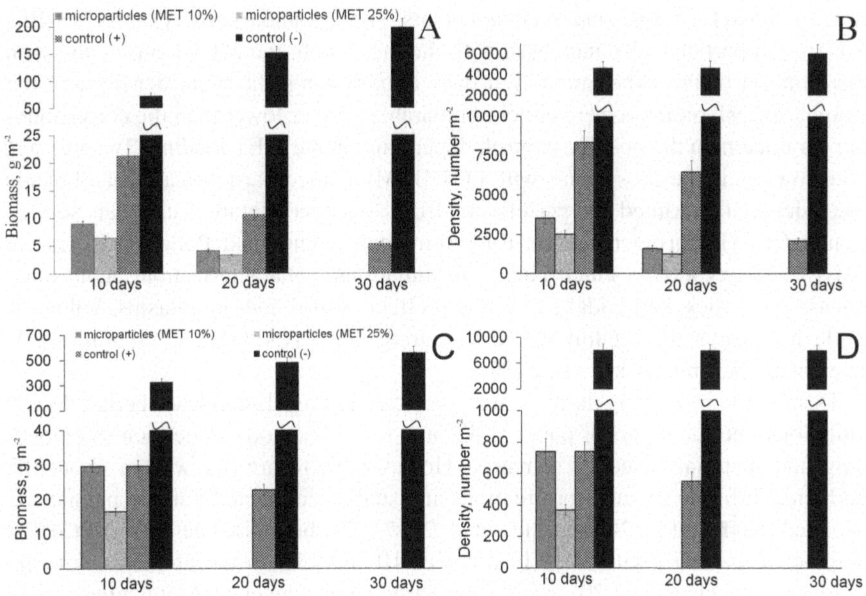


**Fig. 5** The weight of fresh green biomass of *Agrostis stolonifera* (**a**) and *Setaria macrocheata* (**c**) and density of *Agrostis stolonifera* (**b**) and *Setaria macrocheata* (**d**) grown in the laboratory micro-ecosystems with P(3HB)/*MET* microparticles

activity.] (Reprinted by permission from Springer Nature Customer Service Center GmbH: Springer Nature, Environmental Science and Pollution Research, Poly(3-hydroxybutyrate)/metribuzin formulations: characterization, controlled release properties, herbicidal activity, and effect on soil microorganisms, Volova T, Zhila N, Kiselev E, Prudnikova S, Vinogradova O, Nikolaeva E, Shumilova A, Shershneva A, Shishatskaya E, 2016).

[In the positive control, 10 days after sowing, the plant density and the weight of the biomass of *Agrostis stolonifera* were 8333 ± 750 plants $m^{-2}$ and 21.28 ± 1.26 g $m^{-2}$, 20 days after sowing—6481 ± 713 and 10.64 ± 0.84, and 30 days after sowing—2090 ± 187 plants $m^{-2}$ and 5.32 ± 0.32 g $m^{-2}$, respectively. That was almost five to six times lower than in the negative control. For *Setaria macrocheata*, the difference was even more considerable. The inhibitory effect of the experimental P(3HB)/MET microparticles varied depending on the MET loading and the duration of the experiment. Ten days after sowing, the number of *Agrostis stolonifera* plants and their biomass in the experiment with the microparticles were degraded in the soil at the high rate, these parameters were lower by more than a factor of two in comparison with positive control. P(3HB)/MET microparticles with MET loading of 25% had more pronounced herbicidal effects of: 10 days after sowing, the biomass was lower than in the positive control by a factor of 3.3. At Day 20 a considerable number of plants in all treatments were dead, and the green biomass was reduced much more dramatically than in the positive control. At Day 30 all plants were dead in the treatments and positive control. Similar results were obtained for *Setaria macrocheata* plants. The herbicidal activity of the P(3HB)/MET microparticles also increased with the increase in the MET loading and with the duration of the experiment. Ten days after sowing, the plant density and the weight of fresh biomass were either comparable with or lower than the corresponding parameters in the positive control, depending on the MET loading. Twenty days after sowing, in the ecosystems with P(3HB)/MET microparticles, almost all plants were dead.] (Reprinted by permission from Springer Nature Customer Service Center GmbH: Springer Nature, Environmental Science and Pollution Research, Poly(3-hydroxybutyrate)/metribuzin formulations: characterization, controlled release properties, herbicidal activity, and effect on soil microorganisms, Volova T, Zhila N, Kiselev E, Prudnikova S, Vinogradova O, Nikolaeva E, Shumilova A, Shershneva A, Shishatskaya E, 2016).

Despite the increasing number of studies concerning slow-release herbicide formulations, the main part of paper is devoted to the methods of herbicides embedding and materials used as a matrix. However, there are a few data about the herbicidal efficacy of such formulations and studies conducted with crops infested by weed (Kumar et al. 2010b; Zhila et al. 2017). The herbicidal activity of P(3HB)/MET microparticles with MET loadings of 10 and 25% in wheat stands *Triticum aestivum* (cv. Altaiskaya 70) infested by white sweet clover *Melilotus albus* under laboratory conditions was studied (Fig. 6).

The study was compared with negative (untreated) and positive (Sencor Ultra) control. At Day 10 after sowing, the biomass and density of the plants *Melilotus albus* in the negative control reached about 10 g $m^{-2}$ and 6500 plants $m^{-2}$, respectively. These data were considerably higher than the corresponding values in the positive control (5.1 g $m^{-2}$ and 5200 plants $m^{-2}$) and treatments (4100–4900 plants $m^{-2}$), where the plants growth was evidently inhibited. At Day 20 the number of the plants *Melilotus albus* decreased to 1100 and 1350 plants $m^{-2}$ with the treatment of microparticles with MET loadings of 25 and 10%, respectively. The weed density in the positive control was higher (about 2000 plants $m^{-2}$). At the end of the experiment (Day 50), complete suppression of plants *Melilotus albus* was observed in the herbicide-treated ecosystems. Moreover, the density of *Melilotus albus* and the amount of its aboveground biomass were considerably lower in the experiments with microparticles than in the experiment with Sencor Ultra. Effective weed control caused an increase in the productivity of wheat. The aboveground biomass of wheat reached 186–195 g $m^{-2}$ in the experiments with the treatments with P(3HB)/MET microparticles. In the experiments with Sencor Ultra and in the negative control biomass was lower (167 and 136 g $m^{-2}$, respectively).

Thus, these results clearly showed the effectiveness of the P(3HB)/MET microparticles for weed control and also influencing the wheat growth. The activity was significant in comparison with commercial formulations.

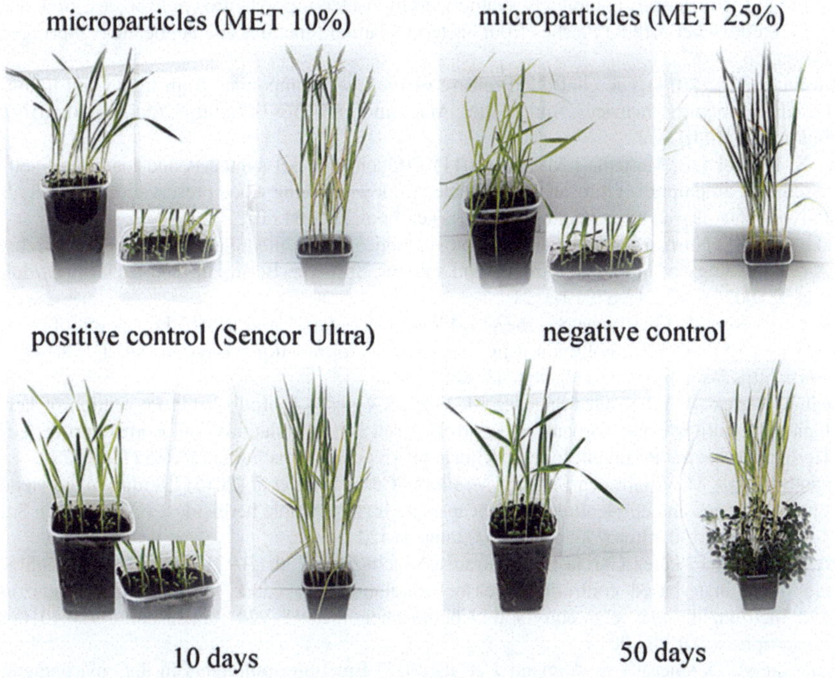


**Fig. 6** Photographs of wheat stands infested with *Melilotus albus* and treated with P(3HB)/*MET* microparticles with MET loadings of 10 and 25%

## 5 Conclusion

The positive results that have been obtained suggest the use of polyhydroxyalkanoates as a biodegradable polymer matrix to construct controlled-release pesticide formulations. Application of such herbicidal and fungicidal formulations has been found to be an effective means of increasing crop productivity and protecting them against pests and pathogens. Moreover, the effect of using these formulations is comparable or superior to the effect of using commercial pesticides. Further research will provide the basis for reducing accumulation and uncontrolled spread of pesticides in the environment and replacing synthetic plastics by biodegradable materials, which can be incorporated in biosphere cycles.

## References

Carvalho FP (2017) Pesticides, environment, and food safety. Food Energ Sec 6(2):48–60. https://doi.org/10.1002/fes3.108

Chanprateep S (2010) Current trends in biodegradable polyhydroxyalkanoates. J Biosci Bioeng 110(6):621–632. https://doi.org/10.1016/j.jbiosc.2010.07.014

Chen G-Q (2010) Plastics completely synthesized by bacteria: polyhydroxyalkanoates. In: Chen G-Q, Steinbüchel A (eds) Plastics from bacteria. Natural functions and applications. Springer, Berlin, pp 17–37

Chowdhury MA (2014) The controlled release of bioactive compounds from lignin and lignin-based biopolymer matrices. Int J Biol Macromol 65:136–147. https://doi.org/10.1016/j.ijbiomac.2014.01.012

Ding X, Richter DL, Matuana LM et al (2011) Efficient one-pot synthesis and loading of self-assembled amphiphilic chitosan nanoparticles for low-leaching wood preservation. Carbohydr Polym 86(1):58–64. https://doi.org/10.1016/j.carbpol.2011.04.002

Fedtke C (1981) Nitrogen metabolism in photosynthetically inhibited plants. In: Borthe H, Trebst A (eds) Biology of inorganic nitrogen and sulphur. Springer, Berlin, pp 260–265. https://doi.org/10.1007/978-3-642-67919-3_21

Fernández-Pérez M, Villafranca-Sánchez M, Flores-Céspedes F et al (2010) Prevention of chloridazon and metribuzin pollution using lignin-based formulations. Environ Pollut 158:1412–1419. https://doi.org/10.1016/j.envpol.2009.12.040

Fernández-Pérez M, Villafranca-Sánchez M, Flores-Céspedes F et al (2011) Prevention of herbicides pollution using sorbents in controlled release formulations. In: Kortekamp A (ed) Herbicides and environment. InTech, Rijeka, pp 157–172. https://doi.org/10.5772/13477

Fernández-Pérez M, Villafranca-Sánchez M, Flores-Céspedes F et al (2015) Lignin-polyethylene glycol matrices and ethylcellulose to encapsulate highly soluble herbicides. J Appl Polym Sci 132:41422–41430. https://doi.org/10.1002/app.41422

Flores-Céspedes F, Pérez-García S, Villafranca-Sánchez M et al (2013) Bentonite and anthracite in alginate-based controlled release formulations to reduce leaching of chloridazon and metribuzin in a calcareous soil. Chemosphere 92:918–924. https://doi.org/10.1016/j.chemosphere.2013.03.001

Gavrilescu M, Demnerová K, Aamand J et al (2015) Emerging pollutants in the environment: present and future challenges in biomonitoring, ecological risks and bioremediation. New Biotechnol 32(1):147–156. https://doi.org/10.1016/j.nbt.2014.01.001

Greene LG, Meyers PA, Springer JT et al (1992) Biological evaluation of pesticides released from temperature-responsive microcapsules. J Agric Food Chem 40:2274–2276. https://doi.org/10.1021/jf00023a044

Grillo R, Pereira AE, Melo S et al (2011) Controlled release system for ametryn using polymer microspheres: preparation, characterization and release kinetics in water. J Hazard Mater 186:1645–1651. https://doi.org/10.1016/j.jhazmat.2010.12.044

Grillo R, Pereira AE, Nishisaka CS et al (2014) Chitosan/tripolyphosphate nanoparticles loaded with paraquat herbicide: an environmentally safer alternative for weed control. J Hazard Mater 278:163–171. https://doi.org/10.1016/j.jhazmat.2014.05.079

Heng KS, Lee YF, Thinagaran L et al (2017) Biosynthesis of PHAs and Their Biomedical Applications. In: Thakur VK (ed) Handbook of composites from renewable materials, vol 5. Scrivener Publishing LLC, Wiley, pp 543–585

Ienczak JL, Schmidell W, de Aragão GMF (2013) High-cell-density culture strategies for polyhydroxyalkanoate production: a review. J Ind Microbiol Biotechnol 40:275–286. https://doi.org/10.1007/s10295-013-1236-z

Kaur G, Roy I (2015) Strategies for large-scale production of polyhydroxyalkanoates. Chem Biochem Eng 29:157–172. https://doi.org/10.15255/cabeq.2014.2255

Keskin G, Kizil G, Bechelany M, Pochat-Bohatier C et al (2017) Potential of polyhydroxyalkanoate (PHA) polymers family as substitutes of petroleum based polymers for packaging applications and solutions brought by their composites to form barrier materials. Pure Appl Chem 89(12):1841–1848. https://doi.org/10.1515/pac-2017-0401

Kumar J, Nisar K, Shakil NA et al (2010a) Controlled release formulations of metribuzin: release kinetics in water and soil. J Environ Sci Health, Part B 45:330–335. https://doi.org/10.1080/03601231003704424

Kumar J, Nisar K, Shakil NA et al (2010b) Residue and bio-efficacy evaluation of controlled release formulations of metribuzin against weeds in wheat. Bull Environ Contam Toxicol 85:357–361. https://doi.org/10.1007/s00128-010-0091-0

Lobo FA, de Aguirre CL, Silva MS et al (2011) Poly(hydroxybutyrate-co-hydroxyvalerate) microspheres loaded with atrazine herbicide: screening of conditions for preparation, physico-chemical characterization, and in vitro release studies. Polym Bull 67(3):479–495. https://doi.org/10.1007/s00289-011-0447-6

Maqueda C, Villaverde J, Sopeña F et al (2008) Novel system or reducing leaching of the herbicide metribuzin using clay-gel-based formulations. J Agric Food Chem 56:11941–11946. https://doi.org/10.1021/jf802364t

McCormick CL (1985) Controlled activity polymers with pendent metribuzin. Effect of structure on hydrolytic release. Ann N Y Acad Sci 446:76–92. https://doi.org/10.1111/j.1749-6632.1985.tb18392.x

Oliveira JL, Campos EVR, Pereira AES et al (2018) Zein nanoparticles as eco-friendly carrier systems for botanical repellents aiming sustainable agriculture. J Agric Food Chem 66(6):1330–1340. https://doi.org/10.1021/acs.jafc.7b05552

Oliveira JL, Campos EVR, Camara MC et al (2019) Highlights in nanosustained release systems aiming agriculture applications. Journal of Siberian Federal University. Biology 12(3):311–328

Ong SY, Sudesh K (2016) Effects of polyhydroxyalkanoate degradation on soil microbial community. Polym Degrad Stab 131:9–19. https://doi.org/10.1016/j.polymdegradstab.2016.06.024

Prudnikova S, Volova T (2012) Ekologicheskaya rol' poligidroksialkanoatov kak analoga sinteticheskikh plastikov: mekhanizmy biodegradatsii v okruzhayushchey srede i vzaimodeystviya s mikroorganizmami. Krasnoyarsk Writer, Krasnoyarsk

Prudnikova SV, Boyandin AN, Kalacheva GS et al (2013) Degradable polyhydroxyalkanoates as herbicide carriers. J Polym Environ 21:675–682. https://doi.org/10.1007/s10924-012-0561-z

Qian K, Shi T, He S et al (2013) Release kinetics of tebuconazole from porous hollow silica nanospheres prepared by miniemulsion method. Micropor Mesopor Mater 169:1–6. https://doi.org/10.1016/j.micromeso.2012.10.017

Rehab A, Akelah A, El-Gamal MM (2002) Controlled-release systems based on the intercalation of polymeric metribuzin onto montmorillonite. J Polym Sci Part A: Polym Chem 40:2513–2525. https://doi.org/10.1002/pola.10326

Roy A, Singh SK, Bajpai J et al (2014) Controlled pesticide release from biodegradable polymers. Centr Eur J Chem 12(4):453–469. https://doi.org/10.2478/s11532-013-0405-2

Sahoo S, Manjaiah KM, Datta SC et al (2014) Kinetics of metribuzin release from bentonite-polymer composites in water. J Environ Sci Health Part B 49:591–600. https://doi.org/10.1080/03601234.2014.911578

Sarkar DJ, Kumar J, Ajoy S et al (2018) Agriculture: polymers in crop production pesticides. In: Mishra M (ed) Encyclopedia of polymer applications. Taylor & Francis, London

Shershneva AM, Murueva AV, Zhila NO et al (2019) Antifungal activity of P3HB microparticles containing tebuconazole. J Environ Sci Health Part B 54(3):196–204. https://doi.org/10.1080/03601234.2018.1550299

Shishatskaya E, Menzianova N, Zhila N et al (2018) By toxic effects of the fungicide tebuconazole on fusarium-infected wheat plants. Plant Physiol Biochem 132:400–407. https://doi.org/10.1016/j.plaphy.2018.09.025

Singh AK, Srivastava JK, Chandel AK et al (2012) Biomedical applications of microbially engineered polyhydroxyalkanoates: an insight into recent advances, bottlenecks, and solutions. Appl Microbiol Biotechnol 103(5):2007–2032. https://doi.org/10.1007/s00253-018-09604-y

Suave J, Dall'Agnol EC, Pezzin APT et al (2010) Biodegradable microspheres of poly(3-hydroxybutyrate)/poly(e-caprolactone) loaded with malathion pesticide: preparation, characterization, and in vitro controlled release testing. J App Polym Sci 117:3419–3427. https://doi.org/10.1002/app.32082

Sudesh K, Hideki A (2010) Practical guide to microbial polyhydroxyalkanoates. Smithers Rapra Technology, Shrewsbury

Undabeytia T, Recio E, Maqueda C et al (2011) Reduced metribuzin pollution with phosphatidylcholine–clay formulations. Pest Manag Sci 67:271–278. https://doi.org/10.1002/ps.2060

Volova TG, Voinova ON, Kalacheva GS et al (2008) The prospects of the use of resorbable polyesters for designing safe pesticides. Dokl Biol Sci 419(1):100–103

Volova TG, Shishatskaya EI, Sinskey AJ (2013) Degradable polymers: production, properties, applications. Nova Science Pub Inc, New York

Volova T, Zhila N, Vinogradova O et al (2016a) Characterization of biodegradable poly-3-hydroxybutyrate films and pellets loaded with the fungicide tebuconazole. Environ Sci Pollut Res 23(6):5243–5254. https://doi.org/10.1007/s11356-015-5739-1

Volova TG, Zhila NO, Vinogradova ON et al (2016b) Constructing herbicide metribuzin sustained-release formulations based on the natural polymer poly-3-hydroxybutyrate as a degradable matrix. J Environ Sci Health Part B 51:113–125. https://doi.org/10.1080/03601234.2015.1092833

Volova T, Zhila N, Kiselev E et al (2016c) Poly(3-hydroxybutyrate)/metribuzin formulations: characterization, controlled release properties, herbicidal activity, and effect on soil microorganisms. Environ Sci Pollut Res 23:23936–23950. https://doi.org/10.1007/s11356-016-7636-7

Volova TG, Prudnikova SV, Zhila NO et al (2017a) Efficacy of tebuconazole embedded in biodegradable poly-3-hydroxybutyrate to inhibit the development of *Fusarium moniliforme* in soil microecosystems. Pest Manag Sci 73(5):925–935. https://doi.org/10.1002/ps.4367

Volova TG, Vinnik YS, Shishatskaya EI, Markelova NM, Zaikov GE (2017b) Natural-based polymers for biomedical applications. Apple Academic Press, Oakville

Volova TG, Prudnikova SV, Zhila NO (2018) Fungicidal activity of slow-release P (3HB)/TEB formulations in wheat plant communities infected by *Fusarium moniliforme*. Environ Sci Pollut Res 25(1):552–561. https://doi.org/10.1007/s11356-017-0466-4

Voss KA, Howard PC, Riley RT et al (2002) Carcinogenicity and mechanism of action of fumonisin B1: a mycotoxin produced by *Fusarium moniliforme* (= F. verticillioides). Cancer Detect Prev 26(1):1–9. https://doi.org/10.1016/S0361-090X(02)00011-9

Yusoff SNM, Kamari A, Aljafree NFA (2016) A review of materials used as carrier agents in pesticide formulations. Int J Environ Sci Technol 13(12):2977–2994. https://doi.org/10.1007/s13762-016-1096-y

Zhang S, Yang G, Zheng Z et al (2009) On-line preconcentration and analysis of metribuzin residues in corn fields by use of a molecularly imprinted polymer. Chromatographia 69:615–619. https://doi.org/10.1365/s10337-008-0862-5

Zhila N, Murueva A, Shershneva A et al (2017) Herbicidal activity of slow-release herbicide formulations in wheat stands infested by weeds. J Environ Sci Health Part B 52(I.10):729–735. https://doi.org/10.1080/03601234.2017.1356668

# Smart Nano-Chitosan for Fungal Disease Control

**Khaidem Aruna Devi, Damyanti Prajapati, Ashok Kumar, Ajay Pal, Deepa Bhagat, Braj Raj Singh, Alok Adholeya, and Vinod Saharan**

**Abstract** The excessive and irrational use of synthetic fungicides has perturbed us with irrevocable soil-water-air contaminations, development of resistance in microbes, and disturbing biosphere. Thus, search for biodegradable/ecofriendly materials has emerged as the main goal to replace/reduce the synthetic fungicides in agriculture for crop protection. Under this scenario, nanobiotechnology seems to be a boon for the synthesis of ecofriendly, biocompatible, and safe fungicides which will not only improve the soil health and the defense system of plants but also help in obtaining healthy food for the continuously growing population. Among the available biomaterials/biopolymers, chitosan is being explored as new generation smart material to be used in agriculture especially for plant protection. This chapter describes various chitosan-based nanomaterials (NMs) which have been used from laboratory to field for control of fungal disease in crops.

**Keywords** Chitosan-based nanomaterials · Essential oil · Antifungal activity Defense system

K. A. Devi · D. Prajapati · A. Kumar · V. Saharan (✉)
Nano Research Facility Laboratory, Department of Molecular Biology and Biotechnology, Rajasthan College of Agriculture, Maharana Pratap University of Agriculture and Technology, Udaipur, Rajasthan, India

A. Pal
Department of Biochemistry, College of Basic Sciences and Humanities, Chaudhary Charan Singh Haryana Agricultural University, Hisar, Haryana, India

D. Bhagat
ICAR-National Bureau of Agricultural Insect Resources (NBAIR), Bangalore, India

B. R. Singh · A. Adholeya
TERI-Deakin Nanobiotechnology Centre, The Energy and Resources Institute (TERI), Gurugram, Haryana, India



L. F. Fraceto et al. (eds.), *Nanopesticides*,
https://doi.org/10.1007/978-3-030-44873-8_2

# 1 Introduction

Past few years have witnessed a tremendous growth in world's total population which is expected to reach up to 8.6 billion by 2030. This sets a great difficulty for the scientists in achieving sustainable agriculture production in view of global warming. To ensure the adequate food supply for growing population, application of synthetic agrochemicals has, therefore, increased many folds. Global application of agrochemicals is ~4.6 million tons, 90% of which gets runs-off into the environment and seep to agricultural products. Pesticides are among these agrochemicals which are being used since long to provide protection against damage caused by severe phytopathogens. Plant pathogens cause significant damage to almost all crops worldwide and this loss compels the farmers to use more and more pesticides to get maximum crop production (Zhang et al. 2011a, b). It is estimated that pesticides are used for one-third of total agricultural production; due to which crop loss declined by 35 to 42% (Pimentel 1997; Liu et al. 2002; Zhang et al. 2011a, b). 32%, 78%, and 54% loss in cereals, fruits, and vegetables, respectively, may be caused if pesticides are not used (Cai 2008).

Global consumption of these pesticides is increasing day by day. The average annual usage of fungicides and bactericides (kg/ha) from 2010 to 2014 in Japan is the greatest (7.934) followed by Mexico (3.275), France (2.162), UK (1.332), Germany (1.194), and Brazil (0.814), which are higher than global average (0.32). The last two countries in the list are USA (0.229) and India (0.058) (Zhang 2018). Although these agrochemicals have significantly contributed to agriculture production, their reckless and non-judicious use has been causing an irreversible damage to the ecosystem due to their nondegradable and toxic nature (Kumaraswamy et al. 2018). Further, most of these agrochemicals are not fully absorbed by plants and seep into the soil/groundwater and eventually get accumulated in living organisms too (Alister and Kogan 2006; Dietz and Herth 2011; Kah 2015; Marutescu et al. 2017). Global pesticide use has also resulted in the loss of biodiversity (Zhang et al. 2011b; Kumar et al. 2013). In addition, pesticide use has led to various human/animal diseases and injured human fecundity and intelligence quotient in past few years (Chen et al. 2004; Zhang 2018). Moreover, the increment of resistance in plant pathogens against these agrochemicals has become a serious issue (Hahn 2014; Xing et al. 2017). Due to this, either new kinds of agrochemicals have been developed or higher doses of the existing ones have been used which in turn has increased the cost and further expedites the resurgence of new plant pathogens.

With the emergence of nanoscience, application of nanotechnological tools has raised hope to deliver new generation agrochemicals which are safe to environment and effective at low doses. New generation pesticides could be comprised of nanostructured materials which act on target in slow/controlled release manner when need arises. Unexplored various bioactive compounds (inorganic and organic) can be used alone or in composite forms through nanotechnology to deliver novel nano-based products for use in agriculture for crop protection especially against fungal disease. Therefore, various inorganic and organic materials for synthesis of nanomaterials (NMs) having biocompatibility, biodegradability, wide biological activities, and ecological safety characteristics are in the forefront list of scientists (Shukla et al. 2013; Kah and Hofmann 2014; Kashyap et al. 2015).

In pursuit of this, chitosan, $\beta$-(1,4)-2-amino-2-deoxy-D-glucose, a hetero-amino-polysaccharide which can easily be obtained from the waste produce of shrimp, crab shells, and cell wall of fungi (Katiyar et al. 2015; Malerba and Cerana 2016), has been in high demand. Chitosan NMs can competently perform many biological applications due to their small size, higher surface area, and cationic nature. Furthermore, they are excellent blending materials for different organic and inorganic molecules due to the availability of functional groups in their structures (Choudhary et al. 2019a, b). Utility of chitosan has been acknowledged in developing chitosan nanoparticles (NPs) either alone or in combination with inorganic and organic substances. The developed chitosan-based nanocomposites could ensure slow, systemic, targeted, and protected release of active ingredients to improve their efficacy and avoid toxicity to environment (Saharan et al. 2015; Saharan and Pal 2016a, b; Choudhary et al. 2017a, b). Chitosan functionalized with various inorganic and organic inputs might ultimately lead to precision farming in a cost-effective manner and can deliver a smart chitosan-based nano-agri-input.

Herein, this chapter highlights various chitosan-based NMs, in-depth, which have potential to protect the plants from fungal diseases (Table 1).

# 2 Chitosan-Based NMs

Chitosan, being an excellent antimicrobial, plant growth regulator and plant elicitor, has been explored in sole as well as functionalized NM forms with other bioactive compounds of inorganic and organic nature. Herein, we have classified chitosan-based NMs in three categories (a) sole chitosan NMs, (b) inorganic based chitosan NMs, and (c) organic based chitosan NMs.

## *2.1 Sole Chitosan NMs*

Since last few years, chitosan NMs have been explored for their diverse biological activities. They have been tested against many plant pathogenic fungi and found to be effective in significantly controlling fungal growth.

Chitosan NPs, in in vitro experiments, at a concentration of 0.6% (w/v), significantly delayed mycelia growth of *Rhizopus* sp. *Colletotrichum capsici*, *Colletotrichum gloeosporioides,* and *Aspergillus niger.* NPs exhibited better tendency as compared with bulk chitosan towards reduction of mycelia growth. In addition, chitosan coated/treated chickpea (*Cicer arietinum*) seeds had higher vigor and very good antifungal activity which could be explained by two facts: (1)

**Table 1** Chitosan NMs used against various fungal pathogens

| Sr. no. | NMs | Size (nm) | Zeta potential (mV) | PDI | Observations | References |
|---|---|---|---|---|---|---|
| 1. | Chitosan NPs | N/A | N/A | N/A | In vitro antifungal activity against *various chili fungal disease* | Chookhongkha et al. (2012) |
| 2. | Chitosan NPs | 192.2 | +45.3 | 0.60 | In vitro antifungal activity against *Alternaria Alternata, Macrophomina. Phaseolina Rhizoctonia solani* | Saharan et al. (2013) |
| 3. | Chitosan NPs | 89.8 | −37 | 0.22 | In vitro antifungal activity against *Pyricularia grisea, Alternaria solani, Fusarium oxysporum,* and promote growth of chickpea seedlings | Sathiyabama and Parthasarathy (2016) |
| 4. | Chitosan NPs | 83.3 | −28 | 0.31 | In vitro and in vivo antifungal activity against rice blast caused by *Pyricularia grisea* | Manikandan and Sathiyabama (2016) |
| 5. | Chitosan NPs | 180.9 | +45.6 | 0.31 | In vitro and greenhouse antifungal activity against wheat head Blight caused by *Fusarium graminearum* | Kheiri et al. (2017) |
| 6. | Cu(II)-chitosan Nanogel | 220 | +40 | 0.20 | In vitro antifungal activity against *Fusarium graminerarium* | Brunel et al. (2013) |
| 7. | Cu–chitosan NPs | 196.4 | +88 | 0.50 | In vitro antifungal activity against *Alternaria alternata, Macrophomina phaseolina* and *Rhizoctonia solani* | Saharan et al. (2013) |
| 8. | Cu–chitosan NPs | 2.5–25 | N/A | N/A | In vitro antifungal activity against *Fusarium solani* | Vokhidova et al. (2014) |

(continued)

**Table 1** (continued)

| Sr. no. | NMs | Size (nm) | Zeta potential (mV) | PDI | Observations | References |
|---|---|---|---|---|---|---|
| 9. | Cu–chitosan NPs | 374.3 | +22.6 | 0.33 | In vitro inhibition of *Alternaria solani* and *Fusarium oxysporum*, and Growth promotion of tomato seedlings | Saharan et al. (2015) |
| 10. | Cu–chitosan NPs | 2–3 | N/A | N/A | In vitro inhibition of *Rhizoctonia solani* and *Sclerotium rolfsii* | Rubina et al. (2017) |
| 11. | Cu–chitosan NPs | 374.3 | +22.6 | 0.33 | In vitro and in vivo antifungal activity against *Curvularia lunata* in maize | Choudhary et al. (2017a, b) |
| 12. | Chitosan–saponin NPs | 373.9 | +31 | 1 | In vitro antifungal activity against *Alternaria alternata, Macrophomina Phaseolina,* and *Rhizoctonia solani* | Saharan et al. (2013) |
| 13. | Oleoyl-chitosan NPs | 296.9 | N/A | N/A | In vitro inhibition of spore germination and mycelia growth of *Verticillium Dahaliae* | Xing et al. (2017) |
| 14. | Zn-chitosan NPs | 200–300 | +34 | 0.22 | In vitro and in vivo antifungal activity against *Curvularia lunata* in maize | Choudhary et al. (2019a, b) |
| 15. | Salicylic acid-loaded chitosan NPs | 368.7 | +34.1 | 0.1 | In vitro and in vivo antifungal activity against post-flowering stalk rot (PFSR) of maize caused by *Fusarium. verticillioides* | Kumaraswamy et al. (2019) |
| 16. | Ag-chitosan NPs | 10–20 | N/A | N/A | In vitro mycelium inhibition of *Rhizoctonia solani, Aspergillus flavus* and *Alternaria .alterneta* | Kaur et al. (2012) |

(continued)

**Table 1** (continued)

| Sr. no. | NMs | Size (nm) | Zeta potential (mV) | PDI | Observations | References |
|---|---|---|---|---|---|---|
| 17. | Ag-chitosan NPs | <100 | N/A | N/A | In vitro botryticidal activity against gray mold (*Botrytis cinerea*) in strawberry | Moussa et al. (2013) |
| 18. | Ag-chitosan NPs | N/A | N/A | N/A | In vitro antifungal activity against *Aspergillus flavus* and *Aspergillus terreus* | Mathew and Kuriakose (2013) |
| 19. | Silica-chitosan NPs | 110 | N/A | N/A | In vitro antifungal activity against *Phomopsis asparagi* | Cao et al. (2016) |
| 20. | Chitosan-peppertree (*Schinus molle*) essential oil (CS-EO) NPs | 754 | N/A | 9.1 ± 1.74 | In vitro antifungal activity against *Aspergillus parasiticus* spores | Luque-Alcaraz et al. (2016) |
| 21. | *Mentha piperita* essential oils in chitosan–cinnamic acid nanogel | N/A | N/A | N/A | In vivo antifungal activity against *Aspergillus flavus* in tomato during post-harvest storage | Beykia et al. (2014) |
| 22. | *Zataria multiflora* essential oils in chitosan nanoparticles | 125–175 | N/A | N/A | In vitro and in vivo botryticidal activity against gray mold (*Botrytis cinerea*) in strawberry at post-harvest stage | Mohammadi et al. (2015) |
| 23. | Chitosan boehmite-alumina nanocomposites films and thyme oil | N/A | N/A | N/A | Inhibited *Monilinia laxa* which caused brown rot infection during post-harvest storage of peaches | Cindi et al. (2015) |
| 24. | Thiadiazole-functionalized chitosan derivatives | N/A | N/A | N/A | In vitro antifungal activity against *Colletotrichum lagenarium*, *Phomopsis asparagi*, and *Monilinia fructicola* | Li et al. (2013) |

(continued)

**Table 1** (continued)

| Sr. no. | NMs | Size (nm) | Zeta potential (mV) | PDI | Observations | References |
|---|---|---|---|---|---|---|
| 25. | Fungicide zineb (Zi) and chitosan-Ag nanoparticles | 4.11 ± 0.37 | N/A | N/A | In vitro antifungal activity against *Neoscytalidium dimidiatum* which caused brown rot disease in dragon fruit during post-harvest storage | Ngoc and Nguyen (2018) |
| 26. | Chitosan-Thyme-oregano, thyme-tea tree and thyme-peppermint EO mixtures | 76.58<br>69.9<br>57.9 | 0.25<br>0.21<br>0.32 | −51<br>−50<br>−53 | In vitro antifungal activity against *Aspergillus niger, Aspergillus flavus, Aspergillus parasiticus*, and *Penicillium chrysogenum*, reducing their growth by 51–77% in rice plant during post-harvest storage | Hossaina et al. (2019) |
| 27. | Chitosan-thymol nanoparticles | 175 ± 21 | 0.4 ± 0.1 | 37 ± 2.7 | In vitro antifungal activity against the mycelial growth of *Botrytis cinerea* in blueberries and tomato cherries during post-harvest storage | Medina et al. (2019) |
| 28. | Chitosan-*Cymbopogon martinii* essential oil | 455–480 | N/A | 39.3–37.2 | In vitro antifungal activity against *Fusarium graminearum*. Which causes Fusarium head blight disease in maize during post-harvest storage | Kalagatur et al. (2018) |

chitosan directly inhibits mycelia growth and (2) seeds treated with chitosan produce more phenolic compounds and lignin (Chookhongkha et al. 2012).

Chitosan NPs, synthesized using ionic gelation method, were investigated against phytopathogenic fungi (*Alternaria alternata, Macrophomina phaseolina*, and *Rhizoctonia solani*) at various concentrations ranging from 0.001 to 0.1% under in vitro conditions. The maximum growth inhibitory effects (87.6%) were found against *Macrophomina phaseolina* at 0.1% concentration. The radial growth of *Rhizoctonia solani* was reduced by all concentrations of chitosan NPs in a dose-dependent manner (Saharan et al. 2013).

In another study, a biological method was used to prepare chitosan NMs using anionic proteins isolated from *Penicillium oxalicum* culture. These biologically synthesized chitosan NMs were significantly found to inhibit the growth of *Pyricularia. grisea, Alternaria solani, Fusarium oxysporum* (Sathiyabama and Parthasarathy 2016). The inhibition rate for *Pyricularia grisea, Fusarium oxysporum ciceri,* and *Alternaria solani* was found to be 92%, 87%, and 72%, respectively. Seed treatment with these NPs exhibited positive morphological effect including enhanced percent germination, vegetative biomass, and seed vigor index of chickpea (*Cicer arietinum*) seedlings. The efficacy of NMs could be attributed to their size as well as highly permeable nature towards biological membranes (Shukla et al. 2013; Saharan et al. 2015). Their small size, lower PDI value, and higher zeta-potential make these NMs more stable and effective against tested phytopathogens.

A 100% suppression of rice (*Oryza sativa*) blast disease symptoms was observed, in vivo, under detached leaf condition, when treated with chitosan NMs prepared using ionic gelation method (Manikandan and Sathiyabama 2016). Chitosans of different molecular weights have also been used to prepare chitosan NMs and check their antifungal property against Fusarium head blight (FHB) in wheat (*Triticum aestivum*) caused by *Fusarium graminearum*. The dynamic light scattering (DLS) study indicated variable z-average size of NMs (180.9, 339.4225.7, and 595.7 nm). Different concentrations of these NMs were tested to evaluate the inhibitory effect on this pathogen, and the maximum growth reduction (77.5%) was found at 5000 ppm. In greenhouse trials, the area under disease progress curve (AUDPC) decreased in plants treated with NMs (Kheiri et al. 2017).

## 2.2 *Inorganic Based Chitosan NMs*

Metals such as copper (Cu), zinc (Zn), and silver (Ag) have been explored in developing chitosan-based NMs as chitosan can easily chelate the metals (Choudhary et al. 2017a). Functionalized chitosan with metals has enabled chitosan NMs more suitable for controlling fungal diseases in plant.

### 2.2.1 Cu–Chitosan NMs

Copper (Cu) is a constituent of many enzymes like ascorbic acid oxidase, laccase, phenolase, cytochrome oxidase, etc., and is therefore vital for photosynthesis, respiration, and carbon-nitrogen balance. Traditionally, it has been used as antifungal agent in many commercially available pesticides (Saharan et al. 2015). Cu, therefore, has been tested for synthesis of smart chitosan-based NMs for controlling fungal disease in plants.

In in vitro model, Saharan et al. (2013) observed 89.5, 63.0, and 60.1% growth inhibition of *Alternaria alternata, Macrophomina phaseolina,* and *Rhizoctonia solani,* respectively, at various concentrations of Cu–chitosan NMs. The higher zeta-potential of chitosan NMs bestowed them a greater binding affinity for negatively charged fungal membrane. In fungi, Cu (II) reduces to Cu (I) which produces toxic $H_2O_2$, resulting in destruction of fungal cell viability. Pure chitosan nanogels were produced to adsorb Cu (II) and assess their antimicrobial activities against *Fusarium graminarium*. Antifungal activity was observed due to the strong synergistic effect between chitosan and Cu. The MIC (Minimum Inhibitory Concentration) of Cu (II) was observed as 250 μg/mL which decreased exponentially upon addition of low amounts of chitosan either in solution or dispersion. Therefore, Cu (II) and chitosan not only seem to be biocompatible and bioactive, but also display a strong synergistic effect in antifungal activities (Brunel et al. 2013).

Porous Cu–chitosan NMs were also examined for their antifungal efficacy in tomato (*Solanum lycopersicum* Mill). DLS, TEM, FTIR, SEM-EDS, and AAS were used for physico-chemical characterization of NMs. In in vitro model, 0.12% concentration caused 70.5 and 73.5% inhibition of mycelia growth and 61.5 and 83.0% inhibition of spore germination in *Alternaria solani* and *Fusarium oxysporum,* respectively. In pots, tomato plants exhibited 87.7% percent efficacy of disease control (PEDC) in early blight, while 61.1% in Fusarium wilt. Cu–chitosan NMs markedly exhibited higher antifungal activity along with only 1–2 mm small black or brown lesions as compared with control plants. At 0.10 and 0.12% concentrations, Cu–chitosan NMs were equally effective on early blight disease as was the commercial fungicide (Saharan et al. 2015).

These NMs were further tested to boost defense responses in *Zea mays* maize crop against Curvularia leaf spot (CLS) disease under in vitro as well as field conditions (Choudhary et al. 2017b). Plants showed significant defense response through higher activities of antioxidant (superoxide dismutase, SOD and peroxidase, POD) and defense enzymes (polyphenol oxidase, PPO and phenylalanine ammonia-lyase, PAL). In NMs treated plants, disease symptoms in the form of small lesions without chlorosis were visualized after 7–8 days of fungal inoculation in pot experiment. PAL activity increased from 46.15 to 66.66%, while PPO activity increased from 3.05 to 16.39%. Application of these NMs increased the activities of POD, PAL, and PPO in plant which further enhanced the production of suberin, melanin, and lignin for cell wall strengthening acting as a mechanical barrier to invading plant pathogen (Kuźniak and Urbanek 2000; Fugate et al. 2016). In pot experiments, at 0.04 to 0.16% concentrations, Cu–chitosan NPs significantly controlled CLS disease while the same effect was observed at 0.12 to 0.16% concentrations of Cu–chitosan NPs in field condition. Study further revealed that these NMs are pH responsive as the Cu release rate increases as pH decreases in plant cell due to fungal infection. The released Cu, therefore, acts smartly on invading fungi (Rubina et al. 2017). Cu–chitosan NMs were prepared using metal vapor synthesis method and their in vitro antifungal effects were checked on hyphal morphology and sclerotia formation in *Sclerotium rolfsii* and *Rhizoctonia solani* AG-4. These NMs were found effective against both the tested fungi in a dose dependent manner (Rubina et al. 2017).

### 2.2.2 Zn-Chitosan NMs

Zinc (Zn) is an essential micronutrient which helps the plants in maintaining their cellular homeostasis. It plays a crucial role during plant's reproductive and grain filling stage and therefore its deficiency or unavailability can result into poor growth and lower grain yield. Zn helps to carry out several biological processes such as electron transport, gene expression, protein and auxin metabolism, structural and functional integrity of biomembranes. It has been found that Zn deficiency in crop also leads to disease suitability.

Zn-chitosan NMs were synthesized and evaluated for their antifungal activity via seed priming and foliar application in maize plants (Fig. 1). These NMs (0.01–0.16%) showed strong in vitro antifungal activity as evident by inhibition of fungal spore germination. The plant immunity was further improved due to enhanced antioxidant and defense enzymes activity, balanced reactive oxygen species (ROS) levels, and more lignin accumulation caused by these NMs. In the field, 0.01–0.16% concentrations were used for seed treatment and foliar application which significantly controlled CLS disease and enriched the grain with Zn micronutrient from 41.27 to 62.21 μg/g DW.

Zn-chitosan NMs displayed high encapsulation efficiency (82%) and exhibited slow release of Zn ions. At acidic pH (from 3 to 1), 20.84–42.80% Zn ions were released rapidly due to protonation of chitosan (Choudhary et al. 2017a, b; Kumaraswamy et al. 2018). It is important as these NMs act strongly when plants are infected with fungi since sudden exposure of Zn (at low pH caused by fungi) creates ions toxicity which averts the growth of fungal cells. Zn-chitosan NMs controlled CLS disease up to 39.5% with significantly higher grain yield. Hence, these

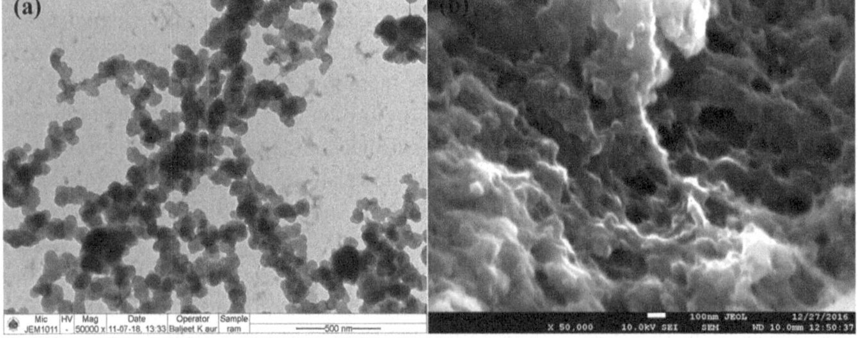


**Fig. 1** TEM and SEM micrograph of Zn-chitosan NMs (Choudhary et al. 2019a, b, Copyright permission from Elsevier)

NPs could be an effective growth promoting, fungal disease controlling, and micronutrient fortifying agent in maize crop (Choudhary et al. 2019a, b).

### 2.2.3 Ag-Chitosan NPs

Silver (Ag)displays multiple modes of inhibitory action against microorganisms (Park et al. 2006). Although metallic Ag is relatively nonreactive, Ag nanoparticles are exceedingly reactive because of their high ability to generate $Ag^+$ ions, which are well known to induce ROS production. ROS are highly detrimental to microbial cells as they can damage surface and interior proteins, lipids, and nucleic acids (Storz and Imlayt 1999; Hwang et al. 2008). Therefore, Ag may be used to prepare NMs as an antifungal treatment for various seed borne plant pathogens. Ag-chitosan NMs exhibited the highest inhibition against *Aspergillus* followed by *Alternaria* and *Rhizoctonia*species. The observed zone of inhibition was 19.66 ± 0.28, 16.33 ± 0.29, 12.66 ± 0.76 against *Aspergillus*, *Alternaria*, and *Rhizoctonia*, respectively. Thus, Ag-chitosan NMs may be used as an alternative to fungicides for controlling seed borne phytopathogens (Kaur et al. 2012).

Nano Ag with irradiated chitosan NMs were investigated along with native chitosan for their ability to hamper the growth of *Botrytis cinerea* Pers, the gray mold of strawberry (*Fragaria ananassa)*, that causes great losses in other agricultural crops too. Ag-irradiated chitosan (IrCTS), as compared with its native fungal chitosan, was found more effective and showed highest antifungal activity at a minimal inhibitory concentration of 125 μg/mL (comprised of 20% Ag and 80% IrCTS). *Botrytis cinerea* treated with the NMs had an obvious alteration in mycelial shape as well as moderate lysis in fungal hyphae. Coating with these NMs led to 90% control of gray mold infection after 7 days of storage and treated fruits still gave fresh-like appearance at the end of storage. Hence, coating with nano Ag-IrCTS solution could be highly recommended regarding its efficiency in prohibiting *Botrytis cinerea* growth, preventing gray mold decay and enhancing the overall quality of coated strawberry fruits (Moussa et al. 2013).

Chitosan was functionalized with 4-((E)-2-(3-hydroxynaphthalen-2-yl) diazen-1-yl) benzoic acid by coupling of hydroxyl functional groups of chitosan with carboxylic acid group of dye by DCC coupling method. The Ag NPs were prepared by sol-gel method while Ag NPs-encapsulated functionalized chitosan was prepared by phase transfer method. The products were characterized by FTIR, UV-VIS, fluorescence, and Nuclear Magnetic Resonance (NMR) spectroscopic methods and by SEM and TEM analysis. The light-fastening properties of the chromophoric system were enhanced when attached to chitosan and they were further improved by the encapsulation of Ag NMs. Their antibacterial analysis was carried out against *Aspergillus flavus* and *Aspergillus terreus* by diffusion plate method and found inhibition zone (20.2 ± 0.15 and 27.0 ± 0.38 mm), showing that NPs can be used for antifungal applications (Mathew and Kuriakose 2013).

### 2.2.4 AgNPs, Chitosan, and Fungicide Zineb (Zi) NMs

Ngoc and Nguyen (2018) examined the synergistic effect of AgNPs, chitosan (CS), and fungicide zineb (Zi) as antifungal materials against *Neoscytalidium dimidiatum* in (*Hylocereus undatus*) dragon fruit. The researchers synthesized Ag@CS by encapsulating AgNPs in CS polymer and then combined with Zi. 4.11 ± 0.37 nm was recorded as diameter of spherical nanoparticles as confirmed by TEM. Ag@CS showed better antifungal ability as compared with each component alone against *N. dimidiatum*. At 5 ppm of Ag@CS, the zone of inhibition was found to be 15.00 ± 0.00 mm which was better than that of Ag alone (13.33 ± 0.58 mm) at 10 ppm . When pure Zi at 500 and 1000 ppm (inhibition zone, 5.00 ± 0.00 mm) was incapable of removing the fungi, the zone of inhibition of Ag@CS-Zi increased to 12.00 ± 0.00 mm, which was nearly equivalent to 5 ppm Ag (12.33 ± 0.58 mm) and much higher than 5000 ppm of Zi (9.00 ± 0.00 mm). Ag@CS-Zi at 2500 ppm of Zi gave inhibition zone of 20.67 ± 0.58 which showed its high antifungal activity as compared with each of individual component.

## 2.3 *Organic Based Chitosan NMs*

Essential oils (EOs) which obtained from plants are aromatic and volatile. They are present in stems, bark, leaves, fruits, etc. (Oussalah et al. 2006). Compounds such as terpenoids and phenolic acids are some of the EOs which are extracted from plants. The food industry used EOs as natural antimicrobials because of their antifungal and antimicrobial properties. (Tassou et al. 1995; Burt 2004; du Plooy et al. 2009).

Many reports have shown that NMs functionalized with essential oils have significant antimicrobial activity because of their chemical stability and solubility, decreased fast evaporation and degradation of EO active components. The controlled and sustained released nature of encapsulated EOs which enhance their bioavailability and efficacy against multidrug-resistant pathogens (Chouhan et al. 2017). As EOs have the property of hydrophobicity, it helps in the partition of lipid present in the cell membrane of the pathogen resulting in the leakage of molecules and ions leading to its death. The activity of essential oils depends on its composition, functional groups present in active components, and their synergistic interactions. Nanoencapsulation of bioactive compounds can be used as an efficient approach to enhance the physical stability of the active ingredient. It can also prevent their interactions with the food components, thus enhancing their bioactivity due to their subcellular size (Donsi et al. 2011). Chitosan, having the properties of biocompatibility, low toxicity, and biodegradability, its encapsulating with EOs is of much interest. (Muzzarelli 2010; Donsi et al. 2011; Harris et al. 2011; Luo et al. 2011).

### 2.3.1 *Zataria multiflora* Essential Oils in Chitosan

Chemical fungicides have been used as a preventive measure of fungal attack during post-harvest storage. However, use of these synthetic fungicides has raised health related questions. So, application of plant EOs at post-harvest stage has been considered as an alternative management to prevent post-harvest decay (Aloui et al. 2014). *Zataria multiflora* Boiss EOs (ZEO) is one of the EOs which appear as potential natural compounds for controlling post-harvest loss in fruits. Quantitatively, the most abundant components in hydro-distilled ZEO are oxygenated monoterpenes (~70%) followed by monoterpene, sesquiterpenes, and oxygenated sesquiterpenes (Sajed et al. 2013). The volatile compounds of EOs are used to maintain fruit quality and decrease fungal decay, but they are easily degraded by high temperature, pressure, light, and oxygen. Furthermore, they are insoluble in water and, for certain applications, a controlled release is required (Martin et al. 2010). Therefore, sustained and controlled released is crucial to obtain maximum benefits of using EOs as antimicrobial agents.

Nano-/microencapsulation technology of these compounds can be a practical and efficient approach to solve some of these problems such as the physical instability. Mohammadi et al. (2015) investigated the nanoencapsulation of ZEO in chitosan nanoparticles (CSNPs) to enhance antifungal activity and stability of the oils against *Botrytis cinerea*, the causal agent of gray mold disease in strawberry. Ionic gelation method was used for encapsulation of ZEO with CSNPs and found an average size of 125–175 nm, as observed by TEM. In vitro release studies also demonstrated a controlled and sustained release of ZEO for 40 days. There was a superior activity of ZEO when encapsulated by CSNPs under both in vitro and in vivo conditions in comparison with unmodified ZEO against *Botrytis cinerea*. At 1500 ppm of encapsulated oils, both disease severity and incidence of *Botrytis*-inoculated strawberries significantly decreased during 7 days of storage at 4 °C followed by 2–3 more days at 20 °C. These findings showed the potential role of CSNPs as a controlled release system for EOs in order to enhance antifungal activities.

### 2.3.2 Chitosan-Thymol Nanoparticles

Thymol (2-isopropyl-5-methylphenol) is the major antimicrobial agent of the aromatic plant thyme (*Thymus vulgaris*). It has a strong antimicrobial property because of its capability of binding bacterial proteins and giving rise to disintegration and permeability of the cell membrane (Juven et al. 1994). Thymol affects energy-generating processes, which makes the cell unable to recover (Ahmad et al. 2011). It, therefore, may be incorporated as a natural antifungal agent in an active packaging to increase shelf-life of foods (Mirdehghan and Valerob 2017).

Medina et al. (2019) conducted the experiment to improve the performance of quinoa protein/chitosan edible films on the extension of post-harvest life of blueberries (*Cyanococcus*) and tomato cherries by addition of chitosan-thymol nanoparticles prepared by ionic gelation method. They obtained NPs with a hydrodynamic diameter (175 ± 21 nm) similar to the diameter measured by TEM (153 ± 42 nm). The PDI and zeta-potential values were 0.4 ± 0.1 and 37 ± 2.7 mV, respectively. Inhibition of radial mycelia growth by chitosan-thymol nanoparticles (CTNPs), chitosan nanoparticles (CNPs), and chitosan/thymol (CT) blend was evaluated in different dilutions added to the potato dextrose agar having the same concentrations of active compounds. CTNPs formulation recorded 100% inhibition for all dilutions (10, 25, and 50%, v/v), whereas CT blend showed total inhibition only at a higher concentration (50% v/v). CNP showed lowest inhibition of mycelia growth (74%) at higher concentration (50% v/v). Therefore, CTNPs was the only treatment that showed inhibitory effect at the lowest dose (10%).

### 2.3.3 Chitosan-Thyme-Oregano, Thyme-Tea Tree and Thyme-Pepper Mint Essential Oils

Bio-nanocomposite based packaging containing plant-derived EOs are presently playing an important role in controlling fungal contamination and proliferation in processed food (Hossain et al. 2017). EOs are more efficiently used in foods when encapsulated in proper delivery systems to overcome dosage limitations and increase the biological stability of active compounds (Van Long et al. 2016). Bioactivities of EOs get enhanced when encapsulated at the nanosize. They pass the cell membranes through passive mechanism or tissue infusion, thereby enabling the reduction of the EOs doses required to ensure antimicrobial activity (Bilia et al. 2014). Flavor, natural aroma, and taste of food maintain the same because of low doses of the bioactive compound applied (Lu et al. 2016).

Hossaina et al. (2019) prepared cellulose nanocrystals (CNCs) reinforced chitosan-based antifungal films by encapsulating EOs nanoemulsion. Chitosan-based nanocomposite films carried with thyme-oregano, thyme-tea tree, and thyme-peppermint EOs mixtures showed reduction of fungal growth by 51–77% against *Aspergillus niger*, *Aspergillus flavus*, *Aspergillus parasiticus*, and *Penicillium chrysogenum* in inoculated rice during 8 weeks of storage at 28 °C. Nanoemulsion prepared with thyme-oregano, thyme-tea tree, and thyme- peppermint have z-averages 76.58, 69.9, 57.9 nm, PDI 0.25, 0.21, 0.32, and zeta-potential −51, −50, −53 mV, respectively. They showed 83.73 ± 2.55, 75.60 ± 1.27, and 87.95 ± 6.81%, respectively, inhibition against *A. niger* after 24 hrs of inoculation. There was a slow release of volatile compounds (26%) and the rice samples packed with bioactive film showed no different change in color, taste, and odor over 12 weeks of storage. CNCs incorporated with chitosan matrix played an important function in stabilizing the physico-chemical and release properties of the nanocomposite films.

### 2.3.4 Chitosan with *Cymbopogon martinii* Essential Oil

*Cymbopogon martinii*, also known as Indian geranium/motia/rosha, is a tropical herbaceous grass belonging to family Poaceae (Duke 1993). Bioactive compounds such as geraniol caryophyllene, humulene, geranyl acetate, linalool, selinenes, limonene, etc. are some of the chemical constituents of its EOs (Rao et al. 2005; Cannon et al. 2013; Verma et al. 2013; Kakaraparthi et al. 2015).

The antifungal activity of *Cymbopogon martini* EOs (CMEOs) was investigated against Fusarium head blight disease in maize, caused by the post-harvest pathogen *Fusarium graminearum* (Kalagatur et al. 2018). They found that the minimum inhibitory concentration and minimum fungicidal concentration of CMEOs were 421.7 ± 27.14 and 618.3 ± 79.35 ppm, respectively. There was a morphological change in vesicles, craters, protuberance, and rough surfaces in macroconidia when exposed with CMEOs as compared with control. ROS content and lipid peroxidation were increased, which induced the death of fungi. Chitosan encapsulated CMEOs nanoparticles (Ce-CMEO-NPs) were synthesized with spherical morphology of size 455–480 nm and zeta-potential of 39.3–37.2 mV. FTIR analysis confirmed that bioactive constituents of CMEOs were well stabilized due to chitosan conjugation and successfully formed Ce-CMEO-NPs. A stabilized complex structure formed between chitosan and CMEOs increased the lifetime antifungal activity of CMEOs by gradual release of antifungal constituents of Ce-CMEO-NPs. Maize grains were used as sample material to check the antifungal and antimycotoxin activities of CMEOs and Ce-CMEO-NPs against *F. graminearum* under laboratory conditions over a storage period of 28 days. Ce-CMEO-NPs and CMEOs reduced fungal growth at 700 ppm and 900 ppm, respectively. Ce-CMEO-NPs offered competent and enhanced antifungal and antimycotoxin activities as compared with CMEO, and it could be due to persistence of antifungal activity by controlled release of antifungal constituents from Ce-CMEO-NPs.

### 2.3.5 Chitosan with Pepper Tree (*Schinus molle*) Essential Oil

*Schinus molle* (Anacardiaceae), also known as pepper tree, has EOs with antimicrobial properties (Lopez et al. 2014). The chemical constituents of EOs, such as ũFC;-pinene, ũFC;-phellandrene, *β*-phellandrene, limonene, monoterpenes, and myrcene, are found in pepper tree. Efficacy of its EOs against the filamentous fungi of *Fusarium solani* has been proved (Rhouma et al. 2009). At 500 ppm of pepper tree EOs, the mycelium inhibition of up to 53.5% was found against *Aspergillus flavus* (Dikshit et al. 1986). It also exhibited substantial antifungal activity against *A. japonicus, A. niger,* and *A. oryzae* (Martins et al. 2014). A minimum inhibitory concentration of >1000 mg/mL of the oil was found against *A. fumigates* (Alanis-Garza et al. 2007).

Luque-Alcaraz et al. (2016) synthesized chitosan nanoparticles, encapsulating pepper tree EOs having the size distribution of 754 ± 7.5 nm and zeta-potential of +9.1 ± 1.74 mV. They tested the effect of different concentrations of chitosan nanoparticles encapsulated pepper tree EOs on the viability of *A. parasiticus* spores. It was found out that all treatments reduced the viability of fungal spores compared with control. These results indicated that the addition of pepper tree EOs in chitosan bionanocomposites is an alternative that preserves the antifungal properties of both components, decreasing the tendency to volatilization of EOs and consequent loss of activity.

### 2.3.6 *Mentha piperita* Essential Oils in Chitosan–Cinnamic Acid Nanogel

Beykia et al. (2014) investigated the encapsulation of *Mentha piperita*EOs in chitosan–cinnamic acid nanogel to increase stability of oils and antimicrobial activity against *Aspergillus flavus*. They found out that because of encapsulation, the extract possessed remarkable antifungal properties against *A. flavus*. The minimum inhibitory concentrations of encapsulated and free EOs against *A. flavus* under sealed condition were 500 and 2100 ppm, respectively. However, when experimented under non-sealed condition, the encapsulated oils performed better result (800 ppm) compared with the free oils which failed to caused complete inhibition within the concentration range tested (up to 3000 ppm). These findings revealed the promising role of chitosan–cinnamic acid nanogel as a carrier for EOs to enhance their antimicrobial properties.

### 2.3.7 Thyme Oil with Chitosan/Boehmite

Cindi et al. (2015) had done their investigation on polyethylene terephthalate (PET) punnets which contained thyme oil (TO sachets) and also packed with chitosan/boehmite nanocomposite lidding films. They found out that, in artificially inoculated peach fruits (cv. Kakawa) (*Prunus persica*) by *Monilinia laxa,* the incidence and severity of brown rot were reduced when stored at 25 °C for 5 days. Moreover, in naturally infected fruits, the brown rot incidence was reduced to 10% when stored at 0.5 °C, 90% RH for 7 days. Active compounds such as thymol (56.43%), β-linalool (37.6%), and caryophyllen (9.47%) were maintained within the punnet. The appearance, taste, and natural peach flavor were remains as such so people preferred fruits packed from commercial punnet containing thyme oil (sachets) and sealed with chitosan/boehmite nanocomposite lidding films.

### 2.3.8 Thiadiazole-Functionalized Chitosan Derivatives

Li et al. (2013) revealed that a group of novel water-soluble chitosan derivatives, such as 1,3,4-thiadiazole (TPCTS), 2-methyl-1,3,4-thiadiazole (MTPCTS), and 2-phenyl-1,3,4-thiadiazole (PTPCTS), had antifungal activities against plant-threatening fungi such as *Colletotrichum lagenarium, Phomopsi asparagi,* and *Monilinia fructicola.* The inhibitory index was recorded as 31.6% at 1.0 mg/mL against the growth of *C. lagenarium*. The antifungal activities of chitosan derivatives were given better result as compared with chitosan. Among the chitosan derivatives tested, MTPCTS gave the best result with the inhibitory indices of 75.3, 82.5, and 65.8% against *C. lagenarium*, *P. asparagi*, and *M. fructicola*, respectively, at 1.0 mg/mL . The length of alkyl substituent in thiadiazole and the hydrophobic moiety tend to affect the antifungal activity of chitosan derivatives.

### 2.3.9 Salicylic Acid-Chitosan NMs

Salicylic acid (SA) is a naturally occurring vital phenolic compound involved in plant's signal transduction pathway for the onset of systemic acquired resistance (SAR) (Raskin et al. 1990; Malamy et al. 1992; Vlot et al. 2009). It is a key element for photosynthesis, vegetative growth, respiration, flower formation, up-regulation of seed germination, senescence, thermogenesis, and cellular redox homeostasis (Khan et al. 2015).

Exogenous application of SA as seed treatment and foliar application induced many metabolic processes in plants and could be an alternative approach for controlling disease, enhancing plant growth and yield. Therefore, SA-chitosan nanoparticles (SA-CS NPs) have been investigated as a biostimulant for promoting plant defense and growth in maize. SA-CS NPs induced significant physiological-biochemical responses under in vitro and in vivo conditions (Fig. 2), as elevated antioxidant-defense enzyme activities (SOD, catalase, peroxidase, etc.), balanced ROS, cell wall reinforcement by lignin deposition, disease control, and plant growth in maize. In field 59.4% and in pots 37.3–49.5% (at 0.01–0.16% concentration)

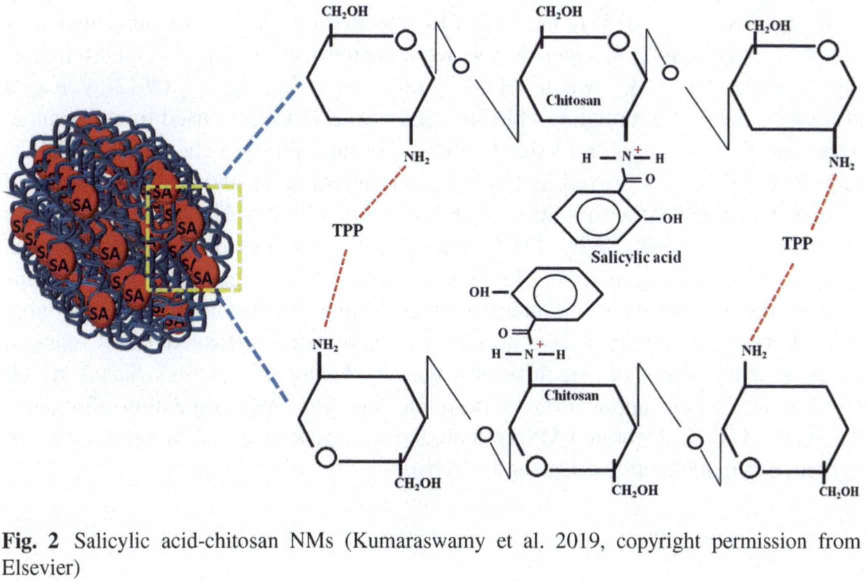


**Fig. 2** Salicylic acid-chitosan NMs (Kumaraswamy et al. 2019, copyright permission from Elsevier)

control of post-flowering stalk rot (PFSR) disease and 57.8% yield enhancement was evident in SA-CS NPs application. NPs at the concentrations of 0.08 and 0.16% significantly evaded in vitro mycelia growth from 62.2 to 100% and spore germination from 48.3 to 60.5%. In NPs treatments (0.01–0.16%), plants endowed reasonably reduced disease severity (25.2 to 33.0%) and higher disease control (PEDC values from 40.5 to 59.4%).

With +34.1 mV zeta-potential, SA-CS NPs were stable in aqueous due to electrostatic repulsion between NPs which averts aggregation and agglomeration of NPs. FTIR study revealed the interaction of –COOH group of SA to primary amide of chitosan. Slow release of SA from SA-CS NPs significantly amended physiological and biochemical responses in maize plant for commendable disease control, plant growth and yield as compared with sole SA application. *Fusarium verticillioides* is an intercellular endophytic pathogen where symptoms appear at flowering stage, so most of the approaches of disease control may not be effective. Thus, application of SA-CS NPs as seed treatment and foliar application before flowering stage can be an effective and preventive approach through boosting plant innate immunity even before the onset of pathogen infection (Kumaraswamy et al. 2019).

### 2.3.10 Silica-Chitosan NMs

Since the discovery of Mobil Crystalline Material 41 (MCM-41), research and development of mesoporous silica nanoparticles (MSNs) has gained worldwide interest due to MSNs' unique properties. These include biocompatibility, low cost, large surface area, tunable pore size for high loading capacity, and ability for targeted and controlled release with surface functionalization and polymer coatings (Wu et al. 2013; Sun et al. 2015).

Bernardos et al. (2015) reported that EOs loaded into MSNs had sustained antifungal activity against *A. niger*. MSNs were synthesized by liquid crystal templating mechanism. A water-soluble chitosan derivative (*N*-(2-hydroxyl) propyl-3-trimethyl ammonium chitosan chloride, HTCC) was used to encapsulate pyraclostrobin (a fungicide)-loaded MSNs. Through physico-chemical and structural analyses, it was proved that electrostatic interactions and hydrogen bonding were responsible for the formation of HTCC-capped MSNs. The loading efficiency of NPs increased to 40.3% by HTCC coating as compared with using bare MSNs as a single encapsulating material (26.7%). Initially, a rapid release of pyraclostrobin-loaded NPs was observed but later it showed a slow and sustained release. Almost same fungicidal activity was expressed by pyraclostrobin-loaded HTCC-capped MSNs with half doses of pyraclostrobin against *Phomopsis asparagi* (Sacc.), which resulted into lower application of pesticide and improved utilization efficiency. Therefore, HTCC-decorated MSNs demonstrated great potential as nanocarriers in agrochemical applications (Cao et al. 2016).

### 2.3.11 Chitosan–Saponin NMs

Saponins are complex glycosidic compounds known for their fungistatic activities (Chapagain et al. 2007). Their self-assembly property in aqueous media has been successfully exploited in chitosan–saponin nanoformulation against cancer cells (Rejinold et al. 2011). But its ability to suppress plant fungal growth was first studied by Saharan et al. (2013), when they synthesized chitosan–saponin NPs to test their synergistic activity against phytopathogenic fungi (*A. alternata*, *M. phaseolina,* and *R. solani*). These NPs were prepared using ionic gelation method by interaction of chitosan, sodium tripolyphosphate, and saponin. Their particle size, polydispersity index, zeta-potential, and structures were confirmed by DLS, FTIR, TEM, and SEM. These NPs inhibited 80.9% of mycelia growth at 0.1% w/v concentration and also showed a dose dependent effect on mycelia growth.

### 2.3.12 Oleoyl-Chitosan NMs

Many scientists have reported about the hydrophobic modifications of chitosan and NP formation by self-aggregation in water. These modifications can introduce hydrophobic groups into chitosan and produce chitosan amphiphilic polymers. Some of these chitosan amphiphilic derivatives can form nanosized self-aggregates in aqueous solution. Derivatives of chitosan having long chain fatty acyl are novel hydrophobic modifications that can form nanoparticles (Xing et al. 2016).

Therefore, oleoyl-chitosan NPs were synthesized using oil-water emulsification method based on O-chitosan, which involved grafting a monounsaturated fatty acid residue, $C_{18}$ oleoyl group, onto the $NH_2$ at C-2 in the chitosan structure (Xing et al. 2016). These NPs were examined for their antifungal activity against *Verticillium dahlia* which causes wilting in woody and herbaceous plants, a problem for which no effective controls have been devised yet. Oleoyl-chitosan NPs dramatically decreased the mycelium growth showing the highest antifungal indexes of 86.81% at 2 mg/mL, and also affected the spore germination and hyphae morphology as crumpled hyphae and spores, thickened cell walls, disappearance of membranous organelles, massive vacuolation of the cytoplasm, and cell wall-plasmalemma separation as observed in SEM and TEM studies. O-chitosan NPs showed inhibitory effect at all tested concentrations which was reversibly concentration-dependent. The dry weight of mycelia was much lower than the control group at pH 4.5 and 5.0. The inactivation of spores by NPs occurred via one of the following mechanisms. Specifically, O-chitosan NPs at lower concentrations could mainly induce an inhibition effect, while at higher concentrations, they primarily led to flocculation. Therefore, the antifungal capability of O-chitosan NPs could restrain the germination and tube growth of conidia. Moreover, these NPs having the characteristics of both coagulants and flocculants could disrupt the dispersion state of spores (Dong et al. 2014; Xing et al. 2017).

## 3 Conclusion

Review of literature confirms that chitosan is a versatile biomaterial that exhibits remarkable fungicidal activity. It can be easily maneuvered through various physical and chemical techniques. Functional groups of chitosan ($-NH_2$ and $-OH$) enable this biopolymer to provide unique platform to make smart fungicides by functionalizing it with inorganic/organic substances to expend its application horizon. In this notation, new generation agrochemicals (like fungicides) can be synthesized which can act smart and timely at lower dose. Chitosan biopolymer has flexible physicochemical properties to convert into smart nano-chitosan product with the help of other bioactive compounds. Therefore, we expect to achieve the following characteristics in new generation fungicides: (a) multi-targeted/multi-mode action to arouse plant immune responses, (b) show direct antifungal activity, (c) slow/controlled release of active component for timely and long lasting effects in crop (Fig. 3). Therefore, chitosan-based NMs have great scope for creation of new generation fungicides which may be economical and ecofriendly, and give minimum chemical load to the biosphere.

**Acknowledgments** Financial support from the DBT-RA Program in Biotechnology and Life Sciences is gratefully acknowledged (Khaidem Aruna Devi). Financial support to Ashok Kumar from ICMR-JRF (ICMR-JRF/2018/30716) and DBT-JRF to Damyanti Prajapati (DBT/2018/MPUAT/1123) also acknowledged.

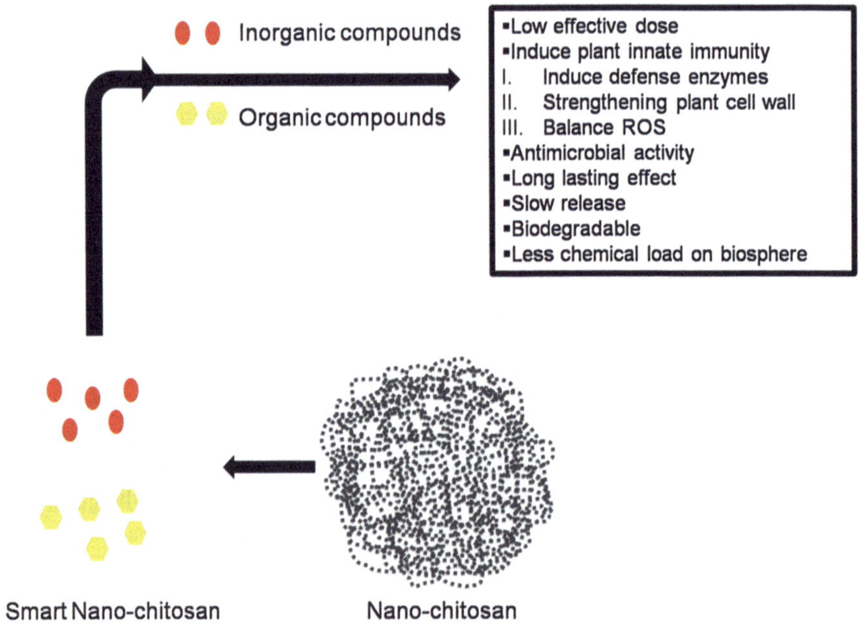


**Fig. 3** Smart Nano-chitosan

## References

Ahmad A, Khan A, Akhtar F, Yousuf S, Xess I, Khan L, Manzoor N (2011) Fungicidal activity of thymol and carvacrol by disrupting ergosterol biosynthesis and membrane integrity against Candida. Eur J Clin Microbiol Infect Dis 30:41–50

Alanis-Garza BA, Gonzalez-Gonzalez GM, Salazar-Aranda R, Waksman de Torres N, Rivas-Galindo VM (2007) Screening of antifungal activity of plants from the northeast of Mexico. J Ethnopharmacol 114(3):468–471

Alister C, Kogan M (2006) ERI: environmental risk index. A simple proposal to select agrochemicals for agricultural use. Crop Prot 25(3):202–211

Aloui H, Khwaldia K, Licciardello F, Mazzaglia A, Muratore G, Hamdi M, Restuccia C (2014) Efficacy of the combined application of chitosan and locust bean gum with different citrus essential oils to control postharvest spoilage caused by *Aspergillus flavus* in dates. Int J Food Microbiol 170:21–28

Bernardos A, Marina T, Žáček P, Pérez-Esteve É, Martínez-Mañez R, Lhotka M, Klouček P (2015) Antifungal effect of essential oil components against *Aspergillus niger* when loaded into silica mesoporous supports. J Sci Food Agric 95(14):2824–2831

Beykia M, Zhaveha S, Khalilib ST, Rahmani-Cheratic T, Abollahic A, Bayatd M, Tabatabaeie M, Mohsenifar A (2014) Encapsulation of *Mentha piperita* essential oils in chitosan–cinnamic acid nanogel with enhanced antimicrobial activity against *Aspergillus flavus*. Ind Crop Prod 54:310–319

Bilia AR, Guccione C, Isacchi B, Righeschi C, Firenzuoli F, Bergonzi MC (2014) Essential oils loaded in nanosystems: a developing strategy for a successful therapeutic approach. In: Evidence-based complement alternative medicine. Oxford University Press, Oxford

Brunel F, Gueddari NEE, Moerschbacher BM (2013) Complexation of copper (II) with chitosan nanogels: toward control of microbial growth. Carbohydr Polym 92:1348–1356

Burt S (2004) Essential oils: their antibacterial properties and potential applications in foods—a review. Int J Food Microbiol 94:223–253

Cai DW (2008) Understand the role of chemical pesticides and prevent misuses of pesticides. Bull Agric Sci Technol 1(6):36–38

Cannon JB, Cantrell CL, Astatkie T, Zheljazkov VD (2013) Modification of yield and composition of essential oils by distillation time. Ind Crops Prod 41:214–220

Cao L, Zhang H, Cao C, Zhang J, Li F, Huang Q (2016) Quaternized chitosan-capped mesoporous silica nanoparticles as nanocarriers for controlled pesticide release. NMs 6(7):126

Chapagain BP, Wiesman Z, Tsror L (2007) In vitro study of the antifungal activity of saponin-rich extracts against prevalent phytopathogenic fungi. Ind Crops Prod 26(2):109–115

Chen JP, Lin G, Zhou BS (2004) Correlation between pesticides exposure and morbidity and mortality of breast cancer. Chin J Public Health-Shenyang 20:289–290

Chookhongkha N, Sopondilok T, Photchanachai S (2012) Effect of chitosan and chitosan nanoparticles on fungal growth and chilli seed quality. In: International conference on post-harvest Pest and disease Management in Exporting Horticultural Crops-PPDM2012 973. Acta Horticulturae, Bangkok, pp 231–237

Choudhary RC, Kumaraswamy RV, Kumari S, Sharma SS, Pal A, Raliya R, Biswas P, Saharan V (2017a) Synthesis, characterization, and application of chitosan NMs loaded with zinc and copper for plant growth and protection. In: Prasad R et al (eds) Nanotechnology. Springer Nature Singapore Pte Ltd., Singapore

Choudhary RC, Kumaraswamy RV, Kumari S, Sharma SS, Pal A, Raliya R, Biswas P, Saharan V (2017b) Cu-chitosan nanoparticle boost defense responses and plant growth in maize (*Zea mays* L.). Sci Rep 7:9754

Choudhary RC, Kumaraswamy RV, Kumari S, Sharma SS, Pal A, Raliya R, Saharan V (2019a) Zinc encapsulated chitosan nanoparticle to promote maize crop yield. Int J Biol Macromol 127:126–135

Choudhary RC, Kumari S, Kumaraswamy RV, Pal A, Raliya R, Biswas P, Saharan V (2019b) Characterization methods for chitosan-based NMs. Plant Nanobionics. Springer, Cham, pp 103–116

Chouhan S, Sharma K, Guleria S (2017) Antimicrobial activity of some essential oils-present status and future perspectives. Medicines 4(3):58

Cindi MD, Shittu T, Sivakumar D, Bautista-Banos S (2015) Chitosan boehmite-alumina nanocomposite films and thyme oil vapour control brown rot in peaches (*Prunus persica* L.) during postharvest storage. Crop Prot 72:127–131

Dietz KJ, Herth S (2011) Plant nanotoxicology. Trends Plant Sci 16(11):582–589

Dikshit A, Nanqvi AA, Husain A (1986) *Schinus molle*: a new source of natural fungitoxicant. Appl Environ Microbiol 51(5):1085–1088

Dong C, Chen W, Liu C (2014) Flocculation of algal cells by amphoteric chitosan-based flocculant. Bioresour Technol 170:239–247

Donsi F, Annunziata M, Sessa M, Ferrari G (2011) Nanoencapsulation of essential oils to enhance their antimicrobial activity in foods. LWT- Food Sci Technol 44:1908–1914

du Plooy W, Regnier T, Combrinck S (2009) Essential oil amended coatings as alternative to synthetic fungicides in citrus postharvest management. Postharvest Biol Technol 53:117–122

Duke JA (1993) CRC handbook of alternative cash crops. CRC Press, Boca Raton

Fugate KK, Ribeiro WS, Lulai EC, Deckard EL, Finger FL (2016) Cold temperature delays wound healing in postharvest sugarbeet roots. Front Plant Sci 7:499

Hahn M (2014) The rising threat of fungicide resistance in plant pathogenic fungi: Botrytis as a case study. J Chem Biol 7(4):133–141

Harris R, Lecumberri E, Mateos-Aparicio I, Mengíbar M, Heras A (2011) Chitosan nanoparticles and microspheres for the encapsulation of natural antioxidants extracted from Ilex paraguariensis. Carbohydr Polym 84:803–806

Hossain F, Follett P, Salmieri S, Vu KD, Jamshidian M, Lacroix M (2017) Perspectives on essential oil loaded nanodelivery packaging technology for controlling stored cereal and grain pests. In: Nollet LML, Rathore HS (eds) Green pesticides handbook essential oils for Pest control. Routledge, London, pp 487–508

Hossaina F, Follettb P, Salmieria S, Vua KD, Fraschinic C, Lacroixa M (2019) Antifungal activities of combined treatments of irradiation and essential oils (EOs) encapsulated chitosan nanocomposite films in *in vitro* and *in situ* conditions. Int J Food Microbiol 295:33–40

Hwang ET, Lee JH, Chae YJ, Kim YS, Kim BC, Sang BI, Gu MB (2008) Analysis of the toxic mode of action of silver nanoparticles using stress-specific bioluminescent bacteria. Small 4(6):746–750

Juven BJ, Kanner J, Schved F, Weisslowicz H (1994) Factors that interact with the antibacterial action of thyme essential oil and its active constituents. J Appl Bacteriol 76:626–631

Kah M (2015) Nanopesticides and nanofertilizers: emerging contaminants or opportunities for risk mitigation? Front Chem 3:64. 13

Kah M, Hofmann T (2014) Nanopesticide research: current trends and future priorities. Environ Int 63:224–235

Kakaraparthi PS, Srinivas KVNS, Kumar JK, Kumar AN, Rajput DK, Anubala S (2015) Changes in the essential oil content and composition of palmarosa (*Cymbopogon martini*) harvested at different stages and short intervals in two different seasons. Ind Crops Prod 69:348–354

Kalagatur NK, Nirmal Ghosh OS, Sundararaj N, Mudili V (2018) Antifungal activity of chitosan nanoparticles encapsulated with Cymbopogon martinii essential oil on plant pathogenic Fungi Fusarium graminearum. Front Pharmacol 9:610

Kashyap PL, Xiang X, Heiden P (2015) Chitosan nanoparticle based delivery systems for sustainable agriculture. Int J Biol Macromol 77:36–51

Katiyar D, Hemantaranjan A, Singh B (2015) Chitosan as a promising natural compound to enhance potential physiological responses in plant: a review. Indian J Plant Physiol 20(1):1–9

Kaur P, Thakur R, Choudhary A (2012) An in vitro study of the antifungal activity of silver/chitosan nanoformulations against important seed borne pathogens. Int J Sci Technol Res 1(7):83–86

Khan MIR, Fatma M, Per TS, Anjum NA, Khan NA (2015) Salicylic acid-induced abiotic stress tolerance and underlying mechanisms in plants. Front Plant Sci 6:462

Kheiri A, Jorf SM, Malihipour A, Saremi H, Nikkhah M (2017) Synthesis and characterization of chitosan nanoparticles and their effect on Fusarium head blight and oxidative activity in wheat. Int J Biol Macromol 102:526–538

Kumar JN, Bora A, Kumar RN, Amb MK, Khan S (2013) Toxicity analysis of pesticides on cyanobacterial species by 16S rDNA molecular characterization. Proc Int Acad Ecol Environ Sci 3(2):101

Kumaraswamy RV, Kumari S, Choudhary RC, Pal A, Raliya R, Biswas P, Saharan V (2018) Engineered chitosan based NMs: bioactivities, mechanisms and perspectives in plant protection and growth. Int J Biol Macromol 113:494–506

Kumaraswamy RV, Kumari S, Choudhary RC, Sharma SS, Pal A, Raliya R, Saharan V (2019) Salicylic acid functionalized chitosan nanoparticle: a sustainable biostimulant for plant. Int J Biol Macromol 123:59–69

Kuźniak E, Urbanek H (2000) The involvement of hydrogen peroxide in plant responses to stresses. Acta Physiol Plant 22(2):195–203

Li Q, Ren J, Dong F, Feng Y, Gu G, Guo Z (2013) Synthesis and antifungal activity of thiadiazole-functionalized chitosan derivatives. Carbohydr Res 373:103–107

Liu CJ, Men WJ, Liu YJ, Zhang H (2002) The pollution of pesticides in soils and its bioremediation. Syst Sci Compr Stud Agric 18(4):295–297

Lopez A, Castro S, Andina MJ (2014) Insecticidal activity of microencapsulated *Schinus molle* essential oil. Ind Crop Prod 53:209–216

Lu W, Kelly AL, Miao S (2016) Emulsion-based encapsulation and delivery systems for polyphenols. Trends Food Sci Technol 47:1–9

Luo Y, Zhang B, Whent M, Yu L, Wang Q (2011) Preparation and characterization of zein/chitosan complex for encapsulation of tocopherol and its in vitro controlled release study. Colloids Surf B: Biointerfaces 85:145–152

Luque-Alcaraz AG, Cortez-Rocha MO, Velázquez-Contreras CA, Acosta-Silva AL, Santacruz-Ortega HDC, Burgos-Hernández A, Argüelles-Monal WM, Plascencia-Jatomea M (2016) Enhanced Antifungal Effect of Chitosan/Pepper Tree (*Schinus molle*) Essential Oil Bionanocomposites on the Viability of *Asppergillus parasiticus* Spores. Journal of Nanomaterials 2016:1–10

Malamy J, Hennig J, Klessig DF (1992) Temperature-dependent induction of salicylic acid and its conjugates during the resistance response to tobacco mosaic virus infection. Plant Cell 4(3):359–366

Malerba M, Cerana R (2016) Chitosan effects on plant systems. Int J Mol Sci 17(7):996

Manikandan A, Sathiyabama M (2016) Preparation of chitosan nanoparticles and its effect on detached rice leaves infected with *Pyricularia grisea*. Int J Biol Macromol 84:58–61

Martin A, Varona S, Navarrete A, Cocero MJ (2010) Encapsulation and co-precipitation processes with supercritical fluids: applications with essential oils. Open Chem Eng J 4(1)

Martins MDR, Arantes S, Candeias F, Tinoco MT, Cruz-Morais J (2014) Antioxidant, antimicrobial and toxicological properties of *Schinus molle* L. essential oils. J Ethnopharmacol 151(1):485–492

Marutescu L, Popa M, Saviuc C, Lazar V, Chifiriuc MC (2017) Botanical pesticides with virucidal, bactericidal, and fungicidal activity. In: New pesticides and soil sensors. Academic Press, London, pp 311–335

Mathew TV, Kuriakose S (2013) Photochemical and antimicrobial properties of silver nanoparticle-encapsulated chitosan functionalized with photoactive groups. Mater Sci Eng C 33(7):4409–4415

Medina E, Caro N, Abugoch L, Gamboa A, Diaz-Dosque M, Tapia C (2019) Chitosan thymol nanoparticles improve the antimicrobial effect and the water vapour barrier of chitosan-quinoa protein films. J Food Eng 240:191–198

Mirdehghan HS, Valerob D (2017) Bioactive compounds in tomato fruit and its antioxidant activity as affected by incorporation of Aloe, eugenol, and thymol in fruit package during storage. Int J Food Prop 20:798–806

Mohammadi A, Hashemi M, Hosseini SM (2015) Nanoencapsulation of *Zataria multiflora* essential oil preparation and characterization with enhanced antifungal activity for controlling *Botrytis cinerea*, the causal agent of gray mould disease. Innov Food Sci Emerg Technol 28:73–80

Moussa SH, Tayel AA, Alsohim AS, Abdallah RR (2013) Botryticidal activity of nanosized silver-chitosan composite and its application for the control of gray mold in strawberry. J Food Sci 78(10):M1589–M1594

Muzzarelli RAA (2010) Chitins and chitosans as immunoadjuvants and nonallergenic drug carriers. Mar Drugs 8:292–312

Ngoc UTP, Nguyen DH (2018) Synergistic antifungal effect of fungicide and chitosan-silver nanoparticles on *Neoscytalidium dimidiatum*. Green Proc Synth 7:132–138

Oussalah M, Caillet S, Saucier L, Lacroix M (2006) Inhibitory effects of selected plant essential oils on the growth of four pathogenic bacteria: E. coli O157:H7, *Salmonella Typhimurium*, *Staphylococcus aureus* and *Listeria monocytogenes*. Food Control 18:414–420

Park HJ, Kim SH, Kim HJ, Choi SH (2006) A new composition of nanosized silica-silver for control of various plant diseases. Plant Pathol J 22(3):295–302

Pimentel D (1997) Pest management in agriculture: techniques for reducing pesticide use: environmental and economic benefits. Wiley, New York, p 456

Rao BR, Kaul PN, Syamasundar KV, Ramesh S (2005) Chemical profiles of primary and secondary essential oils of palmarosa (Cymbopogon martinii (Roxb.) Wats var. motia Burk.). Ind Crops Prod 21:121–127

Raskin I, Skubatz H, Tang W, Meeuse BJ (1990) Salicylic acid levels in thermogenic and non-thermogenic plants. Ann Bot 66(4):369–373

Rejinold NS, Muthunarayanan M, Muthuchelian K, Chennazhi KP, Nair SV, Jayakumar R (2011) Saponin-loaded chitosan nanoparticles and their cytotoxicity to cancer cell lines in vitro. Carbohydr Polym 84(1):407–416

Rhouma A, Daoud HB, Ghanmi S, Salah HB, Romdhane M, Demak M (2009) Antimicrobial activities of leaf extracts of *Pistacia* and *Schinus* species against some plant pathogenic fungi and bacteria. J Plant Pathol 91(2):339–345

Rubina MS, Vasil'kov AY, Naumkin AV, Shtykova EV, Abramchuk SS, Alghuthaymi MA, Abd-Elsalam KA (2017) Synthesis and characterization of chitosan–copper nanocomposites and their fungicidal activity against two sclerotia-forming plant pathogenic fungi. J Nanostruct Chem 7(3):249–258

Saharan V, Pal A (2016a) Chitosan based NMs in plant growth and protection. Springer, Berlin

Saharan V, Pal A (2016b) Current and future prospects of chitosan-based NMs in plant protection and growth. In: Chitosan based NMs in plant growth and protection. Springer, New Delhi, pp 43–48

Saharan V, Mehrotra A, Khatik R, Rawal P, Sharma SS, Pal A (2013) Synthesis of chitosan based nanoparticles and their *in vitro* evaluation against phytopathogenic fungi. Int J Biol Macromol 62:677–683

Saharan V, Sharma G, Yadav M, Choudhary MK, Sharma SS, Pal A, Raliya R, Biswas P (2015) Synthesis and in vitro antifungal efficacy of cu-chitosan nanoparticles against pathogenic fungi of tomato. Int J Biol Macromol 75:346–353

Sajed H, Sahebkar A, Iranshahi M (2013) Zataria multiflora Boiss.(Shirazi thyme)-an ancient condiment with modern pharmaceutical uses. J Ethnopharmacol 145(3):686–698

Sathiyabama M, Parthasarathy R (2016) Biological preparation of chitosan nanoparticles and its in vitro antifungal efficacy against some phytopathogenic fungi. Carbohydr Polym 151:321–325

Shukla SK, Mishra AK, Arotiba OA, Mamba BB (2013) Chitosan-based NMs: a state-of-the-art review. Int J Biol Macromol 59:46–58

Storz G, Imlayt JA (1999) Oxidative stress. Curr Opin Microbiol 2(2):188–194

Sun R, Wang W, Wen Y, Zhang X (2015) Recent advance on mesoporous silica nanoparticles-based controlled release system: intelligent switches open up new horizon. NMs 5(4):2019–2053

Tassou C, Drosinos EH, Nychas GJE (1995) Effects of essential oil of mint (*Mentha piperita*) on *Salmonella enteritidis* and *Listeria monocytogenes* in model food systems at 4°C and 10°C. J Appl Bacteriol 78:593–600

Van Long NN, Joly C, Dantigny P (2016) Active packaging with antifungal activities. Int J Food Microbiol 220:73–90

Verma RS, Padalia RC, Chauhan A (2013) Introduction of *Cymbopogon* distans (Nees ex Steud.) Wats to the sub-tropical India: evaluation of essential oil yield and chemical composition during annual growth. Ind Crops Prod 49:858–863

Vlot AC, Dempsey DMA, Klessig DF (2009) Salicylic acid, a multifaceted hormone to combat disease. Annu Rev Phytopathol 47:177–206

Vokhidova NR, Sattarov ME, Kareva ND, Rashidova SS (2014) Fungicide features of the nanosystems of silkworm (Bombyx mori) chitosan with copper ions. Microbiology 83(6):751–753

Wu SH, Mou CY, Lin HP (2013) Synthesis of mesoporous silica nanoparticles. Chem Soc Rev 42(9):3862–3875

Xing K, Shen X, Zhu X, Ju X, Miao X, Tian J, Qin S (2016) Synthesis and in vitro antifungal efficacy of oleoyl-chitosan nanoparticles against plant pathogenic fungi. Int J Biol Macromol 82:830–836

Xing K, Liu Y, Shen X, Zhu X, Li X, Miao X, Qin S (2017) Effect of O-chitosan nanoparticles on the development and membrane permeability of *Verticillium dahliae*. Carbohydr Polym 165:334–343

Zhang W (2018) A long-term trend of cancer-induced deaths in European countries. Network 3(1–2):1–9

Zhang W, Jiang F, Ou J (2011a) Global pesticide consumption and pollution: with China as a focus. Proc Int Acad Ecol and Environ Sci 1(2):125

Zhang WJ, van der Werf W, Pang Y (2011b) A simulation model for vegetable-insect pest-insect nucleopolyhedrovirus epidemic system. J Environ Entomol 33(3):283–301

# The Use of Nanocarriers to Improve the Efficiency of RNAi-Based Pesticides in Agriculture

**Olivier Christiaens, Marko Petek, Guy Smagghe, and Clauvis Nji Tizi Taning**

**Abstract** RNA interference (RNAi) is a post-transcriptional gene silencing mechanism whereby target gene messenger RNA (mRNA) is neutralized by double-stranded RNA (dsRNA) homologous to the mRNA sequence. The pathway can be exploited for pest and disease control purposes by delivery of exogenous dsRNA targeting a gene essential for the target organism's survival. The most likely dsRNA delivery strategy for invertebrate pest control is through oral uptake, but transdermal dsRNA uptake has been reported to lead to gene silencing in some species as well. To combat plant-pathogenic fungi and viruses, methods that efficiently deliver dsRNAs into plant tissues are needed. While transgenic plants allow for efficient production of such dsRNAs in the plants, non-transgenic spray-based applications are being developed as well. Although RNAi is highly effective in some species, for example, certain beetle species, many insects and nematodes show a variable or lower sensitivity to dietary uptake of dsRNA. In the past decade, several factors and barriers affecting RNAi efficiency in insects have been identified, including dsRNA degradation in the insect body, inefficient cellular uptake of dsRNA, and an impaired endosomal escape into the cytoplasm. Nanocarriers could play a role in enhancing the efficacy of sprayable RNAi-based pesticides by helping to overcome some of these barriers. Several proof-of-concept studies have shown that polymers, liposomes, and peptides, among others could be used in this context but further advances are necessary to optimize these delivery systems. Gathering inspiration from the medical field, where RNAi is also being investigated as a potential therapeutics strategy, could drive forward these agricultural applications. In this chapter, we present an overview of the literature on RNAi barriers and the use of these nanoparticles to increase RNAi efficacy in agricultural pests. Finally, we also discuss a number of biosafety considerations regarding RNAi and the use of these nanoparticle formulations.

---

O. Christiaens (✉) · G. Smagghe · C. N. T. Taning
Department of Plants and Crops, Faculty of Bioscience Engineering, Ghent University, Ghent, Belgium
e-mail: olchrist.christiaens@ugent.be

M. Petek
Department of Biotechnology and Systems Biology, National Institute of Biology, Ljubljana, Slovenia



L. F. Fraceto et al. (eds.), *Nanopesticides*,
https://doi.org/10.1007/978-3-030-44873-8_3

**Keywords** RNA interference · dsRNA delivery · Nanoparticles · Invertebrates RNAi efficiency · Pest control

# 1 RNAi in Crop Protection: An Introduction

The search for novel and environmentally friendly crop protection strategies, allowing us to complement or replace classical synthetic pesticides, is one of the major challenges in agriculture for the coming decades. A promising strategy to combat insects, nematodes, fungi, and viruses is to exploit the natural RNA interference (RNAi) mechanism. This mechanism is present in all eukaryotic organisms and is triggered by cellular uptake of double-stranded RNA (dsRNA) molecules which are homologous to a target gene in the pest or pathogen. These dsRNAs are then processed by the RNAi pathway within the cell which eventually leads to a depletion of messenger RNAs (mRNAs) and proteins which are encoded by that particular target gene. For RNAi-based pest control to occur, the most logical delivery route for dsRNA to the target insect is via feeding, although successful RNAi could also be achieved by topical application of dsRNA on the insects. Regardless of the application or delivery route, dsRNA needs to be taken up inside the cells of the target species, for example, the gut epithelial cells after ingestion, for gene silencing to occur. Once dsRNA is present in the cytoplasm, it is further processed by the cellular RNAi machinery. The cellular RNAi mechanism in insects is illustrated in Fig. 1.

The first proof-of-concept for invertebrate pest control was provided in 2007 by Baum et al. (2007) who showed that dsRNA specific for an essential gene in the Western corn rootworm (WCR; *Diabrotica virgifera virgifera*) expressed by a transgenic maize plant leads to a significant mortality in WCR feeding from the roots of these plants and also leads to a significant protection of the plant and its root system. A decade earlier, however, the first RNAi-based transgenic plants designed for disease control were already on the market. This early success story is the Rainbow papaya variety, which saved papaya culture in Hawaii by providing resistance against the devastating papaya ringspot virus (PRSV) (Gonsalves et al. 2000). Since these first proofs-of-concept, many studies have confirmed the potential of RNAi as control strategy against a wide range of insect pests, but also against plant-parasitic nematodes, fungi, and viruses. Studies have also investigated alternatives for the use of transgenic plants (host-induced gene silencing, HIGS) and have shown that dsRNA can also be sprayed on the field (spray-induced gene silencing, SIGS), leading to an efficient RNAi response (San Miguel and Scott 2016; Zhu et al. 2011).

A very alluring aspect of RNAi, especially from a biosafety point of view, is the sequence-dependent mode of action, as it offers the possibility to design highly species-selective control products. Indeed, Whyard et al. (2009) demonstrated that dsRNAs can be designed to be specific at the species level, able to discriminate, for example, between different species belonging to the *Drosophila* genus. However,

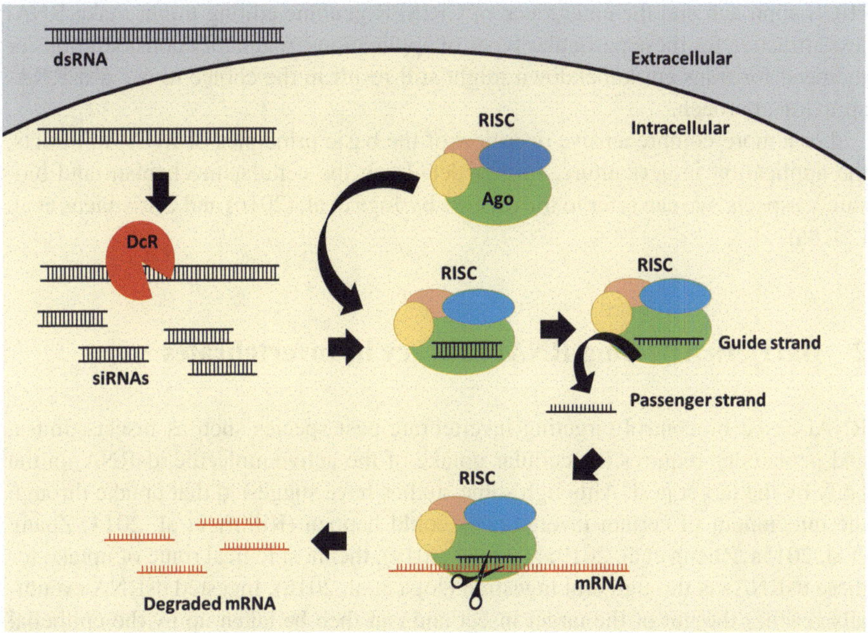


**Fig. 1** Schematic overview of the cellular RNAi mechanism in invertebrates. After successful cellular uptake, long dsRNA is diced, by an enzyme called Dicer (DcR), into small interfering RNAs (siRNAs) which are typically 19–22 nucleotides long. These siRNAs are then taken up in a protein complex called the RNA induced silencing complex (RISC) which includes an Argonaute (Ago) protein. Next, the double-stranded siRNA is separated into a passenger strand and a guide strand. The former is removed from the complex, while the guide strand, which is complementary to the target messenger RNA (mRNA), will then lead this complex towards this mRNA which eventually leads to its degradation

several studies have shown that off-target effects and adverse effects in non-target organisms (NTOs) cannot be excluded and that therefore dsRNA design is crucial (Baum et al. 2007; Christiaens et al. 2018a; Bachman et al. 2013; Haller et al. 2019). Another promising aspect from a biosafety point of view is the limited persistence of dsRNA in the environment, including soil, water, and animals, limiting exposure for NTOs, trophic chain persistence, and so on (Fischer et al. 2017; Parker et al. 2019; Albright III et al. 2017; Dubelman et al. 2014). Besides plant pest and disease control, other intriguing applications of RNAi could find their way to agriculture as well. One example is the use of dsRNA to target pathogens of beneficial insects, such as viruses or parasites. Maori et al. (2009) demonstrated that feeding honeybees with dsRNA specifically targeting an Israeli acute paralysis virus (IAPV) gene led to a lower infection rate and healthier bees. Other studies have confirmed this for other viruses and other bees as well (Piot et al. 2015; Desai et al. 2012). Finally, dsRNAs could theoretically also be used to alter plant characteristics, to increase their pest resistance, or to alter their nutritional composition and post-harvest processes. Even though these applications have so far mainly been explored using a

HIGS approach and the emergence of CRISPR genome editing might make RNAi less attractive for these particular types of applications, regulatory considerations or the need for transient knockdown might still result in the choice to use a dsRNA-spraying approach.

For a more comprehensive overview of the basic principles of RNAi in insects, the application in agriculture, further details on the cellular mechanism and biosafety aspects, we can refer to the reviews by Joga et al. (2016) and Christiaens et al. (2018a).

## 2 Barriers Affecting RNAi Efficacy in Invertebrates

RNAi-based biocontrol targeting invertebrate pest species such as insects, mites, and nematodes requires the cellular uptake of the active molecule, dsRNA, in the body by the target pest. Although some studies have suggested that uptake through the integument of certain invertebrates could happen (Killiny et al. 2014; Zhang et al. 2015a; Zheng et al. 2019; Niu et al. 2019), the most logical route of uptake for these dsRNAs is through oral ingestion (Joga et al. 2016). Ingested dsRNA eventually reaches the gut of the target insect and can then be taken up by the epithelial cells. Although the evolutionary drivers for this are unknown, gut cells of many invertebrates appear to internalize long dsRNA quite efficiently. Certain nematodes, notably *Caenorhabditis elegans*, have highly evolved pathways for the uptake of dsRNA, involving several *sid*-genes which are important in cellular uptake, cytoplasmic release, and also systemic transport of dsRNAs (Winston et al. 2002; Hunter et al. 2006). However, recent research has revealed that *C. elegans* is a special case, showing a strong expansion of RNAi-related effector genes compared to many other nematodes, including plant- and animal-parasitic species (Dalzell et al. 2011). Sid-like homologs have been found in the genomes of most insects, although the number varies depending on the insect order. Furthermore, their importance of these proteins for efficient dsRNA uptake in insects is debated. Certain studies indicate they play a role in efficient RNAi and dsRNA uptake, while other studies suggest they are not involved. What has become clear is that clathrin-mediated endocytosis plays a major role in dsRNA uptake in insects (Cappelle et al. 2016). It is important to note here that cellular dsRNA uptake in invertebrates appears to be variable and could be a major factor explaining variable RNAi efficacy between invertebrate species (Cooper et al. 2019). Furthermore, an impaired cytoplasmic release of dsRNA from late endosomes was also recently reported in lepidopteran Sf9 cells (Shukla et al. 2016; Yoon et al. 2017). For a comprehensive review on (cellular) uptake mechanisms and their influence on RNAi efficacy in invertebrates, we can refer to the recent reviews by Christiaens et al. (2018a) and Cooper et al. (2019).

DsRNAs can be delivered to target pests in different ways. The first RNAi-based pest control product on the market, Bayer's SmartStax PRO maize plant, expresses the insecticidal dsRNA in the plant itself. These dsRNAs, targeting the *snf7* gene in *D. virgifera*, are then ingested by the larvae feeding on the root, taken up in the gut and eventually lead to a high mortality in the exposed larvae. Transgenic plants offer an easy way to achieve a high exposure to the pest insect. However, public concerns, development costs, and regulatory hurdles concerning GMOs have also led to the development of other non-transformative application strategies, recently reviewed by Cagliari et al. (2019). Depending on the target pest, dsRNA may require a plant passage to allow dsRNA to be taken up. This is the case, for example, for phloem or xylem sap sucking insects such as aphids and stinkbugs or for root-feeders. Some proof-of-concept studies for non-transgenic in planta methods, such as stem injection, seed treatment, and root soaking, have been published (Hunter et al. 2012; Taning et al. 2016a; Li et al. 2015). Alternatively, suitably formulated dsRNA which is sprayed on the plants could potentially also be taken up by the plant. Examples of such formulations and applications are described in the next section.

For many insects which feed on the green parts of the plants, spraying can be the ideal non-transformative application strategy, as the practice of spraying is commonplace in agriculture and dsRNA products could be combined with other pesticides. dsRNA which is sprayed on the crop would then be ingested by the target pests, ideally leading to an efficient gene silencing response and lethality. However, as indicated before, dsRNA is a natural molecule which has a relatively short persistence in the environment. And while this is a benefit from a biosafety point of view, it also makes the application and a long-term protection of the crop challenging. Rapid degradation of dsRNA in the target organism after oral uptake has been identified as a major barrier for efficient RNAi-based biocontrol in many invertebrates and pest species, such as aphids, caterpillars, locusts, and beetles (Prentice et al. 2017, 2019; Christiaens et al. 2016, 2014; Garbutt et al. 2013; Castellanos et al. 2019; Guan et al. 2018; Garcia et al. 2017; Allen and Walker III 2012; Wynant et al. 2014). As mentioned previously, additional barriers which have to be taken into account besides dsRNA persistence are cellular uptake and, if dsRNA is taken up, endosomal escape within the cell. Many of these barriers are similar to those that are encountered in the search for RNAi-based therapeutics in humans (Tatiparti et al. 2017). In pharmaceutical research, the use of nanoparticles has proven to be crucial to get siRNA into the target cells and the same might also be the case for applications in agriculture. Some of the above described barriers could be overcome by using nanoparticles, for example, through protection of the dsRNA against nucleolytic degradation or by improving cellular uptake in plants or target pests. In the next two sections of this chapter, we will focus on recent developments using nanoparticles to improve dsRNA stability and delivery in an agricultural context. The next section focuses on invertebrate pests as target species, followed by a section on non-transgenic in planta delivery of nucleic acids and dsRNA in particular.

## 3 Using Nanocarriers to Improve RNAi Efficacy in Invertebrates

Unraveling the different obstacles involved in variable RNAi sensitivity between different insect pest species has clearly indicated that RNAi-based products will have to be formulated to overcome these barriers prior to their field application. A possible solution is the packaging of the dsRNA molecules such that they are protected from degradation by nucleolytic enzymes and can also be easily taken up and released into target cells. Besides RNase enzymes present in the environment and the target species, as discussed above, also UV radiation could affect dsRNA efficacy. To tackle these obstacles, formulations with nanocarriers have been employed to improve RNAi efficiency, mainly by increasing the environmental stability of dsRNA molecules, protecting them against degradation by nucleases and improving target cell delivery of the dsRNA molecules without affecting their ability to silence targeted genes in the pest species. Delivery of dsRNAs into the cytoplasm could be additionally enhanced by using nanomaterials that stimulate endosomal escape or by conjugating nanoparticles with cell-penetrating peptides or viral capsid proteins. It is important to prudently choose a nanocarrier system such that it has no unintended negative effects on non-target organisms, keeping in mind that risk will be assessed on a case-by-case basis. Nevertheless, the potential of exploiting nanocarrier systems to improve RNAi efficacy is indisputable and provides a way to the future development of RNAi-based control products against current RNAi-recalcitrant pest insects. An overview of publications employing nanocarriers to improve RNAi efficiency in pest management is provided in Table 1.

Nanoparticles are generally defined as particles with sizes falling between 1 and 100 nm (Kumar et al. 2018). This range is however somewhat flexible in literature with sizes ranging between 1 and 500 nm. Nanoparticles for nucleic acid delivery can be designed using various types of molecules such as metals, sugars, peptides, cationic polymers, and lipids, with quite a variety of different functions and possible applications (Blanco et al. 2015). In the context of RNAi-based crop protection, nanoparticles have been exploited with the objective to shield dsRNA molecules against UV radiation and nucleases present in the environment and digestive tract of the target pest, which can otherwise degrade unprotected dsRNA molecules. Moreover, some of these nanoparticles have been designed to enhance cell delivery of the dsRNAs molecules once inside the digestive tract of the insect. Some examples of nanoparticles used to improve RNAi efficiency following oral exposure in insects include chitosan, guanylated polymers, core–shell nanoparticles, liposomes, and branched amphiphilic peptide capsules (BAPCs) (Table 1).

Most nanoparticles are designed with a positive charge to enable binding to dsRNA which is negatively charged and are usually either biocompatible and/or biodegradable. The association of the dsRNA molecule to the nanoparticle involves electrostatic interactions between the phosphate groups present in the dsRNA molecule and the cationic groups present in the nanoparticles (for example, amino groups) (Avila et al. 2014). Depending on the nanoparticles, this association process

**Table 1** Overview of nanoparticles used for nucleic acid delivery in the frame of crop protection

| Nanocarrier composition | Particle size (loaded) | Nucleic acid | Target species | Improves cellular uptake | Improves stability | Improves RNAi efficacy | Improves plant uptake | Reference |
|---|---|---|---|---|---|---|---|---|
| *Polymer* | | | | | | | | |
| Chitosan cross-linked to sodium tripolyphosphate | 200 nm | dsRNA | *Aedes aegypti* | – | YES | YES | – | Dhandapani et al. (2019) |
| Chitosan | 15–19 nm | dsRNA | *Caenorhabditis elegans* | YES | – | YES | – | Lichtenberg et al. (2019) |
| Chitosan | 100–200 nm | dsRNA | *Aedes aegypti* | YES | – | YES | – | Kumar et al. (2016) |
| Chitosan | 100–200 nm | siRNA | *Anopheles gambiae, Aedes aegypti* | – | – | YES | – | Zhang et al. (2015b) |
| Chitosan | 15–19 nm | dsRNA | *Aedes aegypti* | – | NO | YES | – | Das et al. (2015) |
| Chitosan | 100–200 nm | siRNA | *Aedes aegypti* | – | – | – | – | Mysore et al. (2014) |
| Chitosan | 100–200 nm | siRNA | *Aedes aegypti* | – | – | – | – | Mysore et al. (2013) |
| Chitosan, quarternized derivative of chitosan (QCH4) | 350–650 nm, 150–350 nm | dsRNA | *Spodoptera frugiperda* Sf9 cell line | – | – | YES | – | Theerawanitchpan et al. (2012) |
| Chitosan | 100–200 nm | dsRNA | *Anopheles gambiae* | – | – | – | – | Zhang et al. (2010) |
| Guanylated polymer | <350 nm | dsRNA | *Spodoptera exigua* | YES | YES | YES | – | Christiaens et al. (2018b) |
| Guanidinium polymer: Poly-[*N*-(3-guanidinopropyl) methacrylamide] (pGPMA) | 240–320 nm | dsRNA | *Spodoptera frugiperda* | YES | – | YES | – | Parsons et al. (2018) |

(continued)

**Table 1** (continued)

| Nanocarrier composition | Particle size (loaded) | Nucleic acid | Target species | Improves cellular uptake | Improves stability | Improves RNAi efficacy | Improves plant uptake | Reference |
|---|---|---|---|---|---|---|---|---|
| Perylenediimide-cored poly(amido amine) (PDI-PAmAm) | <200 nm | dsRNA | *Aphis glycines* | YES | – | YES | – | Zheng et al. (2019) |
| Perylenediimide-cored poly(amido amine) (PDI-PAmAm) | 85 nm, 130 nm | Viral DNA | *Helicoverpa armigera* | YES | – | – | – | Liu et al. (2016) |
| Perylenediimide-cored poly(amido amine) (PDI-PAmAm) | 140 nm | dsRNA | *Ostrinia furnacalis* | YES | – | – | – | Shen et al. (2014) |
| Perylenediimide-cored poly(amido amine) (PDI-PAmAm) | 100–200 nm | dsRNA | *Drosophila melanogaster* S2 cell line and larvae | YES | – | YES | – | Xu et al. (2014) |
| Star polycation (SPc) and detergent | 260 nm | dsRNA | *Aphis glycines* | YES | – | YES | YES | Yan et al. (2019) |
| Star polycation (SPc) | 260 nm | dsRNA | *Agrotis ipsilon* | YES | – | YES | – | Li et al. (2019) |
| *Peptide* | | | | | | | | |
| Branched amphiphilic peptide capsules (BAPCs) | 70–300 nm | dsRNA | *Acyrthosiphon pisum*, *Tribolium castaneum* | YES | – | YES | – | Avila et al. (2018) |
| Cell-penetrating peptide: Peptide transduction domain/ dsRNA binding domain (PTD-DRBD) | 600–2000 nm | dsRNA | *Anthonomus grandis* | YES | YES | YES | – | Gillet et al. (2017) |
| *Liposomal* | | | | | | | | |
| Lipofectamine | N/A | dsRNA | *Euschistus heros* | – | YES | YES | – | Castellanos et al. (2019) |

| | | | | | | | | |
|---|---|---|---|---|---|---|---|---|
| Lipofectamine | N/A | dsRNA | *Spodoptera frugiperda* | – | – | YES | – | Parsons et al. (2018) |
| GenJe plus | N/A | dsRNA | *Blattella germanica* | – | YES | YES | – | Huang et al. (2018) |
| Lipofectamine | N/A | dsRNA | *Drosophila suzukii* | – | – | YES | – | Taning et al. (2016b) |
| Lipofectamine (Cellfectin, DMRIE-C, Transfectin also used in one assay) | N/A | dsRNA | *Drosophila melanogaster, Drosophila sechellia, Drosophila yakuba, Drosophila pseudoobscura* | – | – | YES | – | Whyard et al. (2009) |
| Cellfectin | N/A | dsRNA | *Spodoptera frugiperda* Sf9 cell line | YES | – | YES | – | Theerawanitchpan et al. (2012) |
| Effectene (Qiagen) | N/A | dsRNA | *Aedes aegypti* | – | – | YES | – | Bedoya-Perez et al. (2013) |
| Effectene (Qiagen) | N/A | dsRNA | *Aedes aegypti* | – | – | YES | – | Cancino-Rodezno et al. (2010) |
| *Core–shell nanoparticle* | | | | | | | | |
| Polyethylenimine functionalized carbon quantum dots (CQD-PEI) | 3–5 nm | dsRNA | *Aedes aegypti* | – | YES | YES | – | Das et al. (2015) |
| Amine functionalized silica nanoparticles (ASNP) | 15–18 nm | dsRNA | *Aedes aegypti* | – | NO | NO | – | Das et al. (2015) |
| Fluorescent cationic core–shell nanoparticles | 6 nm | dsRNA | *Drosophila melanogaster* S2 cell line and larvae, *Ostrinia furnacalis* | YES | – | YES | – | He et al. (2013) |
| Cationic perfluorocarbon (PFC) nanoparticles | >200 nm | siRNA | *Acyrthosiphon pisum, Aphis glycines, Schizaphis graminum* | – | – | YES | – | Thairu et al. (2017) |

(continued)

**Table 1** (continued)

| Nanocarrier composition | Particle size (loaded) | Nucleic acid | Target species | Improves cellular uptake | Improves stability | Improves RNAi efficacy | Improves plant uptake | Reference |
|---|---|---|---|---|---|---|---|---|
| Cationic perfluorocarbon (PFC) nanoparticles | >200 nm | siRNA | *Apis mellifera* | – | – | YES | – | Li-Byarlay et al. (2013) |
| *In planta* | | | | | | | | |
| Layered double hydroxy nanosheets (BioClay) | 15–120 nm | dsRNA | *Nicotiana benthamiana, Vigna unguiculata* | – | – | YES | – | Worrall et al. (2019) |
| Layered double hydroxy nanosheets (BioClay) | 15–120 nm | dsRNA | *Arabidopsis thaliana, Nicotiana tabacum, Vigna unguiculata* | YES | YES | YES | YES | Mitter et al. (2017) |
| Single-walled carbon nanotubes (SWNTs) | 20–30 nm | siRNA | *Nicotiana benthamiana* | YES | YES | YES | YES | Demirer et al. (2019a) |

can result in nanoparticle–dsRNA complexes of different sizes, shapes, and structures (de Ilarduya et al. 2010), generally still maintaining the overall positive charge for easy interaction and uptake through the cell membrane that is negatively charged. It is important to note that the size, shape, overall charge, and geometry of the nanoparticle–dsRNA complex can already be affected by physical barriers such as the mesh-like peritrophic membrane that lines the gut of many insects. Additionally, not all nanoparticles are appropriate for all applications, implying that the nanoparticle must be tailor-designed to the biology of the target species in question. For example, carbon quantum dots have been reported to be very effective in dsRNA protection in the strong alkaline midgut of mosquito larvae, while silica-based nanoparticles are inefficient, with dsRNA completely degraded under these conditions (Das et al. 2015). Similarly, lepidopteran insect pests such as *Spodoptera exigua* possess a very alkaline gut environment; hence, nanoparticle stability in this alkaline environment was achieved by modifying cationic polymethacrylate derivatives with protective guanidine side groups (Christiaens et al. 2018b). These high guanidine content nanoparticles protect dsRNA by forming stable complexes at high pH and also enhance cellular uptake by probably imitating arginine-rich cell-penetrating peptides. Like the guanidine containing nanoparticles, the BAPCs nanoparticles also contain amino groups with p*K*a values ranging from 9 to 13. The ϵ-amino group of the lysine in BAPCs nanoparticles will remain protonated up to a pH ~ 10.5, implying that these functional groups will be more stable in alkaline and neutral environments. BAPCs associated with dsRNA targeting an essential gene (glucose regulating protein 78 gene: *BiP*) in the RNAi-recalcitrant insect pest *A. pisum* successfully improved RNAi efficiency through the oral route, causing the premature death of aphids (half time of 4–5 days) compared to exposure to the same amounts of unprotected dsBiP (half time of 11–12 days) (Avila et al. 2018). Another effective group of nanoparticles that has been reported to improve RNAi efficiency in insects is the small cationic core–shell nanoparticle. He et al. (2013) have successfully demonstrated that by complexing cationic core–shell fluorescent nanoparticles (FNP) with dsRNA targeting a midgut-specific *chitinase* gene of the Asian corn borer, RNAi efficiency could be significantly improved, resulting in significant target gene mRNA degradation and subsequently mortality. These examples indicate that the exploitation of nanotechnology in combination with RNAi-based technology to improve RNAi efficiency will play an important role to overcome barriers currently encountered in RNAi-recalcitrant insect pests.

Lipid-based delivery systems have also been exploited to improve RNAi efficiency in insects. Lipid-based transfection reagents are known to naturally form vesicles when brought into an aqueous solution containing dsRNA and these vesicles are commonly known as liposomes. During the formation of liposomes, the negatively charged dsRNA is enveloped by the positively charged lipids resulting in the formation of a lipid bilayer particle which mimics the phospholipid bilayer of the cell membrane (Dalby et al. 2004). Delivery of dsRNA encapsulated in the liposome into the cell then occurs by lipofection. Effectene-micelles encapsulating dsRNA targeting an essential gene in *A. aegypti* have been exploited to improve RNAi efficiency through feeding in this mosquito species. Similarly, by using the commercial transfection agent, Lipofectamine 2000 (Invitrogen), some studies have reported an increase in RNAi efficiency through feeding in different *Drosophila* species (Whyard et al. 2009; Taning et al. 2016b) and the tick species, *Rhipicephalus haemaphysaloides* (Zhang et al. 2018), which are otherwise refractory to RNAi through feeding.

Although still limited in research, the use of carrier proteins in delivering dsRNA into cells has been reported to improve RNAi efficiency in insects. A representative group of protein carriers are cell-penetrating peptides (CPPs), which are short chain cationic peptides consisting of 10–30 amino acids with generally a high prevalence of basic residues, such as arginine and lysine (Durzyńska et al. 2015). CPPs are able to enter cells while transporting a cargo such as dsRNA. The exact mechanism of how CPPs enter the cell is not yet well known; however, it is assumed that endocytosis probably plays a key role (Choi and David 2014). CPPs have been successfully exploited to improve RNAi efficiency through feeding in the cotton boll weevil, *Anthonomus grandis* (Gillet et al. 2017). In the study, a fusion peptide was designed to contain both a dsRNA binding domain (DRBD) and a peptide transduction domain (PTD). The PTD used in the design was an enhanced version of the arginine-rich CPP trans-activating transcriptional activator (TAT) from the human immunodeficiency virus 1, which was modified to have extra properties that can enable the escape of the fusion protein and its cargo from the endosome into the cytoplasm (Vives et al. 1997; Wadia et al. 2004). dsRNA could bind to the DRBD of the CPP, forming a ribonucleoprotein particle (RNP), which could quickly enter the gut cells of *A. grandis* after oral exposure, leading to significant knockdown of the targeted gene compared to the naked dsRNA (Gillet et al. 2017). While CPPs present an intriguing delivery system, more research is required to fully understand how their full potential could be exploited.

Virus-like particles (VLPs) represent another group of nanocarriers which could be exploited to improve RNAi efficiency in insects (Kolliopoulou et al. 2017). VLPs are synthesized by expressing viral capsid proteins in a production platform (bacteria, plants, insects cells, cell-free, and in vitro) (Shirbaghaee and Bolhassani 2016), where they naturally self-assemble into virus-like structures that can incorporate cargoes such as nucleic acids (Aniagyei et al. 2008). VLPs are therefore empty shells that use the same mechanisms as viruses to enter cells and thus can effectively be designed to carry and deliver dsRNAs into the cytoplasm. Depending on the origin of the VLP components, they could be tailored to deliver dsRNA to specific targeted species. These properties of VLPs could be exploited in crop protection to not only improve RNAi efficiency by dsRNA protection and delivery, but also to improve the specificity of the RNAi-based approach in the context of target pest control. Large scale production platforms such as APSE RNA containers (ARCs), established by the start-up company RNAgri, have already exploited this approach to produce nanocontainers which can successfully encapsulate the desired small RNA molecule (Killmer et al. 2019). These ARCs are expected to protect the dsRNA from degradation and also improve delivery into insect cells. Although still limited in research, the potential of exploiting VLPs in agricultural biotechnology for pest control is immense.

## 4 Nanocarriers for Non-transgenic in Planta dsRNA Delivery

Different from animal cells, plant cells have a very tough cellulose-rich cell wall ranging from 0.1 μm to several micrometers in thickness. In addition, the plant cell wall excludes particles larger than approximately 5–20 nm (Schwab et al. 2016), making it a physical barrier for the delivery of biomolecules such as dsRNA. Hence, a number of questions remain unanswered in relation to the delivery of large dsRNA fragments into plant cells, including the mechanism(s) of dsRNA uptake into plant cells and the stability of the topically applied dsRNA to withstand environmental conditions and provide long-term protection. This domain of non-GMO delivery of dsRNA for an RNAi response is still in its infancy, and there is much attention to nanomaterials (Demirer et al. 2019a, b; Landry and Mitter 2019).

One of the first reports of exogenous RNA application into plants for triggering RNAi of a plant gene was in a patent of Monsanto (Sammons et al. 2014). Tobacco plants (*Nicotiana benthamiana*) were pretreated with Silwet L-77 surfactant and sprayed under pressure (with 2.5 bar) with 685-bp-dsRNA and 21-nt-sRNAs. After this initial observation, other people tried infiltration with sRNAs conjugated to a positively charged carrier peptide that combined a copolymer of histidine and lysine with a CPP named Bp100. Interestingly, the conjugated dsRNA molecules could also be absorbed by the roots and they also displayed biological activity throughout the plant (Numata et al. 2014). In continuation, Monsanto developed a line of "BioDirect technology" as a dsRNA spray application. However, there are no details available in the public domain of this technology of siRNA/dsRNA with nanomaterials/co-formulants to realize an efficient plant uptake and RNAi response. Biodirect focuses on bee health applications, targeting honeybee parasites and pathogens, and the control of glyphosate-resistant weeds, tospovirus, canola flea beetles, and Colorado potato beetles (http://nas-sites.org/biotech/files/2016/04/04-Jenkins.pdf).

In most studies investigating these non-transgenic in planta delivery methods, the aim is to protect plants against viral infections. In 2001, Tenllado et al. (2004) were the first to demonstrate the successful exogenous application of dsRNA molecules in plants. Today, virus diseases can be treated in various plants and crops, for instance, maize, papaya, pea, orchid, tobacco (Mitter et al. 2017; Worrall et al. 2019; Gan et al. 2010; Lau et al. 2014; Tenllado et al. 2003). For the success of RNAi, it is essential that the dsRNA remains stable. In an effort to increase the dsRNA stability, the landmark paper of Mitter et al. (2017) demonstrated the binding of dsRNA to layered double hydroxide (LDH) clay nanosheets with an average particle size of 80–300 nm, also named "BioClay." This use of BioClay allowed a sustained release of the dsRNA over time and afforded protection against cucumber mosaic virus (CMV) for at least 20 days when challenged on sprayed leaves and also on newly emerged unsprayed leaves. What is already known on the mechanism is that the BioClay product protects the dsRNA from nucleases, and interestingly the dsRNA/BioClay complex did not wash off even after rigorous rinsing. Extensive analysis by TEM showed that, on the leaf surface, the atmospheric CO2 and moisture resulted in a gradual breakdown of BioClay into a biocompatible residue, and this process released the dsRNA in the plant cell either by passive diffusion or active transport. Today, this product is further developed by the University of Queensland in partnership with their industrial partner Nufarm (https://qaafi.uq.edu.au/article/2018/09/more-sustainable-crops-just-spray-away).

More recently, in 2019, Zhang et al. (2019) established the methodology of biomolecule delivery to plants with DNA nanostructures and detailed the design parameters of importance for uptake in the plant cell. Also, they assessed the impact of DNA nanostructure geometry parameters as size, shape, compactness, and stiffness. Three different DNA nanostructures were used, namely 3D tetrahedrons, 1D hairpin tiles (HT), and 1D nanostrings, to facilitate the delivery and biological action of 21-nt sRNAs. As a model, this work by infiltration was done with leaves of *N. benthamiana.* In detail, each nanostructure can attach a biological cargo to a locus or loci through complementary base pair hybridization. The tetrahedron contains one attachment locus at its apex, the nanostring contains 10 attachment loci at the center of each of its constituent monomers, and the HT monomer contained one attachment locus either at its center (HT-c) or, for a separate construct, an attachment locus at its side (HT-s). Under the confocal microscope, the nanostructure-conjugated sRNAs entered the symplast and silenced GFP expression. Specifically, the sRNAs conjugated to the 3D nanostructures and exhibited mRNA degradation and also a translational arrest of the GFP. Taken all together, the use of carrier compounds increased RNA delivery in plant cells, although it should be remarked that they are still quite expensive and/or difficult to synthesize which may currently hamper their commercial use. It also needs to be noted that depending on the method of RNA application, the efficiency of RNAi fluctuated. For instance, Dalakouras et al. (2018) investigated the delivery of hairpin RNAs and small RNAs into woody and herbaceous plants by trunk injection and petiole absorption. When sRNA was absorbed via the petiole, this was transported through the xylem. Also, when a 499-nt GFP hairpin RNA (hpRNA) was applied on the petiole and/or via trunk injection in grape and apple, this was found in the xylem and apoplast. This may be advantageous to deliver non-processed dsRNA for insect pest control purposes.

## 5 Biosafety and Regulatory Considerations for the Use of Nanoparticles in dsRNA Delivery

Ever-increasing concerns by consumers about pesticidal applications in agriculture have made the biosafety aspect a very important factor in the development of new products. The aim is to make new control strategies more selective and less persistent in the environment, therefore minimizing the potential hazards and exposure to non-target organisms (NTOs). One of the main reasons RNAi-based biocontrols have attracted a lot of attention recently is because they hold great promise for a biosafe and environmentally friendly pest control. The sequence-dependent mechanism allows for the design of highly species-selective dsRNAs, limiting the risk for gene silencing in NTOs (Whyard et al. 2009; Bachman et al. 2013, 2016). Furthermore, the dsRNA has a limited persistence in the environment and is degraded quite rapidly in soil and aquatic environments (Fischer et al. 2017; Parker et al. 2019; Albright III et al. 2017; Dubelman et al. 2014). An additional advantage from a biosafety point of view is the fact that many species, notably vertebrates and humans, have many cellular and extracellular barriers which prevent the uptake of dsRNA from the environment and any biological activity. The risk of cellular uptake in vertebrates after ingestion of dsRNA is generally considered as unlikely, due to the fact that the gastrointestinal tract, the circulatory system, and other biological fluids are very inhospitable environments for nucleic acids. For an extended literature overview of environmental risk assessment and food/feed safety aspects of these dsRNAs, we can refer to two recent reviews which were written for EFSA, as a baseline information source (Christiaens et al. 2018a; Dávalos et al. 2019).

The term "nanoparticles" as used in this chapter refers to a very broad range of chemical or biological compounds which can have very variable effects on the environment and on NTOs. Biosafety aspects of the nanoparticles as such will be discussed elsewhere and are not within the scope of this chapter. However, we would like to draw attention to the fact that the use of nanoparticles to improve dsRNA persistence on the field or to improve delivery in plants or invertebrates could change some of the biosafety aspects which are related to dsRNA. For example, it is a possibility that in future risk assessment of naked dsRNA products, certain NTOs might be excluded from toxicity testing based on their general insensitivity to dsRNAs due to the above discussed barriers. However, if nanocarriers are designed to cross such barriers in target species, they might also lead to silencing effects in these otherwise insensitive NTOs. Another aspect to take into account is the fact that nanocarriers which are designed to improve cellular uptake in insect midgut cells could possibly also do so in mammalian cells. Likewise, nanocarriers designed to withstand nucleolytic degradation in nucleolytic environments (e.g., invertebrate digestive tract) could lead to a longer persistence of dsRNA in vertebrates after ingestion, leading to a more likely cellular exposure. These aspects should be taken into account when considering the regulation of such products and regulators will have to consider whether the active ingredient in such applications is still the dsRNA or whether the dsRNA/nanocarrier complex is the relevant active ingredient.

## References

Albright VC III, Wong CR, Hellmich RL, Coats JR (2017) Dissipation of double-stranded RNA in aquatic microcosms. Environ Toxicol Chem 36:1249–1253

Allen ML, Walker WB III (2012) Saliva of Lygus lineolaris digests double stranded ribonucleic acids. J Insect Physiol 58:391–396. https://doi.org/10.1016/j.jinsphys.2011.12.014

Aniagyei SE, DuFort C, Kao CC, Dragnea B (2008) Self-assembly approaches to nanomaterial encapsulation in viral protein cages. J Mater Chem 18:3763–3774

Avila LA, Lee SY, Tomich JM (2014) Synthetic in vitro delivery systems for plasmid DNA in eukaryotes. J Nanopharmaceutics Drug Deliv 2:17–35
Avila L et al (2018) Delivery of lethal dsRNAs in insect diets by branched amphiphilic peptide capsules. J Control Release 273:139–146
Bachman PM et al (2013) Characterization of the spectrum of insecticidal activity of a double-stranded RNA with targeted activity against Western corn rootworm (Diabrotica virgifera virgifera LeConte). Transgenic Res 22:1207–1222. https://doi.org/10.1007/s11248-013-9716-5
Bachman PM et al (2016) Ecological risk assessment for DvSnf7 RNA: a plant-incorporated protectant with targeted activity against western corn rootworm. Regul Toxicol Pharmacol 81:77–88
Baum JA et al (2007) Control of coleopteran insect pests through RNA interference. Nat Biotechnol 25:1322–1326. https://doi.org/10.1038/nbt1359
Bedoya-Perez LP, Cancino-Rodezno A, Flores-Escobar B, Soberon M, Bravo A (2013) Role of UPR pathway in defense response of Aedes aegypti against Cry11Aa toxin from Bacillus thuringiensis. Int J Mol Sci 14:8467–8478. https://doi.org/10.3390/ijms14048467
Blanco E, Shen H, Ferrari M (2015) Principles of nanoparticle design for overcoming biological barriers to drug delivery. Nat Biotechnol 33:941
Cagliari D et al (2019) Management of Pest Insects and Plant Diseases by non-transformative RNAi. Front Plant Sci 10:1319
Cancino-Rodezno A et al (2010) The mitogen-activated protein kinase p38 is involved in insect defense against cry toxins from Bacillus thuringiensis. Insect Biochem Mol Biol 40:58–63. https://doi.org/10.1016/j.ibmb.2009.12.010
Cappelle K, de Oliveira CFR, Van Eynde B, Christiaens O, Smagghe G (2016) The involvement of clathrin-mediated endocytosis and two Sid-1-like transmembrane proteins in double-stranded RNA uptake in the Colorado potato beetle midgut. Insect Mol Biol 25:315–323. https://doi.org/10.1111/imb.12222
Castellanos NL, Smagghe G, Sharma R, Oliveira EE, Christiaens O (2019) Liposome encapsulation and EDTA formulation of dsRNA targeting essential genes increase oral RNAi-caused mortality in the Neotropical stink bug Euschistus heros. Pest Manag Sci 75:537–548
Choi YS, David AE (2014) Cell penetrating peptides and the mechanisms for intracellular entry. Curr Pharm Biotechnol 15:192–199
Christiaens O, Swevers L, Smagghe G (2014) DsRNA degradation in the pea aphid (Acyrthosiphon pisum) associated with lack of response in RNAi feeding and injection assay. Peptides 53:307–314. https://doi.org/10.1016/j.peptides.2013.12.014
Christiaens O et al (2016) RNA interference: a promising biopesticide strategy against the African Sweetpotato weevil Cylas brunneus. Sci Rep 6:38836
Christiaens O et al (2018a) Literature review of baseline information on RNAi to support the environmental risk assessment of RNAi-based GM plants. EFSA Support Publ 15:1424E
Christiaens O et al (2018b) Increased RNAi efficacy in Spodoptera exigua via the formulation of dsRNA with guanylated polymers. Front Physiol 9:316
Cooper AM, Silver K, Zhang J, Park Y, Zhu KY (2019) Molecular mechanisms influencing efficiency of RNA interference in insects. Pest Manag Sci 75:18–28
Dalakouras A et al (2018) Delivery of hairpin RNAs and small RNAs into woody and herbaceous plants by trunk injection and petiole absorption. Front Plant Sci 9:1253
Dalby B et al (2004) Advanced transfection with Lipofectamine 2000 reagent: primary neurons, siRNA, and high-throughput applications. Methods 33:95–103
Dalzell JJ et al (2011) RNAi effector diversity in nematodes. PLoS Negl Trop Dis 5:e1176. https://doi.org/10.1371/journal.pntd.0001176
Das S, Debnath N, Cui Y, Unrine J, Palli SR (2015) Chitosan, carbon quantum dot, and silica nanoparticle mediated dsRNA delivery for gene silencing in Aedes aegypti: a comparative analysis. ACS Appl Mater Interfaces 7:19530–19535. https://doi.org/10.1021/acsami.5b05232

Dávalos A, Henriques R, Latasa MJ, Laparra M, Coca M (2019) Literature review of baseline information on non-coding RNA (ncRNA) to support the risk assessment of ncRNA-based genetically modified plants for food and feed. EFSA Supporting Publications 16:1688E

de Ilarduya CT, Sun Y, Düzgüneş N (2010) Gene delivery by lipoplexes and polyplexes. Eur J Pharm Sci 40:159–170

Demirer G, Zhang H, Goh N, Chang R, Landry M (2019a) Nanotubes effectively deliver siRNA to intact plant cells and protect siRNA against nuclease degradation. Available at SSRN 3352632

Demirer GS et al (2019b) High aspect ratio nanomaterials enable delivery of functional genetic material without DNA integration in mature plants. Nat Nanotechnol 14:456

Desai SD, Eu YJ, Whyard S, Currie RW (2012) Reduction in deformed wing virus infection in larval and adult honey bees (Apis mellifera L.) by double-stranded RNA ingestion. Insect Mol Biol 21:446–455. https://doi.org/10.1111/j.1365-2583.2012.01150.x

Dhandapani RK, Gurusamy D, Howell JL, Palli SR (2019) Development of CS-TPP-dsRNA nanoparticles to enhance RNAi efficiency in the yellow fever mosquito, Aedes aegypti. *Sci Rep* 9:8775

Dubelman S et al (2014) Environmental fate of double-stranded RNA in agricultural soils. PLoS One 9. https://doi.org/10.1371/journal.pone.0093155

Durzyńska J et al (2015) Viral and other cell-penetrating peptides as vectors of therapeutic agents in medicine. J Pharmacol Exp Ther 354:32–42

Fischer JR et al (2017) Aquatic fate of a double-stranded RNA in a sediment–water system following an over-water application. Environ Toxicol Chem 36:727–734

Gan D et al (2010) Bacterially expressed dsRNA protects maize against SCMV infection. Plant Cell Rep 29:1261–1268

Garbutt JS, Bellés X, Richards EH, Reynolds SE (2013) Persistence of double-stranded RNA in insect hemolymph as a potential determiner of RNA interference success: evidence from Manduca sexta and Blattella germanica. J Insect Physiol 59:171–178. https://doi.org/10.1016/j.jinsphys.2012.05.013

Garcia RA et al (2017) Nucleases as a barrier to gene silencing in the cotton boll weevil, Anthonomus grandis. PLoS One 12:e0189600

Gillet F-X et al (2017) Investigating engineered ribonucleoprotein particles to improve oral RNAi delivery in crop insect pests. Front Physiol 8:256

Gonsalves D, Ferreira S, Manshardt R, Fitch M, Slightom J (2000) Transgenic virus resistant papaya: new hope for controlling papaya ringspot virus in Hawaii. Plant health progress 1:20

Guan R-B et al (2018) A nuclease specific to lepidopteran insects suppresses RNAi. J Biol Chem 293:6011–6021

Haller S, Widmer F, Siegfried BD, Zhuo X, Romeis J (2019) Responses of two ladybird beetle species (Coleoptera: Coccinellidae) to dietary RNAi. Pest Manag Sci 75:2652–2662

He B et al (2013) Fluorescent nanoparticle delivered dsRNA toward genetic control of insect pests. Adv Mater 25:4580–4584

Huang J-H, Liu Y, Lin Y-H, Belles X, Lee H-J (2018) Practical use of RNA interference: Oral delivery of double-stranded RNA in liposome carriers for cockroaches. JoVE-J Vis Exp 135:e57385

Hunter CP et al (2006) Systemic RNAi in Caenorhabditis elegans. Cold Spring Harb Symp Quant Biol 71:95–100. https://doi.org/10.1101/sqb.2006.71.060

Hunter WB, Glick E, Paldi N, Bextine BR (2012) Advances in RNA interference: dsRNA treatment in trees and grapevines for insect Pest suppression. Southwest Entomol 37:85–87

Joga MR, Zotti MJ, Smagghe G, Christiaens O (2016) RNAi efficiency, systemic properties, and novel delivery methods for pest insect control: what we know so far. Front Physiol 7:553

Killiny N, Hajeri S, Tiwari S, Gowda S, Stelinski LL (2014) Double-stranded RNA uptake through topical application, mediates silencing of five CYP4 genes and suppresses insecticide resistance in Diaphorina citri. PLoS One 9:e110536. https://doi.org/10.1371/journal.pone.0110536

Killmer JL, Kumar A, McLaughlin PD, Arhancet JPH (2019) Google Patents

Kolliopoulou A, Taning CN, Smagghe G, Swevers L (2017) Viral delivery of dsRNA for control of insect agricultural pests and vectors of human disease: prospects and challenges. Front Physiol 8:399

Kumar DR et al (2016) Delivery of chitosan/dsRNA nanoparticles for silencing of wing development vestigial (vg) gene in Aedes aegypti mosquitoes. Int J Biol Macromol 86:89–95. https://doi.org/10.1016/j.ijbiomac.2016.01.030

Kumar S et al (2018) Nano-based smart pesticide formulations: emerging opportunities for agriculture. J Control Release 294:131–153

Landry MP, Mitter N (2019) How nanocarriers delivering cargos in plants can change the GMO landscape. Nat Nanotechnol 14:512

Lau S et al (2014) Crude extracts of bacterially-expressed dsRNA protect orchid plants against Cymbidium mosaic virus during transplantation from in vitro culture. J Hortic Sci Biotechnol 89:569–576

Li H, Guan R, Guo H, Miao X (2015) New insights into an RNAi approach for plant defence against piercing-sucking and stem-borer insect pests. Plant Cell Environ 38:2277–2285. https://doi.org/10.1111/pce.12546

Li J et al (2019) A facile-synthesized star Polycation constructed as a highly efficient gene vector in Pest management. ACS Sustain Chem Eng 7:6316–6322

Li-Byarlay H et al (2013) RNA interference knockdown of DNA methyltransferase 3 affects gene alternative splicing in the honey bee. Proc Natl Acad Sci USA 110:12750–12755. https://doi.org/10.1073/pnas.1310735110

Lichtenberg SS, Tsyusko OV, Palli SR, Unrine JM (2019) Uptake and bioactivity of chitosan/double-stranded RNA Polyplex nanoparticles in Caenorhabditis elegans. Environ Sci Technol 53:3832–3840

Liu X et al (2016) Perylenediimide-cored cationic nanocarriers deliver virus DNA to kill insect pests. Polym Chem 7:3740–3746

Maori E et al (2009) IAPV, a bee-affecting virus associated with Colony collapse disorder can be silenced by dsRNA ingestion. Insect Mol Biol 18:55–60. https://doi.org/10.1111/j.1365-2583.2009.00847.x

Mitter N et al (2017) Clay nanosheets for topical delivery of RNAi for sustained protection against plant viruses. Nat plants 3:16207

Mysore K, Flannery EM, Tomchaney M, Severson DW, Duman-Scheel M (2013) Disruption of Aedes aegypti olfactory system development through chitosan/siRNA nanoparticle targeting of semaphorin-1a. PLoS Negl Trop Dis 7:1–12

Mysore K, Andrews E, Li P, Duman-Scheel M (2014) Chitosan/siRNA nanoparticle targeting demonstrates a requirement for single-minded during larval and pupal olfactory system development of the vector mosquito Aedes aegypti. BMC Dev Biol 14:9. https://doi.org/10.1186/1471-213x-14-9

Niu J et al (2019) Topical dsRNA delivery induces gene silencing and mortality in the pea aphid. Pest Manag Sci 75:2873–2881

Numata K, Ohtani M, Yoshizumi T, Demura T, Kodama Y (2014) Local gene silencing in plants via synthetic ds RNA and carrier peptide. Plant Biotechnol J 12:1027–1034

Parker KM et al (2019) Environmental fate of RNA interference pesticides: adsorption and degradation of double-stranded RNA molecules in agricultural soils. Environ Sci Technol 53:3027–3036

Parsons KH, Mondal MH, McCormick CL, Flynt AS (2018) Guanidinium-functionalized interpolyelectrolyte complexes enabling RNAi in resistant insect pests. Biomacromolecules 19:1111–1117

Piot N, Snoeck S, Vanlede M, Smagghe G, Meeus I (2015) The effect of Oral administration of dsRNAcc on viral Replication and Mortality in Bombus terrestris. Viruses-Basel 7:3172–3185. https://doi.org/10.3390/v7062765

Prentice K et al (2017) RNAi-based gene silencing through dsRNA injection or ingestion against the African sweet potato weevil Cylas puncticollis (Coleoptera: Brentidae). Pest Manag Sci 73:44–52

Prentice K, Smagghe G, Gheysen G, Christiaens O (2019) Nuclease activity decreases the RNAi response in the sweetpotato weevil Cylas puncticollis. Insect Biochem Mol Biol 110:80–89

Sammons RD et al (2014) Google Patents

San Miguel K, Scott JG (2016) The next generation of insecticides: dsRNA is stable as a foliar-applied insecticide. Pest Manag Sci 72:801–809. https://doi.org/10.1002/ps.4056

Schwab F et al (2016) Barriers, pathways and processes for uptake, translocation and accumulation of nanomaterials in plants–critical review. Nanotoxicology 10:257–278

Shen D et al (2014) Systemically interfering with immune response by a fluorescent cationic dendrimer delivered gene suppression. J Mater Chem B 2:4653–4659. https://doi.org/10.1039/c4tb00411f

Shirbaghaee Z, Bolhassani A (2016) Different applications of virus-like particles in biology and medicine: vaccination and delivery systems. Biopolymers 105:113–132

Shukla JN et al (2016) Reduced stability and intracellular transport of dsRNA contribute to poor RNAi response in lepidopteran insects. RNA Biol 13:656–669. https://doi.org/10.1080/15476286.2016.1191728

Taning CNT, Andrade EC, Hunter WB, Christiaens O, Smagghe G (2016a) Asian Citrus Psyllid RNAi pathway–RNAi evidence. Sci Rep 6:38082

Taning CNT et al (2016b) Oral RNAi to control Drosophila suzukii: laboratory testing against larval and adult stages. J Pest Sci 89:803–814

Tatiparti K, Sau S, Kashaw SK, Iyer AK (2017) siRNA delivery strategies: a comprehensive review of recent developments. Nano 7:77

Tenllado F, Martinez-Garcia B, Vargas M, Diaz-Ruiz JR (2003) Crude extracts of bacterially expressed dsRNA can be used to protect plants against virus infections. BMC Biotechnol 3:3. https://doi.org/10.1186/1472-6750-3-3

Tenllado F, Llave C, Díaz-Ruíz JR (2004) RNA interference as a new biotechnological tool for the control of virus diseases in plants. Virus Res 102:85–96. https://doi.org/10.1016/j.virusres.2004.01.019

Thairu M et al (2017) Efficacy of RNA interference knockdown using aerosolized short interfering RNAs bound to nanoparticles in three diverse aphid species. Insect Mol Biol 26:356–368

Theerawanitchpan G et al (2012) Chitosan and its quaternized derivative as effective long dsRNA carriers targeting shrimp virus in Spodoptera frugiperda 9 cells. J Biotechnol 160:97–104. https://doi.org/10.1016/j.jbiotec.2012.04.011

Vives E, Brodin P, Lebleu B (1997) A truncated HIV-1 tat protein basic domain rapidly translocates through the plasma membrane and accumulates in the cell nucleus. J Biol Chem 272:16010–16017

Wadia JS, Stan RV, Dowdy SF (2004) Transducible TAT-HA fusogenic peptide enhances escape of TAT-fusion proteins after lipid raft macropinocytosis. Nat Med 10:310

Whyard S, Singh AD, Wong S (2009) Ingested double-stranded RNAs can act as species-specific insecticides. Insect Biochem Mol Biol 39:824–832. https://doi.org/10.1016/j.ibmb.2009.09.007

Winston WM, Molodowitch C, Hunter CP (2002) Systemic RNAi in C-elegans requires the putative transmembrane protein SID-1. Science 295:2456–2459. https://doi.org/10.1126/science.1068836

Worrall EA et al (2019) Exogenous application of RNAi-inducing double-stranded RNA inhibits aphid-mediated transmission of a plant virus. Front Plant Sci 10:265

Wynant N et al (2014) Identification, functional characterization and phylogenetic analysis of double stranded RNA degrading enzymes present in the gut of the desert locust, Schistocerca gregaria. Insect Biochem Mol Biol 46:1–8. https://doi.org/10.1016/j.ibmb.2013.12.008

Xu Z et al (2014) Highly water-soluble perylenediimide-cored poly (amido amine) vector for efficient gene transfection. J Mater Chem B 2:3079–3086

Yan S et al (2019) Spray method application of transdermal dsRNA delivery system for efficient gene silencing and pest control on soybean aphid Aphis glycines. J Pest Sci 93:1–11

Yoon J-S, Gurusamy D, Palli SR (2017) Accumulation of dsRNA in endosomes contributes to inefficient RNA interference in the fall armyworm, Spodoptera frugiperda. Insect Biochem Mol Biol 90:53–60

Zhang X, Zhang J, Zhu KY (2010) Chitosan/double-stranded RNA nanoparticle-mediated RNA interference to silence chitin synthase genes through larval feeding in the African malaria mosquito (Anopheles gambiae). Insect Mol Biol 19:683–693. https://doi.org/10.1111/j.1365-2583.2010.01029.x

Zhang H, Li H, Guan R, Miao X (2015a) Lepidopteran insect species-specific, broad-spectrum, and systemic RNA interference by spraying dsRNA on larvae. Entomol Exp Appl 155:218–228. https://doi.org/10.1111/eea.12300

Zhang Q, Hua G, Adang MJ (2015b) Chitosan/DsiRNA nanoparticle targeting identifies AgCad1 cadherin in Anopheles gambiae larvae as an in vivo receptor of Cry11Ba toxin of Bacillus thuringiensis subsp jegathesan. Insect Biochem Mol Biol 60:33–38. https://doi.org/10.1016/j.ibmb.2015.03.001

Zhang Y et al (2018) Liposome mediated double-stranded RNA delivery to silence ribosomal protein P0 in the tick Rhipicephalus haemaphysaloides. Ticks and Tick-Borne Dis 9:638–644

Zhang H et al (2019) DNA nanostructures coordinate gene silencing in mature plants. Proc Natl Acad Sci 116:7543–7548

Zheng Y et al (2019) A polymer/detergent formulation improves dsRNA penetration through the body wall and RNAi-induced mortality in the soybean aphid Aphis glycines. Pest Manag Sci 75:1993–1999

Zhu F, Xu J, Palli R, Ferguson J, Palli SR (2011) Ingested RNA interference for managing the populations of the Colorado potato beetle, *Leptinotarsa decemlineata. Pest Manage Sci* 67:175–182. https://doi.org/10.1002/ps.2048

# Understanding the Interaction of Nanopesticides with Plants

**Etenaldo F. Santiago, Montcharles S. Pontes, Gilberto J. Arruda, Anderson R. L. Caires, Ian Colbeck, Ronald Maldonado-Rodriguez, and Renato Grillo**

**Abstract** Understanding the interactions between engineered nanomaterials and the environment is essential for unbiased assessments of their agricultural applications. Nano-based pesticides can potentially be safer and/or more efficient than their conventional analogs. However, there is limited information about how nanopesticides influence physiology and metabolism during their interactions with plants, particularly, related to its mode of action. The main question herein is about the interaction between nanopesticides and plants. In this chapter, we start from a theoretical discussion on the complex organization of biological systems, offering a variety of examples showing the effects of nanopesticides from uptake to the mode of action. Moreover, we discuss different examples, how physiological and metabolic responses can help us to understand the behavior of plants exposed to

E. F. Santiago (✉) · M. S. Pontes (✉) · G. J. Arruda
Plant Resources Study Group, Center for Natural Resources Study, Mato Grosso do Sul State University, Dourados, Brazil
e-mail: felipe@uems.br

A. R. L. Caires
Optics and Photonics Group, Institute of Physics, Federal University of Mato Grosso do Sul, Campo Grande, Brazil

School of Life Science, University of Essex, Colchester, UK
e-mail: anderson.caires@ufms.br

I. Colbeck
School of Life Science, University of Essex, Colchester, UK
e-mail: colbi@essex.ac.uk

R. Maldonado-Rodriguez
The Fluoromatics Lab. 10, Geneva, Switzerland
e-mail: info@fluoromatics.com

R. Grillo
Department of Physics and Chemistry, São Paulo State University, Ilha Solteira, Brazil
e-mail: renato.grillo@unesp.br



L. F. Fraceto et al. (eds.), *Nanopesticides*,
https://doi.org/10.1007/978-3-030-44873-8_4

nano-based pesticides. Finally, we demonstrate that prediction models can be used as a routine tool for monitoring and classifying plant response according to their degree of resistance or tolerance to determined nanopesticide, aiming to understand the specific characteristics of nanopesticides into plant systems.

**Keywords** Nanoagrochemicals · Intelligent plant protectant · Smart-agromaterials · Plant systems · Nanopesticides

## 1 Introduction

Plant pathogens and diseases cause significant reductions and losses in crop production. Thus, agrochemicals certainly play an important role in improving modern agricultural practices. However, despite the advances in transgenic materials and their potential benefits for plant management, the accumulation of pesticides and fertilizers in modern agriculture has significantly increased over the last century. This problem may be further intensified by an alarming increase in food demand estimated in the range of 59–98% for total food by 2050 (Conijn et al. 2018).

As such nanotechnology holds emerging promise for addressing these problems in agriculture and food production (White and Gardea-Torresdey 2018). Recently, nanotechnology research on applications in the agrochemical sector has increased, especially in the development and design of new plant-protection products (Kah and Holfman 2014; Zhao et al. 2018a; Gomes et al. 2019; Lowry et al. 2019). Consequently, nano-based plant-protection products can, and will, play an important role in the future of agriculture (Fig. 1).

Despite the great potential of the nanopesticides to partially or totally substitute the conventional agrochemicals by reducing the harmful impact to the environment due to their improved pest control efficacy, shelf-life, solubility, site-specific uptake, and decreased toxicity level for non-target organisms (Grillo et al. 2016; Worral et al. 2018; Zhao et al. 2018a) not many nanomaterials-based plant-protectants products have been commercialized for agricultural crop application (Campos et al. 2018; Worral et al. 2018). In fact, the technical development in the field of nanopesticides has recently led to a new and increased concern over this new class of agrochemical compounds (Zhao et al. 2018a; Li et al. 2019a).

The major goals of recently proposed and novel nanopesticides are to serve as a sustainable amendment, based on the nano-scale properties of the used materials. Thus, nanopesticides are reported to be more potent when compared with analogs and require lower application doses, consequently, reduce the final costs of crop production (Kah and Holfman 2014; Adisa et al. 2019). So, in this chapter, we discuss the recent advances in plant disease management using nanostructured materials themselves as a protectant as well as carriers for agrochemicals such as insecticides, fungicides, and herbicides focused on the holistic understanding of fundamental questions and addressing the scientific gap of plant systems interaction with nanopesticides.

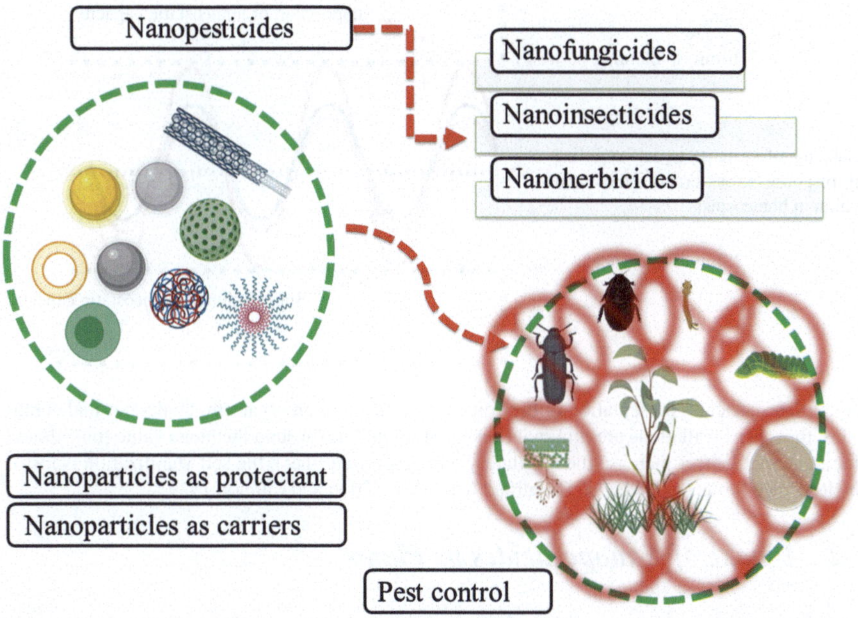


**Fig. 1** Nanomaterials as protectant or delivery systems to provide crop protection. This scheme shows distinct nanomaterials as either protectants or carriers for controlled release of actives such as insecticides, fungicides, and herbicides targeting of a wide range of pests

## *1.1 Nanomaterials Interacting with Plants*

Plants constitute systems in which the cellular flux maintenance depends on complex mechanisms that can be unregulated by environmental stressors such as biotic (herbivores, plant competition, pathogens) and/or abiotic factors (e.g., unfavorable conditions of light, water, mineral availability) including nanopesticides and nanoparticles. The adjustment responses are related to homeostatic ability (Fig. 2).

Negative interactions related to nanomaterials can trigger complex adjustment responses (allosteric responses) from pre-existent information, in a predictive and collaborative way in plants (Sterling 2014). On the other hand, some plants (species and/or variety) when interacting with nanomaterials show metabolic changes that can extrapolate the homeostatic normal range (McEwen and Wingfield 2003; Davis 2016). Thus, the knowledge of these limits can be important to identify tolerant and non-tolerant varieties of nanomaterials to plant management in agricultural systems. Therefore, the tolerance plant responses are the key to the application of nanomaterials in agriculture.

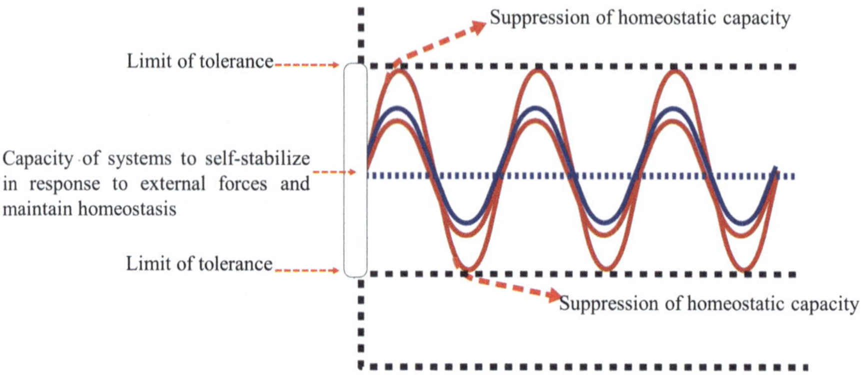


**Fig. 2** Theoretical representation of the homeostatic status of an organism. Under optimal conditions, the inner oscillations are minimal (blue continuum line) around the mean value (blue dotted line), external factors increase the oscillatory amplitude (red lines), the self-stabilization is lost if the tolerance limits are exceeded (Modified from Davis 2016 and Cannon 1929)

## *1.2 Uptake of Nanopesticides by Plants*

Nanopesticides uptake occurs when a particle penetrates the cell walls of plants after application. The cell wall acts as a semipermeable environmental barrier that regulates the trafficking of exogenous materials across the cell membrane through several pores to the plant cell. Therefore, the size, charge, and physicochemical properties of nanopesticides play a critical role in their uptake through these pores. The nanopesticides with a size smaller than the diameter of the cell wall pores can easily penetrate the cell wall and reach the plasma membrane (Navarro et al. 2008). Consequently, nanopesticides may enter plant cells through endocytosis, binding to ion channels or carrier proteins, or by forming complexes associated with the transporters of the plasma membrane (McKnight et al. 2003). Once in the cell cytoplasm, nanopesticides also may interact with cellular organelles, DNA, hormones, proteins, and other cell compounds and affect the plant biochemical, metabolic, or physiological reactions at a particular site.

According to Pérez-de-Luque (2017), important knowledge has been gained about the nanomaterials uptake by plants, although there are still gaps regarding the internalization into plant cells. Additionally, different plant species can differ in their metabolism and physiology, influencing the uptake of nanomaterials as reported by Cifuentes et al. (2010), Larue et al. (2012), and Zhu et al. (2012). Furthermore, the ways of application of the nanopesticides are crucial to determine the efficiency of plant uptake (Fig. 3).

In the case of down-top movement, roots are specialized in the absorption of nutrients and water; it is hypothesized that the macromolecular exudates excreted by root cells might be responsible for the nanopesticides accumulation in the root epidermis. Thus, the nanopesticides can migrate from the epidermal layer to endodermal cells through the apoplastic route (Li et al. 2016a, 2016b). Nanopesticides

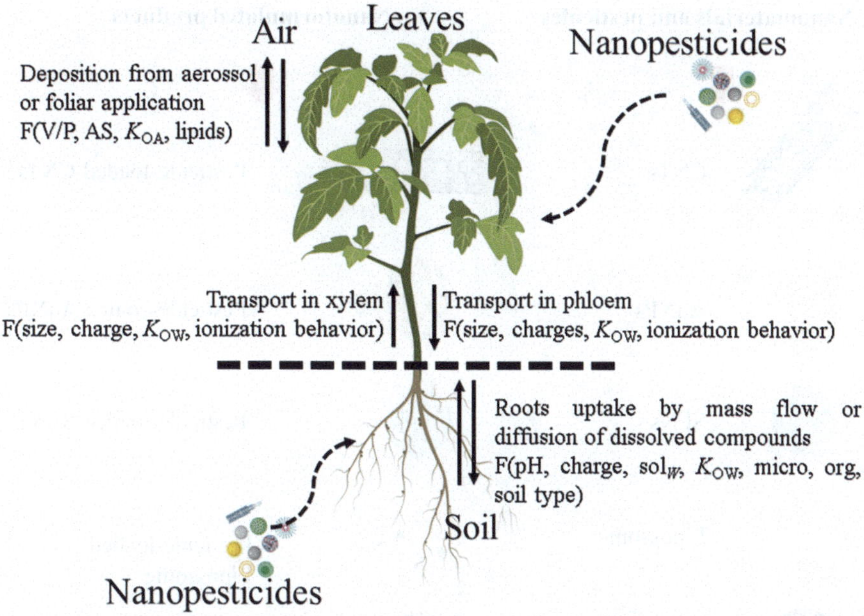


**Fig. 3** Principal pathways that nanopesticides can uptake in plants. The main parameters affecting each pathway are annotated: F—function of; $K_{OA}$—coefficient of octanol–air partition; $K_{OW}$—coefficient of octanol–water partition; micro—soil symbiotic-microbiota; V/P—vapor–particle portioning; AS—plant surface area; size—particle size in nm; $sol_w$—water solubility; org—organic matter content in the soil; lipid—plant lipid content

may also participate in indirect interaction by affecting soil properties and symbiotic-microbiota (Degrassi et al. 2012; Simonin and Richaume 2015). On the other hand, leaves present the stomatal pathway as the most likely route for nanopesticides internalization in case of top-down movement, and the cuticle which hampers penetration of substances (Eichert et al. 2008).

Concerning nanocarrier systems for delivering pesticides, herbicides, and plant growth regulators (Fig. 4), the uptake mechanisms of nanopesticides by plants may be more complex due to the composition of the carrier system. The ability of nanocarriers to protect the active ingredient against degradation and to adsorb to plant surface prolongs the contact time between the agrochemical and the plant surface (e.g., epidermis of leaves or stems) and may be the major factor of efficiency of these nanopesticides compared with the free form of agrochemicals (Pereira et al. 2014). Oliveira and co-workers (2015) hypothesized that the hydrophobic nanocapsules might interact with the leaf cuticle, hence increasing the delivery of an active compound to the plant tissues, while at the same time decreasing the loss of the active element to the environment.

Recently, Bombo et al. (2019) reported that the nanocarriers loaded with atrazine herbicide penetrate the stomata, particularly in the hydathode regions. Hydathode water pores can vary from a few to several microns in size, allowing the nanocarrier

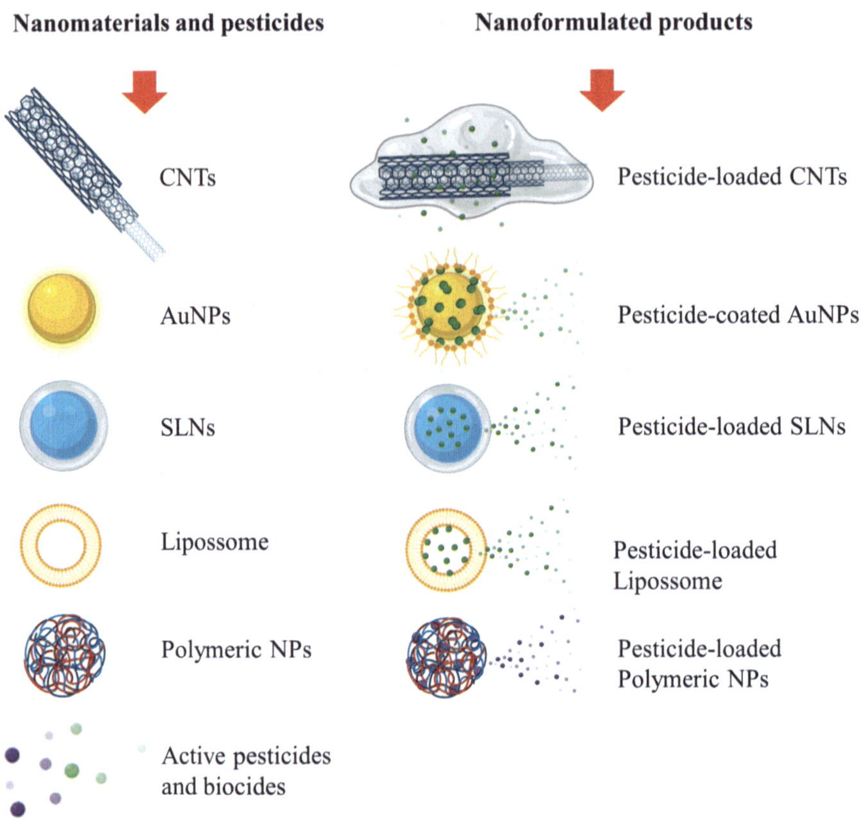


**Fig. 4** Nanocarrier systems for delivering pesticides, herbicides, and plant growth regulators. CNTs: carbon nanotubes; and SLNs: solid lipid nanocarriers

entry and direct access to the vascular system (Martin and von Willert 2000; Nguyen et al. 2014). Depending on the type, composition, coating agent, physicochemical properties, and exposure mode of the nanomaterials, they can be found inside cells in roots, stems, and leaves, via transport in the vascular bundles (Andreotti et al. 2015; Doolette et al. 2015; Bao et al. 2016; Barrios et al. 2016; Bombo et al. 2019). According to Nguyen et al. (2014), both negatively and positively charged nanocarriers can rapidly penetrate the leaf tissues; however, negatively charged nanocarriers had a faster penetration than positively charged ones.

Cellular uptake, targeting, and intracellular trafficking of nanopesticides can be improved by controlling the physicochemical properties of these nanomaterials, such as size (1–100 nm), shape (irregular or geometrically defined), chemical properties, concentration, aggregation, including the target plant sensibility (Hussain et al. 2016; Hayles et al. 2017) and other surface chemical properties (Albanese et al. 2012). Additionally, the nanomaterial uptake and way of interaction also depend on: (1) Mechanical effects related to size and form; (2) Interactions based on affinity; (3) Catalytical and surface effects. Apart from size and shape effects, the last two ways of interactions include responses influenced by the presence and density of ligands able to interact with the receptor target (Albanese et al. 2012), and responses related to the nanoparticle type, its charge and electrostatic interaction as well as the chemical nature of cellular compost, for example, the transitory complex particle-protein (Cerdervall et al. 2007; Yadav et al. 2011). These properties are still not fully understood and are important to reveal the mechanisms involved in the fate and toxicity of the nanopesticides (Fig. 5).

The size of the nanopesticide is crucial to determine the uptake efficiency into the cell wall and, consequently, its permeability into the cell membrane and intracellular trafficking. Understanding the role of nanopesticide size in cellular internalization is a key factor for the design and development of more efficient nanopesticides. After cell wall penetration, nanomaterials in the 120–200 nm size range are internalized via caveolin- or clathrin-mediated endocytosis, while nanomaterials larger than 250 nm show an optimal internalization via phagocytosis (Rejman et al. 2004; Lai et al. 2007; Panariti et al. 2012). Differently, nanomaterials with a size range from 30 to 50 nm can interact with the cell membrane receptors and easy entry into the cytoplasm via receptor-mediated endocytosis (Lu et al. 2009; Wang et al. 2010).

Another critical factor that affects cellular uptake and translocation pathways of nanomaterials is its shape. Nangia and Sureshkumar (2012) using molecular dynamics simulation approaches revealed distinct variations in translocation through cell membranes for cone, cube, rice, rod, pyramid, and sphere-shaped nanoparticles.

Several studies have indicated the shape effects on nanomaterial internalization by biological systems. Up to now, it has been found that there are different pathways

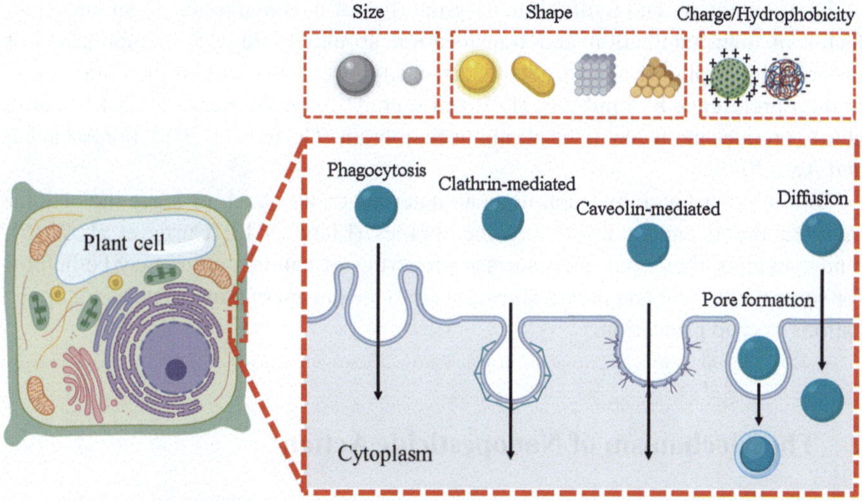


**Fig. 5** Schematic illustration of some physicochemical properties that influence the cellular uptake of nanopesticides and different pathways of nanopesticides translocation across cell membranes. The pathways usually include endocytosis (phagocytosis, clathrin- and caveolin-mediated) and direct penetration (diffusion and pore formation)

as to how nanoparticle shape affects its cell internalization. For instance, Yang and co-worker (2010) reported that the nanoparticle could rotate itself to its sharp edge when penetrating the plasmatic membrane by computationally simulating the interaction of nanoparticles with a lipid bilayer environment. Similar results have also been found by other researchers using specific anisotropic nanomaterials (Titov et al. 2010; Lelimousin and Sansom 2013; Yu et al. 2013). According to Xu et al. (2008), after intracellular internalization, the layered double hydroxide was retained in the cytoplasm, and nanorods were moved towards the nucleus by microtubules.

In addition to size and shape, surface charge and hydrophobicity characteristics of the nanopesticides also play an important role in the cellular internalization and intracellular trafficking pathways. Surface charge of nanomaterials affects translocation across the cell membrane. The translocation time of the nanomaterial into the membrane may increase from milliseconds to hours when the surface charge decreases (Cho et al. 2009; Chu et al. 2014). Furthermore, a charged nanoparticle may induce pore formation on the lipid membrane, and consequently, it may impair the membrane stability and lead to an increase in toxicity (Goodman et al. 2004).

Adsorption or repulsion interactions between nanoparticles and cellular walls with negative charges are dependent on nanoparticle charge (Shomer et al. 2003). The negatively charged cellular walls facilitate the nanoparticle storage and subsequent uptake of positively charged nanostructures through the cell membrane.

Studies have shown that the positively charged nanoparticles have high translocation and internalization than neutral and negatively charged nanomaterials (Marano et al. 2011). Negatively charged nanopesticides may be internalized by clathrin/caveolin-mediated endocytosis, whereas pinocytosis and micropinocytosis are the mechanisms for positively charged (Dausend et al. 2008; Li and Gu 2010).

Hydrophobicity and hydrophilicity properties of nanomaterials are an important factor in their interaction and translocation around biological membranes. For instance, hydrophobic nanostructures can easily insert into cellular membranes due to their preference by lipid tails (D'Rozario et al. 2009; Werner et al. 2012), being able to produce inclusion in the plasmatic membrane (Curtis et al. 2015; Foroozandeh and Aziz 2018).

On the other hand, hydrophilic nanomaterials could attach or leave the cellular membrane and cannot insert into membranes (Liang 2013; Curtis et al. 2015). Understanding these and other surface properties of nanomaterials may be helpful for some theoretical and practical insights into the design of nanoparticles for applications in crop protection.

## 2 The Mechanism of Nanopesticide Action

In last years, an increasing number of publications have emerged concerning the interactions of nanomaterials with plants (Rico et al. 2013; Peralta-Videa et al. 2014; Pošćićć et al. 2016; Oliveira et al. 2018; Pontes et al. 2019). Most of these studies are focused on the potential toxicity of the nanomaterials and both positive and negative effects have been reported.

The nanoparticle's size and shape are related to mechanical effects on plants, restricting the nanoparticles to specific sites in the plant or organ surface; however, the consequences can be manifested throughout the vegetal body resulting in stimuli or inhibition of plant growth. For example, carbon nanotubes (CNTs) of 3 nm can promote the germination of tomato (*Lycopersicon esculentum* Mill.) due to its effects on the entrance and water content in seeds (Khodakovskaya and Dervishi 2009). On the contrary, carbon-based nanoparticles can induce a growth inhibition caused by the mechanical damage due to a perforating effect on cellular or organelles membranes (Hu et al. 2014; Zaytseva and Neumann 2016), involving a primary mechanism of cytotoxicity (Chen and Bothun 2014). In addition, a mechanical blocking induced by nanoparticles can cause clogging of the pores and capillaries, affecting the sap transport (Asli and Neumann 2009; Dietz and Herth 2011).

For instance, colloidal suspensions of clay or titanium oxide nanoparticles can affect the hydraulic conductivity in corn (*Zea mays* L.) roots inhibiting plant growth and transpiration (Asli and Neumann 2009).

Other examples of pesticide effects related to the dust form of nanomaterials are described as damage to the insect's tegument impermeability (Hayles et al. 2017; Benelli 2018). Nanoparticles of silica (SNPs) are related to the death of larval forms of (*Plutella xylostella* L.) Lepidoptera due to tegument abrasion and spiracle blockage (Shoaib et al. 2018). Similar effects are caused by diatomaceous earth on *Prostephanus truncatus* (Horn), a coleoptera plague detected during the corn storage (Kavallieratos et al. 2018). In addition, hydrophilic nano-silica presented high toxicity to *S. littoralis* neonates increasing mortality rates by 80% when tomato plants were experimentally infested with *Spodoptera littoralis* and treated with 350 ppm of nano-silica (El-bendary and El-Helaly 2013).

The effects of nanopesticides on plants are dependent on the physicochemical properties of the nanomaterial. Characteristics such as type, concentration, free charges on the particle surface or formed complexes, among others, can induce different responses in plants, these effects can be categorized as follows: (1) as a positive growth response to the germination process, induction of cellular elongation, tolerance and resistance acquisition against biological and/or abiotic stress in the environment; or (2) as negative growth inhibition related to the direct nanoparticle phytotoxic effect or from the pesticide transported as nanopesticide, resulting in the production of reactive oxygen species (ROS), which changes in the functional cell balance and metabolic destabilization (Hossain et al. 2015; Maruyama et al. 2016).

Tables 1 and 2 summarize some nanomaterials and their effects on growth induction or inhibition of plants. For crop species, for example, low concentrations (300–600 ppm) of zinc oxide nanoparticles (ZnO NPs) stimulate germination and seedling growing of wheat (*Triticum aestivum*) (Mansoor et al. 2019). Similarly, silicon dioxide nanoparticles ($SiO_2$ NPs) at the concentrations 40–400 ppm improved the germination of corn (*Zea mays* L) and bean (*Phaseolus vulgaris* L.) (Sharif-Rad et al. 2016). Enhancement of germination was observed on *Solanum lycopersicum*

**Table 1** Growth stimulus directly and indirectly caused by nanoparticles or nanopesticides on plants

| Vegetal species | Nanoparticle type/ exposition way | Size/shape | Main effects | References |
|---|---|---|---|---|
| *Zea mays* L. | $SiO_2$ NP/imbibition | 10–20 nm/ spherical | Improve germination at 40 ppm | Sharif-Rad et al. (2016) |
| | $CeO_2$ NP/50 mg-Ce per L hydroponic | 2–4 nm/crystal | Changes in plant photosynthesis and gas exchange | Spielman-sun et al. (2019) |
| *Lycopersicon esculentum* Mill. | Carbon (CNTs)/ seeds in culture medium | –/Nanotubes | 10–40 μg/mL accelerate seed germination, increase vigor | Khodakovskaya and Dervishi (2009) |
| | $SiO_2$NP/Petri dishes | 12 nm | 8 g L improves seed germination | Siddiqui and Al-Whaibi (2014) |
| | $CeO_2$NP/seeds, plants irrigation in plastic pots with sand | 30–75 nm/ spherical | 20 mg/L beneficial for growth (dry mass) | Singh et al. (2019) |
| | AgNPs+graphene oxide/spray on plants | ~18 nm/ spherical | Antibacterial activity at 16 ppm | Ocsoy et al. (2013) |
| *Eleusine coracana* Gaertn | Chitosan nanoparticles (CNPs)/seeds imbibition | 20–50 nm/ spherical | Antifungal activity at CTZ solution (1%) | Sathiyabama and Manikandan (2016) |
| *Oryza sativa* L. | CNPs/leaves treated with CNPs solution | 20–50 nm/ spherical | Antifungal activity at (0.1% (w/v), 500 μL/leaf) | Manikandan and Sathiyabama (2016) |
| | Si and $TiO_2$ NPs/foliar spray | –/– | Improve antioxidant enzyme activities, decrease Cd in tissues | Rizwan et al. (2019) |
| *Zea mays* L. | Au NPs/seeds imbibition | 10–30 nm/ spherical or near-spherical | Germination increase and seedling growth improvement at 10 ppm | Mahakham et al. (2016) |
| *Hyssopus officinalis* L. *Nigella sativa* L. | $SiO_2$ NP/seeds imbibition | 10–20 nm/ spherical | Improved seed germination at 400 mg $L^{-1}$ | Sharif-Rad et al. (2016) |

(continued)

**Table 1** (continued)

| Vegetal species | Nanoparticle type/ exposition way | Size/shape | Main effects | References |
|---|---|---|---|---|
| *Spinacia oleracea* L. | $TiO_2$ NPs/seeds imbibition | – | 0.25–4% increase of germination and vigor indexes at 2,5% | Zheng et al. (2005) |
| | $CeO_2$ NP/foliar spray | ~4–7 nm/– | Metabolic reprogramming in leaves and roots at 0.3 mg per plant | Zhang et al. (2019) |
| | $TiO_2$ NP/seeds soaked | –/– | Increase the antioxidant enzyme activities at 0.25% | Hong et al. (2005) |
| *Brassica napus* L. | $TiO_2$ NPs/roots or leaves exposition | 14 nm/crystal anatase | 10–100 mg/L induction of root elongation | Larue et al. (2012) |
| *Triticum aestivum* L. | ZnO NPs/seed imbibition for 4 h | 30–40 nm/ sphere-crystal | 300–600 ppm shoot length, shoot weight, and vigor index increase | Mansoor et al. (2019) |
| *Avena sativa* L. | ZnO NPs | ~20–50 nm/ sphere | 750 mg/kg improve germination percentage | Maity et al. (2018) |
| | $TiO_2$ NPs | ~30–60 nm/ crystal/sphere | | |
| | CuO NPs | ~50–80 nm/ sphere | | |
| | AgNPs | ~5–15 nm crystal | | |
| *Eruca sativa* Mill. | AgNPs/seeds soaked | 5–17.5 nm/– | Root length induction at 10–20 mg $L^{-1}$ | Vannini et al. (2013) |
| *Brassica juncea* (L.) Czern. | AuNPs/spray on leaves | 10–20 nm/ spherical | Seed germination induction (25 ppm), plant growth, and chlorophyll content | Arora et al. (2012) |
| *Cicer arietinum* L. | $TiO_2$ NPs/spray on leaves | ~5–20 nm/– | Reduction of membrane damage during cold stress treatment at 5 ppm | Mohammadi et al. (2013) |
| *Glycine max* L. | $CeO_2$ NPs/substrate addiction | 10–30 nm/ spherical-quadrilateral | Stimulated plant growth at 100 mg kg | Cao et al. (2017) |

L. (*Lycopersicon esculentum* Mill.) seeds; $SiO_2$ NPs also improved the germination and seedling vigor (Siddiqui and Al-Whaibi 2014).

The effects of nanomaterials on plants, related to the doses or nanoparticles size distribution, are well exemplified by Youssef and Elamawi (2018). So, they described that lower concentrations of ZnO NPs (10 and 25 ppm) enhanced seed germination

**Table 2** Growth inhibition directly and indirectly caused by nanoparticles and nanopesticides on plants

| Vegetal species | Nanoparticle type/ exposition way | Size/shape | Main effects | References |
|---|---|---|---|---|
| *Allium cepa* L. | ZnO NPs/bulb base immersion | 10–80 nm/– | Root growth inhibition, cytotoxicity and genotoxicity at 5 or 50 μg/mL | Spielman-sun et al. (2019) |
| | AgNPs/seeds in solution | 10, 20, 51, and 73 nm/ spherical | Cytotoxicity and genotoxicity NP size related at 10 ppm | Scherer et al. (2019) |
| *Vicia faba* L. | ZnO NPs | 30 nm/– | Cytotoxicity and genotoxicity (100–200 mg/L) | Youssef and Elamawi (2018) |
| *Brassica sp.* | Chitosan NPP + Paraquat (NP-CS:PQ)/foliar spray | ~300 nm/– | Leaf necrosis; dry matter reduction (2 kg/ ha) | Grillo et al. (2014) |
| *Brassica rapa ssp.* | CuO NPs/seed imbibition | 25–55 nm/– | Seedling growth decrease; lipid peroxidation, genotoxicity at 250 and 500 mg/L | Chung et al. (2019) |
| *Zea mays L.* | $CeO_2$ NPs/seed inoculation | 10 nm/– | Increase accumulation of $H_2O_2$ at 400 and 800 mg/kg | Zhao et al. (2012) |
| *Amaranthus retroflexus* L. *Taraxacum officinale* F. H. Wigg | $SiO_2$ NPs/seeds imbibition | 10–20 nm/ spherical | Germination, biomass, photosynthetic pigments, and total protein decreasing at 400–4000 mg L | Sharif-Rad et al. (2016) |
| *Triticum aestivum* L | $TiO_2$ NPs/seeds imbibition | 40 nm/ tetragonal crystal | Inhibition of germination and P (dependent on cultivar and concentration) at 1000 mg kg | Zahra et al. (2019) |
| *Linum usitatissimum* L., cv. Electra, *Lolium perenne* L., cv. Tove *Hordeum vulgare* L. | Zero-valent iron NP (nZVI)/seeds imbibition in different substrates | – | >500 mg L Toxic effects on germination | El-Temsah and Joner (2012) |
| *Oryza sativa* L. cv. KDML 105 | AgNPs/seeds imbibition | 20–150 nm/ spherical | 0.1–1000 mg/L, seed germination and seedling growth inhibition | Thuesombat et al. (2014) |
| | MWCNTs/cell suspension | 10–30 nm | Accumulation of ROS and cell viability decrease at 20 mg/L | Tan et al. (2009) |

(continued)

**Table 2** (continued)

| Vegetal species | Nanoparticle type/ exposition way | Size/shape | Main effects | References |
|---|---|---|---|---|
| *Glycine max* (L.) Merr. | AgNP/irrigation in plastic pots | 60 nm/ spherical | Damage in leaves; lipid peroxidation and changes in catalase activity | Galazzi and Arruda (2018) |
| *Cynodon dactylon* (L.) Pers | NPP (purified diatomite (PDE) + $Fe_3O_4$ NP + *glyphosate*)/ *spray on leaves* | ~40 nm ($Fe_3O_4$NP) | Weed (*Cynodon dactylon*) mortality under pH 5 | Xiang et al. (2017) |
| *Bidens pilosa* L. | Chitosan NPP (NP-CS/ TPP) | –/Aggregated | Plant growth reduction (concentrations of 75 g/L of imazapic and 25 g/L of imazapyr) | Maruyama et al. (2016) |
| | Atrazine-ATZ containing PCL nanocapsules | –/– | Pre-emergent herbicidal activity, growth reduction at 200 g $ha^{-1}$ | Preisler et al. (2020) |
| *Amaranthus viridis* L. | NPP ATZ (PCL)/seeds imbibition | Mean diameter 260 nm/– | Post-emergent herbicidal activity, decrease PSII activity at 2000 g $ha^{-1}$ | Sousa et al. (2018) |
| *Raphanus raphanistrum* L. | NPP ATZ + AMZ/ substrate or plant spraying | 111–178 nm/– spherical | Pre- and post-emergent herbicidal activity growth reduction at 1:10 v/v (0.3 kg/ha) | Oliveira et al. (2015) |

and improved seedling growth of faba bean (*Vicia faba* L.), while higher concentrations of NPs (100 and 200 ppm) resulted in phytotoxicity effect. Also, Scherer et al. (2019), in its study with onion seeds (*Allium cepa* L.), showed that AgNPs with different sizes (10, 20, 51 and 73 nm) had distinct cytotoxicity and genotoxicity responses.

## 2.1 Nanoparticles and Seed Germination

Seed germination involves a sequence of events configuring in a critical stage of plant growth in which the benefits by application of nanopesticides or nanoparticles to protection against pathogens are hardly dose-dependent (Fig. 6). In seeds, the uptake of nanopesticides into tegument (seed coat) may increase germination in the seed priming stage due to its protectant activity. Some papers report an increase in seed germination when seeds are exposed to nanomaterials, with or without negative outcomes in seed embryo development, seedling growth, or plant survival. For

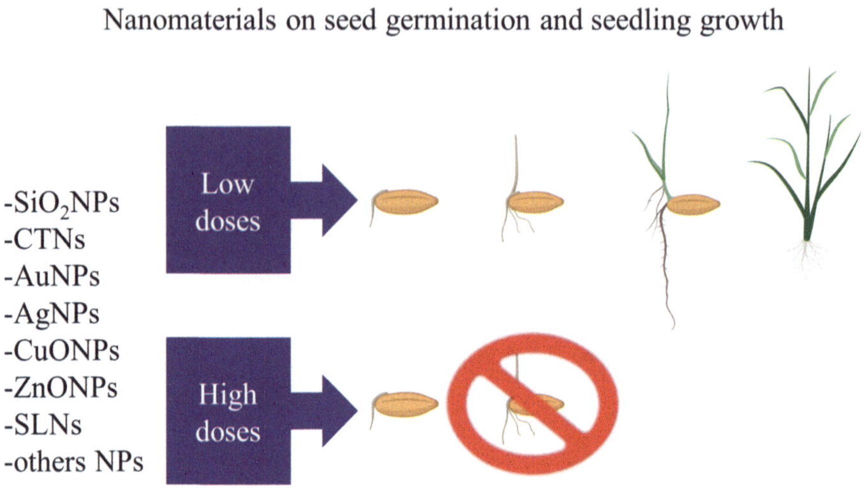


**Fig. 6** Nanomaterials on seed germination and seedling growth. Nanoparticles or nanopesticides at low doses (varying according to the NP characteristics and plant species) can induce or inhibit seedling growth

instance, Raja et al. (2019) have reported an improvement in the germination of *Vigna mungo* seeds exposed to biogenic zinc and copper nanoparticles.

In this way, chitosan nanoparticles have presented a promising effect on seed germination and seedling growth of wheat at a lower concentration (5 μg/mL compared with 50 μg/mL) and they can stimulate the growth of wheat seedlings by up-regulating indole-3-acetic acid (IAA) synthetic genes and down-regulating metabolic genes (Li et al. 2019a, b). On the contrary, silver nanoparticles have been reported to have had negative effects on seed germination and seedling growth of rice (*Oryza sativa* L.). Additionally, increased penetration was reported for the smaller silver nanoparticles (20 nm) into roots (Thuesombat et al. 2014).

Uptake and translocation of nanopesticides are dependent on the physicochemical, biochemical, and physiological properties of nanomaterials as well as the presence of an ion transporter in plant body or tissue (Goodman et al. 2004; Foroozandeh and Aziz 2018).

For instance, titanium oxide nanoparticles ($TiO_2$ NPs) <14 nm induced root elongation in rapeseed (*Brassica napus* L.) and wheat (*Triticum aestivum* L.), which the high reactivity of the nanoparticle surface was responsible for increasing the hydromineral flux in the roots (Larue et al. 2012). The $TiO_2$ NP size was also determinant to the nanoparticle absorption and potential interference on rice (*Oryza sativa* L.) fluxes (Cai et al. 2017). Adverse physicochemical effects on the hydraulic conductivity of roots can vary according to the nanoparticle's characteristics, the vegetal species, and the plant development stage (Margenot et al. 2018).

Furthermore, additional investigations have shown that there are dependencies of dose–response, particle charge, coating agent, and nanoparticle size in the modulation of biochemical, metabolic, and physiologic pathways, gene expression, and developmental responses of plants, such as seed germination, root and shoot growth, plant height, and biomass partitioning (Bao et al. 2016; Xiong et al. 2017; Li et al. 2019a; Nath et al. 2019).

Enhancement in the germination of seeds treated with nanoparticles, especially metallic nanoparticles, usually occurs at low concentrations (the value may vary according to the NP, size, charge plant species, variety, among others). Therefore, it is difficult to generalize these findings since studies of mechanisms of action, particularly when the aim is the utilization of the nanoparticle as a nanopesticide, have still not been elucidated. For example, the seeds germination stimuli and vigor observed on spinach (*Spinacia oleracea* L.) as described by Zheng et al. (2005) are related to the antioxidant activity of $TiO_2$ NPs at low concentrations (0.25–4%) as well as the reduction of free radicals during the seed storage induced by $TiO_2$ NPs. So it happens due to the nanoparticles' effects on the mechanisms of anions superoxide and hydroxide generation.

## 2.2 Nanopesticides and Reactive Oxygen Species (ROS) Generation

The mechanisms of oxidative stress and plant growth inhibition related to nanoparticles or nanopesticides involve different pathways, depend on the nanomaterial properties or the pesticide mode of action (Stark 2011; Hossain et al. 2015). The reactive oxygen species (ROS) includes several natural products of cellular oxidative metabolism, such as the free radical superoxide ($O_2^{-}*$), hydroxyl radicals and ions ($*OH$ and $OH^{-}$), and nonradicals, such as hydrogen peroxide ($H_2O_2$). However, the cellular levels of ROS must be low to maintaining cell homeostasis. Otherwise, ROS can be very toxic for plants at high concentrations and excess ROS may induce DNA damage, lipid peroxidation, enzyme inhibition, and cell death (Blokhina and Fagerstedt 2010; Heyno et al. 2011; Sharma et al. 2012).

Although the high levels of ROS are related to several negative effects on cell metabolism of plants, these molecules may also participate in complex regulatory mechanisms of integration plant-environment (Czarnocka and Karpiński 2018). The positive role of ROS in plant regulation depends on the balance between ROS generation and the neutralization of their excess by cellular antioxidant agents (Dayem et al. 2017). For instance, metallic nanoparticles are particularly able to induce the intracellular ROS generation, the oxidants and free radicals located on the nanoparticle surface can produce ROS by Fenton reactions (Dayem et al. 2017; Hou et al. 2019). Hence, the subcellular nanoparticles targets can be organelles such as mitochondria causing structural damages and inducing stress responses related to the endoplasmic reticulum (Wang et al. 2019) or peroxidation of membranes (Chung et al. 2019). Figure 7 shows the schematic representation of plant metabolic self-defense in response to nanopesticides.

When metallic nanoparticles are utilized as nanopesticide this nanomaterial can suppress crop pests and pathogens by directly acting on target-site through a large

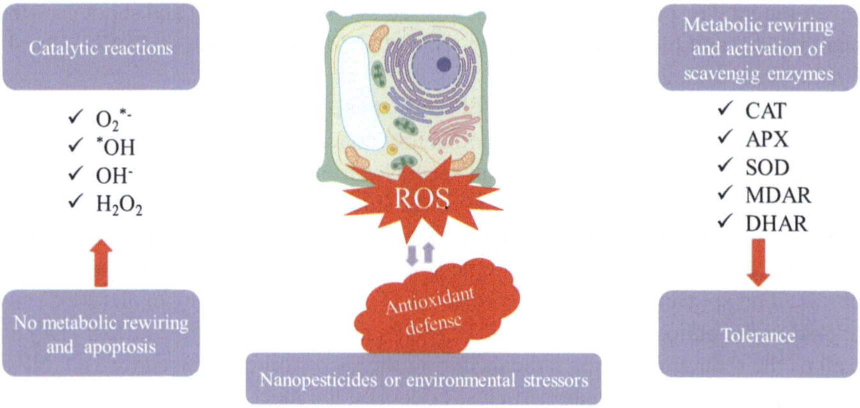


**Fig. 7** Schematic representation of plant metabolic self-defense in response to nanopesticides. Superoxide radical ($O_2$), hydroxyl radical (OH, $OH^-$), hydrogen peroxide ($H_2O_2$), CAT: catalase; APX: ascorbate peroxidase; SOD: superoxide dismutase; MDAR: monodehydroascorbate reductase; DHAR: dehydroascorbate reductase

variety of mechanisms. In plant–pathogen systems, copper nanopesticides ($Cu(OH)_2$) on *Cucumis sativus* L., for instance, were correlated with the expression of genes encoding catalase, peroxidase, amonialyase, superoxide dismutase, polyphenol oxidase, and others (Sathiyabama and Manikandan 2018; Adisa et al. 2019; Zhao et al. 2018b). Also, the potential increase in the cold stress tolerance described as a reduction of the membrane damage was observed on foliar tissues of chickpea (*Cicer arietinum* L.) treated with low concentrations (5 ppm) of $TiO_2$ NPs (Mohammadi et al. 2013).

Zhao et al. (2016a) demonstrated the impacts of foliar exposure to the $Cu(OH)_2$ nanopesticides in the metabolism of lettuce plants. Notably, plants exposed to nanopesticides triggered the generation of ROS and induced metabolites serving as ROS scavenger significantly reduced. A decrease in dehydroascorbic acid and cis-caffeic acid (two important antioxidants) was also observed, suggesting that the antioxidant defense system was impaired. However, an increased detoxification behavior was observed in this study, reported by increases in nicotianamine, amino acids, and polyamines. Nicotianamine is a copper sequestering agent (chelator), its enhanced values may represent a plant detoxification mechanism.

In the same way, polyamines may mitigate oxidative stress and enhance tolerance. Zhao and co-workers (2016b) revealed that cucumber plant roots exudate metabolomics and that nanocopper treated plants up-regulated some amino acids which bind ionic and nanoparticulate copper, likely to plant detoxification.

Despite the agricultural potential, copper nanoparticles are related to deleterious effects in plants, such as structural changes in roots and shoot tissues by increased lignification, deformation of stomata and chloroplasts, reduction in number of thylakoids per grana, and decrease in chlorophyll and carotenoids content (Perreault et al. 2010; Shi et al. 2013; Nair and Chung 2015; Da Costa and Sharma 2016; Xiong et al. 2017).

Furthering the work, Zhao et al. (2018b) also reported that nanopesticides induced metabolic reprogramming in cucumber (*Cucumis sativus*) and maize (*Zea mays* L.) plants. In maize plants, up-regulation of intermediate metabolites of the glycolysis pathway and tricarboxylic acid cycle (TCA) suggests the activation of energy metabolism by $Cu(OH)_2$ nanopesticides, and in addition, the antioxidant defense-related pathway was enhanced as revealed by the increase in levels of aromatic compounds (4-hydroxycinnamic acid and 1,2,4-benzenetriol) and their precursors (phenylalanine, tyrosine) probably indicating the activation of shikimate–phenylpropanoid biosynthesis. On the other hand, in cucumber plants, arginine and proline metabolic pathways were the most altered pathway.

By the way, a large number of nanomaterials showed plant-defense properties. However, deleterious effects have been reported. For example, silver (Ag) nanoparticles have been largely considered the most favorable nanopesticide, attributed to their high bactericidal and viricidal efficacy and low toxicity. Despite adverse effects being reported in several plant species, cytotoxic and genotoxic effects in *Allium cepa* roots increased with decreasing silver nanoparticles diameter as reported by Scherer et al. (2019).

The existing literature physiologic effects of the silver nanoparticles on plants report changes in germination, plant growth, development, and operation of photosystem II (Rizwan et al. 2017). Furthermore, silver nanoparticles can bind with chlorophyll molecules and form a novel hybrid system, which produces around 10-times higher excited electrons due to plasmon resonance effect and fast electron–hole separation, which subsequently promotes photosynthesis process (Govorov and Carmeli 2007; Queiroz et al. 2016). Additionally, Falco et al. (2019) have recently demonstrated that AgNPs also can alter the $CO_2$ assimilation rate, stomatal conductance, and photochemical efficiency of photosystem II of *Vicia faba* when internalized into leaves.

The pesticide effects of silver nanoparticles were also described to tomato (*Lycopersicon esculentum* Mill.), and the antibacterial activity of AgNPs on graphene oxide (GO) solution (Ag@dsDNA@GO) showed antibacterial effects towards *Xanthomonas perforans* (Ocsoy et al. 2013). Growth responses of roots and biomass accumulation are described as nanoparticle interference on the cellular structure of underground organs or functional aspects of the photosynthetic complex. For instance, an increase in the growth of roots was observed in rocket salad (*Eruca sativa* Mill.) when treated with silver nanoparticle (AgNPs), and it was observed that the growth response is mediated by the expression and accumulation of proteins related to sulfur metabolism (Vannini et al. 2013).

### 2.2.1 Ionic Release and Binding Affinity of Nanopesticides

Studies have shown that small nanomaterials may penetration into the plant tissue and thus increase their toxicity. Also, the balance between size, concentration, and biodegradability of the NPs are the key factors to the nanopesticide toxicity (Scherer et al. 2019).

The physicochemical properties of nanopesticides influence how they interact with plant cells and, consequently, their overall potential toxicity. The most common cause of the toxic effects of metal oxides nanopesticides into cells is the dissolution and release of toxic ions contained in their cores, upon oxidation by environmental agents. Understanding these toxic properties can lead to the development of safer nanopesticides. Some studies have shown that the degradation of metal-based nanomaterials causes the gradual release of metal ions (Soenen et al. 2015; Sukhanova et al. 2018). The behavior of nanopesticides can be altered by modifying the surface. The storage, temperature, pH, and functionalization play a key role in nanostructure dissolution and release of toxic ions (Kittler et al. 2010; Soenen et al. 2013).

Several nanomaterials have potential applications as a plant protectant, for example; Sabella et al. (2014) suggest that a wide class of NPs (such as metallic, metal oxide, and semiconductor NPs) are able to release toxic ions in cell when they come into contact with cytosol. Also, these authors suggest that ionic release is a major responsible for intracellular toxicity profiles of these nanomaterials. In fact, we cannot exclude the possibility of nanopesticide toxicity be reported in non-target species induced by released ions, might also be responsible for other toxic mechanisms, such as lysosomal damage (Xia et al. 2008; Stern et al. 2012). After cellular internalization, nanopesticides may interact with plant organelles (such as mitochondria, peroxisomes, and chloroplast), DNA, proteins, and pigments (Kathiravan et al. 2009; Saptarshi et al. 2013; Queiroz et al. 2016; Ahsan et al. 2018).

The binding affinity of a nanopesticide with target and non-target binding sites helps us to understand its bioavailability, distribution, toxicity, and elimination from the plant cell. Proteins possess a broad range of structural and functional properties, including cellular signaling, molecular recognition, catalysis, metabolite production, and ligand binding. The binding of a protein to nanopesticide (such as CuO, ZnO, $TiO_2$, $SiO_2$, or FeO) can result in minor conformational changes or protein denaturation (Saptarshi et al. 2013; Chibber and Ahmad 2016).

Also, change in conformation and mobility of the genomic DNA induced by nanoparticles are reported (Bhar et al. 2018; Ma et al. 2018). A large range of studies reports cytogenetic abnormalities induced by potential nanopesticides such as AgNPs, ZnONPs, and $TiO_2$NPs (Kumari et al. 2009; Ghosh et al. 2010; Lòpez-Moreno et al. 2010). The Figure 8 illustrate the balance between size, concentration, and biodegradability as a key factor to the nanopesticide toxicity.

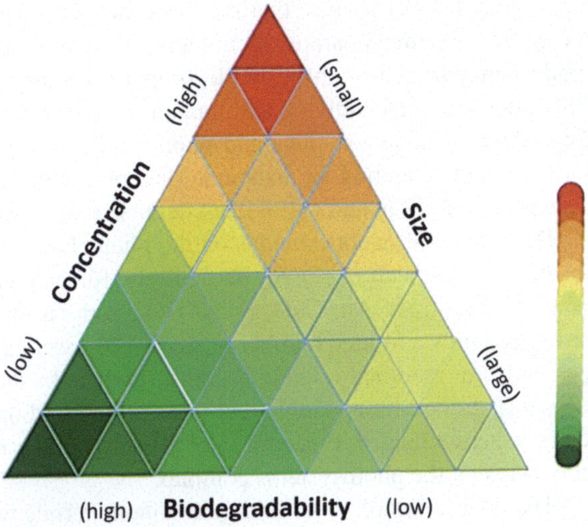


**Fig. 8** Balance between concentration, size, and biodegradability affecting the nanopesticide toxicity. Cold colors (green) represent low toxicity, while hot colors (red) high

### 2.2.2 Nanopesticide on the Photosynthetic Apparatus

Oxygenic photosynthesis is a key process to maintaining life on earth and is known to be very sensitive to exogenous stimuli. Oxidation–reduction reactions of photosynthesis are a key phase of plant metabolism, the process can be grouped into two phases: the first phase is the photochemical or Hill's reaction, and the second biochemical phase is named Calvin and Benson cycle (Taiz and Zeiger 2002).

Light-dependent photochemical reactions take place in the thylakoidal membranes of chloroplasts. Hence, light (photons) energy supplies the driving force for oxygen evolution (water photolysis), and thus nicotinamide adenine dinucleotide phosphate (NADP+) is reducted to NADPH and there is adenosine triphosphate (ATP) formation (Trebst 1994). The protein–pigment complexes involved in electron transport chain (ETC) into thylakoid membranes are two photosystems (PSI and PSII), and a range of peripheral polypeptides attached to pigments and redox systems from ETC, responsible for connection between PSII and cytochrome $b_6f$ complex (Cyt $b_6f$) at the ETC (Trebst 1994; Dekker and Van-Grondelle 2000; Nelson and Yocum 2006). On the Calvin and Benson cycle step, biochemical reactions occur in the stroma side of chloroplasts by the fixation and reduction of $CO_2$ and formation of carbohydrates (Sharkey and Weise 2016).

Nanopesticides can affect the photosynthetic apparatus in photochemical or biochemical phases, depending on the dosage or physicochemical properties of nanomaterials and consequently can affect crop productivity. However, the effects of metal oxide nanopesticides (for antimicrobial and insecticidal applications), nanocapsules, and nanomaterials for controlled release of agrochemicals on the chlorophyll content and photosynthetic photochemistry are still not fully understood.

In the cell cytoplasm, nanopesticides may interact with chloroplasts, and consequently, affect the plant photosynthetic reactions at that particular site by binding to photosynthetic apparatus and impair their functioning. In a recent review, Tighe-

Neira et al. (2018) suggest that the most metallic nanoparticles may be harmful to the photosynthetic apparatus by inducing structural and functional damages. Metal oxide nanoparticles with pesticide properties alter the photosynthetic efficiency, PSII photochemical activity, and quantum yield in plants. Thus, knowledge on the interaction of nanopesticides and photosynthetic light harnessing events can provide an understanding of nanomaterials-photoinduced oxidative stress, electron transport inhibition, and antioxidant defense system in plants.

The impact of nanopesticides on the photochemical reactions of photosynthetic machinery has been explored, particularly, for nanoparticles with some pesticidal properties (Ag, Al, Au, CuO, $SiO_2$, $TiO_2$, and Zn nanomaterials). However, there have been only a limited number of studies concerning the photosynthetic impact of pesticides in modified release systems, produced using polymeric or lipidic nanocapsules, carbon-based single and multiwalled carbon nanotubes, and other nanocarriers active-loaded. Figure 9 shows the schematic representation of the electron flux between the photosystems complex.

The interaction of silver and gold nanomaterials with photosynthetic apparatus has been widely studied for two related competing effects—the enhanced light absorption by chlorophyll molecule due to the surface plasmon resonance effects of nanoparticle and the decrease in quantum efficiency of photosystems due to the

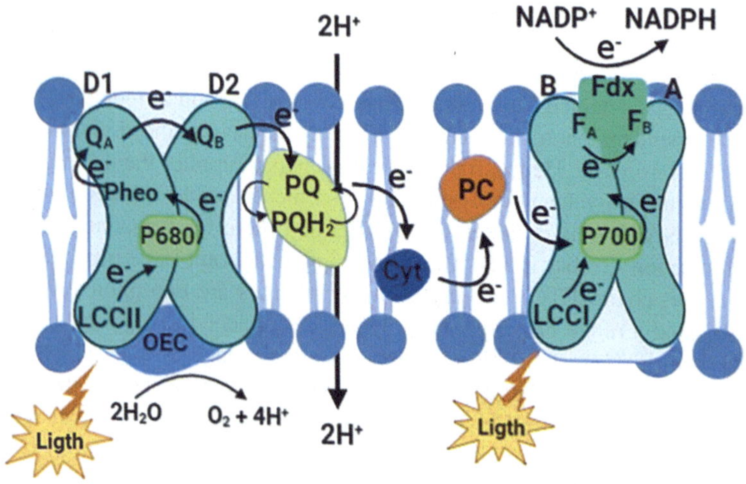


**Fig. 9** Schematic representation of the electron flux involving two photosystems: PSII reaction center (RC) containing a chlorophyll P680 molecule and PSI RC (containing a chlorophyll P700 molecule anchored, respectively, by the structural membrane proteins (D1, D2, B, and A). First, photosynthetically active radiation is absorbed by Light Harvesting Complex (LHCI and LHCII) and trigger the electron transport reactions to the PSII RC. Water photolysis are mediated by Oxygen-Evolving Complex (OEC) with release of protons (H+) and O2., also known as donor side of PSII. Pheophytin (Pheo) is the primary electron acceptor, and transfer electrons to the Quinones (QA and QB). Followed by a sequence of energy transport through Plastoquinone (PQ), Cytochrome (Cyt) complex, and Plastocyanin (PC) to PSI RC. At PSI from P700 to Ferredoxin (Fdx) complex and consequently NADPH formation. Modified from Buchanan et al. (2000), Kerbauy (2004)

energy transfer from excited chlorophyll to nanoparticles (Barazzouk et al. 2005; Nieder et al. 2010). Falco et al. (2019) demonstrated that silver nanoparticles (AgNPs) could also induce overproduction of ROS, causing a decrease in the photosynthetic activity (Fv/Fm) of the *Vicia faba* leaves associated. In addition, they also showed that AgNPs caused an increase of the nonphotochemical quenching (NPQ), possibly due to the transfer of the excited electrons of the chlorophylls to the metal surface of the AgNPs. Also, growth responses of plants (height, stem diameter, number of branches, number of pods) of Indian mustard (*Brassica juncea* (L.) Czern.) treated with gold nanoparticles (AuNPs) are attributed to an increase of photosynthetic rates (Arora et al. 2012).

Silver nanoparticles are reported to increase the content of non-reducing center of $Q_B$ in *Chlamydomonas reinhardtii*, decrease the amount of plastoquinone and PSII efficiency in *Spirodela polyrhiza*, also reported to decrease the photosynthetic activity in *Arabidopsis thaliana*, *Glycine max,* and *Vicia faba* plants (Matorin et al. 2013; Falco et al. 2015; Sosan et al. 2016; Shabnam et al. 2016; Queiroz et al. 2016). On the other hand, $TiO_2$ nanoparticles, a potential delivery system for active compounds, induced an increase in the chlorophyll content, light absorbance, photolysis of water, and electron transport in spinach leaves (Hong et al. 2005; Mingyu et al. 2007; Yang et al. 2007).

The effects of $TiO_2$ NPs on the photochemical reactions are also described for spinach (*Spinacia oleracea* L.) (Zheng et al. 2005) and chloroplast resistance to aging by the reduction of free radicals' production and the induction of the antioxidant enzymes activities (Hong et al. 2005).

As an example of a nano-based carrier system for agrochemicals, we cite the encapsulation of atrazine herbicide described firstly by Grillo et al. (2012). Atrazine is a $Q_B$-binding inhibitor herbicide, blocking the electron flow through the PSII. Poly(epsilon-caprolactone) nanocapsules loaded with atrazine herbicide effectively increases its pre- and post-emergence herbicidal activity, and the inhibition of PSII photochemistry was more intense in leaves treated with undiluted nanocapsules carrying atrazine (Oliveira et al. 2015). Also, Sousa et al. (2018) used the same poly(epsilon-caprolactone) nanocapsules loaded atrazine against *Amaranthus viridis* (slender amaranth) and *Bidens pilosa* (hairy beggarticks) weeds. Thus, they observed a greater decrease in the PSII activity for both species (above 50% inhibition relative to the control) than the commercial atrazine formulation at the same concentration (around 40% inhibition).

In a bioassay with *Chlamydomonas reinhardtii* as organism model exposed to nanoatrazine, Scognamiglio et al. (2019) report changes in the redox state of the electron carriers within the PSII reaction center, such as the accumulation of $Q_A$ in the reduced state due to $Q_B$-binding and the further filling up of the membrane plastoquinone pool with electrons. Also, Preisler et al. (2020) suggest that nanoatrazine could be applied for efficient weed control without additional phytotoxicity to susceptible crops compared with non-nanoatrazine, provided that a safe application interval is respected.

## 2.3 Nanoencapsulated Pesticides and Plant Responses

When pesticide molecules are temporally trapped by nanoparticles, the complexes formed assume new properties with high interest for agricultural uses. The viability of pesticide transport as nanoforms (Ashitha and Mathew 2019) opens a vast field with several possibilities to develop new products. Other nanomaterials with a large potential for nanopesticide are multiwalled carbon nanotubes (MWCNTs). They have been exposed to several plants such as zucchini, corn, tomato, and soybean with no apparent toxic effects (De La Torre-Roche et al. 2013), suggesting their possible application as delivery system for controlled release of pesticides.

Encapsulation of the fungicide zineb (zinc ethylene bis-dithiocarbamate) into carbon nanotubes (CNT-g-PCA hybrid material with 20–40 nm) results in a water-soluble nanopesticide (NPP) more effective than the bulk zineb to reduce the fungi (*Alternaria alternata*) growth (Sarlak et al. 2014). Also, Fan et al. (2018) reported the modulatory and protective effect of MWCNT on paraquat toxicity in *Arabidopsis* plants on the root surface area, in which these results may be explained by the extent of paraquat adsorption on the surface of MWCNT and to stimulation of photosynthesis, and antioxidant protection.

Xiang et al. (2017) described interesting data about the properties of a controllable nanopesticide system with magnetic collectability, the authors incorporated the herbicide glyphosate in a magnetic nanocarrier the micro-nano pores of diatomite/$Fe_3O_4$, promoting high weed (*Cynodon dactylon* (L.) Pers) mortality. Also, Maruyama et al. (2016) evaluated a nanopesticide based on chitosan as carriers, demonstrated that the encapsulated herbicides were more effective to reduce the hairy beggarticks (*Bidens Pilosa* L) growth than the free imidazolinone form because the pesticide was released more slowly, enabling the use of lower dosages. Another advantage was that the results of cytotoxicity assays indicated low toxicity and genotoxicity to the nanopesticide. In addition, the growth inhibition on *B. Pilosa* plants can be achieved by the atrazine nanoencapsulation, the nanopesticide shows pre- and post-emergent herbicidal activity leading to very high mortality rates of the *B. pilosa* seedlings (Sousa et al. 2018; Preisler et al. 2020).

Observations on weed control have also been evaluated in target and non-target plant systems. Grillo and co-authors (2014) reported reduced toxicity to non-target organisms compared with the conventional herbicide using chitosan/tripolyphosphate (CS/TPP) nanocarriers loaded with paraquat herbicide while keeping the herbicidal activity against *Brassica* sp. Also, nanoencapsulation of atrazine herbicide with poly(lactic-co-glycolic-acid), solid lipid nanoparticles, or poly(epsilon-caprolactone) nanocapsules showed promising results by effectively pre- and post-emergence plant control (Oliveira et al. 2015; Schnoor et al. 2018; Sousa et al. 2018), suggesting that nanoencapsulation could potentially long-term reduce the residual effect of herbicides in agricultural lands, due to the enhanced efficiency of lower dosage applied.

It has been reported that these herbicides inhibit the photosynthetic electron transport flow, paraquat blocking the electron transport on photosystem I (PSI) level (Donaldson 2013), and atrazine blocking the electron transfer on the acceptor side of photosystem II (PSII) (Hess 2000). However, the mechanisms of action of a large variety of nanoherbicides remain poorly understood. In another investigation, Grillo et al. (2015) studied the influence of aquatic humic substances on paraquat-loaded CS/TPP nanoparticles. *Allium cepa* genotoxicity studies and ecotoxicity assays with *Pseudokirchneriella subcapitata* revealed that aquatic humic substances decreased the toxicity of paraquat. In this way, polymeric nanoparticles containing ametryn, atrazine, or simazine were slightly less genotoxic human lymphocyte and *Allium cepa* cell cultures than its free form (Grillo et al. 2012; Lima et al. 2012; Clemente et al. 2014).

Despite the rapid progress in the study of plant response of nanopesticides in the past years, we look forward to furthering developments of the plant–nanopesticides interaction studies based on physiological, proteomic, genomic, and metabolomic studies. It will be helpful to better understand the mechanisms involved in the interactions of the current and new nanopesticides with plants. Hence, it is not possible to generalize because most interactions of plants with nanomaterials tend to be species-specific, and the effect of nanopesticide on crop plant (target and non-target species) is correlated and dependent on the plant species, growth stages, type of tissue of application, environmental conditions, time exposure, dose and method of application, among others factors (Fraceto et al. 2016; Rizwan et al. 2017; Pérez-de-Luque (2017); Grillo et al. 2018).

The continuum uses of chemical pesticides and the evolutionary race result in pest populations more tolerant creating a vicious circle were the pesticide doses need to be incremented resulting in environmental and health risks. However, alternative strategies are required to promote sustainable crop protection, as the combination of different practices together (Integrated Pest Management—IPM), overcoming the shortcomings of individual practices (Chandler et al. 2011). Also, nanotechnology in agriculture (especially for nanopesticides) has emerged as a new tool to improve crop productivity (Fig. 10).

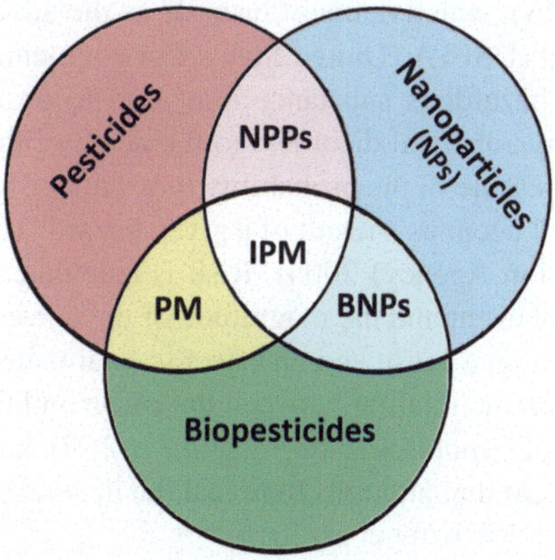


**Fig. 10** Interrelations between agricultural productivity components involving three main elements (pesticides, nanoparticles, and biopesticides) and their derivations (pest management—PM, nanopesticides—NPPs, and bionanopesticides—BNPs) resulting in the integrated pest management—IPM

Also, we would like to stress that to improve the understanding of the interaction of nanopesticide with plants, we need a more holistic view of studying the integration of different levels of biological organization in response to nanopesticides application. Surprisingly, little information is available yet on the regulation of homeostatic responses of plants (target and non-target species) to nanopesticides using a systemic approach.

## 2.4 A Brief Review of Computational Tools for Nanopesticide Risk Assessment in Plants

For understanding the dynamics of interactions between plants and nanopesticides, "in vivo" and "in vitro" methods are necessary, but for the sake of nanotoxicity assessment for non-target organisms, those methods are time-consuming and expensive, and the approval of bioethical committee may be required.

Computational "in silico" tools (chemometrics, bioinformatics, machine learning, statistics, among others) are used to predict nanoparticles properties and their interaction with protein complexes and other biomolecular structures. In vivo and in vitro methods for testing phytonanotoxicity are quite expensive and time-consuming. For instance, a risk assessment study for a single chemical structure may cost into the millions of dollars and take 3–4 years to accomplish. The importance of computational tools for evaluating environmental and human health safety of engineered nanomaterials is manifested by the fast-growing amount of research publications on computational nanotoxicology (Hastings et al. 2015a). Nevertheless, "in silico" methods are expected to help to reduce the number of sacrificed laboratory animals (Pacheco and Buzea 2018; Zuverza-Mena et al. 2017).

The terms "risk" and "hazard" may be defined in many ways. For the purpose of this chapter, we follow the definitions adopted by the United States Environmental Protection Agency (EPA), which defines "hazard" as the adverse effect or inherent toxicity of a compound (USEPA (United States Environmental Protection Agency) 2004). Exposure to a hazardous substance may lead to an adverse health effect, varying from minor physiological disorders to the death of the exposed individual. Risk is defined as "A measure of the probability that damage to life, health, property and/or environment will occur as a result of a given hazard" (USEPA (United States Environmental Protection Agency) 2004). Risk is calculated as a function of the probability of a harmful event and the magnitude of the consequences of that event. When the relation between a cause and an effect is established, we speak of *known* or *identified risks*. When the relation between the cause and the damage is not well established, we speak of *hypothetical* or *potential risk*. Risk assessment is a multidisciplinary research field that attempts to reveal the general principles and rules of nanomaterials toxicity (Hristozov et al. 2016).

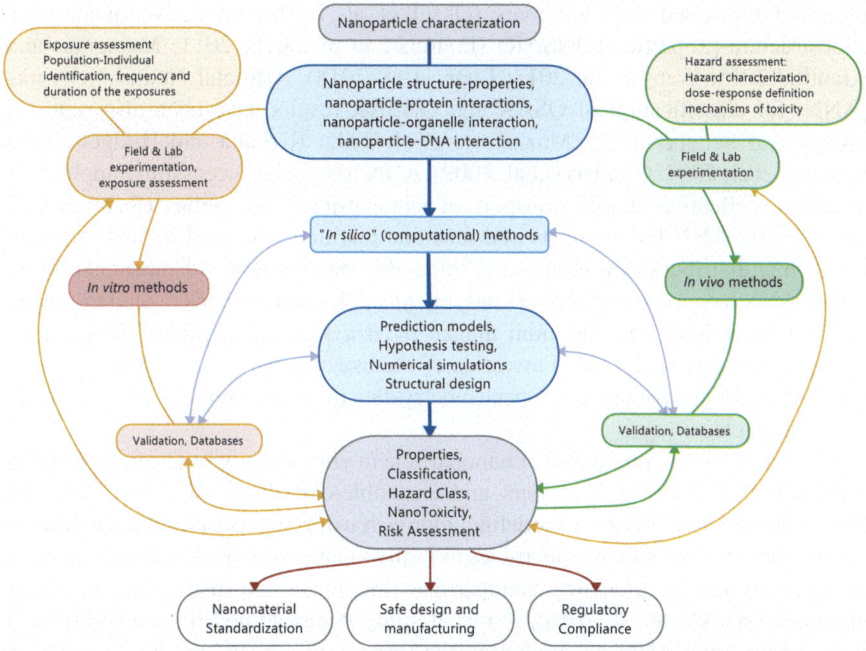


**Fig. 11** Scheme of a typical risk assessment procedure for engineered nanomaterials and nanopesticides

A typical risk assessment procedure (see Fig. 11) is performed in three basic steps (Holden et al. 2016):

1. Exposure assessment that refers to the identification and characterization of the populations exposed and determines the magnitude, frequency, and duration of the exposures;
2. Hazard assessment, which involves two main phases:
   (a) Hazard characterization: nanodose–response determination for critical target organs, tissues, cells, subcellular structures as well as possibly involved mechanisms of toxicity;
   (b) Hazard identification: identification of nanochemical properties that may cause adverse effects;
3. Final risk assessment based on exposure and hazard assessments.

Computational models play a complementary role in allowing rapid prediction of potential toxicities of new and modified nanomaterials (Qu et al. 2013; Yanamala et al. 2013; Kleandrova et al. 2014; Concu et al. 2017; Kovalishyn et al. 2018; Sinha et al. 2019).

Several computational approaches are being implemented for molecular description of interactions of nanoparticles with proteins and other biomolecules (Villaverde et al. 2017; Banu et al. 2018; Selvaraj et al. 2018; Deringer et al. 2019), genetic programming-based decision trees (Oksel et al. 2016), predictive quantitative nanostructure-properties (QNSAR) (Fourches et al. 2010, 2011; Melagraki and Afantitis 2015; Wang et al. 2017; Luan et al. 2018). Artificial Neural Networks (ANN) for classification of QSAR and QNSAR results have been also reported (Wiese and Schaper 1993; Mazzatorta et al. 2005; Rizkalla and Hildgen 2005; Stenemo et al. 2007; Goudarzi et al. 2009). ANN have been successfully applied for modeling collective charge transport in nanoparticles assemblies (Suvakov and Tadic 2010). ANN-based supervised machine learning was used for investigating the atomic distribution in mono- and bimetallic nanoparticles (Timoshenko et al. 2019). ANN have been applied for studying physiological changes related to cerium oxide nanoparticles and cadmium uptake by *Brassica napus* plants (Rossi et al. 2019). Lazarovits et al. (2019) investigated the use of supervised machine learning using ANN for processing mass spectrometry data in order to predict the in vivo fate of nanoparticles.

Machine learning prediction of nanoparticle in vitro toxicity has been reported in a comparative study of classifiers and ensemble-classifiers using the Copeland Index (Furxhi et al. 2019). A modeling approach using supervised ANN enabled to successfully predict $TiO_2$ nanoparticles mobility in intact soil media (Fazeli-Sangani et al. 2019) and for modeling nanoparticles biouptake and distribution in a host organism (Winkler et al. 2014). A recent study demonstrated the capability of a Back-propagating ANN for predicting the toxicity of 17, 36, and 72 data sets of metal oxide nanoparticles (Fjodorova et al. 2017).

Current chemometrics or computational methods for molecular modeling are able to predict electronic configuration and conformational properties, specific reactivity, and mechanisms of actions for molecular systems, ranging from small molecules to nanomolecules and up to large biomolecules (Khan et al. 2019; Mikolajczyk et al. 2019; Villaverde et al. 2017; Yan et al. 2019). Quantum chemical calculations and molecular dynamics (MD) simulations are already well introduced as a routine tool for evaluation of potential human and environmental risks associated with nanomaterials (Slater et al. 2017; Kovalishyn et al. 2018). Results from such theoretical calculations provide researchers and experimentalists with huge data that are further classified and analyzed with machine learning paradigms like statistical tools and ANN (Gajewicz et al. 2018).

The knowledge gained from computational studies involving interactions of nanoparticles with biological systems helps to build algorithms for assessing the likelihood of toxicity in a variety of natural environmental scenarios (Richarz et al. 2015; Villaverde et al. 2018). Computer simulations are used to evaluate the structure of surfaces and for identifying new properties even with the smallest variation of atoms positions at edges, corners, surface steps, and defects (Zeng et al. 2018; Lamon et al. 2019). The mechanisms of the interatomic interactions between nanoparticles and biological molecules are not well understood. Comprehension of the mechanisms of such interactions will aid the safe production and utilization of the nanomaterials. Computational studies are helpful to understand the precise nature of interparticle interactions (Puzyn et al. 2018), the structure of the interface, and the packing of arrays and superstructures that are difficult to probe experimentally (Banares et al. 2017; Hastings et al. 2015b; Chen and Riviere 2017; Wang et al. 2017). However, similar to experimentalists who face several issues, computational nanoscientists have also various challenges; for example, poor nanotoxicity data (Tong et al. 2017; Gajewicz 2017), simulations involving many nanoparticles are computationally too intensive and not feasible using advanced ab initio or Density Functional Theory (DFT) approaches; convergence problems often occur in dealing with large molecules.

In silico models centered on quantitative nanostructure–activity/toxicity relationships (QnSAR/QnSTR) are valuable computational tools for supporting risk assessment (Toropova and Toropov 2018; Burello and Worth 2011a, 2011b, 2011c; Peter et al. 2019), by rationalizing the search for safer nanomaterials (Sizochenko et al. 2019; Lamon et al. 2019). In a recent study (Concu et al. 2017), a unified QSTR-perturbation model based on artificial neural networks was developed for simultaneously predicting general toxicity profiles of nanomaterials under diverse experimental conditions.

The construction of QnSAR models (Fig. 12) requires (1) the integration of expertise of nanomaterial scientists, chemists, (eco)-toxicologists, and modelers from academia, regulatory agencies, and industry, (2) collaborative databases to support the development of computational methods for toxicological risk assessment of nanopesticides. Among such initiatives, we may enumerate several European modeling and database Projects: (NanoPUZZLES, ModENPTox, PreNanoTox, MembraneNanoPart, MODERN, eNanoMapper, and EU COST TD1204 MODENA) as well as to create synergies within the European NanoSafety Cluster (Banares et al. 2017). The EU-funded eNanoMapper project (Hastings et al. 2015a) proposes a computational infrastructure for toxicological data management of engineered nanomaterials based on open standards, ontologies, and an interoperable design to enable a more effective, integrated approach to European research in nanotechnology (Jeliazkova et al. 2015). The lazar framework for read-across predictions was expanded for the prediction of nanoparticle toxicities, and a new methodology for calculating nanoparticle descriptors from core and coating structures was implemented. Nano-lazar provides a flexible and reproducible framework for downloading data and ontologies from the open eNanoMapper infrastructure, developing and validating nanoparticle read-across models, open-source code, and a free graphical interface for nanoparticle read-across predictions (Helma et al. 2017; Ambure et al. 2019).

Scientific research for evaluating the hazards of nanopesticides on the environmental burden and human health faces three main challenges: (1) integration and evaluation of scientific evidence, toxicity data and computational models, (2) categorization and labeling of nanomaterial-based raw materials and marketed products, and (3) establishing hazard threshold levels for environmental and human health. Computational toxicology must become a priority.

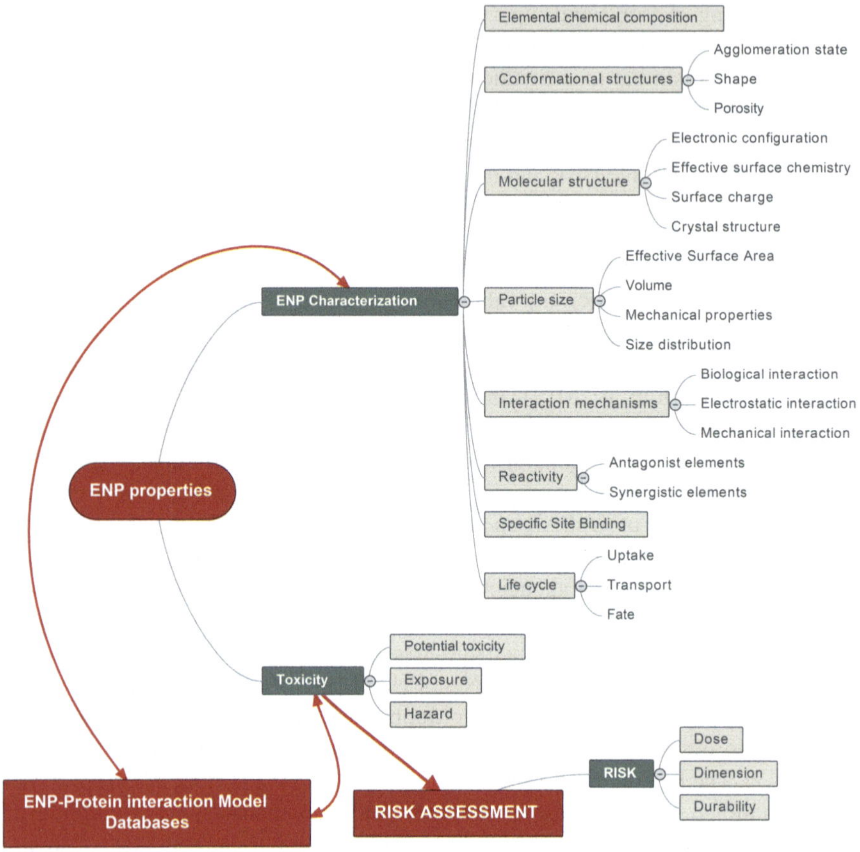


**Fig. 12** Schematic illustration for construction of a quantitative nanostructure-activity/toxicity relationship model (QnSAR/QnSTR)

## 3 Conclusions

Our growing knowledge of new technologies involving nanoparticles and nanopesticides has generated expectations as to the possible reduction in the use of pesticides around the world. However, the extension of these benefits needs to be observed with attention since similar speculations about the effects of transgenics on the pesticides reduction were also not confirmed.

Despite emergent research groups developing studies with nanocapsules for agricultural applications our knowledge of nanopesticides and their effects on plants in particular on weeds is still lacking. Although most of the data with nanoparticles involve cultivated plants, macrophytes and model species for crop studies show promising results, some responses are species-specific and interpretations must be done with prudence.

Related to pesticides, one of the most significant challenges to modern agriculture is the use of products with high efficiency, with low cost but commercially viable that are less dangerous to human and environmental health. For this purpose, some nanopesticides appear to be more harmful to the non-target organism, which justifies further studies in this area. Hence, new ecotoxicity protocols should be validated in order to understand the real risk assessment of the nanopesticides concerning the commercial one to target and non-target organisms.

Finally, the use of nanopesticides in agriculture based on sustainable concepts needs to assemble researches, producers, governor, and other social actors, including conscious consumers able to influence the market laws.

**Acknowledgement** R.G. would like to thank the Conselho Nacional de Desenvolvimento Científico e Tecnológico (CNPQ)—Brazil (grant numbers: 427498/2018-0 and 454417/2017-9). A.R.L.C. also acknowledges the financial support provided by the CAPES-PrInt funding program (grant numbers: 88887.353061/2019-00 and 88887.311920/2018-00) and the National Institute of Science and Technology of Basic Optics and Optics Applied to Life Science (grant number: 465360/2014-9). This study was financed in part by the Coordenação de Aperfeiçoamento de Pessoal de Nível Superior—Brasil (CAPES)—Finance Code 001 and in part by the Universidade Estadual de Mato Grosso do Sul (UEMS)—PIBAP/UEMS.

## References

Adisa IO, Pullagurala VLR, Peralta-Videa JR, Dimkpa CO, Elmer WH, Gardea-Torresdey JL, White JC (2019) Recent advances in nano-enabled fertilizers and pesticides: a critical review of mechanisms of action. Environ Sci Nano 6:2002–2030

Ahsan SM, Rao CM, Ahmad MF (2018) Nanoparticle-protein interaction: the significance and role of protein corona. Adv Exp Med Biol 1048:175–198

Albanese A, Tang PS, Chan WC (2012) The effects of nanoparticles size, shape, and surface chemistry on biological systems. Annu Rev Biomed Eng 14:1–16

Ambure P, Halder AK, Gonzalez-Diaz H, Cordeiro M (2019) QSAR-Co: an open source software for developing robust multitasking or multitarget classification-based QSAR models. J Chem Inf Model 59(6):2538

Andreotti F, Mucha AP, Caetano C, Rodríguez P, Gomes CR, Almeida CMR (2015) Interactions between salt marsh plants and Cudr nanoparticles–effects on metal uptake and phytoremediation processes. Ecotoxicol Environ Saf 120:303–309

Arora S, Sharma P, Kumar S, Nayan R, Khanna PK, Zaidi MGH (2012) Gold-nanoparticle induced enhancement in growth and seed yield of *Brassica juncea*. Plant Growth Regul 66:303–310

Ashitha A, Mathew J (2019) Characteristics and types of slow/controlled release of pesticides. In: Rakhimol KR, Thomas S, Volova T, Jayachandran K (eds) Controlled release of pesticides for sustainable agriculture. Springer, Cham, pp 141–153

Asli S, Neumann PM (2009) Colloidal suspensions of clay or titanium dioxide nanoparticles can inhibit leaf growth and transpiration via physical effects on root water transport plant. Cell Environ 32:577–584

Banares MA, Haase A, Tran L, Lobaskin V, Oberdorster G, Rallo R, Leszczynski J, Hoet P, Korenstein R, Hardy B, Puzyn T (2017) CompNanoTox2015: novel perspectives from a European conference on computational nanotoxicology on predictive nanotoxicology. Nanotoxicology 11(7):839–845

Banu H, Joseph MC, Nisar MN (2018) In-silico approach to investigate death domains associated with nano-particle-mediated cellular responses. Comput Biol Chem 75:11–23
Bao D, Oh ZG, Chen Z (2016) Characterization of silver nanoparticles internalized by *Arabidopsis* plants using single particle ICP-MS analysis. Front Plant Sci 7:32
Barazzouk S, Kamat PV, Hotchandani S (2005) Photoinduced electron transfer between chlorophyll a and gold nanoparticles. J Phys Chem B 109:716–723
Barrios AC, Rico CM, Trujillo-Reyes J, Medina-Velo IA, Peralta-Videa JR, Gardea-Torresdey JL (2016) Effects of uncoated and citric acid coated cerium oxide nanoparticles, bulk cerium oxide, cerium acetate, and citric acid on tomato plants. Sci Total Environ 563:956–964
Benelli G (2018) Mode of action of nanoparticles against insects. Environ Sci Pollut Res 25: 12329–12341
Bhar R, Kaur G, Mehta SK (2018) Experimental validation of DNA interactions with nanoparticles derived from metal coupled amphiphiles. J Biomol Struct Dyn 36(14):3614–3622
Blokhina O, Fagerstedt KV (2010) Reactive oxygen species and nitric oxide in plant mitochondria: origin and redundant regulatory systems. Physiol Plant 138(4):447–462
Bombo AB, Pereira AES, Lusa MG, Oliveira EM, Oliveira JL, Campos EVR, Jesus MB, Oliveira HC, Fraceto LF, Mayer JLS (2019) A mechanistic view of interactions of a nanoherbicide with target organism. J Agric Food Chem 67(16):4453–4462
Buchanan BB, Gruissem W, Jones RL (2000) Biochemistry and molecular biology of plants, 1st edn. Courier Companies, Maryland, 1367p
Burello E, Worth A (2011a) Computational nanotoxicology: predicting toxicity of nanoparticles. Nat Nanotechnol 6(3):138–139
Burello E, Worth AP (2011b) QSAR modeling of nanomaterials. Wiley Interdiscip Rev Nanomed Nanobiotechnol 3(3):298–306
Burello E, Worth AP (2011c) A theoretical framework for predicting the oxidative stress potential of oxide nanoparticles. Nanotoxicology 5(2):228–235
Cai F, Wu X, Zhang H, Shen X, Zhang M, Chen W, Gao Q, White JC, Tao S, Wang X (2017) Impact of $TiO_2$ nanoparticles on lead uptake and bioaccumulation in rice (*Oryza sativa* L.). NanoImpact 5:101–108
Campos EVR, Oliveira JL, Fraceto LF, Grillo R (2018) Global market of nanomaterials and colloidal formulations for agriculture: an overview. In: Singh HB, Faceto LF, Lima R, Mishra S (eds) Emerging trends in agri-nanotechnology: fundamental and applied aspects, 1st edn. Centre for Agriculture and Biosciences International (CABI), Wallingford, pp 1–312
Cannon W (1929) The wisdom of the body. Physiol Rev 9:399–431
Cao Z, Stowers C, Rossi L, Zhang W, Lombardini L, Ma X (2017) Physiological effects of cerium oxide nanoparticles on the photosynthesis and water use efficiency of soybean (*Glycine max* (L.) Merr.). Environ Sci Nano 4:1086–1094
Cerdervall T, Lynch I, Lindman S, Berggärd T, Thulin E, Nilsson H, Dawson KA, Linse S (2007) Understanding the nanoparticle–protein corona using methods to quantify exchange rates and affinities of proteins for nanoparticles. Proc Natl Acad Sci U S A 104:2050–2055
Chandler D, Bailey AS, Tatchell GM, Davidson G, Greaves J, Grant WP (2011) The development, regulation and use of biopesticides for integrated pest management. Philos Trans R Soc Lond B Biol Sci 366:1987–1998
Chen KL, Bothun GD (2014) Nanoparticles meet cell membranes: probing nonspecific interactions using model membranes. Environ Sci Technol 48:873–880
Chen R, Riviere JE (2017) Biological and environmental surface interactions of nanomaterials: characterization, modeling, and prediction. Wiley Interdiscip Rev Nanomed Nanobiotechnol 9(3):e1440. https://doi.org/10.1002/wnan.1440
Chibber S, Ahmad I (2016) Molecular docking, a tool to determine interaction of CuO and $TiO_2$ nanoparticles with human serum albumin. Biochem Biophys Rep 6:63–67
Cho EC, Xie J, Wurm PA, Xia Y (2009) Understanding the role of surface charges in cellular adsorption versus internalization by selectively removing gold nanoparticles on the cell surface with a I2/KI etchant. Nano Lett 9(3):1080–1084

Chu ZQ, Zhang SL, Zhang BK, Zhang CY, Fang CY, Rehor I, Cigler P, Chang HC, Lin G, Liu RB, Li Q (2014) Unambiguous observation of shape effects on cellular fate of nanoparticles. Sci Rep 4:4495
Chung I, Rekha K, Venkidasamy B, Thiruvengadam M (2019) Effect of copper oxide nanoparticles on the physiology, bioactive molecules, and transcriptional changes in *Brassica rapa* ssp. rapa seedlings. Water Air Soil Pollut 230:48
Cifuentes Z, Custardoy L, de la Fuente JM, Marquina C, Ibarra MR, Rubiales D, Perez-de-Luque A (2010) Absorption and translocation to the aerial part of magnetic carbon-coated nanoparticles through the root of different crop plants. J Nanobiotechnology 8(1):26–28
Clemente Z, Grillo R, Jonsson C, Santos NZP, Feitosa LOL, Fraceto RLF (2014) Ecotoxicological evaluation of poly(epsilon-caprolactone) nanocapsules containing triazine herbicides. J Nanosci Nanotechnol 14:4911–4917
Concu R, Kleandrova VV, Speck-Planche A, Cordeiro M (2017) Probing the toxicity of nanoparticles: a unified in silico machine learning model based on perturbation theory. Nanotoxicology 11(7):891–906
Conijn JG, Bindraban PS, Schröder JJ, Jongschaap REE (2018) Can our global food system meet food demand within planetary boundaries? Agric Ecosyst Environ 251:244–256
Curtis EM, Bahrami AH, Weikl TR, Hall CK (2015) Modeling nanoparticle wrapping or translocation in bilayer membranes. Nanoscale 7(34):14505–14514
Czarnocka W, Karpiński S (2018) Friend or foe? Reactive oxygen species production, scavenging and signaling in plant response to environmental stresses. Free Radic Biol Med 122:4–20
D'Rozario RSG, Wee CL, Wallace EJ, Sansom MSP (2009) The interaction of C60 and its derivatives with a lipid bilayer via molecular dynamics simulations. Nanotechnology 20(11):115102
Da Costa MVJ, Sharma PK (2016) Effect of copper oxide nanoparticles on growth, morphology, photosynthesis, and antioxidant response in *Oryza sativa*. Photosynthetica 54:110–119
Dausend J, Musyanovych A, Dass M, Walther P, Schrezenmeier H, Landfester K et al (2008) Uptake mechanism of oppositely charged fluorescent nanoparticles in HeLa cells. Macromol Biosci 8(12):1135–1143
Davis KJ (2016) Adaptive homeostasis. Mol Asp Med 49:1–7
Dayem AA, Hossain MK, Lee SB, Kim K, Saha SK, Yang G, Choi HY, Cho S (2017) The role of reactive oxygen species (ROS) in the biological activities of metallic nanoparticles. Int J Mol Sci 18(1):E120
De La Torre-Roche R, Hawthorne J, Deng Y, Xing B, Cai W, Newman LA, Wang Q, Ma X, Hamdi H, White JC (2013) Multiwalled carbon nanotubes and c60 fullerenes differentially impact the accumulation of weathered pesticides in four agricultural plants. Environ Sci Technol 47:12539–12547
Degrassi G, Bertani I, Devescovi G, Fabrizi A, Gatti A, Venturi V (2012) Response of pant-bacteria interaction models to nanoparticles. J Environ Qual 8:39–50
Dekker JP, Van-Grondelle R (2000) Primary charge separation in photosystem II. Photosynth Res 63:195–208
Deringer VL, Caro MA, Csanyi G (2019) Machine learning interatomic potentials as emerging tools for materials science. Adv Mater 31(46):e1902765
Dietz KJ, Herth S (2011) Plant nanotoxicology. Trends Plant Sci 11(16):582–589
Donaldson O (2013) Paraquat. In: Peterson ME, Talcott PA (eds) In small animal toxicology. Elsevier, St Louis, pp 731–739
Doolette CL, McLaughlin MJ, Kirby JK, Navarro DA (2015) Bioavailability of silver and silver sulfide nanoparticles to lettuce (*Lactuca sativa* L.): effect of agricultural amendments on plant uptake. J Hazard Mater 300:788–795
Eichert T, Kurtz A, Steiner U, Goldbach HE (2008) Size exclusion limits and lateral heterogeneity of the stomatal foliar uptake pathway for aqueous solutes and water-suspended nanoparticles. Physiol Plant 134:151–160
El-bendary HM, El-Helaly AA (2013) First record nanotechnology in agricultural: silica nanoparticles a potential new insecticide for pest control. Appl Sci Rep 4(3):241–246

El-Temsah YS, Joner EJ (2012) Impact of Fe and Ag nanoparticles on seed germination and differences in bioavailability during exposure in aqueous suspension and soil. Environ Toxicol 27(1):42–49

Falco WF, Queiroz AM, Fernandes J, Botero ER, Falcão EA, Guimarães FEG, M'Peko JC, Oliveira SL, Colbeck I, Caires ARL (2015) Interaction between chlorophyll and silver nanoparticles: a close analysis of chlorophyll fluorescence quenching. J Photochem Photobiol A Chem 299: 203–209

Falco WF, Scherer MD, Oliveira SL, Wender H, Colbeck I, Lawson T, Caires ARL (2019) Phytotoxicity of silver nanoparticles on *Vicia faba*: evaluation of particle size effects on photosynthetic performance and leaf gas exchange. Sci Total Environ 701:134816

Fan X, Xu J, Lavoie M, Peijnenburg WJGM, Zhu Y, Lu T, Fu Z, Zhu T, Qian H (2018) Multiwall carbon nanotubes modulate paraquat toxicity in *Arabidopsis Thaliana*. Environ Pollut 233: 633–641

Fazeli-Sangani M, Owens G, Nazari B, Astaraei A, Fotovat A, Emami H (2019) Different modelling approaches for predicting titanium dioxide nanoparticles mobility in intact soil media. Sci Total Environ 665:1168–1181

Fjodorova N, Novic M, Gajewicz A, Rasulev B (2017) The way to cover prediction for cytotoxicity for all existing nano-sized metal oxides by using neural network method. Nanotoxicology 11(4):475–483

Foroozandeh P, Aziz AA (2018) Insights into cellular uptake and intracellular trafficking of nanoparticles. Nanoscale Res Lett 13:339

Fourches D, Pu D, Tassa C, Weissleder R, Shaw SY, Mumper RJ, Tropsha A (2010) Quantitative nanostructure-activity relationship modeling. ACS Nano 4(10):5703–5712

Fourches D, Pu D, Tropsha A (2011) Exploring quantitative nanostructure-activity relationships (QNAR) modeling as a tool for predicting biological effects of manufactured nanoparticles. Comb Chem High Throughput Screen 14(3):217–225

Fraceto LF, Grillo R, Medeiros GA, Scognamiglio V, Rea G, Bartolucci C (2016) Nanotechnology in agriculture: which innovation potential does it have? Front Environ Sci 4:20

Furxhi I, Murphy F, Mullins M, Poland CA (2019) Machine learning prediction of nanoparticle in vitro toxicity: a comparative study of classifiers and ensemble-classifiers using the Copeland Index. Toxicol Lett 312:157–166

Gajewicz A (2017) What if the number of nanotoxicity data is too small for developing predictive Nano-QSAR models? An alternative read-across based approach for filling data gaps. Nanoscale 9(24):8435–8448

Gajewicz A, Puzyn T, Odziomek K, Urbaszek P, Haase A, Riebeling C, Luch A, Irfan MA, Landsiedel R, van der Zandle M, Bouwmeester H (2018) Decision tree models to classify nanomaterials according to the DF4nanoGrouping scheme. Nanotoxicology 12(1):1–17

Galazzi RM, Arruda MAZ (2018) Evaluation of changes in the macro and micronutrients homeostasis of transgenic and non-transgenic soybean plants after cultivation with silver nanoparticles through ionomic approaches. J Trace Elem Med Biol 48:181–187

Ghosh M, Bandyopadhyay M, Mukherjee A (2010) Genotoxicity of titanium dioxide ($TiO_2$) nanoparticles at two trophic levels: plant and human lymphocytes. Chemosphere 81:1253–1262

Gomes SIL, Scott-Fordsmand J, Campos EVR, Grillo R, Fraceto LF, Amorin MJB (2019) On the safety of nanoformulations to non-target soil invertebrates – an atrazine case study. Environ Sci Nano 6:1950–1958

Goodman CM, McCusker CD, Yilmaz T, Rotello VM (2004) Toxicity of gold nanoparticles functionalized with cationic and anionic side chains. Bioconjug Chem 15(4):897–900

Goudarzi N, Goodarzi M, Araujo MC, Galvao RK (2009) QSPR modeling of soil sorption coefficients (K(OC)) of pesticides using SPA-ANN and SPA-MLR. J Agric Food Chem 57(15): 7153–7158

Govorov AO, Carmeli I (2007) Hybrid structures composed of photosynthetic system and metal nanoparticles: plasmon enhancement effect. Nano Lett 7:620–625

Grillo R, Santos NZP, Maruyama CR, Rosa AH, Lima R, Fraceto LF (2012) Poly(ε-caprolactone) nanocapsules as carrier systems for herbicides: physico-chemical characterization and genotoxicity evaluation. J Hazard Mater 231:1–9

Grillo R, Pereira AES, Nishisaka CS, de Lima R, Oehlke K, Greiner R (2014) Chitosan/tripolyphosphate nanoparticles loaded with paraquat herbicide: an environmentally safer alternative for weed control. J Hazard Mater 278:163–171

Grillo R, Clemente Z, Oliveira JL, Campos EV, Chalupe VC, Jonsson CM, Lima R, Sanches G, Nishisaka CS, Rosa AH, Oehlke K, Greiner JL, Fraceto LF (2015) Chitosan nanoparticles loaded the herbicide paraquat: the influence of the aquatic humic substances on the colloidal stability and toxicity. J Hazard Mater 286:562–572

Grillo R, Abhilash PC, Fraceto LF (2016) Nanotechnology applied to bio-encapsulation of pesticides. J Nanosci Nanotechnol 16:1231–1234

Grillo R, Jesus MB, Fraceto LF (2018) Editorial: environmental impact of nanotechnology: analyzing the present for building the future. Front Environ Sci 6:1–3

Hastings J, Jeliazkova N, Owen G, Tsiliki G, Munteanu CR, Steinbeck C, Willighagen E (2015a) eNanoMapper: harnessing ontologies to enable data integration for nanomaterial risk assessment. J Biomed Semantics 6(1):10

Hastings J, Jeliazkova N, Owen G, Tsiliki G, Munteanu CR, Steinbeck C, Willighagen E (2015b) eNanoMapper: harnessing ontologies to enable data integration for nanomaterial risk assessment. J Biomed Semantics 6(10):2–15

Hayles J, Johnson L, Worthley C, Losic D (2017) Nanopesticides: a review of current research and perspectives. In: New pesticides and soil sensors. Academic Press, Cambridge, pp 193–225

Helma C, Rautenberg M, Gebele D (2017) Nano-lazar: read across predictions for nanoparticle toxicities with calculated and measured properties. Front Pharmacol 8:377

Hess DF (2000) Light-dependent herbicides: an overview. Weed Sci 48:160–170

Heyno E, Mary V, Schopfer P, Krieger-Liszkay A (2011) Oxygen activation at the plasma membrane: relation between superoxide and hydroxyl radical production by isolated membranes. Planta 234(1):35–45

Holden PA, Gardea-Torresdey JL, Klaessig F, Turco RF, Mortimer M, Hund-Rinke K, Hubal EAC, Avery D, Barceló D, Behra R, Cohen Y, Deydier-Stephan L, Ferguson PL, Fernandes TF, Harthorn BH, Henderson M, Hoke RA, Hristozov D, Johnston JM, Kane AB, Kaputska L, Keller AA, Lenihan HS, Lovel W, Murphy CJ, Nisbet RM, Peterson EJ, Salinas ER, Scheringer M, Sharma M, Speed DE, Sultan Y, Westerhoff P, White JC, Wiesner MR, Wong EM, Xing B, Horan MS, Godwin HA, Nel AE (2016) Considerations of environmentally relevant test conditions for improved evaluation of ecological hazards of engineered nanomaterials. Environ Sci Technol 50(12):6124–6145

Hong F, Zhou J, Liu C, Yang F, Wu C, Zheng L, Yang P (2005) Effect of nano-$TiO_2$ on photochemical reaction of chloroplasts of spinach. Biol Trace Elem Res 105:269–279

Hossain Z, Mustafa G, Komatsu S (2015) Plant responses to nanoparticle stress. Int J Mol Sci 16:26644–26653

Hou H, Huang X, Wei G, Xu F, Wang Y, Zhou S (2019) Fenton reaction assisted photodynamic therapy for cancer with multifunctional magnetic nanoparticles. ACS Appl Mater Interfaces 11(33):29579–29592

Hristozov D, Gottardo S, Semenzin E, Oomen A, Bos P, Peijnenburg W, van Tongeren M, Nowack B, Hunt N, Brunelli A, Scott-Fprdsmand JJ, Tran L, Marcomini A (2016) Frameworks and tools for risk assessment of manufactured nanomaterials. Environ Int 95:36–53

Hu X, Lu K, Mu L, Kang J, Zhou Q (2014) Interactions between graphene oxide and plant cells: regulation of cell morphology, uptake, organelle damage, oxidative effects and metabolic disorders. Carbon 80:665–676

Hussain I, Singh NB, Singh A, Singh H, Singh SC (2016) Green synthesis of nanoparticles and its potential application. Biotechnol Lett 38:545–560

Jeliazkova N, Chomenidis C, Doganis P, Fadeel B, Grafstrom R, Hardy B, Hastings J, Hegi M, Jeliazkov V, Kochev N, Kohonen P, Munteanu CR, Sarimveis H, Smeets B, Sopasakis P, Tsiliki

G, Vorgrimmler D, Willighagen E (2015) The eNanoMapper database for nanomaterial safety information. Beilstein J Nanotechnol 6:1609–1634

Kah M, Holfman T (2014) Nanopesticides research; current trends and future priorities. Environ Int 63:224–235

Kathiravan A, Chandramohan M, Renganathan R, Sekar S (2009) Photoinduced electron transfer from phycoerythrin to colloidal metal semiconductor nanoparticles. Spectrochim Acta Part A: Mol Biomol Spectrosc 72:496–501

Kavallieratos NG, Athanassiou CG, Peteinatos GG, Boukouvala MC, Benelli G (2018) Insecticidal effect and impact of fitness of three diatomaceous earths on different maize hybrids for the eco-friendly control of the invasive stored-product pest *Prostephanus truncatus* (Horn). Environ Sci Pollut Res 25:10407–10417

Kerbauy GB (2004) Fisiologia vegetal. Guanabara Koogan, Rio de Janeiro, 452p

Khan MI, Alsaedi A, Hayat T, Khan NB (2019) Modeling and computational analysis of hybrid class nanomaterials subject to entropy generation. Comput Methods Prog Biomed 179:104973

Khodakovskaya M, Dervishi E (2009) Carbon nanotubes are able to penetrate plant seed coat and dramatically affect seed germination and plant growth. ASC Nano 10(3):3221–3227

Kittler S, Greulich C, Diendorf J, Koeller M, Epple M (2010) Toxicity of silver nanoparticles increase during storage because of slow dissolution under release of silver ions. Chem Mater 22(16):4548–4554

Kleandrova VV, Luan F, Gonzalez-Diaz H, Ruso JM, Speck-Planche A, Cordeiro MN (2014) Computational tool for risk assessment of nanomaterials: novel QSTR-perturbation model for simultaneous prediction of ecotoxicity and cytotoxicity of uncoated and coated nanoparticles under multiple experimental conditions. Environ Sci Technol 48(24):14686–14694

Kovalishyn V, Abramenko N, Kopernyk I, Charochkina L, Metelytsia L, Tetko IV, Peijnenburg W, Kustov L (2018) Modelling the toxicity of a large set of metal and metal oxide nanoparticles using the OCHEM platform. Food Chem Toxicol 112:507–517

Kumari M, Mukherjee A, Chandrasekaran N (2009) Genotoxicity of silver nanoparticles in *Allium cepa*. Sci Total Environ 407:5243–5246

Lai SK, Hida K, Man ST, Chen C, Machamer C, Schroer TA, Hanes J (2007) Privileged delivery of polymer nanoparticles to the perinuclear region of live cells via a non-clathrin, non-degradative pathway. Biomaterials 28(18):2876–2884

Lamon L, Asturiol D, Vilchez A, Ruperez-Illescas R, Cabellos J, Richarz A, Worth A (2019) Computational models for the assessment of manufactured nanomaterials: development of model reporting standards and mapping of the model landscape. Comput Toxicol 9:143–151

Larue C, Veronesi G, Flank AM, Surble S, Herlin-Boime N, Carrière M (2012) Comparative uptake and impact of TiO2 nanoparticles in wheat and rapeseed. J Toxicol Environ Health A 75:722–734

Lazarovits J, Sindhwani S, Tavares AJ, Zhang Y, Song F, Audet J, Krieger J, Syed AM, Stordy B, Chan WCW (2019) Supervised learning and mass spectrometry predicts the *in vivo* fate of nanomaterials. ACS Nano 13(7):8023–8034

Lelimousin M, Sansom MS (2013) Membrane perturbation by carbon nanotube insertion: pathways to internalization. Small 9:3639–3646

Li Y, Gu N (2010) Thermodynamics of charged nanoparticle adsorption on charge-neutral membranes: a simulation study. J Phys Chem B 114(8):2749–2754

Li H, Ye X, Guo X, Geng Z, Wang G (2016a) Effects of surface ligands on the uptake and transport of gold nanoparticles in rice and tomato. J Hazard Mater 314:188–196

Li J, Hu J, Ma C, Wang Y, Wu C, Huang J, Xing B (2016b) Uptake, translocation and physiological effects of magnetic iron oxide (γ-Fe2O3) nanoparticles in corn (*Zea mays* L.). Chemosphere 159:326–334

Li L, Xu Z, Kah M, Lin D, Filser J (2019a) Nanopesticides: a comprehensive assessment of environmental risk is needed before widespread agricultural application. Environ Sci Technol 53(14):7923–7924

Li R, He J, Xie H, Wang W, Bose SK, Sun Y, Hu J, Yin H (2019b) Effects of chitosan nanoparticles on seed germination and seedling growth of wheat (*Triticum aestivum* L.). Int J Biol Macromol 126:91–100

Liang Q (2013) Penetration of polymer-grafted nanoparticles through a lipid bilayer. Soft Matter 9:5594–5601

Lima R, Feitosa LO, Grillo R, Pereira AES, Fraceto LF (2012) Evaluation of the genotoxicity of polymeric microparticles containing ametryn herbicide. J Environ Sci Eng 4:553–565

Lòpez-Moreno ML, de la Rosa G, Hernàndez-Viezcas JA, Castillo-Michel H, Botez CE, Peralta-Videa JR, Gardea-Torresdey JL (2010) Evidence of the differential biotransformation and genotoxicity of ZnO and CeO2 nanoparticles on soybean (*Glycine max*) plants. Environ Sci Technol 44:7315–7320

Lowry GV, Avellan A, Gilbertson LM (2019) Opportunities and challenges for nanotechnology in the Agri-tech revolution. Nat Nanotechnol 14:517–522

Lu F, Wu SH, Hung Y, Mou CY (2009) Size effect on cell uptake in well-suspended, uniform mesoporous silica nanoparticles. Small 5(12):1408–1413

Luan F, Tang L, Zhang L, Zhang S, Monteagudo MC, Cordeiro M (2018) A further development of the QNAR model to predict the cellular uptake of nanoparticles by pancreatic cancer cells. Food Chem Toxicol 112:571–580

Ma W, Saccardo A, Roccatano D, Aboagye-Mensah D, Alkaseen M, Jewkes M, Di Nezza F, Baron M, Soloviev M, Ferrari E (2018) Modular assembly of proteins on nanoparticles. Nat Commun 9:1489

Mahakham W, Theerakulpisut P, Maensiri S, Phumying S, Sarmah AK (2016) Environmentally benign synthesis of phytochemicals-capped gold nanoparticles as nanopriming agent for promoting maize seed germination. Sci Total Environ 573:1089–1102

Maity A, Natarajan N, Vijay D, Srinivasan R, Pastor M, Malaviya DR (2018) Influence of metal nanoparticles (NPs) on germination and yield of Oat (*Avena sativa*) and Berseem (*Trifolium alexandrinum*). Proc Natl Acad Sci India Sect B Biol Sci 88(2):595–607

Manikandan A, Sathiyabama M (2016) Preparation of chitosan nanoparticles and its effects on detached rice leaves infected with Pyricularia grisea. Int J Biol Macromol 84:58–61

Mansoor N, Younus A, Jamil Y, Shahid M (2019) Impact of nanosized and bulk ZnO on germination and early growth response of *Triticum aestivum*. Pak J Agric Sci 56(4):879–884

Marano F, Hussain S, Rodrigues-Lima F, Baeza-Squiban A, Boland S (2011) Nanoparticles: molecular targets and cell signaling. Arch Toxicol 85(7):733–741

Margenot AJ, Rippner DA, Dumlao MR, Nezami S, Green PG, Parikh SJ, McElrone AJ (2018) Copper oxide nanoparticle effects on root growth and hydraulic conductivity of two vegetable crops. Plant Soil 431:333–345

Martin CE, von Willert DJ (2000) Leaf epidermal hydathodes and the ecophysiological consequences of foliar water uptake in species of *Crassula* from the Namib Desert in Southern Africa. Plant Biol 2(2):229–242

Maruyama CR, Guilger M, Pascoli M, Bileshy-José M (2016) Nanoparticles based on chitosan as carriers for the combined herbicides imazapic and imazapyr. Sci Rep 6:19768

Matorin DN, Todorenko DA, Seifullina NK, Zayadan BK, Rubin AB (2013) Effect of silver nanoparticles on the parameters of chlorophyll fluorescence and P700 reaction in the green alga *Chlamydomonas reinhardtii*. Microbiology 82:809–814

Mazzatorta P, Smiesko M, Lo Piparo E, Benfenati E (2005) QSAR model for predicting pesticide aquatic toxicity. J Chem Inf Model 45(6):1767–1774

McEwen BS, Wingfield JC (2003) The concept of allostasis in biology and biomedicine. Horm Behav 43:2–15

McKnight TE, Melechko AV, Griffin GD, Guillorn MA, Merkulov VI, Serna F, Hensley DK, Doktycz MJ, Lowndes DH, Simpson ML (2003) Intracellular integration of synthetic nanostructures with viable cells for controlled biochemical manipulation. Nanotechnology 14:551

Melagraki G, Afantitis A (2015) A risk assessment tool for the virtual screening of metal oxide nanoparticles through Enalos InSilicoNano platform. Curr Top Med Chem 15(18):1827–1836

Mikolajczyk A, Sizochenko N, Mulkiewicz E, Malankowska A, Rasulev B, Puzyn T (2019) A chemoinformatics approach for the characterization of hybrid nanomaterials: safer and efficient design perspective. Nanoscale 11(24):11808–11818

Mingyu S, Fashui H, Chao L, Xiao W, Xiaoqing L, Liang C, Fengqing G, Zhongrui L (2007) Effects of nano-anatase $TiO_2$ on absorption, distribution of light, and photoreduction activities of chloroplast membrane of spinach. Biol Trace Elem Res 118:120–130

Mohammadi R, Maali-Amiri R, Abbasi A (2013) Effect of $TiO_2$ nanoparticles on chickpea response to cold stress. Biol Trace Elem Res 152:403–410

Nair PMG, Chung IM (2015) Biochemical, anatomical and molecular level changes in cucumber (*Cucumis sativus*) seedlings exposed to copper oxide nanoparticles. Biologia 70:1575–1585

Nangia S, Sureshkumar R (2012) Effects of nanoparticle charge and shape anisotropy on translocation through cell membranes. Langmuir 28(51):17666–17671

Nath A, Molnár MA, Das A, Bánvölgyi S, Márki E, Vatai G (2019) Agrochemicals from nanomaterials – synthesis, mechanisms of biochemical activities and application. In: Verma SK, Das AK (eds) Analysis, fate, and toxicity of engineered nanomaterials in plants, vol 84. Elsevier, Amsterdam, pp 263–312

Navarro E, Baun A, Behra R, Hartmann NB, Filser J, Miao AJ, Quigg A, Santschi PH, Sigg L (2008) Environmental behavior and ecotoxicity of engineered nanoparticles to algae, plants, and fungi. Ecotoxicology 17:372–386

Nelson N, Yocum CF (2006) Structure and function of photosystem I and II. Annu Rev Plant Biol 57:521–565

Nguyen MH, Lee JS, Hwang IC, Park HJ (2014) Evaluation of penetration of nanocarriers into red pepper leaf using confocal laser scanning microscopy. Crop Prot 66:61–66

Nieder JB, Bittl R, Brecht M (2010) Fluorescence studies into the effect of plasmonic interactions on protein function. Angew Chem Int Ed Engl 49(52):10217–10220

Ocsoy I, Paret ML, Ocsoy MA, Kunwar S, Chen T, You M, Tan W (2013) Nanotechnology in plant disease management: DNA directed silver nanoparticles on graphene oxide as an antibacterial against *Xanthomonas Perforans*. ACS Nano 7:8972–8980

Oksel C, Winkler DA, Ma CY, Wilkins T, Wang XZ (2016) Accurate and interpretable nanoSAR models from genetic programming-based decision tree construction approaches. Nanotoxicology 10(7):1001–1012

Oliveira HC, Stolf MR, Martinez CB, Grillo R, de Jesus MB, Fraceto LF (2015) Nanoencapsulation enhances the post-emergence herbicidal activity of atrazine against mustard plants. PLoS One 10:e0132971

Oliveira JL, Campos EVR, Pereira AES, Pasquoto T, Lima R, Grillo R, Andrade DJ, Santos FA, Faceto LF (2018) Zein nanoparticles as eco-friendly carrier systems for botanical repellents aiming sustainable agriculture. J Agric Food Chem 66(6):1330–1340

Pacheco I, Buzea C (2018) Nanoparticle uptake by plants: beneficial or detrimental? In: Faisal M, Saquib Q, Alatar A, Al-Khedhairy A (eds) Phytotoxicity of nanoparticles. Springer, Basel, p 60

Panariti A, Miserocchi G, Rivolta I (2012) The effect of nanoparticle uptake on cellular behavior: disrupting or enabling functions? Nanotechnol Sci Appl 5:87–100

Peralta-Videa JR, Hernandez-Viezcas JA, Zhao L, Dias BD, Ge Y, Priester JH, Holden PA, Gardea-Torresdey JL (2014) Cerium dioxide and zinc oxide nano-particles alter the nutritional value of soil cultivated soybean plants. Plant Physiol Biochem 80:128–135

Pereira AE, Grillo R, Mello NF, Rosa AH, Fraceto LF (2014) Application of poly(epsilon-caprolactone) nanoparticles containing atrazine herbicide as an alternative technique to control weeds and reduce damage to the environment. J Hazard Mater 268:207–215

Pérez-de-Luque A (2017) Interaction of nanomaterials with plants: what do we need for real applications in agriculture? Front Environ Sci 5:1–12

Perreault F, Oukarroum A, Pirastru L, Sirois L, Gerson Matias W, Popovic R (2010) Evaluation of copper oxide nanoparticles toxicity using chlorophyll fluorescence imaging in *Lemna gibba*. J Bot 2010:763142

Peter SC, Dhanjal JK, Malik V, Radhakrishnan N, Jayakanthan M, Sundar D (2019) Quantitative structure-activity relationship (qsar): modeling approaches to biological applications. Encycl Bioinform Comput Biol 2:661–676

Pontes MS, Grillo R, Graciano DE, Falco WF, Lima SM, Caires ARL, Andrade LHC, Santiago EF (2019) How does aquatic macrophyte *Salvinia auriculata* respond to nanoceria upon an increased $CO_2$ source? A Fourier transform-infrared photoacoustic spectroscopy and chlorophyll *a* fluorescence study. Ecotoxicol Environ Saf 180:526–534

Pošćićć F, Mattiello A, Fellet G, Miceli F, Marchiol L (2016) Effects of cerium and titanium oxide nanoparticles in soil on the nutrient composition of barley (*Hordeum vulgare* L.) kernels. Int J Environ Res Public Health 13:577–592

Preisler AC, Pereira AES, Campos EVR, Dalazen G, Fraceto LF, Oliveira HC (2020) Atrazine nanoencapsulation improves pre-emergence herbicidal activity against *Bidens pilosa* without enhancing long-term residual effect on *Glycine max*. Pest Manag Sci 76(1):141–149

Puzyn T, Jeliazkova N, Sarimveis H, Marchese-Robinson RL, Lobaskin V, Rallo R, Richarz AN, Gajewicz A, Papadopulos MG, Hastings J, Cronin MTD, Benfenati E, Fernandez A (2018) Perspectives from the NanoSafety Modelling Cluster on the validation criteria for (Q)SAR models used in nanotechnology. Food Chem Toxicol 112:478–494

Qu ZG, He XC, Lin M, Sha BY, Shi XH, Lu TJ, Xu F (2013) Advances in the understanding of nanomaterial-biomembrane interactions and their mathematical and numerical modeling. Nanomedicine 8(6):995–1011

Queiroz AM, Mezacasa AV, Graciano DE, Falco WF, M'Peko JC, Guimarães FEG, Lawson T, Colbeck I, Oliveira SL, Caires ARL (2016) Quenching of chlorophyll fluorescence induced by silver nanoparticles. Spectrochim Acta Part A Mol Biomol Spectrosc 168:73–77

Raja K, Sowmya R, Sudhagar R, Moorthy PS, Govindaraju K, Subramanian KS (2019) Biogenic ZnO and Cu nanoparticles to improve seed germination quality in blackgram (*Vigna mungo*). Mater Lett 235:164–167

Rejman J, Oberle V, Zuhorn I, Hoekstra D (2004) Size-dependent internalization of particles via the pathways of clathrin-and caveolae mediated endocytosis. Biochem J 377:159–169

Richarz A-N, Madden JC, Marchese-Robinson RL, Lubiński Ł, Mokshina E, Urbaszek P, Kuz'min VE, Puzyn T, Cronin MTD (2015) Development of computational models for the prediction of the toxicity of nanomaterials. Perspect Sci 3(1–4):27–29

Rico CM, Morales MI, Barrios AC, McCreary R, Hong J, Lee WY, Nunez J, Peralta-Videa JR, Gardea-Torresdey JL (2013) Effect of cerium oxide nano-particles on the quality of rice (*Oryza sativa* L.) grains. J Agric Food Chem 61:11278–11285

Rizkalla N, Hildgen P (2005) Artificial neural networks: comparison of two programs for modeling a process of nanoparticle preparation. Drug Dev Ind Pharm 31(10):1019–1033

Rizwan M, Ali S, Qayyum MF, Ok YS, Adrees M, Ibrahim M, Zia-ur-Rehman M, Farid M, Abbas F (2017) Effect of metal and metal oxide nanoparticles on growth and physiology of globally important food crops: a critical review. J Hazard Mater 322:2–16

Rizwan M, Ali S, Rehman MZ, Malik S, Adress M, Qayyum MF, Alamri SA, Alyemeni MN, Ahmad P (2019) Effect of foliar applications of silicon and titanium dioxide nanoparticles on growth, oxidative stress, and cadmium accumulation by rice (*Oryza sativa*). Acta Physiol Plant 41(3):35

Rossi L, Bagheri M, Zhang W, Chen Z, Burken JG, Ma X (2019) Using artificial neural network to investigate physiological changes and cerium oxide nanoparticles and cadmium uptake by *Brassica napus* plants. Environ Pollut 246:381–389

Sabella S, Carney RP, Brunetti V, Malvindi MA, Al-Juffali N, Vecchio G, James SM, Bakr OS, Cingolani R, Stellacci F, Pompa PP (2014) A general mechanism for intracellular toxicity of metal-containing nanoparticles. Nanoscale 6(12):7052–7061

Saptarshi SR, Duschi A, Lopata AL (2013) Interaction of nanoparticles with proteins: relation to bio-reactivity of the nanoparticle. J Nanobiotechnology 11(26):1–12

Sarlak N, Taherifar A, Salehi F (2014) Synthesis of nanopesticides by encapsulating pesticide nanoparticles using functionalized carbon nanotubes and application of new nanocomposite for plant disease treatment. J Agric Food Chem 62(21):4833–4838

Sathiyabama M, Manikandan A (2016) Chitosan nanoparticle induced defense responses in finger millet plants against blast disease caused by *Pyricularia grisea* (Cke.) Sacc. Carbohydr Polym 154:241–246

Sathiyabama M, Manikandan A (2018) Application of copper-chitosan nanoparticles stimulate growth and induce resistance in finger millet (*Eleusine coracana* Gaertn.) plants against blast disease. J Agric Food Chem 66(8):1784–1790

Scherer MD, Sposito JCV, Falco WF, Grisolia AB, Andrade LHC, Lima SM, Machado G, Nascimento VA, Gonçalves DA, Wender H, Oliveira SL, Caires ARL (2019) Cytotoxic and genotoxic effects of silver nanoparticles on meristematic cells of *Allium cepa* roots: a close analysis of particle size dependence. Sci Total Environ 660:459–467

Schnoor B, Elhendawy A, Joseph S, Putman M, Chácon-Cerdas R, Flores-Mora D, Bravo-Moraga F, Gonzalez-Nilo F, Salvador-Morales C (2018) Engineering atrazine loaded poly(lactic-co-glycolic acid) nanoparticles to ameliorate environmental challenges. J Agric Food Chem 66:7889–7898

Scognamiglio V, Antonacci A, Arduini F, Moscone D, Campos EV, Fraceto LF, Palleschi G (2019) An eco-designed paper-based algal biosensor for nanoformulated herbicide optical detection. J Hazard Mater 373:483–492

Selvaraj C, Sakkiah S, Tong W, Hong H (2018) Molecular dynamics simulations and applications in computational toxicology and nanotoxicology. Food Chem Toxicol 112:495–506

Shabnam N, Sharmila P, Pardha-Saradhi P (2016) Impact of ionic and nanoparticle speciation states of silver on light harnessing photosynthetic events in *Spirodela polyrhiza*. Int J Phytoremediation 19:80–86

Sharif-Rad J, Sharif-Rad M, Teixeira da Silva JA (2016) Morphological, physiological and biochemical responses of crops (*Zea mays* L., *Phaseolus vulgaris* L.), medicinal plants (*Hyssopus officinalis* L., *Nigella sativa* L.), and weeds (*Amaranthus retroflexus* L., *Taraxacum officinale* F. H. Wigg) exposed to $SiO_2$ nanoparticles. J Agric Sci Technol 18:1027–1040

Sharkey TD, Weise SE (2016) The glucose 6-phosphate shunt around the Calvin–Benson cycle. J Exp Bot 67:4067–4077

Sharma P, Jha AB, Dubey RS, Pessarakli M (2012) Reactive oxygen species, oxidative damage, and antioxidative defense mechanism in plants under stressful conditions. J Bot 2012:217037

Shi J, Peng C, Yang Y, Yang J, Zhang H, Yuan X, Chen Y, Hu T (2013) Phytotoxicity and accumulation of copper oxide nanoparticles to the Cu-tolerant plant *Elsholtzia splendens*. Nanotoxicology 8:179–188

Shoaib A, Elabasy A, Waqas M, Lin L, Cheng X, Zhang Q, Shi Z (2018) Entomotoxic effect of silicon dioxide nanoparticles on *Plutella xylostella* (L.) (Lepidoptera: Plutellidae) under laboratory conditions. Toxicol Environ Chem 1(100):80–91

Shomer I, Novacky AJ, Pike SM, Yermiyahu U, Kinraide TB (2003) Electrical potentials of plant cell walls in response to the ionic environment. Plant Physiol 133:411–422

Siddiqui MH, Al-Whaibi MH (2014) Role of nano-SiO2 in germination of tomato (*Lycopersicum esculentum* seeds Mill). Saudi J Biol Sci 21(1):13–17

Simonin M, Richaume A (2015) Impact of engineered nanoparticles on the activity, abundance, and diversity of soil microbial communities: a review. Environ Sci Pollut Res Int 22:13710–13723

Singh S, Hussain I, Singh NB, Singh H (2019) Uptake, translocation and impact of green synthesized nanoceria on growth and antioxidant enzymes activity of *Solanum lycopersicum* L. Ecotoxicol Environ Saf 182:109410

Sinha M, Dhawan A, Parthasarathi R (2019) In Silico approaches in predictive genetic toxicology. Methods Mol Biol 2031:351–373

Sizochenko N, Syzochenko M, Fjodorova N, Rasulev B, Leszczynski J (2019) Evaluating genotoxicity of metal oxide nanoparticles: application of advanced supervised and unsupervised machine learning techniques. Ecotoxicol Environ Saf 185:109733

Slater R, Stratonovitch P, Elia J, Semenov MA, Denholm I (2017) Use of an individual-based simulation model to explore and evaluate potential insecticide resistance management strategies. Pest Manag Sci 73(7):1364–1372

Soenen SJ, Demeester J, De Smedt SC, Braechmans K (2013) Turning a frown upside down: exploiting nanoparticle toxicity for anticancer therapy. Nano Today 8(2):121–125

Soenen SJ, Parak WJ, Rejman J, Manshian B (2015) (Intra)cellular stability of inorganic nanoparticles: effects on cytotoxicity, particle functionality and biomedical applications. Chem Rev 115(5):2109–2135

Sosan A, Svistunenko D, Straltsova D, Tsiurkina K, Smolich I, Lawson T, Subramaniam S, Golovko V, Anderson D, Sokolik A, Colbeck I (2016) Engineered silver nanoparticles are sensed at the plasma membrane and dramatically modify the physiology of *Arabidopsis thaliana* plants. Plant J 85:245–257

Sousa GFM, Gomes DG, Campos EVR, Oliveira JL, Fraceto LF, Stolf-Moreira R, Oliveira HC (2018) Post-emergence herbicidal activity of nanoatrazine against susceptible weeds. Front Environ Sci 6:1–6

Spielman-sun E, Avellan A, Bland GD, Tappero RV, Acerbo AS, Unrine JM, Giraldo JP, Lowry GV (2019) Nanoparticle surface charge influences translocation and leaf distribution in vascular plants with contrasting anatomy. Environ Sci Nano 6:2508–2519

Stark WJ (2011) Nanoparticles in biological systems. Angew Chem Int Ed 50:1242–1258

Stenemo F, Lindahl AM, Gardenas A, Jarvis N (2007) Meta-modeling of the pesticide fate model MACRO for groundwater exposure assessments using artificial neural networks. J Contam Hydrol 93(1–4):270–283

Sterling P (2014) Homeostasis vs Allostasis implications for brain function and mental disorders. Clin Rev Educ 71(10):1192–1193

Stern ST, Adiseshaiah PP, Crist RM (2012) Autophagy and lysosomal dysfunction as emerging mechanisms of nanomaterial toxicity. Part Fibre Toxicol 9(20):1–15

Sukhanova A, Bozrova S, Sokolov P, Berestovoy M, Karaulov A, Nebiev I (2018) Dependence of nanoparticle toxicity on their physical and chemical properties. Nanoscale Res Lett 13(44):1–21

Suvakov M, Tadic B (2010) Modeling collective charge transport in nanoparticle assemblies. J Phys Condens Matter 22(16):163201

Taiz L, Zeiger E (2002) Plant physiology, 3rd edn. Sinauer Associates, Sunderland, 672p

Tan X, Lin C, Fugetsu B (2009) Studies on toxicity of multi-walled carbon nanotubes on suspension rice cells. Carbon 47:3479–3487

Thuesombat P, Hannongbua S, Akasit S, Chadchawan S (2014) Effect of silver nanoparticles on rice (*Oryza sativa* L. cv. KDML 105) seed germination and seedling growth. Ecotoxicol Environ Saf 104:302–309

Tighe-Neira R, Carmora E, Recio G, Nunes-Nesi A, Reyes-Diaz M, Alberdi M, Rengel Z, Inostroza-Blancheteau C (2018) Metallic nanoparticles influence the structure and function of the photosynthetic apparatus in plants. Plant Physiol Biochem 130:408–417

Timoshenko J, Wrasman CJ, Luneau M, Shirman T, Cargnello M, Bare SR, Aizenberg J, Friend CM, Frenkel AI (2019) Probing atomic distributions in mono- and bimetallic nanoparticles by supervised machine learning. Nano Lett 19(1):520–529

Titov AV, Kral P, Pearson R (2010) Sandwiched graphene-membrane superstructures. ACS Nano 4:229–234

Tong X, Zhang B, Fan Y, Chen Y (2017) Mechanism exploration of ion transport in nanocomposite cation exchange membranes. ACS Appl Mater Interfaces 9(15):13491–13499

Toropova AP, Toropov AA (2018) Use of the index of ideality of correlation to improve models of eco-toxicity. Environ Sci Pollut Res Int 25(31):31771–31775

Trebst A (1994) Dynamics in photosystem II structure and function. In: Schulze ED, Caldwell MM (eds) Ecophysiology of photosynthesis. Springer, Berlin, 579p

USEPA (United States Environmental Protection Agency) (2004) Air quality criteria for particulate matter (external review draft). National Center for Environmental Assessment, Washington, DC. http://cfpub2.epa.gov/ncea/cfm/partmatt.cfm

Vannini C, Domingo G, Onelli E, Prinsi B, Marsoni M, Espen L, Bracale M (2013) Morphological and proteomic responses of *Eruca sativa* exposed to silver nanoparticles or silver nitrate. PLoS One 8(7):e68752

Villaverde JJ, Lopez-Goti C, Alcami M, Lamsabhi AM, Alonso-Prados JL, Sandin-Espana P (2017) Quantum chemistry in environmental pesticide risk assessment. Pest Manag Sci 73(11): 2199–2202
Villaverde JJ, Sevilla-Moran B, Lopez-Goti C, Alonso-Prados JL, Sandin-Espana P (2018) Considerations of nano-QSAR/QSPR models for nanopesticide risk assessment within the European legislative framework. Sci Total Environ 634:1530–1539
Wang S-H, Lee C-W, Chiou A, Wei P-K (2010) Size-dependent endocytosis of gold nanoparticles studied by three-dimensional mapping of plasmonic scattering images. J Nanobiotechnology 8(33):1–13
Wang W, Sedykh A, Sun H, Zhao L, Russo DP, Zhou H, Bin Y, Zhu H (2017) Predicting nano-bio interactions by integrating nanoparticle libraries and quantitative nanostructure activity relationship modeling. ACS Nano 11(12):12641–12649
Wang M, Wang J, Liu Y, Wang J, Nie Y, Si B, Liu Y, Wang Y, Chen S, Hei TK, Wu L, Zhao G, Xu A (2019) Subcellular targets of zinc oxide nanoparticles during the aging process: role of cross-talk between mitochondrial dysfunction and endoplasmic reticulum stress in the genotoxic response. Toxicol Sci 171(1):159–171
Werner M, Sommer JU, Baulin VA (2012) Homo-polymers with balanced hydrophobicity translocate through lipid bilayers and enhance local solvent permeability. Soft Matter 8:11714–11722
White JC, Gardea-Torresdey J (2018) Achieving food security through the very small. Nat Nanotechnol 13:621–629
Wiese M, Schaper KJ (1993) Application of neural networks in the QSAR analysis of percent effect biological data: comparison with adaptive least squares and nonlinear regression analysis. SAR QSAR Environ Res 1(2–3):137–152
Winkler DA, Burden FR, Yan B, Weissleder R, Tassa C, Shaw S, Epa VC (2014) Modelling and predicting the biological effects of nanomaterials. SAR QSAR Environ Res 25(2):161–172
Worral EA, Hamid A, Mody KT, Mitter N, Pappu HR (2018) Nanotechnology for plant disease management. Agronomy 285(8):1–24
Xia T, Kovochich M, Zink JI, Nel AE (2008) Cationic polystyrene nanosphere toxicity depends on cell-specific endocytic and mitochondrial injury pathways. ASC Nano 2(1):85–96
Xiang Y, Zhang G, Chi Y, Cai D, Wu Z (2017) Fabrication of a controllable nanopesticide system with magnetic collectability. Chem Eng J 328:320–330
Xiong T, Dumat C, Dappe V, Vezin H, Schreck E, Shahid M, Pieral A, Sobanska S (2017) Copper oxide nanoparticle foliar uptake, phytotoxicity, and consequences for sustainable urban agriculture. Environ Sci Technol 51:5242–5251
Xu ZP, Niebert M, Porazik K, Walker TL, Cooper HM, Middelberg APJ, Gray PP, Bartlett PF, Lu GQM (2008) Subcellular compartment targeting of layered double hydroxide nanoparticles. J Control Release 130(1):86–94
Yadav KS, Jacob S, Sachdeva G, Sawant KK (2011) Intracellular delivery of etoposide loaded biodegradable nanoparticles: cytotoxicity and cellular uptake studies. J Nanosci Nanotechnol 11:6657–6667
Yan X, Sedykh A, Wang W, Zhao X, Yan B, Zhu H (2019) In silico profiling nanoparticles: predictive nanomodeling using universal nanodescriptors and various machine learning approaches. Nanoscale 11(17):8352–8362
Yanamala N, Kagan VE, Shvedova AA (2013) Molecular modeling in structural nanotoxicology: interactions of nano-particles with nano-machinery of cells. Adv Drug Deliv Rev 65(15):2070–2077
Yang K, Ma YQ (2010) Computer simulation of the translocation of nanoparticles with different shapes across a lipid bilayer. Nat Nanotechnol 5:579–583
Yang F, Liu C, Gao F, Su M, Wu X, Zheng L, Hong F, Yang P (2007) The improvement of spinach growth by nano-anatase $TiO_2$ treatment is related to nitrogen photoreduction. Biol Trace Elem Res 119:77–88
Youssef MS, Elamawi RM (2018) Evaluation of phytotoxicity, cytotoxicity, and genotoxicity of ZnO nanoparticles in *Vicia faba*. Environ Sci Pollut Res Int:1–13. https://doi.org/10.1007/s11356-018-3250-1

Yu YS, Lv M, Xiu P, Huynh T, Zhang M, Castelli M, Liu ZR, Huang Q, Fan CH, Fang HP, Zhou RH (2013) Destructive extraction of phospholipids from *Escherichia coli* membranes by graphene nanosheets. Nat Nanotechnol 8:594–601

Zahra Z, Ali A, Parveen A, Kim E, Khokhar MF, Baig S, Hina K, Choi H-K, Arshad M (2019) Exposure-response of wheat cultivars to $TiO_2$ nanoparticles in contrasted soils. Soil Sediment Contam 28(2):184–199

Zaytseva O, Neumann G (2016) Carbon nanomaterials: production, impact on plant development, agricultural and environmental applications. Chem Biol Technol Agric 3:17

Zeng X, Zhu L, Zheng X, Cecchini M, Huang X (2018) Harnessing complexity in molecular self-assembly using computer simulations. Phys Chem Chem Phys 20(10):6767–6776

Zhang H, Lu L, Zhao X, Zhao S, Gu X, Du W, Wei H, Ji R, Zhao L (2019) Metabolomics reveals the "invisible" responses of spinach plants exposed to $CeO_2$ nanoparticles. Environ Sci Technol 53(10):6007–6017

Zhao L, Peng B, Hernandez-Viezcas JA, Rico C, Sun Y, Peralta-Videa JR, Tang X, Niu G, Jin L, Varela-Ramirez A, Zhang J-y, Gardea-Torresdey JL (2012) Stress response and tolerance of *Zea mays* to $CeO_2$ nanoparticles: cross talk among $H_2O_2$, heat shock protein, and lipid peroxidation. ACS Nano 11(6):9615–9622

Zhao L, Ortiz C, Adeleye AS, Hu Q, Zhou H, Huang Y, Keller AA (2016a) Metabolomics to detect responses of lettuce (*Lactuca sativa*) to $Cu(OH)_2$ nanopesticides: oxidative stress response and detoxification mechanisms. Environ Sci Technol 20(17):9697–9707

Zhao L, Huang Y, Hu J, Zhou H, Adeleye AS, Keller AA (2016b) 1H NMR and GCMS based metabolomics reveal defense and detoxification mechanism of cucumber plant under nanoCu stress. Environ Sci Technol 50(4):2000–2010

Zhao X, Cui H, Wang Y, Sun G, Cui B, Zeng Z (2018a) Development strategies and prospect of nano-based smart pesticide formulation. J Agric Food Chem 66(26):6504–6512

Zhao L, Huang Y, Keller AA (2018b) Comparative metabolic response between cucumber (*Cucumis sativus*) and corn (*Zea mays*) to a $Cu(OH)_2$ nanopesticide. J Agric Food Chem 66 (26):6628–6636

Zheng L, Hong F, Lu S, Liu C (2005) Effect of nano-$TiO_2$ on strength of naturally aged seeds and growth of spinach. Biol Trace Elem Res 104:83–92

Zhu ZJ, Wang H, Yan B, Zheng H, Jiang Y, Miranda OR, Rotello VM, Xing B, Vachet RW (2012) Effect of surface charge on the uptake and distribution of gold nanoparticles in four plant species. Environ Sci Technol 46:12391–12998

Zuverza-Mena N, Martinez-Fernandez D, Du W, Hernandez-Viezcas JA, Bonilla-Bird N, Lopez-Moreno ML, Komárek M, Gardea-Torresdey JL (2017) Exposure of engineered nanomaterials to plants: insights into the physiological and biochemical responses-a review. Plant Physiol Biochem 110:236–264

# Methods for Understanding the Fate of Nanopesticides in Soil and Water

**Rocío López-Cabeza**

**Abstract** The interest in the application of nanotechnology in the formulation of plant protection products has increased in recent years. Currently, the methods used for understanding the behavior and fate of nanopesticides in soils are the same as those used for conventional formulations of pesticides. Nevertheless, the assessment of environmental risk of nanopesticides requires some modifications of the conventional methodologies to adapt them to specific properties of nanoformulations. Thus, the nano-character itself is reasonably expected to bring some novel features which modify the environmental fate and may influence the applicability of the conventional methodological concepts. This chapter reviews the most widely used methods to evaluate the different processes to which nanopesticides are subjected in the soil (sorption, persistence, and leaching). The advantages and disadvantages of each method are discussed and its applicability for nanoformulations of pesticides is assessed, focusing mainly on nanopesticides constituted by an active ingredient associated with a nanocarrier.

**Keywords** Nanopesticides · Nanoformulation · Soil · Fate · Durability · Release Sorption · Persistence · Leaching

## 1 Introduction

The definition of nanopesticide is usually used to any nanoformulation of pesticide that includes an active ingredient or some engineered structure that improves the pesticidal properties in the nanometer size range and/or a formulation that has novel

R. López-Cabeza (✉)
RECETOX, Faculty of Science, Masaryk University, Brno, Czech Republic
e-mail: rocio.lopez-cabeza@recetox.muni.cz



L. F. Fraceto et al. (eds.), *Nanopesticides*,
https://doi.org/10.1007/978-3-030-44873-8_5

properties associated with the small size of these components (Kah et al. 2013; Kookana et al. 2014). The development of these nanoformulations has received a lot of attention in recent years (Kah and Hofmann 2014). As in the case of a new active ingredient (AI) or conventional formulation, the registration and commercial production of nanopesticides would require the assessment of their efficacy, physicochemical properties, behavior, environmental fate, transformation, and toxicity by following standardized testing guideline by the Organization for Economic Co-operation and Development (OECD) or other standard protocols. Nevertheless, these guidelines are designed for conventional formulations of pesticides, thus it may not be appropriate for nanopesticides (Li et al. 2019a). Therefore, due to the structure and particular properties of nanoformulations, the applicability of the conventional methodological concepts is questioned and the procedures demand certain modifications and checks in order to be applied to nanopesticides. Firstly, the nano-character itself is reasonably expected to bring some novel features which modify the environmental fate including the behavior during the measurement protocol. In addition, most often, a nanopesticide is composed of a nanocarrier and a pesticide active ingredient. The properties of both, and their mutual interaction and association significantly influence the fate and behavior in soils. For example, in conventional methodologies, all studied processes are well derived from the measurements of the mass of active ingredient per volume of liquid or per mass of solid (AI concentrations). However, in the case of pesticide nanoformulations, other parameters have to be considered and become pivotal, such as particle number concentration, particle size distribution, and relation between the free AI and the AI associated with the nanocarrier (encapsulation efficiency), among others (Kookana et al. 2014). Thereby, the applicability of conventional protocols to nanopesticides must be confirmed before implementation (Kah et al. 2014). This applicability depends on the similarities between the nanoformulation and conventional formulation in relation to their behavior in soils. For this reason, before establishing the best methodologies to understand and predict the fate of nanopesticides in soils, it is necessary to know the role of the nanoparticles in the nanoformulation of pesticides and the expected behavior derived from this role. Three situations can be considered depending on the function performed by the nanoparticles:

(a) The nanoformulation function is limited to increase the apparent solubility/dispersion of the active ingredient or protect it from the degradation. This group includes the nano-emulsions and nano-dispersions. In this case, it is recommended that the product be treated as a conventional pesticide in the risk assessment process (Kah et al. 2013; Kookana et al. 2014). Therefore, the standard methodologies used for chemicals should be sufficient to evaluate the fate of these nanopesticides.
(b) The nanoformulation is composed of an active ingredient bound to a nanocarrier that maintains the structure of complex after spraying (and/or dilution) and the AI is released in a slow/targeted manner. In this case, the fate of the pesticide associated with the nanoformulation depends on the durability of the nanopesticide, i.e., the time during which the nanocarrier and the active ingredi-

ent are associated together. This group includes a wide variety of nanoformulations such as some polymer-based nanomaterials (nanocapsules, nanospheres, micelles, and nanogels), lipid-based nanomaterials (liposomes, solid lipid nanoparticles, and nanostructured lipid carriers), and clay-based nanomaterials (clays, layered double hydroxides) (Nuruzzaman et al. 2016).

(c) The nanoparticle has pesticidal properties. Silica based plant growth regulators, nanometals such as silver, copper or aluminum bactericides or fungicides, and oxides as $TiO_2$ are the major examples of this group (Kah and Hofmann 2014; Athanassiou et al. 2018).

While everything mentioned above is relevant for the measurement of the nanopesticide fate in the environment in general, there are also some specific problems related to their study in soils. The detection and characterization of nanoparticles in complex matrices such as soil is extremely challenging, for example, due to the presence of natural colloids that are almost indistinguishable from the nanoparticles under study (Hassellöv et al. 2008). Therefore, the endpoints of the methods that are available for the study of nanopesticide fate in soils still depend mostly on AI concentration (in soil, in solution, in biota, etc.). Probably, the only way forward to evaluate the specific processes with nano-character is not their direct measurements but the involvement of some additional steps that help to identify what really happens in the system. For example, by knowing the release kinetics outside the soil, one may estimate the durability of nanoformulations in soil solution or water from soil pores, and leachate might be subjected to the approaches separating bound and free pesticides. Also, the behavior of nanoformulations can be addressed indirectly from analysis of the coupled and measurable processes.

In this chapter, the most common methodologies for studying the behavior and fate of nanoformulated pesticides in soils found in the literature are described (Fig. 1 and Table 1), mainly focusing on slow/targeted release nanoformulations. In addition, the methodologies for assessing the fate of nanometals and nanometal oxides have also been briefly discussed since these engineered nanoparticles have very promising pesticidal properties.

## 2 Durability of Nanocarrier-AI Association

As mentioned in the previous section, in the case of controlled release nanopesticides, in order to set up the methodologies for evaluating the fate of the AI in soils is necessary to know the durability of the nanocarrier-AI complex after its application in the environment. Kah and Hofmann (2014) introduced the concept of durability based on the desorption kinetics of pollutants from soil colloids, which could be considered equivalent to the release of the AI from the nanocarrier. These authors defined three possible situations (Kah and Hofmann 2014):

- *Rapid release of the AI from the nanocarrier material (short durability).* If the complete release of the AI from the nanocarrier is much faster than the environ-

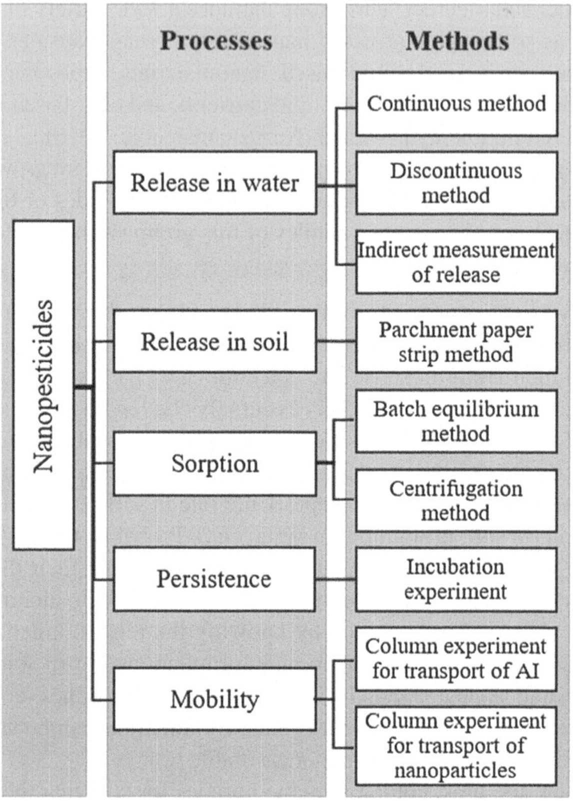


**Fig. 1** Summary of the methodologies described in this chapter

mental process of interest (degradation, leaching, runoff, etc.) the exposure of the AI is similar to that of conventional pesticides formulations. Therefore, the effect of the nanoformulation on the behavior of the AI is negligible and the environmental fate of nanopesticides is the same as the pure AI.

- *Intermediate release of the AI from the nanocarrier material.* Release kinetics of nanopesticide is required to assess the AI transfer from the nanocarrier to environment. In this case, the environmental processes studied will depend on the release rate, the properties of the released AI, and the properties of the AI in the nanocarrier-AI form (Kah and Hofmann 2014). Therefore, it is expected that the nanoformulation can influence the fate of AI in the soil and a more complex exposure assessment will be necessary (Kah et al. 2018).
- *Slow or no release of the AI from the nanocarrier material (long durability).* The fate of the nanopesticides will depend on the properties of the released AI and colloidal properties of the nanocarrier independently.

It is worth to mention that the durability of the nanocarrier-AI association can be strongly influenced by various factors such as AI concentration and/or properties of

**Table 1** Summary of the methodology applied to various nanoformulations of pesticides, nanometals and nanometal oxides reported in the literature

| Nanopesticide | Methods described in the reference | Reference |
|---|---|---|
| Chitosan- and iron(III)-modified smectites loaded with imazamox | – Release in water: sample-and-separate method<br>– Mathematic model of release: Korsmeyer–Peppas<br>– Mobility: column experiment for transport of AI | Cabrera et al. (2016) |
| Poly(epsilon-caprolactone) loaded with carbendazim and tebuconazole. Solid lipid nanoparticles loaded with carbendazim and tebuconazole | – Release in water: sample-and-separate method<br>– Mathematic model of release: Zero order, First order and Higuchi | Campos et al. (2015) |
| Mesoporous silica loaded with 2,4-dichlorophenoxy acetic acid | – Release in water: dialysis bag method<br>– Mobility: column experiment for transport of AI | Cao et al. (2018) |
| Chitosan/tripolyphosphate loaded with hexaconazole | – Release in water: parchment paper strip method<br>– Release in soil: parchment paper strip method<br>– Mathematic model of release: Korsmeyer–Peppas | Chauhan et al. (2017) |
| Rice husk biochar loaded with 2,4-dichlorophenoxyacetic acid | – Release in water: sample-and-separate method<br>– Release in soil: parchment paper strip method<br>– Mathematic model of release: Korsmeyer–Peppas<br>– Mobility: column experiment for transport of AI | Evy Alice Abigail (2019) |
| Lignin-polyethylene glycol coated with ethylcellulose loaded with imidacloprid | – Release in water: sample-and-separate method<br>– Mathematic model of release: Korsmeyer–Peppas | Flores-Céspedes et al. (2012) |
| Chitosan/tripolyphosphate loaded with paraquat | – Release in water: two-compartment method<br>– Mathematic model of release: Korsmeyer–Peppas<br>– Sorption: batch method<br>– Mathematic model of sorption kinetics: pseudo-first order and pseudo-second order | Grillo et al. (2014) |
| Poly(epsilon-caprolactone) loaded with atrazine | – Sorption: batch and centrifugation methods<br>– Persistence: soil incubation experiment | Kah et al. (2014) |

(continued)

**Table 1** (continued)

| Nanopesticide | Methods described in the reference | Reference |
|---|---|---|
| Three polymer-based nanoformulations of bifenthrin | – Release: indirect method<br>– Sorption: batch method<br>– Persistence: soil incubation experiment | Kah et al. (2016) |
| Three polymer-based nanoformulations of clothianidin | – Sorption: batch and centrifugation methods | Kah et al. (2018) |
| Poly(vinylpyrrolidone)-coated silver nanoparticles | – Release in water: dialysis bag method<br>– Mathematic model of release: modified first order | Kittler et al. (2010) |
| Alginate/chitosan loaded with acetamiprid | – Release in water: parchment paper strip method<br>– Release in soil: parchment paper strip method<br>– Mathematic model of release: Korsmeyer–Peppas | Kumar et al. (2015) |
| Citrate-stabilized silver nanoparticles | – Release in water: sample-and-separate method | Liu and Hurt (2010) |
| Carboxymethyl-β-cyclodextrin-$Fe_3O_4$ magnetic nanoparticle loaded with diuron | – Release in water: sample-and-separate method | Liu et al. (2014) |
| Silver nanoparticles | – Mobility: column experiment for transport of nanoparticles | Mahdi et al. (2018) |
| Poly(citric acid)/poly(ethylene glycol)/poly(citric acid) loaded with imidacloprid | – Release in water: dialysis bag method | Memarizadeh et al. (2014) |
| $TiO_2$ and ZnO nanoparticles | – Release in water: sample-and-separate method | Miller et al. (2010) |
| CuO nanoparticles | – Release in water: dialysis bag method<br>– Mathematic model of release: modified first order | Misra et al. (2012) |
| Chitosan coated beeswax solid lipid nanoparticles loaded with deltamethrin | – Release in water: sample-and-separate method | Nguyen et al. (2012) |
| ZnO nanoparticles | – Release in water: sample-and-separate method<br>– Mathematic model of release: First order | Peng et al. (2011) |
| Poly(epsilon-caprolactone) loaded with atrazine | – Release in water: two-compartment method<br>– Mathematic model of release: Korsmeyer–Peppas<br>– Mobility: column experiment for transport of AI | Pereira et al. (2014) |

(continued)

**Table 1** (continued)

| Nanopesticide | Methods described in the reference | Reference |
|---|---|---|
| Poly(methacrylic acid-ran-butylmethacrylate) loaded with bifenthrin | – Mobility: column experiment for transport of nanoparticles | Petosa et al. (2017) |
| Nano-size calcium carbonate loaded with validamycin | – Release in water: sample-and-separate method | Qian et al. (2011) |
| Porous hollow silica nanospheres loaded with tebuconaloze | – Release in water: sample-and-separate method<br>– Mathematic model of release: Korsmeyer–Peppas | Qian et al. (2013) |
| Pectin/chitosan/tripolyphosphate loaded with paraquat | – Release in water: two-compartment method<br>– Mathematic model of release: Korsmeyer–Peppas<br>– Sorption: batch method<br>– Mathematic model of sorption kinetics: pseudo-first order and pseudo-second order<br>– Mobility: column experiment for transport of AI | Rashidipour et al. (2019) |
| Silver nanoparticles | – Mobility: column experiment for transport of AI | Sagee et al. (2012) |
| Chitosan/pectin loaded with carbendazim | – Release in water: dialysis bag method | Sandhya et al. (2017) |
| Poly(ethylene glycol)/aliphatic and aromatic diacids loaded with thiamethoxam | – Release in soil: parchment paper strip method<br>– Mathematic model of release: Baker–Lonsdale, Hixson–Crowell, Higuchi, First order and Korsmeyer–Peppas | Sarkar et al. (2012) |
| Alginate/chitosan loaded with paraquat | – Release in water: two-compartment method<br>– Mathematic model of release: Korsmeyer–Peppas<br>– Sorption: batch method<br>– Mathematic model of sorption kinetics: pseudo-first order, pseudo-second order and intraparticle diffusion | dos Santos Silva et al. (2011) |
| ZnO nanoparticles | – Mobility: column experiment for transport of AI | Sun et al. (2015) |
| mPEG-PLGA loaded with metolachlor | – Release in water: dialysis bag method<br>– Mathematic model of release: Korsmeyer–Peppas | Tong et al. (2017) |
| Silver nanoparticles | – Sorption: batch method | Torrent et al. (2019) |

(continued)

**Table 1** (continued)

| Nanopesticide | Methods described in the reference | Reference |
|---|---|---|
| Ag and $CeO_2$ nanoparticles | – Sorption: batch method | Van Koetsem et al. (2018) |
| Silver nanoparticles | – Sorption: batch method<br>– Mathematic model of equilibrium isotherm: Langmuir and Freundlich | Wang et al. (2018) |
| Poly(lactic acid) loaded with abamectin | – Release in water: dialysis bag method<br>– Mathematic model of release: First order | Yu et al. (2017) |

the surrounding environment, including temperature, polarity, ionic strength, etc. The dilution of nanopesticides below the solubility of their particular AIs could be expected to result in rapid release in the case of most nanoformulations, since the principle of the nanocarrier-AI association often consists in a hydrophobic interaction that is relatively weak.

Currently, no standard protocols to measure the durability of nanopesticides have been proposed by regulatory agencies. The most common approach reported in the literature consists in measuring the relative rate of the AI release in water under infinite sink conditions.

## *2.1 Release Experiments in Water*

The most commonly used methods for measuring the release rate of nanopesticides in water can be grouped in two categories: continuous methods (membrane isolation methods) and discontinuous methods (sampling-and-separate methods).

### 2.1.1 Continuous Methods: Dialysis Methods

In dialysis methods, the nanoparticles loaded with the AI are contained in a compartment (donor compartment) that is physically separated from compartment with the release medium (acceptor compartment) by a semipermeable membrane. Pesticides non-bound to the nanoparticles cross through the membrane to the acceptor compartment, while the pesticide associated with the nanoparticles is unable to penetrate the membrane. In this method, it is assumed that the diffusion rate through the membrane and within the acceptor compartment is very high compared to the release kinetics and, therefore, does not limit the release process. The most popular dialysis system for the determination of release kinetics of nanopesticides is based on the addition of the nanoformulation into a bag made of cellulose semipermeable membrane. Then, the bag is sealed thoroughly and immersed into a vessel containing the release medium in sink conditions. The release medium is selected following several criteria such as the AI solubility (Sandhya et al. 2017), or the stability of the nanoformulation in different media (Tong et al. 2017), at different pH (Memarizadeh et al. 2014; Sandhya et al. 2017; Cao et al. 2018), or at different ionic strength values (Cao et al. 2018). Volume of release medium must be six to tenfold greater than that is inside the dialysis bag, which provides the driving force for the transport of the AI to the outside and also allows maintaining the sink conditions (D'Souza and DeLuca 2006). Pesticide diffusion from the dialysis bag to the vessel can be favored by agitation, since unstirred water layer effect is minimized (D'Souza and DeLuca 2006). Periodically, samples are taken from release medium and analyzed in order to determine the total amount of pesticide released over time. After each sampling, the volume of release medium taken is replaced with an equivalent amount of fresh dissolution medium to ensure a constant total solution volume (Cao et al. 2018). An alternative setup is a two-compartment models in which a glass vessel that contains the nanopesticide is covered with a dialyzing membrane and introduced in a chamber containing the release medium (dos Santos Silva et al. 2011; Grillo et al. 2014; Pereira et al. 2014). The dialysis method has also been used to study the release (dissolution) of metal ions from metal nanoparticles (Kittler et al. 2010; Misra et al. 2012). However, these studies are scarcely found in the literature.

The main drawback of dialysis is the possible sorption of the AI on the membrane. Nevertheless, when the sorption is low, this problems could be solved by performing dialysis experiment with a control unformulated AI together with the nanoformulation under study. Sometimes, the sorption of the AI in the membrane is very high due to the low water solubility of the AI, and the comparison is not possible. In this case, the addition of a co-solvent that increases the solubility of the AI in the release medium are recommended. In the extreme event that the affinity between the AI and the membrane is so high that the use of a co-solvent cannot avoid the high sorption, the use of dialysis method must be discarded (D'Souza and DeLuca 2006).

### 2.1.2 Discontinuous Methods: Sample-and-Separate Methods

This method consists in introducing the nanopesticide into a vessel containing the release medium and release is evaluated over time. Unlike dialysis methods, in samples-and-separate methods, nanocarrier-AI complex is in direct contact with the bulk medium. The medium is mainly selected according to the solubility of the AI (Qian et al. 2011, 2013), and the pH values of this medium tested are chosen for the purpose to evaluate the stability of the nanoformulation at different pH (Qian et al. 2011). The volume of release medium should be sufficient to maintain sink conditions without compromising the sensitivity of the assay studied (D'Souza and DeLuca 2006). After the incorporation of the nanoparticles to the release medium, the system must be subjected to continuous or intermittent agitation during the experiment (Qian et al. 2013; Cabrera et al. 2016). Periodically, samples are taken from the bulk medium and the nanoparticles are separated by filtration (Nguyen et al. 2012; Liu et al. 2014) or centrifugation (Qian et al. 2013) and then analyzed. After every sampling, the release medium removed must be replaced with an equal volume of fresh release medium than that withdrawn to maintain sink conditions during all the experiment (Flores-Céspedes et al. 2012). An alternative setup of this method was carried out by Campos et al. (2015), in which, after the addition of a nanoformulation suspension sample to the release medium (water), aliquots of this mixture were added to Falcon tubes which were closed and agitated at a room temperature. At certain intervals of time, one of the tubes was removed from the shaker and centrifuged. The supernatant was filtered and measured to determine the pesticide release at each time.

This procedure is also used to assess the dilution rates (release rate) of metal ions from nanoparticles of the corresponding metal such as, for example, Ag (Liu and Hurt 2010) or ZnO nanoparticles (Miller et al. 2010; Peng et al. 2011). The ion release is determined by dilution of the nanometal stock solution in the release medium to desired concentration. Then, at selected times, aliquots of the suspension are withdrawn and the supernatant containing the dissolved metal ions are separated from the nanoparticles by ultrafiltration or centrifugation.

The main advantage of sample-and-separate method is obtaining the amount of AI released from the nanocarriers directly. However, this method has a number of disadvantages such as the possible overestimation of the AI released because of the forces of filtration and/or centrifugation that could compress or crash the nanoparticles in the case of AI-nanocarrier complexes. Another limitation of this method is the inability to accurately quantify the amount of AI released in real-time due to time delay from sampling to analysis (Zhou et al. 2016). Therefore, comparing the two methods, the dialysis would be the most suitable method to study the AI release from nanopesticides in which the AI is associated with a nanocarrier, as long as no AI retention in the membrane occurs. On the other hand, in the case of nanometals, discontinuous method has been shown to be appropriate.

It is worth noting that, although release experiments in water has been employed widely for assessing the nanopesticide durability, they are usually performed in unrealistically conditions of high concentration of AI and in ionized water. Thus, under these conditions, the results obtained are not very representative of the real conditions of pH, ionic strength, and dilution factor to which the nanoformulation is subjected when it is diluted in the tank before its application in the field (Kah et al. 2018).

### 2.2 Release Experiment in Soil

Measuring the release rate of the AI from the nanocarrier in soils is critical, since it allows estimating the durability of nanopesticides after application in the field. However, the design of this experiment is a challenge due to the difficulty of obtaining measurements at realistic soil-solution ratios (Kah et al. 2018). In several works, a similar procedure to evaluate the release of the AI in soils has been proposed

(Sarkar et al. 2012; Kumar et al. 2015; Chauhan et al. 2017). The nanopesticide tested is wrapped in parchment papers and placed inside the soil samples (25–50 g) contained in beakers. Then, water is added to the soil to bring around 60% water holding capacity. The beakers are covered with parafilm that is drilled to allow air exchange, and incubated at 30 °C in biochemical oxygen demand incubator. Periodically, beakers are taken out from the incubator, and the parchment paper strips are removed from the soil. The soil samples are extracted with an appropriate procedure using organic solvents (depending on the AI) and the AI is analyzed.

## *2.3 Mathematical Models for Nanopesticide Release*

Both in water and in soils, the data from release experiment can be fitting with several mathematical models including zero order (Campos et al. 2015), first order (Sarkar et al. 2012; Campos et al. 2015; Yu et al. 2017; Li et al. 2019b), Higuchi (Sarkar et al. 2012; Campos et al. 2015) or Korsmeyer–Peppas (Qian et al. 2011; dos Santos Silva et al. 2011; Flores-Céspedes et al. 2012; Sarkar et al. 2012; Pereira et al. 2014; Kumar et al. 2015; Cabrera et al. 2016; Chauhan et al. 2017; Tong et al. 2017; Evy Alice Abigail 2019; Rashidipour et al. 2019). The latter is the most used and allows elucidating the type of release mechanism. The Korsmeyer–Peppas model is described by the equation (Korsmeyer et al. 1983):

$$\frac{M_t}{M_\infty} = k \cdot t^n \tag{1}$$

where $M_t/M_\infty$ is the fraction of compound released in time $t$, $k$ is the characteristic kinetic constant of the Nanocarrier-AI system, and $n$ is the release exponent, which indicates the type of release mechanism.

This model is appropriate to describe the release of compounds from nanocarriers when the preponderant mechanism is not well known, or when two mechanisms apparently independent are involved: diffusion of the AI through the polymer and transport controlled by swelling-relaxation of the polymeric chains. In this last mechanism, the dissolution medium penetrates the matrix causing the swelling of the polymer that adopts a rubbery state that allows the AI contained in it to diffuse outwards (Langer and Peppas 1981). Ritger and Peppas 1987 proposed that the value of n characterizes the mechanism of release, establishing three different situations in the case of spherical polymer particles:

- *$n < 0.43$ (case I transport)* indicates that the process is mainly regulated by diffusion and, thus, release mechanism follows Fick's Laws.
- *$n > 0.85$ (case II transport)* implies that the mechanism is regulated by swelling and relaxation processes of the polymer.
- *$0.43 < n < 0.85$ (intermediate values)* suggests anomalous behavior with non-Fickian release kinetic in which a combination of diffusion and relaxation of the polymeric chain occurs.

## 2.4 Indirect Estimation of Durability

Sometimes, the direct determination of the durability is not possible due to the release experiments could present some artifact such us the sorption of the AI to the membranes. In these cases, an indirect estimation of durability of nanopesticides based on other soil processes such as persistence or leaching could be feasible. Kah et al. (2016) proposed an indirect approach to estimate the durability of several nanoformulations of bifenthrin-loaded to polymer nanocarriers based on the degradation kinetics of the free AI and the AI associated with the nanocarriers. The assumptions behind this approach are that (1) only the portion of AI released is available to be degraded, (2) at time zero, all the insecticide is associated with the nanocarrier, and (3) both release and degradation process can be described by first-order kinetics. The authors proposed that the concentrations of formulated (Nanocarrier-AI), released (AI), and degraded bifenthrin (AI') could be described by a sequential first-order model:

$$\text{Nanocarier-AI} \xrightarrow{k_1} \text{AI} \xrightarrow{k_2} \text{AI}'$$

where Nanocarrier-AI is the nanoformulation, AI is the active ingredient released from the nanocarrier, AI' is the active ingredient degraded, $k_1$ is the release rate of AI from nanocarrier, and $k_2$ is the degradation rate of the released AI. Then, the results obtained in the degradation experiment of the nanoformulations and the pure AI were fitted to sequential first-order model and the value of $k_1$ and $k_2$ were obtained. Thus, $k_1$ was calculated from the degradation experiment of the nanoformulation and $k_2$ was determined from the degradation curve of the pure bifenthrin. Finally, the release-half-lives ($R_{50}$) were determined from $k_1$ as the time necessary for half the bifentrin to be released from the nanocarrier.

# 3 Methods for Measuring Sorption of Active Ingredients in Soils

Two methods have been proposed in the literature to assess the sorption of AIs from nanopesticides in soils: (1) batch equilibrium method and (2) centrifugation method.

## 3.1 Batch Equilibrium Method

The batch equilibrium approach (standardized by OECD guideline 106; OECD 2000) is the most widely used method for evaluating the sorption of pesticides (and other chemicals) in soils. Logically, it is frequently used also in the studies that determine the effect of the nanoformulation on the sorption of AIs in soils (dos Santos Silva et al. 2011; Grillo et al. 2014; Kah et al. 2014, 2016, 2018; Rashidipour et al. 2019). According to the OECD guidelines (OECD 2000), soil samples are shaken with 0.01 M $CaCl_2$ solution which is spiked with the pesticide studied (or nanopesticide) until equilibrium of the distribution of the pesticide between soil and solution. Both the soil/solution ratio and the time to achieve equilibrium are determined in preliminary experiments (OECD 2000). The aqueous phase is separated by centrifugation and analyzed to obtain the pesticide concentration. The sorbed mass of the pesticide at equilibrium is calculated indirectly from the difference between the concentration of the initial solution and the solution in equilibrium with the soil. It is also possible to determine the amount of the pesticide sorbed on the soil directly by extracting with an organic solvent (OECD 2000; Kah et al. 2016). Figure 2 depicts schematically how the batch equilibrium method is performed.

The benefit of the OECD approach is that, if wished, the supernatants could be filtered in order to differentiate the fate of the AI loaded onto the nanocarriers from those released within the time frame of the experiment (Kah et al. 2014). On the other hand, it has some limitations because it is designed for conventional formulations of pesticides. For instance, in this method it is necessary to use large volume of solution and a vigorous shaking that could alter the structure of the nanopesticide

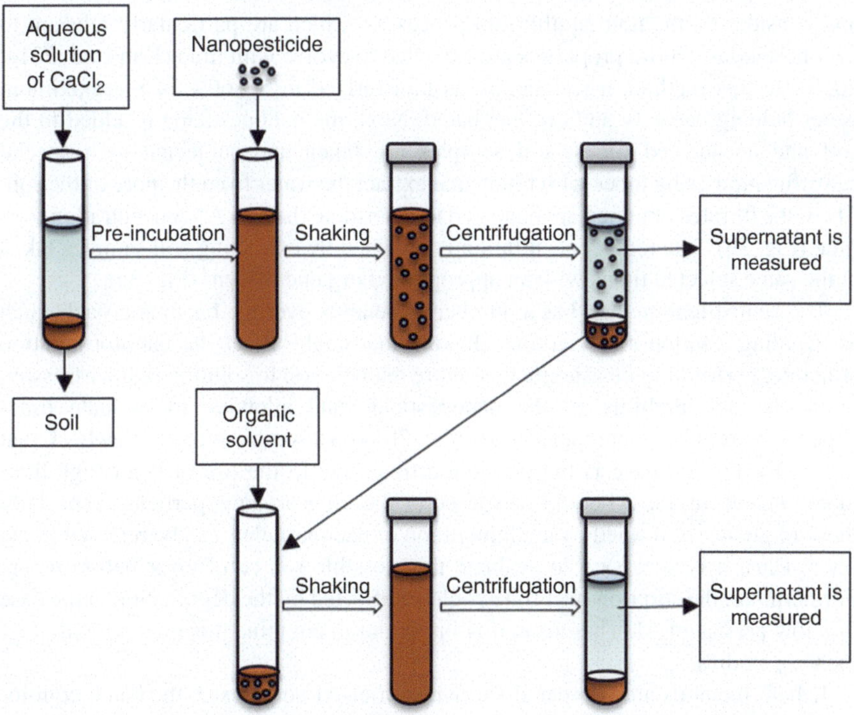


**Fig. 2** Schematic diagram of batch equilibrium method

and the interaction between the AI and the nanocarrier, accelerating the release (Kah et al. 2014).

The batch sorption experiment is the most widely method used in the literature to evaluate the retention of metal ions and metal nanoparticles in soils (Van Koetsem et al. 2018; Wang et al. 2018; Torrent et al. 2019). However, due to their colloidal properties, this method would not be recommended to assess the sorption of nanopesticides whose nanoparticles have pesticidal properties (as the aforementioned metals) in soils. This is because this type of nanopesticides behaves as the similar way that the soil colloids, since they have colloidal characteristics such as aggregation, settling, and interaction with surface and matrix effect, among others. Therefore, the sorption of theses nanopesticides in soils is a dynamic process and a phase partitioning between environmental matrices cannot be assumed under these non-equilibrium conditions (Westerhoff and Nowack 2013; Kookana et al. 2014).

## 3.2 Centrifugation Method

The centrifugal method has been proposed as an alternative to batch equilibrium method in order to determine sorption of pesticides at a realistic soil-solution ratio and considering the non-equilibrium processes, which are particularly relevant to nanopesticides, whose properties are expected to evolve with time (Kah et al. 2014, 2018). In this method, a soil sample is moisturized to 50–60% of the maximum water holding capacity and pre-incubated. Next, the nanopesticide is added to the soil and at selected times, soil samples are taken and subjected to a special centrifugation using tubes with filters that extract the water from the pore of the soil. Then, the filtrates obtained are analyzed to determine the free AI concentration over time (Fig. 3b). The total concentration is obtained by extracting soil samples taken at the same selected times with an appropriated organic solvent (Fig. 3a).

The centrifugal method has a number of benefits over the batch approach, such as avoiding dilution and vigorous shaking that could affect the nanoformulation structure, performing measurement at more realistic soil to solution radio, assessing indirectly the durability of the nanopesticide and allowing to evaluate time-dependent sorption phenomena (Kah et al. 2014). However, a major drawback that this method could have is the possible artifact due to the use of centrifuge filter tubes. Therefore, the AI could be sorbed of the filter or some particles, especially those of great size formed by agglomeration of nanoparticles, can be retained in the filter. Thus, it is necessary to evaluate this possible source of error before using. Furthermore, the sorption over time could be affected by the degradation in the case of a low persistent AI. Therefore, it is important to combine this method with degradation studies.

If both methods are compared for nanocarrier-AI complexes, the batch equilibrium method could be considered the most appropriate one for regulatory assessment of AIs because this method is more consistent and reproducible in comparison with the centrifugal method. Nevertheless, due to the particular properties of these

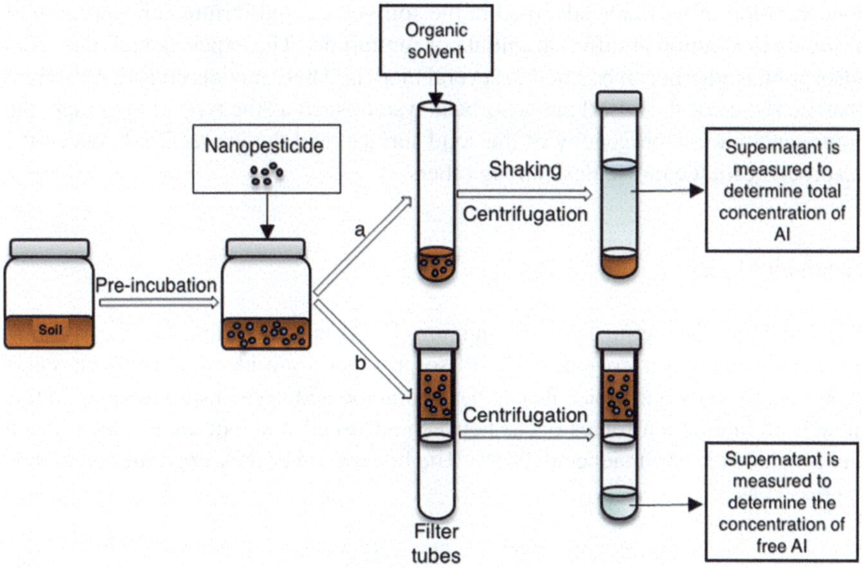


**Fig. 3** Schematic illustration of incubation experiment where total concentration of AI (**a**) and concentration of free AI (**b**) are determined over the time

nanopesticides, the use of centrifugal technique can provide more accurate results and more information about the durability of nanopesticides, and thus, about the time during which the nanoformulation can influence on the active ingredient behavior. In the case of nanometals, although batch equilibrium method is frequently used, this approach would not be adequate, since the retention of nanoparticles in the soil is determined by non-equilibrium processes due to their colloidal characteristics.

## 3.3 *Mathematical Models Used in Sorption Experiments*

Several mathematical models of the equilibrium isotherms and sorption kinetics have been proposed to study the sorption of pesticides in soils. These same models could be also used to assess the sorption of AIs from nanopesticides in soil. The most common models are discussed below.

### 3.3.1 Mathematical Models of Equilibrium Isotherms

Adsorption isotherms are used to describe the equilibrium when the relation between the amount of pesticide adsorbed and the amount of pesticide in equilibrium is not linear. In this case, the adsorption equilibrium is evaluated at isothermal temperature and different pesticide concentrations. The isotherm is obtained by plotting concentration of pesticide adsorbed in the soil versus equilibrium concentration of pesticide in solution at different initial concentrations. The experimental data from adsorption isotherm can be fitted to several models. These models consider different characteristics of the adsorbate-adsorbent system such as the type of coverage, the homogeneity or heterogeneity of the solid surface, and the interaction between the adsorbed pesticide molecules, among others.

### Langmuir Model

The Langmuir isotherm theory (Langmuir 1918) considers that (1) the sorption involves a monolayer coverage, (2) all sorption sites are identical (uniform energies), and (3) only one molecule can be accommodated in each sorption site. In this model, no interaction forces occur between adsorbed pesticide molecules (Ghosh and Singh 2012; Al-Smadi et al. 2019). The linear form of the Langmuir equation is expressed as:

$$\frac{C_e}{q_e} = \frac{1}{b \cdot Q_0} + \frac{C_e}{Q_0} \tag{2}$$

where $C_e$ is the equilibrium concentration, $q_e$ is the amount of AI adsorbed per unit mass of adsorbent at equilibrium, $Q_0$ is the theoretical monolayer capacity, and $b$ is the sorption equilibrium constant related to the energy of sorption. From the plot of $C_e/q_e$ versus $C_e$, the values of $Q_0$ and $b$ can be calculated.

### Freundlich Model

The Freundlich model is an empirical model used for heterogeneous systems. In the model, infinity surface coverage is assumed and an extremely strong interaction between adsorbed molecules occurs. Thus, the greater the adsorbate concentration in solution, the greater concentration of adsorbate on the surface of the adsorbent (Ghosh and Singh 2012), and therefore, the sorption can be described by an exponential equation:

$$q_e = K_f \times C_e^{\frac{1}{n}} \tag{3}$$

where $q_e$ is the amount of AI adsorbed per mass unity of adsorbent, $C_e$ is the equilibrium concentration of pesticide, $K_f$ and $n$ are the Freundlich constants representing the adsorption capacity and the adsorption intensity, respectively.

The lineal form of Freundlich expression is

$$\log q_e = \log K_f + n \cdot \log C_e. \tag{4}$$

The values of $K_f$ and $n$ can be determined by plotting log $q_e$ versus log $C_e$.

### 3.3.2 Mathematical Models of Sorption Kinetics

The main sorption kinetics models found in the literature for nanopesticides are pseudo-first order, pseudo-second order, and intraparticle diffusion (dos Santos Silva et al. 2011; Grillo et al. 2014; Rashidipour et al. 2019).

Lagergren Pseudo-First-Order Model

Pseudo-first-order model is usually used to describe reversible reactions in which an equilibrium is established between liquid and solid phases (Al-Smadi et al. 2019). The linearized form of this model is as follows:

$$\log(q_e - q_t) = \log q_e - \frac{k_1}{2.303} \cdot t \tag{5}$$

where $q_e$ is the amount of pesticide adsorbed per unit mass of sorbent at equilibrium, $q_t$ is the amount of pesticide adsorbed per unit mass of sorbent at time $t$, and $k_1$ is the pseudo-first-order adsorption constant.

Pseudo-Second-Order Model

In the pseudo-second-order model, the rate-limiting factor is the chemical adsorption, i.e., the interaction of the pesticide molecules with the adsorption sites by chemical bonding (Al-Smadi et al. 2019). This model considers that the adsorption capacity is directly proportional to the number of available active soil sites (Cáceres et al. 2010). The lineal form of the pseudo-second-order equation is

$$\frac{\mathrm{t}}{q_t} = \frac{1}{k_2 \cdot q_e^2} + \frac{1}{q_e} \cdot t \tag{6}$$

where $q_e$ is the amount of pesticide adsorbed per unit mass of sorbent at equilibrium, $q_t$ is the amount of pesticide adsorbed per unit mass of sorbent at time $t$, and $k_2$ is the pseudo-second-order adsorption constant.

Weber and Morris's Intraparticle Diffusion Model

This model describes the sorption processes that are diffusion-controlled, i.e., the sorption rate depends on the velocity at which sorbate (pesticide) diffuses into the sorbent (soil surface) and in the sorbate solution (Cáceres et al. 2010). The equation that describes this model is

$$q_t = k_{\mathrm{int}} \cdot t^{1/2} + C \tag{7}$$

where $k_{int}$ is the intraparticle diffusion rate constant, $C$ is a constant that provides an indication of boundary layer effect, and $q_t$ is the amount of pesticide adsorbed per unit mass of sorbent at time $t$.

## 4 Method for Assessing the Persistence of Active Ingredients in Soils

Incubation method according to OECD guideline 307 (OECD 2002) is recommended to evaluate the persistence of an AI associated with a nanocarrier in soils (Kah et al. 2014, 2016; Kookana et al. 2014). Fresh soil is pre-incubated and then, the nanoformulation is added to it. Periodically, soil aliquots are sampled, extracted with organic solvent, and analyzed to determine the concentration of AI that remains in the soil over the time (Fig. 3a). The protocol followed for nanopesticides is the same than those used for conventional pesticides (or other compound). It is recommended to carry out the incubation experiment with pure AI in parallel with the nanoformulation of AI, since the differences observed between both experiments could reveal essential information about the effect of the nanocarrier-AI association on the persistence and release of AIs (Kookana et al. 2014). For instance, if the degradation rate of the AI is faster in the pure AI experiment than in the nanoformulation experiment, it could mean that the release of the AI from the nanocarrier is very slow and the amount of AI bioavailable to be degraded is very low. On the other hand, similar degradation rates for the pure AI and the nanopesticide would indicate that the release of the AI occurs rapidly in comparison with degradation kinetics. It must be kept in mind that this approach requires assuming that only the AI released from the nanopesticide is bioavailable to be degraded, while the AI associated with the nanocarrier is protected from degradation. Nevertheless, this assumption does not always have to be true, and thus more research on bioavailability are necessary to establish a realistic estimation of release based on the differences between degradation kinetics of pure AIs and nanopesticides. Thus, for example, in the case described above, the reason for a similar degradation might not be a rapid release but a degradation of the AI associated with the nanocarrier due to abiotic or biotic degradation (Kah et al. 2014).

## 5 Methods for Evaluating the Mobility of Nanopesticide in Soils

The use of column experiments has been suggested to study the mobility of nanopesticides in soils, both of the AI and of nanoparticles. These experiments allow assessing the leaching rate of nanopesticides, as well as its horizontal distribution in the soils. A schematic illustration of the column experiments described in this section is shown in Fig. 4.

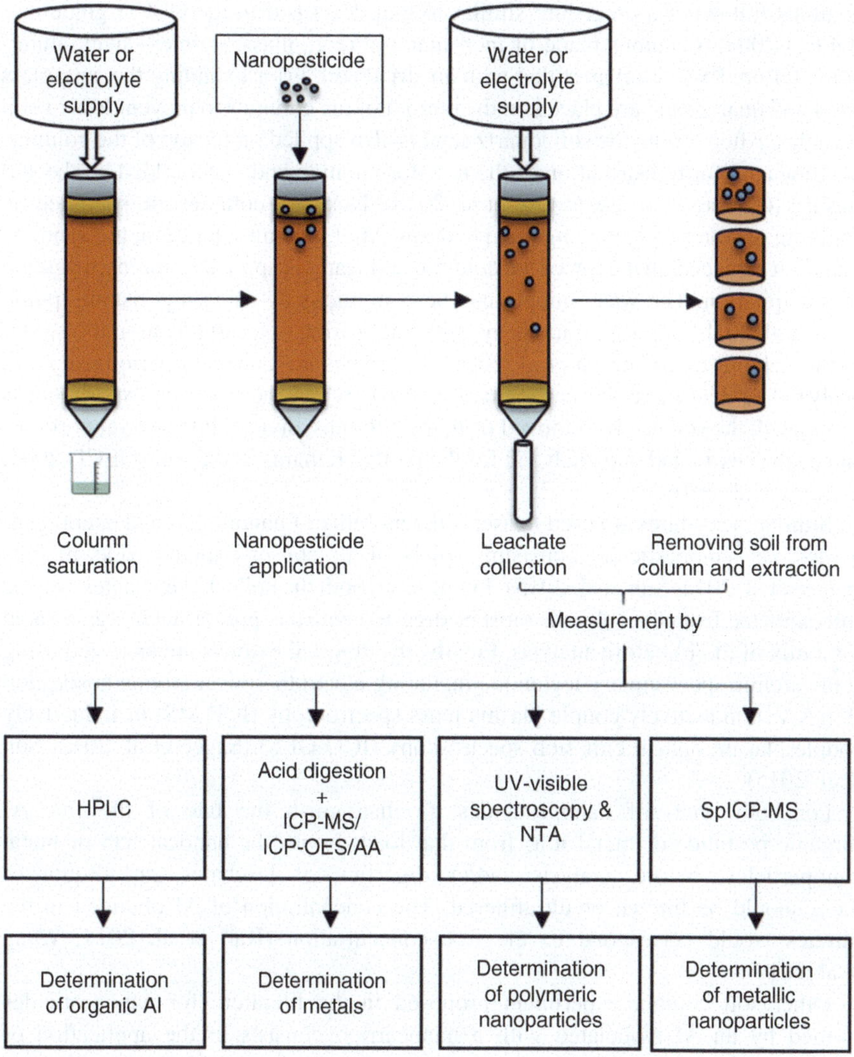


**Fig. 4** Schematic illustration of the column experiments described in the Sect. 5

## 5.1 *Column Experiments for Evaluating the Mobility of Active Ingredients in Soils*

Column experiments used in the studies of mobility in soils for conventional pesticides have also been recommended to study the AI transport of nanopesticides in soils. Different designs of column experiments for nanopesticides are found in the literature. One of these designs is the leaching experiment of AIs through soil columns following a procedure similar to that described in the OECD guidelines (OECD 2004). Columns made of inert material (e.g., glass, stainless steel, aluminum, Teflon, PVC, etc.) are filled with air-dried soil. Prior to adding the soil, glass wool and quartz sand are placed on the bottom of the columns to prevent loss of soil particles. After adding the soil, quartz sand is also applied on the top of the columns to allow a uniform distribution of the nanoformulation and water added on the soil surface (Cabrera et al. 2016; Cao et al. 2018). Next, the columns are saturated by applying of water and allowing them to drain. After column saturation, the nanoformulation is placed on the top of the column, and water is applied imitating irrigation or precipitation. The water supply can be continuous by using a peristaltic pump (Cao et al. 2018) or discontinuous by daily application of a certain amount of water to the column top (Cabrera et al. 2016). Leachates are collected periodically and analyzed to determine the concentration of AI. When the leaching experiment is completed, the soil can be removed from the columns, divided into several portions, and each is extracted and analyzed for the AI that remains in the soil (OECD 2004; Cabrera et al. 2016).

Similar procedures are used to assess the mobility of nanometals and nanometal-oxides, i.e. nanopesticides consisting solely of a metal or a metal oxide, in soils (Sagee et al. 2012; Sun et al. 2015). In this case, both the collected leachates and the soil extracted from the columns must be digested with an appropriate inorganic acid (or a mix of them) before analysis. Finally, the digested extracts are analyzed using some atomic spectrometry technique including electrothermical atomic absorption (ET-AAS), inductively couple plasma mass spectroscopy (ICP-MS) or inductively couple plasma optical emission spectroscopy (ICO-OES) (Sagee et al. 2012; Sun et al. 2015).

For both types of nanopesticides, to distinguish the fate of the free AI (organic pesticide or metal ion) from that loaded onto the nanocarriers or metal nanoparticles, before analysis and/or digestion of leachates, an aliquot of them should be filtered or ultrafiltered. The concentration of AI obtained in the filtrates would correspond to the free concentration (Kah et al. 2014; Wang et al. 2014).

Other soil column experiment proposed in the literature for nanopesticides formed by an AI associated with a nanocarrier, consists in the application of nanopesticides to soil columns constructed by joining several PVC rings to assess the distribution at different soil depths (Pereira et al. 2014; Evy Alice Abigail 2019; Rashidipour et al. 2019). After the nanopesticide addition, precipitation/irrigation is simulated by addition of water at desired time intervals. Next, the column is split into individual rings and the AI concentration is quantified in each ring separately. This method does not distinguish between the free AI and the AI associated with the nanocarrier.

## 5.2 *Column Experiments for Evaluating the Mobility of Nanoparticles in Soils*

Detection of nanopesticide in "particle form" in soils and leachates is extremely complex due in part to the presence of natural colloids that are almost identical to the nanoparticles to be quantified (Hassellöv et al. 2008). Therefore, studies on the mobility of nanoparticles in the soil, especially for those non-metallic nanocomponents (e.g., polymers), are very scarce in the literature. Petosa et al. (2013) conducted a leaching experiment using polymeric nanoparticles. In this work, the authors proposed a method to study the vertical transport of engineered nanoparticles (polyacrylic acid-coated cerium dioxide and polyacrylic acid nanocapsule) using glass columns filled with quartz sand (artificial soil) and loamy sand (natural soil). Firstly, both quartz sand and loamy sand had to be equilibrated. The quartz sand was preconditioned by soaking in electrolyte solution and then, it was wet-packed into the glass columns. In the case of loamy sand columns, the glass columns were filled with the dried-soil and next, the soil columns were conditioned with $CaCl_2$ solution to stabilize the soil colloids, followed by an electrolyte solution. After equilibrating the columns, at least three pore volumes of engineered nanoparticle suspensions were applied to them, followed by particle-free electrolyte solution. Then, influent and effluent particle concentrations were monitored using a UV-visible spectrophotometer. Petosa et al. (2017) repeated this procedure for evaluating the mobility of four polymeric nanocapsules, a polymeric nanocapsule loaded with bifenthrin and a commercial formulation of this insecticide (Capture® LFR). In this study, all the hollow nanocarriers, the bifenthrin nanoformulation and the commercial formulation were monitored by a UV-visible spectrophotometer. Additionally, nanopesticide elution was also monitored by nanoparticle tracking analysis. This study was performed at different pH values and in presence and absence of fertilizer. The different scenarios shown in these experiments allowed the authors to obtain an insight of the interaction between the polymer delivery system and the model soil used, as well as elucidated the mechanism governing the nanoparticle transport.

Assessment of the mobility of nanometals in their nanoparticle forms using columns is also possible. Mahdi et al. (2018) evaluated the transport of Ag-nanoparticles (AgNPs) using polyethylene hydraulic soil columns filled with three natural soils. The columns were filled three quarter and saturated with water and left equilibrate for 24 h. Then, the Ag-nanoparticle solution were diluted in Milli-Q water, mixed with a portion of water-saturated soil and incorporated to the column. Next, rainwater was simulated by applying to the top of the columns water in pulses (one pulse per day). Leaching and distribution at different depths of Ag nanoparticles were evaluated after applying different volumes of water. For this purpose, three series of columns were prepared for each soil by adding to one, two, three pulses of water, respectively. After applying the desired number of pulses of each series, the leachates were collected and the soil was removed from the columns and divided into four layer. In the leachates, Ag-Nanoparticles were extracted by sedimentation and dilution to remove the leached soil particles and organic matter. In the case of the soil layer, the Ag-nanoparticles were extracted following the aqueous extraction procedure detailed in Mahdi et al. (2017). After extractions, soil and leachate extracts were measured for concentration and particle size by the single particle inductively couple mass spectrometry method (spICP-MS) that allows the determination of very low nanoparticle concentration in soils. A critical limitation of this technique is the difficulty and even the inability to determine the Ag nanoparticles in the presence of dissolved species of the monitored metallic element (Laborda et al. 2016). However, a new generation of spICP-MS instruments has been developed that presents certain improvements that allow solving this issue (Montaño et al. 2014). Therefore, this quantification technique has a great potential for the detection of metallic nanoparticles in soil and water samples.

The combined use of the leaching experiments described in both section for each type of nanopesticide could give a wide information on their mobility through the soil. For instance, in the case of nanopesticides formed by an AI loaded onto a nanocarrier, comparing the leaching rate of the AI when it is applied as pure form or associated with a nanocarrier could provide information on the release rate of the AI from the nanopesticide. Thus, if a delay in the leaching of the AI or a lower percentage of AI leached is observed in the column where the nanopesticide is added, it may be indicative of a slow release of the AI. In the case of the study of distribution of the AI at different soil depths, the extent of the AI filtration could be evaluated when the nanopesticide is applied. Finally, the method proposed in Petosa et al. (2017) allows assessing the transport of the nanoparticles (not AI) through the soil and how different soil properties as pH and presence of fertilizer could affect on this transport.

## 6 Conclusion

Interest in the use of nanopesticides has increased in recent years. The procedures followed to assess the fate of these new nanoformulations in soils are practically the same as those used for conventional pesticide formulations. However, nanopesticides have a number of physicochemical properties associated with their nano-size and colloidal character that make that their behavior in soils differs from that of conventional pesticides. For this reason, slight modifications have gradually been made to these procedures to adapt them to nanoformulations. Even so, understanding fate of nanopesticides using these modified methods has some limitations. Nevertheless, due to the growing interest in the application of nanotechnology in the formulation of plant protection products, the development of new methodologies that overcome the limitations of conventional methods is expected.

**Acknowledgments** The work of the author on nanopesticide topic is supported by the grant from the Czech Science Foundation (GACR) 18-19324S. The valuable review and feedback on the chapter from dr. Melanie Kah and dr. Jakub Hofman are appreciated.

## References

Al-Smadi BM, Al Oran EH, Abu Hajar HA (2019) Adsorption-desorption of cypermethrin and chlorfenapyr on Jordanian soils. Arab J Geosci 12:6–13. https://doi.org/10.1007/s12517-019-4614-7

Athanassiou CG, Kavallieratos NG, Benelli G et al (2018) Nanoparticles for pest control: current status and future perspectives. J Pest Sci 91:1–15. https://doi.org/10.1007/s10340-017-0898-0

Cabrera A, Celis R, Hermosín MC (2016) Imazamox-clay complexes with chitosan- and iron(III)-modified smectites and their use in nanoformulations. Pest Manag Sci 72:1285–1294. https://doi.org/10.1002/ps.4106

Cáceres L, Escudey M, Fuentes E, Báez ME (2010) Modeling the sorption kinetic of metsulfuron-methyl on Andisols and Ultisols volcanic ash-derived soils: kinetics parameters and solute transport mechanisms. J Hazard Mater 179:795–803. https://doi.org/10.1016/j.jhazmat.2010.03.074

Campos EVR, De Oliveira JL, Da Silva CMG et al (2015) Polymeric and solid lipid nanoparticles for sustained release of Carbendazim and Tebuconazole in agricultural applications. Sci Rep 5:1–14. https://doi.org/10.1038/srep13809

Cao L, Zhou Z, Niu S et al (2018) Positive-charge functionalized mesoporous silica nanoparticles as nanocarriers for controlled 2,4-dichlorophenoxy acetic acid sodium salt release. J Agric Food Chem 66:6594–6603. https://doi.org/10.1021/acs.jafc.7b01957

Chauhan N, Dilbaghi N, Gopal M et al (2017) Development of chitosan nanocapsules for the controlled release of hexaconazole. Int J Biol Macromol 97:616–624. https://doi.org/10.1016/j.ijbiomac.2016.12.059

D'Souza SS, DeLuca PP (2006) Methods to assess in vitro drug release from injectable polymeric particulate systems. Pharm Res 23:460–474. https://doi.org/10.1007/s11095-005-9397-8

dos Santos Silva M, Cocenza DS, Grillo R et al (2011) Paraquat-loaded alginate/chitosan nanoparticles: preparation, characterization and soil sorption studies. J Hazard Mater 190:366–374. https://doi.org/10.1016/j.jhazmat.2011.03.057

Evy Alice Abigail M (2019) Biochar-based nanocarriers: fabrication, characterization, and application as 2,4-dichlorophenoxyacetic acid nanoformulation for sustained release. 3 Biotech 9:1–9. https://doi.org/10.1007/s13205-019-1829-y

Flores-Céspedes F, Figueredo-Flores CI, Daza-Fernández I et al (2012) Preparation and characterization of imidacloprid lignin-polyethylene glycol matrices coated with ethylcellulose. J Agric Food Chem 60:1042–1051. https://doi.org/10.1021/jf2037483

Ghosh RK, Singh N (2012) Sorption of metolachlor and atrazine in fly ash amended soils: comparison of optimized isotherm models. J Environ Sci Health Part B Pestic Food Contam Agric Wastes 47:718–727. https://doi.org/10.1080/03601234.2012.669288

Grillo R, Pereira AES, Nishisaka CS et al (2014) Chitosan/tripolyphosphate nanoparticles loaded with paraquat herbicide: an environmentally safer alternative for weed control. J Hazard Mater 278:163–171. https://doi.org/10.1016/j.jhazmat.2014.05.079

Hassellöv M, Readman JW, Ranville JF, Tiede K (2008) Nanoparticle analysis and characterization methodologies in environmental risk assessment of engineered nanoparticles. Ecotoxicology 17:344–361. https://doi.org/10.1007/s10646-008-0225-x

Kah M, Hofmann T (2014) Nanopesticide research: current trends and future priorities. Environ Int 63:224–235. https://doi.org/10.1016/j.envint.2013.11.015

Kah M, Beulke S, Tiede K, Hofmann T (2013) Nanopesticides: state of knowledge, environmental fate, and exposure modeling. Crit Rev Environ Sci Technol 43:1823–1867. https://doi.org/10.1080/10643389.2012.671750

Kah M, Machinski P, Koerner P et al (2014) Analysing the fate of nanopesticides in soil and the applicability of regulatory protocols using a polymer-based nanoformulation of atrazine. Environ Sci Pollut Res 21:11699–11707. https://doi.org/10.1007/s11356-014-2523-6

Kah M, Weniger AK, Hofmann T (2016) Impacts of (nano)formulations on the fate of an insecticide in soil and consequences for environmental exposure assessment. Environ Sci Technol 50:10960–10967. https://doi.org/10.1021/acs.est.6b02477

Kah M, Walch H, Hofmann T (2018) Environmental fate of nanopesticides: durability, sorption and photodegradation of nanoformulated clothianidin. Environ Sci Nano 5:882–889. https://doi.org/10.1039/c8en00038g

Kittler S, Greulich C, Diendorf J et al (2010) Toxicity of silver nanoparticles increases during storage because of slow dissolution under release of silver ions. Chem Mater 22:4548–4554. https://doi.org/10.1021/cm100023p

Kookana RS, Boxall ABA, Reeves PT et al (2014) Nanopesticides: guiding principles for regulatory evaluation of environmental risks. J Agric Food Chem 62:4227–4240. https://doi.org/10.1021/jf500232f

Korsmeyer RW, Gurny R, Doelker E et al (1983) Mechanisms of solute release from porous hydrophilic polymers. Int J Pharm 15:25–35. https://doi.org/10.1016/0378-5173(83)90064-9

Kumar S, Chauhan N, Gopal M et al (2015) Development and evaluation of alginate-chitosan nanocapsules for controlled release of acetamiprid. Int J Biol Macromol 81:631–637. https://doi.org/10.1016/j.ijbiomac.2015.08.062

Laborda F, Bolea E, Cepriá G et al (2016) Detection, characterization and quantification of inorganic engineered nanomaterials: a review of techniques and methodological approaches for the analysis of complex samples. Anal Chim Acta 904:10–32. https://doi.org/10.1016/j.aca.2015.11.008

Langer RS, Peppas NA (1981) Present and future applications of biomaterials in controlled drug delivery systems. Biomaterials 2:201–214. https://doi.org/10.1016/0142-9612(81)90059-4

Langmuir I (1918) The adsorption of gases on plane surfaces of glass, mica, and platinum. J Am Chem Soc 40:1361–1403. https://doi.org/10.1021/ja02242a004

Li L, Xu Z, Kah M et al (2019a) Nanopesticides: a comprehensive assessment of environmental risk is needed before widespread agricultural application. Environ Sci Technol 53:7923–7924. https://doi.org/10.1021/acs.est.9b03146

Li X, Ding X, Bian C et al (2019b) Hydrophobic drug adsorption loss to syringe filters from a perspective of drug delivery. J Pharmacol Toxicol Methods 95:79–85. https://doi.org/10.1016/j.vascn.2018.12.001

Liu J, Hurt RH (2010) Ion release kinetics and particle persistence in aqueous nano-silver colloids. Environ Sci Technol 44:2169–2175. https://doi.org/10.1021/es9035557

Liu W, Yao J, Cai M et al (2014) Synthesis of a novel nanopesticide and its potential toxic effect on soil microbial activity. J Nanopart Res 16:2677. https://doi.org/10.1007/s11051-014-2677-7

Mahdi KNM, Peters RJB, Klumpp E et al (2017) Silver nanoparticles in soil: aqueous extraction combined with single-particle ICP-MS for detection and characterization. Environ Nanotechnol Monit Manag 7:24–33. https://doi.org/10.1016/j.enmm.2016.12.002

Mahdi KNM, Peters R, van der Ploeg M et al (2018) Tracking the transport of silver nanoparticles in soil: a saturated column experiment. Water Air Soil Pollut 229:334. https://doi.org/10.1007/s11270-018-3985-9

Memarizadeh N, Ghadamyari M, Adeli M, Talebi K (2014) Preparation, characterization and efficiency of nanoencapsulated imidacloprid under laboratory conditions. Ecotoxicol Environ Saf 107:77–83. https://doi.org/10.1016/j.ecoenv.2014.05.009

Miller RJ, Lenihan HS, Muller EB et al (2010) Impacts of metal oxide nanoparticles on marine phytoplankton. Environ Sci Technol 44:7329–7334. https://doi.org/10.1021/es100247x

Misra SK, Dybowska A, Berhanu D et al (2012) Isotopically modified nanoparticles for enhanced detection in bioaccumulation studies. Environ Sci Technol 46:1216–1222. https://doi.org/10.1021/es2039757

Montaño MD, Badiei HR, Bazargan S, Ranville JF (2014) Improvements in the detection and characterization of engineered nanoparticles using spICP-MS with microsecond dwell times. Environ Sci Nano 1:338–346. https://doi.org/10.1039/c4en00058g

Nguyen HM, Hwang IC, Park JW, Park HJ (2012) Enhanced payload and photo-protection for pesticides using nanostructured lipid carriers with corn oil as liquid lipid. J Microencapsul 29:596–604. https://doi.org/10.3109/02652048.2012.668960

Nuruzzaman M, Rahman MM, Liu Y, Naidu R (2016) Nanoencapsulation, nano-guard for pesticides: a new window for safe application. J Agric Food Chem 64:1447–1483. https://doi.org/10.1021/acs.jafc.5b05214

OECD (2000) Test no. 106 adsorption - desorption using a batch equilibrium method. In: OECD guidelines for the testing of chemicals. OECD, Paris, pp 1–44. https://doi.org/10.1787/9789264069602-en

OECD (2002) Test no 307: Aerobic and anaerobic transformation in soil. In: OECD guidelines for the testing of chemical. OECD, Paris, pp 1–17. https://doi.org/10.1787/2074577x

OECD (2004) Test no. 312: leaching in soil columns. In: OECD guidelines for the testing of chemicals. OECD, Paris, pp 1–15. https://doi.org/10.1787/2074577x

Peng X, Palma S, Fisher NS, Wong SS (2011) Effect of morphology of ZnO nanostructures on their toxicity to marine algae. Aquat Toxicol 102:186–196. https://doi.org/10.1016/j.aquatox.2011.01.014

Pereira AES, Grillo R, Mello NFS et al (2014) Application of poly(epsilon-caprolactone) nanoparticles containing atrazine herbicide as an alternative technique to control weeds and reduce damage to the environment. J Hazard Mater 268:207–215. https://doi.org/10.1016/j.jhazmat.2014.01.025

Petosa AR, Öhl C, Rajput F, Tufenkji N (2013) Mobility of nanosized cerium dioxide and polymeric capsules in quartz and loamy sands saturated with model and natural groundwaters. Water Res 47:5889–5900. https://doi.org/10.1016/j.watres.2013.07.006

Petosa AR, Rajput F, Selvam O et al (2017) Assessing the transport potential of polymeric nanocapsules developed for crop protection. Water Res 111:10–17. https://doi.org/10.1016/j.watres.2016.12.030

Qian K, Shi T, Tang T et al (2011) Preparation and characterization of nano-sized calcium carbonate as controlled release pesticide carrier for validamycin against Rhizoctonia solani. Microchim Acta 173:51–57. https://doi.org/10.1007/s00604-010-0523-x

Qian K, Shi T, He S et al (2013) Release kinetics of tebuconazole from porous hollow silica nanospheres prepared by miniemulsion method. Microporous Mesoporous Mater 169:1–6. https://doi.org/10.1016/j.micromeso.2012.10.017

Rashidipour M, Maleki A, Kordi S et al (2019) Pectin/chitosan/tripolyphosphate nanoparticles: efficient carriers for reducing soil sorption, cytotoxicity, and mutagenicity of paraquat and enhancing its herbicide activity. J Agric Food Chem 67:5736–5745. https://doi.org/10.1021/acs.jafc.9b01106

Ritger PL, Peppas NA (1987) A simple equation for description of solute release II. Fickian and anomalous release from swellable devices. J Control Release 5:37–42. https://doi.org/10.1016/0168-3659(87)90035-6

Sagee O, Dror I, Berkowitz B (2012) Transport of silver nanoparticles (AgNPs) in soil. Chemosphere 88:670–675. https://doi.org/10.1016/j.chemosphere.2012.03.055

Sandhya KS, Kumar D, Dilbaghi N (2017) Preparation, characterization, and bio-efficacy evaluation of controlled release carbendazim-loaded polymeric nanoparticles. Environ Sci Pollut Res 24:926–937. https://doi.org/10.1007/s11356-016-7774-y

Sarkar DJ, Kumar J, Shakil NA, Walia S (2012) Release kinetics of controlled release formulations of thiamethoxam employing nano-ranged amphiphilic PEG and diacid based block polymers in soil. J Environ Sci Health Part A 47:1701–1712. https://doi.org/10.1080/10934529.2012.687294

Sun P, Shijirbaatar A, Fang J et al (2015) Distinguishable transport behavior of zinc oxide nanoparticles in silica sand and soil columns. Sci Total Environ 505:189–198. https://doi.org/10.1016/j.scitotenv.2014.09.095

Tong Y, Wu Y, Zhao C et al (2017) Polymeric nanoparticles as a metolachlor carrier: water-based formulation for hydrophobic pesticides and absorption by plants. J Agric Food Chem 65:7371–7378. https://doi.org/10.1021/acs.jafc.7b02197

Torrent L, Marguí E, Queralt I et al (2019) Interaction of silver nanoparticles with mediterranean agricultural soils: lab-controlled adsorption and desorption studies. J Environ Sci (China) 83:205–216. https://doi.org/10.1016/j.jes.2019.03.018

Van Koetsem F, Woldetsadik GS, Folens K et al (2018) Partitioning of Ag and CeO2 nanoparticles versus Ag and Ce ions in soil suspensions and effect of natural organic matter on CeO2 nanoparticles stability. Chemosphere 200:471–480. https://doi.org/10.1016/j.chemosphere.2018.02.133

Wang D, Su C, Zhang W et al (2014) Laboratory assessment of the mobility of water-dispersed engineered nanoparticles in a red soil (Ultisol). J Hydrol 519:1677–1687. https://doi.org/10.1016/j.jhydrol.2014.09.053

Wang R, Du H, Wang Y et al (2018) Retention of silver nanoparticles and silver ion to natural soils: effects of soil physicochemical properties. J Soils Sediments 18:2491–2499. https://doi.org/10.1007/s11368-018-1918-2

Westerhoff P, Nowack B (2013) Searching for global descriptors of engineered nanomaterial fate and transport in the environment. Acc Chem Res 46:844–853. https://doi.org/10.1021/ar300030n

Yu M, Yao J, Liang J et al (2017) Development of functionalized abamectin poly(lactic acid) nanoparticles with regulatable adhesion to enhance foliar retention. RSC Adv 7:11271–11280. https://doi.org/10.1039/c6ra27345a

Zhou Y, He C, Chen K et al (2016) A new method for evaluating actual drug release kinetics of nanoparticles inside dialysis devices via numerical deconvolution. J Control Release 243:11–20. https://doi.org/10.1016/j.jconrel.2016.09.031

# Interactions of Nanoenabled Agrochemicals with Soil Microbiome

**Catherine Santaella and Barbara Plancot**

**Abstract** Soil is a dynamic, physically, spatially, and temporally heterogeneous but well-organized, three-dimensional porous matrix mixing mineral and organic matter and living organisms. Among them, soil microbiota constitutes a reservoir in which plants select a specific microbiome, contributing to their growth and their health. Microbes in soil also contribute to many ecosystemic services in agrosystems, as the recycling of major nutrients in the soil ecosystem (carbon, nitrogen, phosphorus, sulfur…).

Nanoagrochemicals are active substances based on nanotechnologies and nanoformulations to improve the characteristics and properties of active molecules as pesticides for agronomy purposes, e.g., biocides, herbicides but also nutrients. Nanotechnologies have burst into agronomy with a potential for innovation in order to improve the efficiency of pesticides, nutrients, their delivery and thus contribute to the reduction of inputs in agriculture. However, the impact of these nanopesticides on the soil microbiota as non-target organism remains underestimated up to now.

The chapter reviews the approaches and trends in the evaluation of nanopesticides implications on soil microbiota, focusing on copper- and silver-based nanoparticles as pesticides or on formulation or nanocarriers of conventional pesticides. By confronting the current knowledge and comparing methodologies, the potential and the pitfalls to overcome are discussed, together with future directions.

**Keywords** Nanoagrochemicals · Non-target organisms · Soil microbiota · Metallic nanoparticles · Polymer-based pesticide nanoformulations

---

C. Santaella (✉) · B. Plancot
Aix-Marseille University, Laboratory of Microbial Ecology of the Rhizosphere, Bioscience and Biotechnologies Institute of Aix-Marseille, St Paul lez Durance, France
e-mail: catherine.santaella@cea.fr



L. F. Fraceto et al. (eds.), *Nanopesticides*,
https://doi.org/10.1007/978-3-030-44873-8_6

## 1 Introduction

Agrochemicals, also known as phytopharmaceuticals products or pesticides, are substances used in agriculture to increase crop yield and to control pests, such as plant pathogens (fungi and bacteria), herbs, and nematodes. Nanoagrochemicals or nanopesticides are active substances based on nanotechnologies and nanoformulation to improve their characteristics and properties. Nanoenabled agrochemicals encompass nanofertilizers, nanopesticides, soil enhancer, and more recently nanosensors (Parisi et al. 2015; Fraceto et al. 2016; Baker et al. 2017; Adisa et al. 2019). Unlike nanomaterials, which are defined as materials with at least one dimension between 1 and 100 nm, nanopesticides encompass a range of heterogeneous products in terms of particle size: most of nanopesticides exceed the 100 nm size threshold. However, the nanoscale dimension usually provides particles with new chemical and physical properties, and is source of innovation in agricultural sector. The outcomes of nanotechnologies applied to pesticides are smart objects, endowed with increased efficacy, due to the reduction of losses and controlled delivery of the active ingredient, together with potential reduction of doses (Kah et al. 2018).

This reduction in the quantities of pesticides, used to increase agricultural productivity, could be particularly welcome in a paradoxical context that confronts the injunction for a more sustainable agriculture to preserve the earth's resources, feeding an increasing world population expected to reach from 7.7 billion to 9.7 billion in 2050 (https://population.un.org/wpp/), and fluctuating yields due to global warming and climatic events (drought, flooding heatwave events, etc.).

Pesticides and nanopesticides, sprayed on plants and soils or used as seed coating, can interact with the soil ecosystem, with potential consequences for the soil microbiomes, the soil fertility, and ecosystemic services.

Soil microbiota, encompasses a community of microorganisms, bacteria, archaea, fungi, viruses, and protists, associated to this environment. Soil microbiota plays a fundamental role in the cycle of elements, especially carbon but also nitrogen, phosphorus, sulfur, and other elements, the recycling of organic matter, the degradation of pollutants and the soil formation, by water and microbial alteration of rocks. Hence, soil organisms are key drivers for relevant ecosystem services in agricultural landscapes, such as nutrient cycling, soil structure, pest control, and biodiversity.

But more importantly for agriculture, the soil is a reservoir of microorganisms, in which the plant selects a specific microbiome, which contributes to the growth of the plant and its health. Thus, via the selected microbiome, the plant acclimates more quickly to stress, whether abiotic (drought, flooding, chemical toxics) or biotic (plant pathogens). The role of the plant microbiome is often compared to that of the intestinal microbiome for humans (Schlaeppi et al. 2014). Soil microbiome is considered as the second genome of the plant and the agricultural potential of the soil. Some microbiomes associated to soils can be suppressors of plant pathogens and naturally help controlling plant diseases.

Thus, understanding the interactions of nanopesticides with soil and plant microbiomes is essential in order to develop smart nanoagrochemicals that associate efficiency and eco-compatibility, in order to preserve the microbial diversity of the soil.

Before jumping into the nanoworld of pesticides, we would like to highlight the fact that the impact on soil microbiome as non-target organisms of regular (non-nano) pesticides is not so well described, even if they are currently used on agroecosystems at a rate of billions of tons. Pesticide risk assessment on soil microorganisms is certainly sidelined when considering the effects on non-target organisms. In Europe, as far as environmental risk assessment is concerned on non-target soil microorganisms, obtaining a marketing authorization from EFSA (European Food Safety Authority) only requires to evaluate the effect of the active substance on nitrogen mineralization (OCDE 216 2000; Thiour-Mauprivez et al. 2019). However, the European Food Safety Authority (EFSA) Panel on Plant Protection Products and their Residues recently proposed specific protection goals and testing strategies (e.g., functional assays based on soil respiration, exoenzyme activity and potential ammonium oxidation, PAO, test), which take into account the relevant exposure routes for in-soil organisms and the potential direct and indirect effects.

Many pesticides are systemic in plant and may act on a target that both can be found in plants and in microorganisms, as it for herbicides (Thiour-Mauprivez et al. 2019). Pesticides that control biotic plant disease can indiscriminately affect microorganisms pathogenic or beneficial to the soil ecosystem and to the plant. Regular use of organophosphates or pesticides reduces the microbial community and soil fertility though pesticides are not always toxic for microbial communities (Lo 2010). Some effects can be transient, e.g. the modulation of soil enzymatic activities by biopesticides (Shao and Zhang 2017). Pesticides can be both a felicity or a curse to soil microbial community (Karpouzas et al. 2016). Indeed, some pesticides are used as source energy for microbes and can challenge and select some specific and competitive microbial communities. However, whether these selected microbes are friendly or not is a main concern. As example, glyphosate, one of the most used herbicide in the world, enhances the resistance to chloramphenicol and kanamycin in *E. coli* and *S. typhimurium* (Kurenbach et al. 2017). Thus, crossed-resistances to herbicides and antibiotics could be a major concern, as exposure of bacteria to non-antibiotic chemicals such as herbicides could promote the resistance to antibiotics (Rangasamy et al. 2018; Van Bruggen et al. 2018).

Thus understanding the impact of pesticides and nanopesticides on non-target organisms and the resilience of the soil ecosystem is an evidence and an open question, and the approaches are still debated.

This chapter analyses the interactions and impacts of nanopesticides on soil microbial communities. It is not an exhaustive review but rather an illustration of the knowledge in the field, the gaps and future prospects.

Before getting into the details of microbial nanopesticides interactions and their impacts on soil life, it is necessary to understand: (1) the complexity of soil and plant-soil-microbial system, indeed, the nanoform of pesticides may alter their fate and diffusion in the soil matrix, and (2) the main methods to characterize impacts on soil activity and soil microbiota.

# 2 Soil-Plant-Microbiota: A Complex System

## 2.1 *Soil Is Complex and Heterogeneous Matrix*

Soil is biomaterial and the support for microbial communities that form the foundation of trophic food webs, supporting terrestrial life. A fertile soil contains up to $10^{12}$ bacteria and 25 km of fungi. However, as cells cluster together, only about a tiny fraction of the soil surface area ($10^{-6}$%) is covered by soil microbes (Young et al. 2008).

Soil is a dynamic, physically, spatially, and temporally heterogeneous but well-organized, three-dimensional porous matrix made from mineral and organic matter, different physical matter states (solids, liquids, and gases) and living organisms. There is a complex feedback between the chemistry of the matter and the biology of microorganisms living in soil habitat (Fig. 1). At a local scale, soil is a three-dimensional hierarchical network based on aggregates and on pores that are periodically connected during wetting events. Aggregates are the functional unit of a soil ecosystem (Wilpiszeski et al. 2019). Organo-mineral associations drive the formation of clusters (2–20 μm) through electrostatic and hydrophobic interactions between clays and organic matter, especially extracellular polymeric substances (Santaella et al. 2008) forming hutches for bacteria and fungi (Totsche et al. 2018; Watteau and Villemin 2018). The formation of

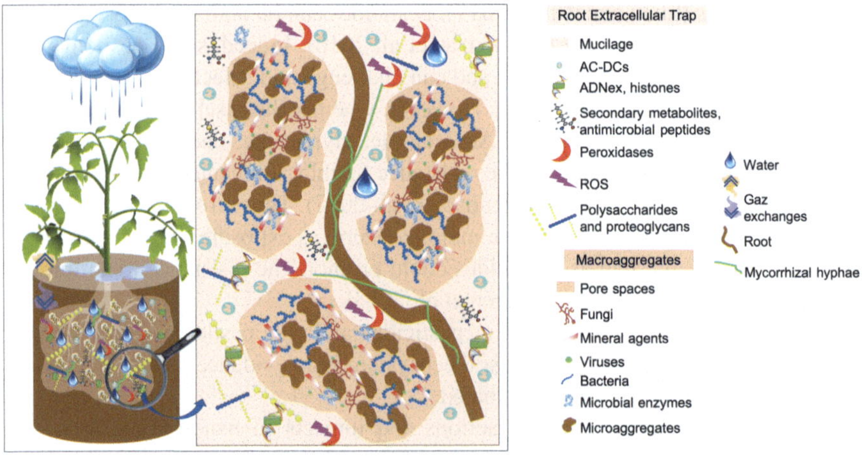


**Fig. 1** Soil matrix, a complex system (adapted from Wilpiszeski et al. 2019 and Driouich et al. 2019). The microstructure of soil aggregates hosts different soil communities and functional diversity. Pore spaces within microaggregates (1–2 μm) and interaggregates (10–30 μm) allow gases, water and nutrients to diffuse. Diffusion of gas, water and nutrients is modulated according to the diameter of pore spaces from 10–30 μm in interaggregates to 1–2 μm within intraaggregates. At the root tip, a network of polysaccharides and proteoglycans embeds cap-derived cells (AC-DCs) and exudates

stable clusters is stimulated at the interface between the plant root and the soil, the rhizosphere, as plant exudates and desquamated cells promote hot spots of bacteria (Watteau and Villemin 2018). These clusters assemble into microaggregates (<250 μm) cementing mineral agents (oxides, hydroxides, and oxyhydroxides of iron, manganese, aluminum, silicon, aluminosilicates, and carbonates) and entangling organic matter (Totsche et al. 2018). Temporary binding through hyphae from fungi or actinomycetes, roots, proteins, and extracellular polymeric substances gathers microaggregates into macroaggregates (>250 μm) and pores. This architecture creates a variable flow of water and nutrients that can be accessed by soil organisms (Wilpiszeski et al. 2019). Proteins with enzymatic activities can be everywhere, inside cells, inside or at the surface of microaggregates, and macroaggregates and even in pores during a waterlogging event. Soil and especially clays, organic matter, and minerals can sorb chemical compounds circulating in the pore water solution (the so-called cation exchange capacity) and interact with microorganisms.

The microstructure of soil aggregates directly impacts soil communities and functional diversity. The diffusion gas, water, and nutrients are modulated according to the diameter of pore spaces from 10 to 30 μm in inter-aggregates to 1–2 μm within intra-aggregates. Soil microstructure offers micro-niche for microorganisms. As example, nitrogen cycle relies on communities inhabiting distinct portions or the soil structure. Nitrifiers are most abundant and active in 2- to 20-μm microaggregates, while nitrogen-fixing bacteria were most abundant in the <2 μm clay fraction (references in Wilpiszeski et al. (2019)).

Recently, Driouich et al. (2019) described a new structure, the Root Extracellular Trap (RET), expected to set in soil the interactions and relations between plant and rhizosphere microorganisms. At the tip of the root, cap-derived cells (AC-DCs, Driouich et al. 2019) are released in the rhizosphere as single cells (border cells, Hawes et al. 2000) or files of cells still attached together (border-like cells, Vicré et al. 2005). These two types of cells are implicated in the root defense (Hawes et al. 2012, 2016; Plancot et al. 2013). At the scale of a root system, root cap-derived cells and their secretions form a cloudy network of "sticky" mucilage between the soil and the roots, composed of cells and defense-related compounds released into the surrounding soil environment and, consist mainly of glycan-containing molecules (i.e., proteoglycans and polysaccharides), antimicrobial compounds including proteins, peptides and secondary metabolites, histones, and extracellular DNA (Hawes et al. 2016; Ropitaux et al. 2019; Driouich et al. 2019), regulating interactions and relations of the plant with rhizosphere microorganisms.

All these architectural structure in the soil controls the interactions between plants, microbes, and also pollutants. This is why understanding the interactions between the soil matrix and nanopesticides will be so important.

## 2.2 *Microbiome vs Microbiota*

There is some confusion and quite a controversy in the use of these two words, supported by the semantic analysis of the word stem as "microbiome" or "micro-biome" (Lederberg and McCray 2001). According to the author of that word, "microbiome" refers to "the ecological community of commensal, symbiotic, and pathogenic microorganisms that literally share our body space and have been all but ignored as determinants of health and disease" (Lederberg and McCray 2001). However, this definition overlaps with that of the microbiota, quite equivalent to the microflora in the gut, defined as the microbial communities that inhabit our gastrointestinal tract. In the dynamic trend of omes and omics, microbiomes could tend to define a population of microorganisms and their genetic potential while microbiota defines the collection of microbes. The composition of a microbial community as described by high throughput sequencing approaches (see the next paragraph) refers to a microbiome, while a fecal microorganism transplantation refers to a microbiota.

## 2.3 *How to Analyze the Impacts of Nanopesticides on Soil Microbiota*

### 2.3.1 Microbiome Analysis

Microbiome analysis relies on metagenomics and more generally omics (transcriptomics, proteomics, and metabolomics), which allow microorganisms to be studied in their environment without the need for a culture step. Microbial communities can be characterized by their composition (who is there), abundances (how many of them?), their activities, e.g., RNA, proteins, and metabolites (what are they doing).

One approach to characterize microbiome is amplicon sequencing or "DNA metabarcoding." DNA metabarcoding is based on high throughput sequencing of amplicons of taxonomic markers, such as ribosomal RNA genes (16S rRNA for bacteria and archaea, 18s RNA for eucaryota) or Internal Transcribed Spacer (ITS, for fungi), as universal barcode sequences of the microorganism identity (Caporaso et al. 2011; Shokralla et al. 2012). 16S rRNA and 18S rRNA genes code ribosomal RNA, a non-coding RNA (not translated to protein) that is part of the small subunit of the ribosome, responsible for the translation of messenger RNAs (mRNAs) into proteins. These genes are not submitted to lateral gene transfer, and contained conserved and variable regions termed V1–V9.

This allows to analyze the composition and the abundance of taxa, which are groups of closely related organisms, using a sequence similarity criterion. A deeper investigation of microbiomes can be reached by metagenomics, which analyzes the whole set of genes present, leading to the composition, but also to the whole set of functions potentially displayed by the microbiome.

How to interpret changes in the abundance of specific taxa, drifts in microbial community profiles or potential alteration in microbial functions? Hugerth and Andersson (2017) provide a comprehensive analysis of how sequencing data are obtained and processed for microbial community analysis.

Next-generation-sequencing data are usually interpreted in terms of alpha- and betadiversity. Alphadiversity will refer to the diversity within a single type of sample based on replicates (Whittaker 1960). This diversity is characterized by an estimation of the richness (number of sequence, Chao 1 estimator) or as richness and evenness (e.g., the Shannon diversity index). Evenness corresponds to the regularity of the presence of a taxon in a community. Apart from the fact that it is difficult to correctly estimate alphadiversity, the interpretation of this data is hampered by the preconceived idea that higher diversity is better. The temptation to conclude to drama is great when comparing the richness of a control sample to that of a treatment. Shade (2017) advises to consider these data as a starting point for further inquiry of ecological mechanisms rather than an "answer" to community outcomes.

Betadiversity measures the extent to which two samples are different. For this purpose, different tools based on metrics allow to measure the distance between microbiomes, based on OTU (operational taxonomic unit) abundance and/or on phylogenetic distance. Changes in defined taxa and shifts in community profiles can be detected. However, it is difficult to interpret the meaning and consequences of these changes on soil functioning. The role and importance of taxa in an ecosystem are not always related to their abundance. As example, rare microorganisms with an abundance less than 0.1%, could act as a reservoir to rapidly respond to environmental changes and contribute to community stability (Shade et al. 2014). Moreover, inferring functional role of a microbial community based on 16S rRNA partial gene sequence is unsatisfactory.

Beyond diversity patterns, interaction networks of ecological or functional associations between taxa are essential drivers of ecological community structure and dynamics. Keystone microbes are those whose interactions cascade through the community (Berry and Widder 2014). Some highly connected keystone taxa can be good predictors of whole-community compositional change under environmental disturbance (Herren and McMahon 2018).

### 2.3.2 Microbial Enzymatic Activities

The interaction between soil and pesticides may result in altered biochemical processes driven by microorganisms. Soil contain many enzymes, as free, immobilized, and extracellular or intracellular entities. Soil enzyme activities are soil quality indicators, playing many roles in nutrient element cycling and organic matter decomposition (García-Ruiz et al. 2008; Karaca et al. 2011). Thus, soil enzyme activities are good biological responses to analyze the soil response to a stress such as pesticides.

Soil enzymes have a crucial role in element cycling such as C cycle (glycosyl hydrolases, oxidases, and peroxidases), N cycle (proteases, peptidases, urease, and chitinase), P cycle (phosphatases), and S cycle (arylsulfatase). Dehydrogenase are intracellular enzymes found in all living organisms that are involve in energy transfer in microbial metabolic reactions and biological oxidation of soil organic matter. They are widely used as an indicator of overall soil microbial activity (Wolińska et al. 2015).

# 3 Impact of Nanopesticides on Non-target Soil Microorganisms and Microbiomes

Most of nanopesticides are systemic and are intended to be active inside the plant. However, as nanopesticides are disseminated in the environment, soil microbiota and microfauna, and plants become non-target organisms, and exposed to the impacts of these bioactive molecules.

Regarding nanopesticides impact, the standpoint of non-target organisms is still not already set in the literature. A Web of Science (WOS) bibliometric analysis (October 2019) of (nanopesticide∗ AND non-target) yields 23 references.

As non-target organisms, plant or microbes are not viewed with the same importance. The search for keywords nanopesticide∗ AND soil∗ AND microb∗ in WOS (October 2019) returned 12 references while (nanopesticide∗ AND plant∗) yielded 106 references. The importance of soil microorganisms for ecosystem functioning remains greatly underestimated.

The chapter will focus on microbiome and microbiota as non-target organisms of nanopesticides.

Different types of pesticides have been formulated as nanopesticides, including nanoformulations of conventional pesticides or nanomaterials as pesticides, many of them being metallic and metal oxide nanoparticles.

## *3.1 Nanopesticides Based on Metal and Metal Oxide Nanoparticles*

### 3.1.1 Copper

The impacts of metal and metal oxide nanoparticles, on the microbiome of the soil and rhizosphere, have been widely studied, mainly with the envision of environmental pollution effects (Anjum et al. 2013; Simonin and Richaume 2015; Tian et al. 2019; Rajput et al. 2020). Among the most investigated in toxicity studies, nanoparticles based on $TiO_2$, Ag, ZnO, Cu, and Fe rule the ranking.

Currently, two types of nanomaterials have resulted in nanoenabled commercial agrochemicals, available on the market: copper nanoparticles as fungicides to control diseases on fruit tree, vegetables, and crops, and colloidal silver to treat fungal pathogens on seeds, tubers, and vegetative plants (He et al. 2019).

We will focus on reports of the impacts of Cu- and Ag-based nanomaterials on soil microbial communities, especially those for which the doses tested were compatible with applications in agriculture, as nanofertilizers or nanopesticides.

Copper is both an essential nutrient for living organisms as plants and microorganisms, and a renowned biocide since ages. Some copper-based pesticides are currently authorized in organic farming as fungicides and bactericides on grapes, trees, and fruits. Initially used as lime neutralized copper sulfate in the Bordeaux mixture to cure grapes infected with downy mildew (Millardet et al. 1933), copper-based pesticides can exist as copper hydroxide, cuprous oxide, copper oxychloride, copper ammonium carbonate, and copper octanoate. Indeed, as the solubility of copper sulfate favors phytotoxicity and decreases the persistence on the plant/tree leaves and fruits, and fungicide activity, less soluble forms known as fixed-coppers have been developed (e.g., copper hydroxide, copper oxychloride, basic copper sulfate cuprous oxide, etc.). These fixed-coppers are particles whose size determines coverage and adherence to plant leaves, and release of copper ions. Initially marketed as micronized particles, copper nanosized particles have rapidly been developed and commercialized to improve the coverage of the plant fruits or leaves, and to control the release of Cu ions. Currently, at least two nanosized copper formulations are available: Kocide® 3000 (DuPont) and NANOCU (Bio Nano Technology) (He et al. 2019).

(Simonin et al. 2018a) assessed the impact of nanosized bare CuO (~50 nm, specific surface area 23 $m^2 g^{-1}$, 0.1, 1, and 100 mg $kg^{-1}$ dry soil) vs Cu ions ($CuSO_4$) in five agricultural soils with contrasting properties (pH between 6.4 and 8.21), to take into account soil biological complexity and physico-chemical diversity. Soil moisture was adjusted to the water holding capacity specific to each soil, and soil microcosms were incubated in the dark at 28 °C, over 90 days. At the highest concentration (100 mg $kg^{-1}$ dry soil), in the five soils tested, CuONPs cause significant reductions that worsen over time, on soil microbial activities involved in carbon and nitrogen cycles, respiration, nitrification, and denitrification. Lowest doses show limited effects, mostly at 90 days, with decreases of respiration in the sandy-loam soil from 1 mg $kg^{-1}$, and in denitrification at 1 mg $kg^{-1}$ in the loamy soil. Globally, denitrification is the most sensitive microbial activity to CuONPs in most soil types, while soil respiration and nitrification are mainly impacted in coarse soils. CuONPs and ionic Cu show distinct impact on soil microbial activities, likely explained by the low dissolution of CuONPs, less than 2% in soil solution, over time. Thus at low and agricultural-relevant concentrations, CuONPs have limited effects on soil microbial activities involved in carbon and *N* cycles. Occasionally, coarse soil texture with low organic matter or clays contents are more likely to be affected.

In this type of soil (loamy soil with low clay content), potentially more sensitive, enhanced with CuONPs (1 and 100 mg $kg^{-1}$), Simonin et al. (2018a) grew winter wheat (*Triticum aestivum*) over 50 days in climatic chambers. The plant exudates stimulate heterotrophic microbial activities as microbial respiration and denitrification. However, this does not counterbalance or even worsen (e.g., 1 mg $kg^{-1}$ CuONPs for microbial respiration) the effects of CuONPs on these enzymatic activities. Thus the plant influences the microbial response to CuONPs exposure but does not mitigate the negative effects of CuONPs.

VandeVoort and Arai (2012) confirmed the toxicity of Cu-based NPs to nitrifiers and the very different behavior between CuONPs and $Cu^{2+}$ ions in terms of $Cu^{2+}$ release, adsorption, and impact on nitrification in batch nitrification kinetic experiments.

Asadishad et al. (2018) investigated the impact of nanosized CuO and Cu ions on soil enzyme activity and microbial community composition of a biosolid-amended agricultural soil, over 30 days. Surface soil (pH 6.7) was sampled from an agricultural site at the Macdonald campus of McGill University (most likely sandy loams, loamy sands or clay soils based on Collaborative Geographic Information Systems, Authors' note) amended with a biosolid from a waste water treatment plant, was enhanced with bare CuONPs (40 nm) at 1, 10, and 100 mg total CuNPs $kg^{-1}$ soil. In soil solution, CuONPs dissolution occurs within the first 2 h (70%) and remains stable up to 30 days, likely because of soil dissolved organic matter binding to reactive sites on the NP surface.

The activities of five soil extracellular microbial enzymes involved in C, N, and P nutrient cycling were measured in the soil amended with biosolids and exposed to bare CuONPs or Cu ions at 2 h, and 30 days after treatment with the NPs suspensions or ionic solutions. After some transient inhibitory at 2 h, no significant enzyme inhibition is observed for the soil-biosolids slurry exposed to CuONPs after 30 days. CuONPs and $Cu^{2+}$ show similar effects on soil enzyme activities at short term but CuONPs tends to stimulate some enzyme activity at longer exposure time, suggesting a specific nanoeffect. Over 70% of the CuONPs was dissolved at 2 h, and this dissolution increased to 77% in 30 days suggesting that most of the CuONPs ended up as $Cu^{2+}$ or Cu organic complexes explaining their similar trends for some of the enzymes. The initial decrease in enzyme activity observed at 2 h may be linked to the antimicrobial activity of $Cu^{2+}$ and CuONPs. Nonetheless, these data shows that the activity of the five extracellular soil enzymes generally recovers after 30 days of exposure to CuONPs.

Kocide® 3000 (Dupont) is fungicide/bactericide based on copper hydroxide, approved by the US EPA for citrus, conifers, field crops, small fruits, tree crops, vegetables, vines, and some other fruits. Kocide® 3000 contains micronized particles made from nanosheets of $Cu(OH)_2$ embedded in a carbon-based matrix that promptly dissociates in water (Adeleye et al. 2014).

Simonin et al. (2018b) designed outdoor terrestrial mesocosms with a sandy-clay-loam soil (57.7% sand, 20.5% clay, 21.9% silt, 4% organic matter, pH = 5.8) seeded with seven forage crops composed of forbs, graminoids, and legumes as representatives of the three main plant functional groups. To assess the environmental impacts of sequential applications under low-input or conventional farming scenarios, the nanopesticide was applied alongside three different mineral fertilization levels (Ambient, Low, and High). The foliage of forage was sprayed with the Kocide® 3000 suspension (6.68 mg $L^{-1}$ in water, 30 mg $m^{-2}$, at Day 0, 75, and 155, and 15 days before each subsequent plant harvest). The mean particle size was 38.7 ± 8.2 nm (TEM) and an average hydrodynamic diameter of 120 ± 30 nm in the dosing water with a secondary peak with particles size greater than 700 nm (Simonin et al. 2018c). The authors monitored enzymatic activities involved in C, N, P, and S cycling, soil $N_2$ fixation rates (conversion of molecular $N_2$ in the air to ammonia or nitrogenous compounds available to the plant) and mycorrhizal colonization of plant roots, over a year. The authors report no detrimental effects on the forage biomass and mycorrhizal association with plant roots. However, they evidence a dual, beneficial or negative, interactive effects between nanopesticide and fertilization treatments on extracellular microbial enzymatic activities. In the Ambient fertilization, Kocide® 3000 applications transiently inhibited enzyme activities at short term (15 days) and decreased P and C cycling at long term (6 months after the last Kocide® 3000 applications), while positive effects on plant biomass and enzyme activities occurred in the high fertilization treatment. In Ambient fertilization, the authors hypothesize that at short term, nutrient limitation combined to the copper biocide activity could decrease the ability of microbial community to cope with the stress. At long term, the decrease of enzymatic activities could be related to responses to Kocide® 3000 driven by seasonal effects and low water availability.

At long term, Kocide® 3000 treatment stimulated or unaffected enzyme activities in the Ambient and high fertilizations. This could arise from the adaptation of the microbial community to Cu, with the selection of Cu-tolerant species, and the depletion of resources in soil, with a nutritional effect of Kocide® 3000 and contained micronutrients.

The authors conclude on limited or positive effects of repeated Kocide® 3000 applications on forage production and soil microbial processes in conventional farming with high fertilization rates, but they warn about detrimental effects on microbially mediated soil processes involved in C and P cycling and on forage production in the context of lower-intensity fertilization (e.g., organic farming). This study of the impact of Cu-based nanopesticide on the microbial compartment is certainly the most complete, examining the impact of sequential applications over a growing season in an outdoor mesocosm. However, it would be interesting to verify the last conclusions in soils under organic farm, using fertilizers suited for this mode of cultivation. Here the soil was supplemented with an inorganic fertilizer, while in organic farming, fertilizers are usually derived from animal and vegetable matters or agricultural practices.

Zhang et al. (2019) applied a commercial $Cu(OH)_2$ nanopesticide formulation, the active ingredient of this formulation, the synthesized $Cu(OH)_2$ nanotubes with comparable morphology to the active ingredient, and $CuSO_4$ to a silty soil (pH 8.17, organic content 3.4%) at 0.5, 5 and 50 mg $kg^{-1}$, followed by an application of neonicotinoid thiacloprid, an insecticide, after an interval of 21 d. The overall pattern of soil bacterial community composition shows that $Cu(OH)_2$ nanopesticides at 50 mg $kg^{-1}$ significantly decreased the alpha-diversity of bacteria in soil and drastically altered the community composition. The relative abundance of *Gemmatimonas* decreased by ~30% in soil with $Cu(OH)_2$ nanopesticides 50 mg $kg^{-1}$ as compared to control. Their relative abundance showed a significant positive correlation ($r = 0.89$, $p < 0.05$) with the degradation rate constant of thiacloprid. The $Cu(OH)_2$ nanopesticides reduced nitrile hydratase activity and downregulated thiacloprid-degradative *nth* gene abundance that contributes to the mitigation of thiacloprid degradation. The authors suggest to reconsider the use of nanopesticides based on $Cu(OH)_2$. However, in this study, the authors used a concentration of $Cu(OH)_2$ that is tenfold the recommended dose of this nanopesticide (5 mg $kg^{-1}$). Moreover, the Cu applied (50 mg $kg^{-1}$) was high as compared to the Cu background (4.1 mg $kg^{-1}$), while in Simonin et al. (2018b) the Cu amount applied to the mesocosms (5.43 mg/mesocosm containing 81 kg of soil) was much lower than the background concentration (90.5 mg $kg^{-1}$). The presence of background Cu in soils may select tolerant communities, which would be less affected by the additional addition of Cu.

Assessing how CuNPs may interact with pollutants and pesticides in soil, Parada et al. (2019) incubated CuNPs (40–60 nm) at 0.05 and 0.15% w/w and ATZ (3 mg $kg^{-1}$) in an Andisol (a soil rich in organic matter) for 30 days. Microbial community profiles assessed by PCR-Denaturing Gradient Gel Electrophoresis (PCR-DGGE) on bacteria, fungi, and nitrifying bacteria, remained relatively stable throughout the experiment. However, CuNPs at 0.15% w/w caused a significant decrease in ATZ dissipations showing an increase in the persistence of ATZ in soil. This persistence was mostly associated to physical-chemical interaction with soil particles.

Paddy soils are typical soils agricultural soils in China, and are under periodical flood–dry water management, constantly changing redox potential in the soil environment. Shi et al. (2018) exposed two paddy soils (organic content 4.1 and 8.01%) to CuONPs (hydrodynamic diameter in water 240.0 nm) and CuO bulk particles (BP, average particles size of 1346 nm) at 10, 100, and 1000 mg $kg^{-1}$ for CuONPs and 1000 mg $kg^{-1}$ for CuOBPs. The authors show differentiated behavior between NPs and BPs in paddy soils and a role for the organic matter. Microbial available Cu was higher for CuONPs than for CuOBPs. In the low organic matter soil, CuONPs changed the soil properties by increasing the pH and Eh, accelerated the degradation or mineralization of the organic matter, as well as the Fe reduction process, by increasing the Fe(II) content by 293% after flooding for 60 days. The microbial biomass carbon in both soils was severely inhibited by CuONPs and to a minor extent by BPs at 100 mg $kg^{-1}$. The organic matter could partly mitigate the negative effects of CuONPs.

For a complete review of copper-based nanoparticles implication on terrestrial and aquatic environment, see Rajput et al. (2020).

### 3.1.2 Silver

Silver is known as a biocide since ages. Silver-based nanopesticides show antimicrobial/biocidal properties against a broad of classes of microorganisms, e.g., bacteria, fungi, and virus (Durán et al. 2016).

Some silver-based nanopesticides are already patented and commercial in the technology of plant protection, the processing of seed material, and the enhancement of plant development. Some examples are WA-CV-WA13B, WA-AT-WB13R, and WA-PR-WB13R (Bio-Plus Co.Ltd., Pohang, Korea), and Zerebra® agro,

Zeroxxee®, Silver leaf, Zeromix® (AgroKhimProm Group, Russia and Commonwealth of Independent States, Grand Harvest Research Innovation Company). Even if these nanopesticides are claimed to effectively inhibit phytopathogen diseases in a broad set of plants, to strengthen the plant immune system, and to reduce stress reduction (Jung et al. 2010; Parada et al. 2019), most of the published knowledge on the impact of silver-based nanopesticides on non-target microbes and microbiomes originates from studies on the environmental impact of AgNPs.

Hund-Rinke et al. (2019) amended a loamy, acidic sand (73% sand, 22% silt and 5% clay; pH 5.6, low organic matter content, 1.1%) with biosolids and AgNPs (NM-300K dispersed in a mixture of a stabilizing agents, particle size of ~ 15 nm, 99%) to achieve a target concentration of 0.19–15 mg $kg^{-1}$ soil. Soil samples amended with biosolids and AgNPs or standard ionic solutions were kept static in the dark at 22 °C for up to 30 days. The impact of AgNPs was assessed by soil respiration (Micro-Resp test), exoenzyme activity, potential ammonium oxidation (PAO) test, and next-generation sequencing to survey bacterial diversity by sequencing the 16S rRNA gene. The four tests showed similar sensitivity towards the silver nanomaterial, with significant effects at AgNPs concentrations from at least 1.67 mg $kg^{-1}$. The authors evidenced no differences in the Shannon index or evenness as indicators of alphadiversity. However, next generation sequencing evidenced a different sensitivity of bacterial orders, and shift in the microbial community, with an enrichment of Proteobacteria (*Caulobacterales*, *Burkholderiales*, and *Xanthomonadales*), *Cytophagales,* and *Sphingobacteriales*. The adverse impact on some nitrifiers (*Nitrosomonadales*) matched the inhibition of PAO activity.

Examining the long-term effect of these AgNPs (140 days) on ammonium oxidizing bacteria (AOB), Schlich et al. (2018) incubated AgNPs (NM-300K, diameter~15 nm, and a small proportion at ~95 nm) and $AgNO_3$ added to a sandy loam soil (pH 5.61, 0.93% organic content) at 0.56, 1.67, and 5 mg $kg^{-1}$ dry matter soil. At 1.67 and 5 mg $kg^{-1}$ AgNPs, they show a relative inhibition of AOB starting from day 14, which increases up to 140 days, while inhibition occurs from day one and increases over time, even at the lowest dose (0.56 mg $kg^{-1}$) in the case of silver ionic form.

Vitali et al. (2019) analyzed the effects of AgNPs on the phyllosphere and rhizosphere associated microbiota of a black poplar tree. Nanopowder, amorphous-carbon-coated with Ag nanoparticles (1 mg $L^{-1}$, average particle size ~25 nm, specific surface area 23 $m^2$ $g^{-1}$, dispersed in water with a soap surfactant) were chronically supplied at leaf and root level of three-year-old poplar trees (3 m, 15 L pots filled with soil fertilizer mixture) over 10 weeks (4 weeks with single supply followed by 6 weeks with twice supply). The final concentration exposure of plants to AgNPs was 16 mg $L^{-1}$ (volume not indicated) in both leaf and root treatments (surface of the pot estimated to 615 $cm^2$, Authors' note). The soil was protected during foliar exposure, and no fertilizer was added during the time of the experiment. The author used next generation sequencing of the V3–V4 region of 16S rRNA and the ITS 1 region to analyze the bacterial and fungal microbiomes, respectively. Leaf AgNPs treatment increased bacteria and fungi evenness and determined a significant reduction in both microbial groups, while root AgNPs treatment reduced the bacterial and fungal biodiversity. Bioinformatics functional analysis showed that AgNPs treatment reduced the aerobic and stimulated facultative anaerobic and oxidative stress-tolerant bacteria. However, in this study, the AgNPs treatments mimicked a polluted environment and not an agricultural treatment with Ag nanopesticide. As example, Ag concentration in Zerebra® Agro, a commercial silver-based nanopesticide, is 0.5 g $L^{-1}$ and the recommended dose for plant treatment is 0.1 L $t^{-1}$ in seed, and 0.1 L $ha^{-1}$ (50 mg $ha^{-1}$) for application in vegetation period on agricultural crops from 1 to 3 times, instead of 20 g $ha^{-1}$ in Vitali et al. (2019) study (assuming at least 100 ml were used).

Asadishad et al. (2018) investigated the impact of AgNPs (50 nm citrate-coated AgNPs) and their dissolved ions on soil enzyme activity and microbial community composition of a biosolid-amended agricultural soil. Surface soil (~35 cm depth, pH 6.7) was collected at the Macdonald campus of McGill University amended with a biosolid from a municipal wastewater treatment plant (soil/biosolid weight ratio 50/1). AgNPs were added at 1, 10, and 100 mg total AgNP $kg^{-1}$ soil. Dissolution occurred within the first 2 h and remained stable up to 30 days. At short term (2 h), AgNs showed no effect at 1 and 10 mg $kg^{-1}$ extracellular enzymatic activities implicated in P, C, and N cycling. At 100 mg $kg^{-1}$, AgNPs moderately impacted these enzymatic activities as compared to $Ag^+$, likely because only 37% of the AgNPs was dissolved at 2 h. The microbial community of the soil was analyzed by 16S rRNA gene amplicon sequencing after 2 h and 30 days of exposure. The relative abundance of the *Gammaproteobacteria* class was significantly higher for $Ag^+$ ions and AgNPs at 100 mg $kg^{-1}$ soil than in all other treatments. The *Alphaproteobacteria* community responded differently to dissolved Ag and AgNPs, with a decrease in the relative abundance $Ag^+$ 100 mg $kg^{-1}$ soil.

Also focusing on long-term experiments, Grün et al. (2018, 2019) incubated at 15 ± 4.5 °C over a period of one year, AgNPs, (BAM-N001 AgPure) with concentrations ranging from 0.01 to 1 mg AgNPs $kg^{-1}$ soil from an arable field cultivated with wheat. The soil was classified as a loamy soil (pH 7.1 in $CaCl_2$, clay content of 17–30%, total organic content 2.8%). The toxicity of AgNPs to the microbiota was indicative of the time-dependent reactivity in the complex physicochemical soil system. Over time, AgNPs (0.01 mg $kg^{-1}$) have short-term (1 day and 1 week) stimulatory effects on *Acidobacteria, Actinobacteria,* and *Bacteroidetes*. After 1 month, *Actinobateria* are negatively impacted. The relative abundance of *beta-Proteobacteria* is decreased from the first day of incubation until to the end of the experiment (1 year). On the average, for the three concentrations tested, the negative effects were the highest for *beta-Proteobacteria* and *Bacteroidetes. Actinobacteria* and *alpha-Proteobacteria* were statistically unaffected by AgNPs treatments after 1-year exposure.

Globally, the author report fluctuations of positive and negative effects over time with a strong toxicity event at 90 days and a decline of silver toxicity on some bacterial phyla at day 28, 180, and 365 at the different concentrations tested. These trends are likely explained by potential transformations such as changes in aggregation and oxidation state, dissolution, sulfidation, sorption of inorganic and organic species that result in a transient pattern of dissolution or stability of AgNPs. In response to these events, the bacterial community showed transient resistance and resilience mechanisms.

Grün et al. (2018) show that one year of exposure to 0.01 mg AgNPs $kg^{-1}$, negatively impacted the microbial soil community involved in nitrogen, with a decrease in the abundance of AOB (*amoA* gene copy numbers), the leucine aminopeptidase activity (N substrate turnover), and the abundance of nitrogen fixing microorganisms (*nifH* gene copy numbers).

Guilger et al. (2017) biogenically synthesized silver nanoparticles using the fungus *Trichoderma harzianum.* The AgNPs (spherical nanoparticles size distribution between 20 and 30 nm by scanning electron microscopy, $0.15 \times 10^{12}$ and $0.31 \times 10^{12}$ NPs $mL^{-1}$) were incubated $0.15 \times 10^{12}$ and $0.31 \times 10^{12}$ NPs $mL^{-1}$ in an agricultural soil (pH 6.8, 14% organic content) at 25 °C for 6 months. The authors quantitatively followed overtime the distribution and abundance of several genes involved in the nitrogen cycle (Fig. 2): *nifH* (nitrogen fixation), *amoA* (nitrification), *nirK*, *nirS*, and *narG* (first stage of denitrification), and *cmorB* and *nosZ* (second stage of denitrification).

Over time, the authors evidence a sequential modulation of the abundance of bacteria and genes involved in N cycle in the samples exposed to the biogenic AgNPs. During the first 30 days, a higher increase in the abundance of bacteria in the samples exposed to AgNPs than in the control sample is observed, but the distribution of genes stay comparable. Over time, this increase in the abundance of bacteria still happens, which could traduce a stimulation of bacteria involved in N cycle in the samples exposed to biogenic AgNPs. At 90 days, differences do occur in the distribution of genes, with decreases in the bacteria producing nitrate reductases (*narG*) that persists up to 180 days, and reduction nitrogenase reductase enzymes (*nifH*) and oscillations in the proportions of *nifH* and up to 180 days. Bacteria that presented the *cmorB* nitrate reductase genes increased up to 90 days post-exposure and decreased after this period, while the bacteria that presented the nitrous oxide reductase gene (*nosZ*) oscillated in the opposite way, increasing for the first two periods and decreasing for the last two periods (90 and 180 days). The coating of the

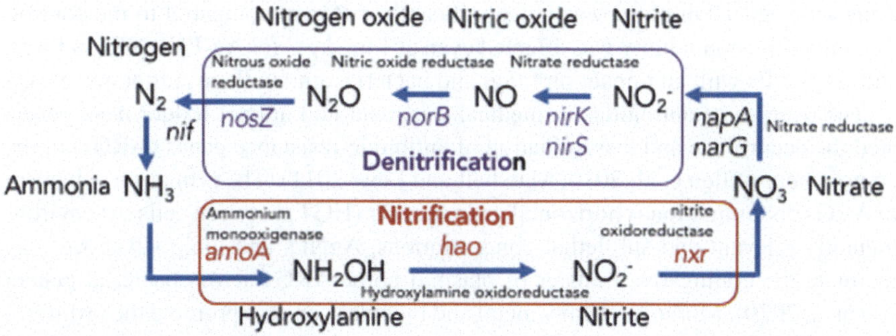


**Fig. 2** Nitrogen cycle

nanoparticles may have retarded the release of $Ag^+$, which could explain possessed a coating, which could have delayed the release of $Ag^+$ and explain the latency phase observed in the changes in abundance of bacteria and genes involved in the nitrogen cycle. Thus the impact of the biogenic AgNPs tends to show a stimulation of bacteria involved in N cycle together with some cycles of impact and recovery of the community.

VandeVoort et al. (2014) incubated AgNPs (PVP coated 50 nm and 15 nm) at 1, 10, and 100 mg $L^{-1}$ in a Toccoa soil (AgNPs display near 100% sorption onto Toccoa soil surfaces at all concentrations used for the denitrification experiments). PVP coated 50 nm AgNPs did not show significant differences in $NO_3$ depletion rate from the control condition at any concentration, while the smallest PVP coated 15 nm AgNPs showed the greatest differences from the control condition in the reaction rate and a concentration dependent inhibition. At 1 mg $L^{-1}$ the depletion rate was not significantly different than that of the control, and it took 68 h to achieve 90% $NO_3$ depletion, while at 10 mg $L^{-1}$ and 100 mg $L^{-1}$ it took 111 h and 194 h, respectively. The dissolution of 15 nm AgNPs was an order of magnitude greater than the larger AgNPs and they displayed a better colloidal stability. Phase transformation readily occurred in 15 nm AgNPs as ~ 75% of Ag(0) speciation in pAg15 was changed to $Ag_2S$ and Ag(I) sorbed humic acid during the incubation period. The Ag speciation changed to a much lesser extent 50 nm AgNPs. These results show designing the NPs characteristics and the dose, denitrification can be unaffected by AgNPs.

AgNPs can undergo phase transformation in the aquatic environment and in soil, especially sulfidation (Hashimoto et al. 2017). Judy et al. (2015) investigated the impact of AgNPs, focusing on different Ag speciation and NPs coating. They exposed a biosolids-amended sandy loam soil (pH 6.8) to 1, 10, or 100 mg $Ag_2S$ NPs, polyvinylpyrrolidone (PVP) coated AgNPs and $Ag^+$. The soil mixture was inoculated with a commercial inoculum or an arbuscular mycorrhizal fungi (AMF), prior to sewing tomato seeds (*Solanum lycopersicum*). The authors monitored the colonization of tomato roots by the fungi, the microbial community structure in biosolids-amended soil, and ammonium nitrate extractable Ag concentrations. Except for three treatments (100 mg $kg^{-1}$ for Ag-PVP NPs and $Ag^+$ and 10 mg $kg^{-1}$ for AgS NPs), mycorrhizal colonization of tomato roots for all Ag treatments at 1 mg $kg^{-1}$ and 10 mg $kg^{-1}$ was not significantly different compared to the control. The microbial community was affected even at 1 mg $kg^{-1}$ for Ag-PVP NPs and $Ag^+$, and $Ag_2S$ NPs with an impact on fungi and bacteria, among them Actinomycetes.

The overuse of antibiotics in medical treatment and animal fodder have generated the occurrence and dissemination of antibiotic resistance genes (ARGs) in the environment (Allen et al. 2010; Marshall and Levy 2011). The primary mechanism of ARGs dissemination is horizontal gene transfer (HGT) between cells. At environmentally relevant and sub-lethal concentrations, AgNPs and ionic silver $Ag^+$ can facilitate the conjugative transfer of plasmid-borne ARGs across bacterial genera (Lu et al. 2020). Moreover, heavy metal and biocides can also promote the proliferation of ARGs via co-selection (Seiler and Berendonk 2012; Zhu et al. 2013; Baker et al. 2017). This prompted to investigate the potential ecological risks of environmental levels of AgNPs as an abiotic pressure to co-select antibiotic resistance genes (ARGs) or promote plasmid transfer between bacteria by horizontal transfers. Chen et al. (2019) used high throughput quantitative PCR to analyze the effect of AgNPs (100 ppm) on the co-selection pressure of ARGs in the rhizosphere and phyllosphere of 3 months aged *Coriandrum sativum* L. growing on a soil (pH 6.69) containing Cr, Cu, Zn, and Pb, and exposed to (~20 nm and ~50 nm) AgNPs. The exposure to AgNPs did not induce any significant increases in the total abundance of ARGs in either the rhizosphere or phyllosphere. However, the overall pattern of resistome was shifted following AgNPs application, with a significance increase in the relative abundance of efflux pumps genes, which is an important mechanism for co-selection of antibiotic resistance genes by heavy metals.

### 3.1.3 Others Nanopesticides Based on Inorganic Nanomaterials

Other nanopesticides are envisaged, based on nanomaterials of $TiO_2$, ZnO, $CeO_2$, Si NPs and even carbon nanotubes. For reviews on environmental impacts on microbiota of these NPs see Liné et al. (2017) and Tian et al. (2019).

### 3.1.4 Tentative Conclusion on Ag- and Cu-Based NPs in Agriculture

Altogether these data could tend to underline that Cu- and AgNPs can drastically shift the composition of microbial communities, and alter the activities of extracellular enzymes involved in element cycling. However, except one (Simonin et al. 2018b), many of these studies were dedicated to environmental impact of NPs and not to evaluate the impact of Cu- and Ag-based nanopesticides on off-target soil microbiota. At agronomical relevant concentrations and use, Kocide® 3000 ($Cu(OH)_2$) and CuONPs (0.1 mg $kg^{-1}$, (Simonin et al. 2018a; Simonin et al. 2018b) showed limited effects on soil microbial activities involved in carbon and nitrogen cycles.

For AgNPs, based on commercial AgNPs nanopesticides as Zerebra Agro® (Patent of the Russian Federation 2,419,439 as of 27.05.2011), the concentration targeted for agronomical applications is estimated to 0.2 mg $kg^{-1}$ (assuming a dispersion of AgNPs on a bulk soil density of 1.2 mg $cm^{-3}$, and a soil depth of 20 cm). At concentration close to this operational concentration, Grün et al. (2018, 2019) evidenced some long-term impact on proteobacteria and bacteria involved in N cycle. Note that AgNPs used in this study are AgPure®, which are designed for the antimicrobial functionalization of surfaces and bulk materials. Zerebra Agro® is composed of silver NPs modified with polyhexamethylene biguanidine, a polymer also endowed with biocide properties.

The behavior and fate of Cu- and AgNPs in soil depend on variables inherent to the NPs, e.g., particle size, surface charge, isoelectric point (pH at which the NPs carry not net electrical charge) and extrinsically related to the properties of the complex soil matrix. The shape of nanoparticles is a big player in governing the dissolution, and the interactions with cells. The properties of AgNPs, and NPs in general, can differentially affect the composition and functions of microbial communities depending on the level of exposure (Zhai et al. 2016).

Globally, the NPs can experience dissolution, transformation (oxidation and reduction), aggregation with soil colloids, adsorption especially on clays, (for a review, see Anjum et al. (2013) and reference inside). Important parameter that control the fate of Ag and Cu-based NPs, are the soil texture, clays are key players in the retention of NPs (Cornelis et al. 2014), pH, organic content, divalent cations, etc. High soil pH value increase the number of negatively charged sites and enhance Ag-sorption, while low pH tends to promote the dissolution of AgNPs. As shown by Schlich and Hund-Rinke (2015) in a variety of soils, AgNPs toxicity towards microbial activities such as substrate-induced respiration and to ammonia oxidizing bacteria declined with increasing clay content and increasing pH. Simonin et al. (2018a) also conclude on the same line about occasional impacts of CuONPs at agricultural relevant concentration, on coarse soil texture with low organic matter or clays contents. For the record, acidic soils occupy approximately 30% of the world's ice free land area but only about 4.5% of the acid soil area is used for arable crops (von Uexküll and Mutert 1995). The use of acidic soil can favors the dissolution of Cu- and AgNPs with the release of free ions, that can enhance the short-term impact of the nanos. In many studies commented in this chapter, the soils used were acid, and contained low clay contents, which make them worse case studies.

An interesting result from the literature is that the ionic or nanoform of the pesticide can show differentiated impacts, likely related to the fraction of ions released (e.g., Asadishad et al. 2018). Some authors already pointed that ionic and nanoforms of a metal may show similarities and differences, in the mode of antibacterial activity (Kędziora et al. 2018) or in the impact on a microbial community extracted from a soil and exposed in vitro to AgNPs (Zhai et al. 2016).

In long-term studies, the toxicity of NPs is kinetic and seems related to dissolution or transformation events in the soil, that lead to transient adjustment and adaptation of the microbial community. As evidenced by VandeVoort et al. (2014), tuning the surface properties of NPs could help to control the dissolution and phase changes, and likely to reduce the toxicity towards microbial cells.

As shown by Guilger et al. (2017), promising direction probably relies on biogenic nanoparticles, that show minimal impact on human cells, and denitrification, but strong activity toward a set of plant pathogenic fungi.

Among microbial activities that may be affected by NPs, denitrification ranks first. At a microscale level, soil structural organization provides diverse niches that are favorable to bacteria with different needs (aerobic or anaerobic) and lifestyle. Examining the localization on denitrifiers in soils, Lensi et al. (1995) showed that the <2 μm fractions contains a moderate density of denitrifiers, with high specific activity while the 20–2 μm fraction contained microaggregates and exhibited the highest microbial biomass C and organic N content and a high density of denitrifiers, with a moderate specific denitrifying activity. The main factors positively influencing denitrification are the absence of oxygen, the availability of nitrate and carbon (source of electrons) (Zumft 1997). Denitrifiers are also sensitive to pH

(Šlmek and Cooper 2002), and hydric conditions (Szukics et al. 2010). Denitrification is favored at lower soil redox potential values, which in turn is related to soil texture (Kunickis et al. 2010). Sandy textured soils generally show redox values too high for denitrification, while clayey textured soils provided lower redox values that were within the range for this biological transformation. VandeVoort and Arai (2012) related negative impacts on denitrification to the silver nanoparticle affinity for soil surfaces and to the physicochemical properties e.g., size, coating, sedimentation rate, solubility, surface charge properties, dispersibility (VandeVoort et al. 2014), showing that AgNPs properties could be tuned to avoid impact on denitrification. Hence, the biogenic AgNPs synthetized by green process (Guilger et al. 2017) did not show dramatic impact on the nitrogen cycle.

An understanding of the microbiome interactions with NPs at a microscale level could support a better design of the structure and properties of the NPs.

## 3.2 *Impact of Nanopesticides Based on Nanoformulation of Pesticides*

Nanotechnology has the potential to positively impact the agrifood sector, minimizing adverse problems of agricultural practices on environment and human health, improving food security and productivity (Fraceto et al. 2016). In this context, nanocapsulated formulations, nanoemulsion, nanogel of conventional pesticides, and metal and metal oxide nanoparticles have been designed in order to control the release of the active ingredient, favor adsorption on plant leaves and reduce the dose, protect the active molecule (Fraceto et al. 2016; Chhipa 2017). While the impact of metal and metal oxide nanoparticles on the soil microbiota as non-target organism has been addressed in the literature, those of nanoformulation of pesticides still stay poorly documented.

Liu et al. (2014) synthesized a new nanopesticide CM-β-CD-MNPs-Diuron (average diameter of 25 nm by TEM) from an inclusion complex of cyclodextrin-$Fe_3O_4$ magnetic nanoparticles as host and diuron as guest molecules. Their potential toxic effects on soil microbiota was evaluated by microcalorimetry, measure of urease enzyme activity and qPCR. By recording heat flow rate of microbial growth, microcalorimetry provides information on microbial biochemical processes and evaluate the metabolism of microbial biomass in soil. Soil samples (1.00 g) in ampoules were spiked with glucose and ammonium sulfate were exposed to different concentrations of CM-β-CD-MNPs-Diuron (5.00, 20.0, 80.0, and 150 mg $g^{-1}$ in dry soil samples) at 28 °C. CM-β-CD-MNPs-Diuron leads to the inhibition of the metabolic activity of microorganism in soil. Urease catalyzes the conversion reaction of urea to carbon dioxide and ammonia, leading to available nitrogen for plants. There was a significant effect ($p < 0.05$) of CM-β-CD-MNPs-Diuron on the urease enzyme activity at 7, 14, and 21 days of incubation. Real-time qPCR and universal probes were used to quantify the impact of different concentration (0.00, 5.00, 20.0,

80.0, and 150 mg $g^{-1}$) of CM-β-CD-MNPs-Diuron on population size of the microorganism community in soil for 21 days. The abundance the soil bacterial community decreases when the dose of CM-β-CD-MNPs-Diuron increases while for *Actinobacteria*, the population does not change significantly at the different doses. Diuron has a negative effect on the microbial population (Prado and Airoldi 2001) and iron-based nanoparticles are toxic to bacterial community of soil due to the generation of reactive oxygen species (He et al. 2019; Guilger et al. 2017). Altogether, these results show CM-β-CD-MNPs-Diuron exerts a stress on soil microorganism and that encapsulation of Diuron did not help to decrease the toxicity.

As a counter-example, Maruyama et al. (2016) decreased the toxicity of Imazapic and Imazapyr herbicides by formulating them in alginate/chitosan and chitosan/tripolyphosphate nanoparticles (average size of 400 nm). These systemic herbicides are used to control weeds in many crops, and are used as combination to bypass the resistance of plants. An agricultural soil was sampled and exposed to the herbicides using doses equivalent to the application rates employed in the field.

The impact of the formulations on soil was assessed by qPCR of genes involved in nitrogen cycle. Quantification of *nifH*, *nirk*, *nirS*, *narG*, *norB*, and *nosZ*, bacterial genes in the soil samples treated with the nanoparticles for 7 and 30 days showed that the encapsulated herbicides were less toxic, compared to the free Imazapic and Imazapyr compounds.

Essential oils are a promising option for substituting the synthetic pesticides used in agriculture. Neem oil is effective against a wide range of pests, exhibiting a broad spectrum of action due to its systemic and transmembrane activities. However, its use in the field is limited by its short persistence in the environment (Shah et al. 2017; Kumar et al. 2019). Pascoli et al. (2019) formulated neem oil loaded zein nanoparticles as spherical particles of 100–200 nm (atomic force microscopy). Zein is a corn protein. The impact of these NPs on soil microbiota was also assessed by qPCR of specific genes from nitrogen cycle bacteria: *nifH*, *amoA* (encoding ammonia monooxygenase, nitrification enzyme: conversion of ammonia to hydroxylamine), *haO* (encoding hydroxylamine oxidase, nitrification enzyme: oxidation of hydroxylamine to nitrite), *narG*, *nirK* and *nirS*, *cnorB,* and *nosZ*. No change in the number of genes which encode nitrogen-fixing enzymes and denitrifying enzymes was detected, suggesting no effect on soil bacteria involved in nitrogen cycle. The encapsulation in zein nanoparticles reduced the genotoxicity of neem oil to *Allia cepa* and nullified the toxicity in *Caenorhabditis elegans*. Thus encapsulation of the herbicides could improve their mode of action and reduced their toxicity.

Hexaconazole is a pest control and a plant growth regulator. In order to reduce its adverse effects in some plants (Kumar et al. 2015) have developed nanoparticles of hexaconazole using polyethyleneglycol-400 (PEG) as the surface stabilizing agent. The nanoparticles show an average size of 100 nm (SEM). The impact of hexaconazole NPs on non-target soil microbiota was assessed by measure of the quantities of ammonium, nitrite, and nitrate-nitrogen as indicators of soil nitrification activity. Ammonia and nitrite oxidizing bacteria are unaffected by hexaconazole NPs, and commercial formulation of hexaconazole.

### *3.3 Different Nanovectors of Pesticides and Different Impacts*

The different examples discussed above show that in most cases, the nanoformulation of pesticides and herbicides using organic polymers, improved or did not worsen the impact of the active ingredient on non-target soil microbiota. Compared the inorganic metal and metal oxide nanoparticles discussed in the first part, the average size of these pesticides encapsulated in polymers is far higher than those of the inorganic nanoparticles and these organic formulations seem safer toward nitrogen cycle (nitrification and denitrification).

Regarding nitrogen cycle, many studies focus on the abundance of nitrification and denitrification genes, using qPCR. Taking advantage of the diversity revealed by metagenomic in microbial functional groups, Ma et al. (2019) reevaluated the coverage of existing DNA primers for denitrification functional genes, using in silico approach. They confirm that the existing primers reveal a partial vision of the denitrifiers community. As example, the non-specific coverage of fungi lead to underestimation of the potential importance of fungal denitrifiers.

## 4 Conclusion and Future Directions

Nanotechnology sounds promising to decrease pesticides impact on non-target soil microorganisms. There is a great potential in modulating the surface of the NPs, to tune their properties, their interaction and fate in soil. Encapsulation of active ingredients in polymers, formulation of biogenic NPs, and designing the properties of NPs to reduce their impact appear as promising opportunities.

From a futuristic perspective, but already under development, smart nanoparticulate vectors of pesticides can be designed in order to deal with controlled and targeted release, taking advantage of environment stimuli responsive nanopesticides (Camara et al. 2019). All these smart-devices should rely on green-technologies and biocompatible materials.

An important prerequisite to the development of nanopesticides is to assess their innocuity on soil microorganisms in order to preserve the soil ecosystem, and to control the diffusion of nanopesticides. In the soil matrix, to avoid contamination of the water compartment. Soil depth targeted release could be envisioned using as synthetic virus-based model nanopesticides those mobility whose mobility allows them to reach different depths in soil (Chariou et al. 2019).

Currently, research is focusing on the search for microbiota that allow plants to defend themselves against abiotic (drought, flooding, etc.) or biotic (plant pathogens) stresses or to improve the growth and yield of field crops. Nanopesticides must fit in this scheme, and allow combined uses of both approach, in preventive (seed treatment, disinfection or stimulation of seedling transplanting) and curative (plant treatment) treatments.

Regarding the impact on microbiome, the methology focuses on diversity revealed by sequencing an amplicon of 16S rRNA, to the impact on the bacterial community present. A broader approach would address the diversity of bacteria, together with those of archaea, fungi, protozoa, etc., allowing to examine how nanopesticides are disrupting the networks of interactions between these communities. A sharper advance could focus on the active communities (complementary DNA) and the expression of genes. Indeed, extracellular DNA can persist in soil, and hide some changes in the community.

Interaction between microorganisms and macroorganisms should be deciphered, especially addressing how nanopesticides present in the soil or systemic in the plant may modify the microbiota recruited in plant roots and shoots, which is currently poorly documented. Some organisms inhabiting soils, such as nematodes, can also modify the impact of nanopesticides on soil microbiota. Recently, Bart et al. (2019) evidenced that nematodes can mitigate the toxicity of pesticides on soil microbial enzymatic activities.

Going back to soil, which is the basic matrix for agronomy, the microstructure of soil aggregates directly impacts soil communities and functional diversity, and likely the implications of nanoagrochemicals. To overcome the complexity of soil matrix, microfluidic techniques provide new ways of studying soil microbial ecology by allowing simulation and manipulation of chemical conditions and physical structures at the microscale in soil model habitats (Aleklett et al. 2018).

As final conclusion, soil must be considered as a super-organism (Lovelock 1993). In order to design smart solutions for agronomy, the soil ecosystem has to be addressed globally and in interaction with the air and water compartment.

## References

Adeleye AS, Conway JR, Perez T, Rutten P, Keller AA (2014) Influence of extracellular polymeric substances on the long-term fate, dissolution, and speciation of copper-based nanoparticles. Environ Sci Technol 48:12561–12568

Adisa IO, Pullagurala VLR, Peralta-Videa JR, Dimkpa CO, Elmer WH, Gardea-Torresdey JL, White JC (2019) Recent advances in nano-enabled fertilizers and pesticides: a critical review of mechanisms of action. Environ Sci Nano 6:2002–2030

Aleklett K, Kiers ET, Ohlsson P, Shimizu TS, Caldas VE, Hammer EC (2018) Build your own soil: exploring microfluidics to create microbial habitat structures. ISME J 12:312–319

Allen HK, Donato J, Wang HH, Cloud-Hansen KA, Davies J, Handelsman J (2010) Call of the wild: antibiotic resistance genes in natural environments. Nat Rev Microbiol 8:251–259

Anjum NA, Gill SS, Duarte AC, Pereira E, Ahmad I (2013) Silver nanoparticles in soil–plant systems. J Nanopart Res 15:1896

Asadishad B, Chahal S, Akbari A, Cianciarelli V, Azodi M, Ghoshal S, Tufenkji N (2018) Amendment of agricultural soil with metal nanoparticles: effects on soil enzyme activity and microbial community composition. Environ Sci Technol 52:1908–1918

Baker S, Volova T, Prudnikova SV, Satish S, Prasad MNN (2017) Nanoagroparticles emerging trends and future prospect in modern agriculture system. Environ Toxicol Pharmacol 53:10–17

Bart S, Pelosi C, Barraud A, Péry ARR, Cheviron N, Grondin V, Mougin C, Crouzet O (2019) Earthworms mitigate pesticide effects on soil microbial activities. Front Microbiol 10:1535. https://www.frontiersin.org/articles/10.3389/fmicb.2019.01535/full. Accessed 22 Jan 2020
Berry D, Widder S (2014) Deciphering microbial interactions and detecting keystone species with co-occurrence networks. Front Microbiol 5:219
Camara MC, Campos EVR, Monteiro RA, Do Espirito Santo Pereira A, De Freitas Proença PL, Fraceto LF (2019) Development of stimuli-responsive nano-based pesticides: emerging opportunities for agriculture. J Nanobiotechnol 17:100
Caporaso JG, Lauber CL, Walters WA, Berg-Lyons D, Lozupone CA, Turnbaugh PJ, Fierer N, Knight R (2011) Global patterns of 16S rRNA diversity at a depth of millions of sequences per sample. Proc Natl Acad Sci 108:4516–4522
Chariou PL, Dogan AB, Welsh AG, Saidel GM, Baskaran H, Steinmetz NF (2019) Soil mobility of synthetic and virus-based model nanopesticides. Nat Nanotechnol 14:712–718
Chen Q-L, Zhu D, An X-L, Ding J, Zhu Y-G, Cui L (2019) Does nano silver promote the selection of antibiotic resistance genes in soil and plant? Environ Int 128:399–406
Chhipa H (2017) Nanofertilizers and nanopesticides for agriculture. Environ Chem Lett 15:15–22
Cornelis G, Hund-Rinke K, Kuhlbusch T, Brink N, Van den Nickel C (2014) Fate and bioavailability of engineered nanoparticles in soils: a review. Crit Rev Environ Sci Technol 44:2720–2764
Driouich A, Smith C, Ropitaux M, Chambard M, Boulogne I, Bernard S, Follet-Gueye M-L, Vicré M, Moore J (2019) Root extracellular traps versus neutrophil extracellular traps in host defence, a case of functional convergence? Biol Rev Camb Philos Soc 94:1685–1700
Durán N, Durán M, De Jesus MB, Seabra AB, Fávaro WJ, Nakazato G (2016) Silver nanoparticles: a new view on mechanistic aspects on antimicrobial activity. Nanomed Nanotechnol Biol Med 12:789–799
Fraceto LF, Grillo R, De Medeiros GA, Scognamiglio V, Rea G, Bartolucci C (2016) Nanotechnology in agriculture: which innovation potential does it have? Front Environ Sci 4:20. https://www.frontiersin.org/articles/10.3389/fenvs.2016.00020/full. Accessed 16 Jan 2020
García-Ruiz R, Ochoa V, Hinojosa MB, Carreira JA (2008) Suitability of enzyme activities for the monitoring of soil quality improvement in organic agricultural systems. Soil Biol Biochem 40:2137–2145
Grün A-L, Straskraba S, Schulz S, Schloter M, Emmerling C (2018) Long-term effects of environmentally relevant concentrations of silver nanoparticles on microbial biomass, enzyme activity, and functional genes involved in the nitrogen cycle of loamy soil. J Environ Sci 69:12–22
Grün A-L, Manz W, Kohl YL, Meier F, Straskraba S, Jost C, Drexel R, Emmerling C (2019) Impact of silver nanoparticles (AgNP) on soil microbial community depending on functionalization, concentration, exposure time, and soil texture. Environ Sci Eur 31:15
Guilger M, Pasquoto-Stigliani T, Bilesky-Jose N, Grillo R, Abhilash PC, Fraceto LF, Lima R (2017) Biogenic silver nanoparticles based on trichoderma harzianum: synthesis, characterization, toxicity evaluation and biological activity. Sci Rep 7:44421
Hashimoto Y, Takeuchi S, Mitsunobu S, Ok Y-S (2017) Chemical speciation of silver (Ag) in soils under aerobic and anaerobic conditions: Ag nanoparticles vs. ionic Ag. J Hazard Mater 322:318–324
Hawes MC, Gunawardena U, Miyasaka S, Zhao X (2000) The role of root border cells in plant defense. Trends Plant Sci 5:128–133
Hawes MC, Curlango-Rivera G, Xiong Z, Kessler JO (2012) Roles of root border cells in plant defense and regulation of rhizosphere microbial populations by extracellular DNA 'trapping'. Plant Soil 355:1–16
Hawes M, Allen C, Turgeon BG, Curlango-Rivera G, Minh TT, Huskey DA, Xiong Z (2016) Root border cells and their role in plant defense. Annu Rev Phytopathol 54:143–161
He X, Deng H, Hwang H (2019) The current application of nanotechnology in food and agriculture. J Food Drug Anal 27:1–21
Herren CM, McMahon KD (2018) Keystone taxa predict compositional change in microbial communities. Environ Microbiol 20:2207–2217

Hugerth LW, Andersson AF (2017) Analysing microbial community composition through amplicon sequencing: from sampling to hypothesis testing. Front Microbiol 8:1561

Hund-Rinke K, Hümmler A, Schlinkert R, Wege F, Broll G (2019) Evaluation of microbial shifts caused by a silver nanomaterial: comparison of four test systems. Environ Sci Eur 31:86

Judy JD, McNear Jr DH, Chen C, Lewis RW, Tsyusko OV, Bertsch PM, Rao W, Stegemeier J, Lowry GV, McGrath SP, Durenkamp M, Unrine J (2015) Nanomaterials in biosolids inhibit nodulation, shift microbial community composition, and result in increased metal uptake relative to bulk/dissolved metals. Environ Sci Technol 49(14):8751–8758

Jung J-H, Kim S-W, Min J-S, Kim Y-J, Lamsal K, Kim KS, Lee YS (2010) The effect of nano-silver liquid against the white rot of the green onion caused by Sclerotium cepivorum. Mycobiology 38:39–45

Kah M, Kookana RS, Gogos A, Bucheli TD (2018) A critical evaluation of nanopesticides and nanofertilizers against their conventional analogues. Nat Nanotechnol 13:677–684

Karaca A, Cetin SC, Turgay OC, Kizilkaya R (2011) Soil enzymes as indication of soil quality. In: Shukla G, Varma A (eds) Soil enzymology. Springer, Berlin, p 119–148 https://doi.org/10.1007/978-3-642-14225-3_7. Accessed 16 Jan 2020

Karpouzas DG, Tsiamis G, Trevisan M, Ferrari F, Malandain C, Sibourg O, Martin-Laurent F (2016) "LOVE TO HATE" pesticides: felicity or curse for the soil microbial community? An FP7 IAPP Marie Curie project aiming to establish tools for the assessment of the mechanisms controlling the interactions of pesticides with soil microorganisms. Environ Sci Pollut Res Int 23:18947–18951

Kędziora A, Speruda M, Krzyżewska E, Rybka J, Łukowiak A, Bugla-Płoskońska G (2018) Similarities and differences between silver ions and silver in nanoforms as antibacterial agents. Int J Mol Sci 19:pii: E444

Kumar R, Nair KK, Alam MI, Gogoi R, Singh PK, Srivastava C, Gopal M, Goswami A (2015) Development and quality control of nanohexaconazole as an effective fungicide and its biosafety studies on soil nitifiers. J Nanosci Nanotechnol 15:1350–1356

Kumar S, Nehra M, Dilbaghi N, Marrazza G, Hassan AA, Kim K-H (2019) Nano-based smart pesticide formulations: emerging opportunities for agriculture. J Control Release 294:131–153

Kunickis SH, Gilliam JW, Evans RO, Dukes M (2010) Soil characteristics and their role in developing conditions favorable for denitrification. In: Proceedings of the 19th World Congress of Soil Science: Soil solutions for a changing world, Division Symposium 22 Management of landscapes for the future, Brisbane, 1–6 August 2010, pp 34–37

Kurenbach B, Gibson PS, Hill AM, Bitzer AS, Silby MW, Godsoe W, Heinemann JA (2017) Herbicide ingredients change Salmonella enterica sv. Typhimurium and Escherichia coli antibiotic responses. Microbiology 163:1791–1801

Lederberg J, McCray AT (2001) 'Ome Sweet 'Omics–a genealogical treasury of words. Scientist 15(7):8. https://link.galegroup.com/apps/doc/A73535513/AONE?sid=lms. Accessed 16 Jan 2020.

Lensi R, Clays-Josserand A, Jocteur ML (1995) Denitrifiers and denitrifying activity in size fractions of a mollisol under permanent pasture and continuous cultivation. Soil Biol Biochem 27:61–69

Liné C, Larue C, Flahaut E (2017) Carbon nanotubes: impacts and behaviour in the terrestrial ecosystem—a review. Carbon 123:767–785

Liu W, Yao J, Cai M, Chai H, Zhang C, Sun J, Chandankere R, Masakorala K (2014) Synthesis of a novel nanopesticide and its potential toxic effect on soil microbial activity. J Nanopart Res 16:2677

Lo C-C (2010) Effect of pesticides on soil microbial community. J Environ Sci Health Part B 45:348–359

Lovelock JE (1993) The soil as a model for the Earth. Geoderma 57:213–215

Lu J, Wang Y, Jin M, Yuan Z, Bond P, Guo J (2020) Both silver ions and silver nanoparticles facilitate the horizontal transfer of plasmid-mediated antibiotic resistance genes. Water Res 169:115229

Ma Y, Zilles JL, Kent AD (2019) An evaluation of primers for detecting denitrifiers via their functional genes. Environ Microbiol 21:1196–1210

Marshall BM, Levy SB (2011) Food animals and antimicrobials: impacts on human health. Clin Microbiol Rev 24:718–733

Maruyama CR, Guilger M, Pascoli M, Bileshy-José N, Abhilash PC, Fraceto LF, de Lima R (2016) Nanoparticles based on chitosan as carriers for the combined herbicides imazapic and imazapyr. Sci Rep 6:1–15

Millardet A, Gayon U, Schneiderhan FJ (1933) The discovery of Bordeaux mixture. Three papers: I. Treatment of mildew and rot. II. Treatment of mildew with copper sulphate and lime mixture. III. Concerning the history of the treatment of mildew with copper sulphate. American Phytopathological Society, Ithaca

OCDE 216 (2000) Test No. 216: Soil microorganisms: nitrogen transformation test. https://doi.org/10.1787/9789264070226-en

Parada J, Rubilar O, Sousa DZ, Martínez M, Fernández-Baldo MA, Tortella GR (2019) Short term changes in the abundance of nitrifying microorganisms in a soil-plant system simultaneously exposed to copper nanoparticles and atrazine. Sci Total Environ 670:1068–1074

Parisi C, Vigani M, Rodríguez-Cerezo E (2015) Agricultural nanotechnologies: what are the current possibilities? Nano Today 10:124–127

Pascoli M, Jacques MT, Agarrayua DA, Avila DS, Lima R, Fraceto LF (2019) Neem oil based nanopesticide as an environmentally-friendly formulation for applications in sustainable agriculture: an ecotoxicological perspective. Sci Total Environ 677:57–67

Plancot B, Santaella C, Jaber R, Kiefer-Meyer MC, Follet-Gueye M-L, Leprince J, Gattin I, Souc C, Driouich A, Vicré-Gibouin M (2013) Deciphering the responses of root border-like cells of arabidopsis and flax to pathogen-derived elicitors. Plant Physiol 163:1584–1597

Prado AG, Airoldi C (2001) The effect of the herbicide diuron on soil microbial activity. Pest Manag Sci 57:640–644

Rajput V, Minkina T, Ahmed B, Sushkova S, Singh R, Soldatov M, Laratte B, Fedorenko A, Mandzhieva S, Blicharska E, Musarrat J, Saquib Q, Flieger J, Gorovtsov A (2020) Interaction of copper-based nanoparticles to soil, terrestrial, and aquatic systems: critical review of the state of the science and future perspectives. In: De Voogt P (ed) Reviews of environmental contamination and toxicology, vol 252. Springer, Cham, pp 51–96

Rangasamy K, Athiappan M, Devarajan N, Samykannu G, Parray JA, Aruljothi KN, Shameem N, Alqarawi AA, Hashem A, Abd_Allah E.F. (2018) Pesticide degrading natural multidrug resistance bacterial flora. Microb Pathog 114:304–310

Ropitaux M, Bernard S, Follet-Gueye M-L, Vicré M, Boulogne I, Driouich A (2019) Xyloglucan and cellulose form molecular cross-bridges connecting root border cells in pea (Pisum sativum). Plant Physiol Biochem 139:191–196

Santaella C, Schue M, Berge O, Heulin T, Achouak W (2008) The exopolysaccharide of Rhizobium sp. YAS34 is not necessary for biofilm formation on Arabidopsis thaliana and Brassica napus roots but contributes to root colonization. Environ Microbiol 10:2150–2163

Schlaeppi K, Dombrowski N, Oter RG, Van Themaat EVL, Schulze-Lefert P (2014) Quantitative divergence of the bacterial root microbiota in Arabidopsis thaliana relatives. Proc Natl Acad Sci 111:585–592

Schlich K, Hund-Rinke K (2015) Influence of soil properties on the effect of silver nanomaterials on microbial activity in five soils. Environ Pollut 196:321–330

Schlich K, Hoppe M, Kraas M, Schubert J, Chanana M, Hund-Rinke K (2018) Long-term effects of three different silver sulfide nanomaterials, silver nitrate and bulk silver sulfide on soil microorganisms and plants. Environ Pollut 242:1850–1859

Seiler C, Berendonk TU (2012) Heavy metal driven co-selection of antibiotic resistance in soil and water bodies impacted by agriculture and aquaculture. Front Microbiol 3:399

Shade A (2017) Diversity is the question, not the answer. ISME J 11:1–6

Shade A, Jones SE, Caporaso JG, Handelsman J, Knight R, Fierer N, Gilbert JA (2014) Conditionally rare taxa disproportionately contribute to temporal changes in microbial diversity. mBio 5:e01371-01314

Shah FM, Razaq M, Ali A, Han P, Chen J (2017) Comparative role of neem seed extract, moringa leaf extract and imidacloprid in the management of wheat aphids in relation to yield losses in Pakistan. PLoS One 12:e0184639

Shao H, Zhang Y (2017) Non-target effects on soil microbial parameters of the synthetic pesticide carbendazim with the biopesticides cantharidin and norcantharidin. Sci Rep 7:1–12

Shi J, Ye J, Fang H, Zhang S, Xu C (2018) Effects of copper oxide nanoparticles on paddy soil properties and components. Nanomaterials 8(10):839. https://www.ncbi.nlm.nih.gov/pmc/articles/PMC6215298/. Accessed 23 Jan 2020

Shokralla S, Spall JL, Gibson JF, Hajibabaei M (2012) Next-generation sequencing technologies for environmental DNA research. Mol Ecol 21:1794–1805

ŠImek M, Cooper JE (2002) The influence of soil pH on denitrification: progress towards the understanding of this interaction over the last 50 years. Eur J Soil Sci 53:345–354

Simonin M, Richaume A (2015) Impact of engineered nanoparticles on the activity, abundance, and diversity of soil microbial communities: a review. Environ Sci Pollut Res 22:13710–13723

Simonin M, Cantarel AAM, Crouzet A, Gervaix J, Martins JMF, Richaume A (2018a) Negative effects of copper oxide nanoparticles on carbon and nitrogen cycle microbial activities in contrasting agricultural soils and in presence of plants. Front Microbiol 9:3102. https://www.frontiersin.org/articles/10.3389/fmicb.2018.03102/full. Accessed 23 Jan 2020

Simonin M, Colman BP, Tang W, Judy JD, Anderson SM, Bergemann CM, Rocca JD, Unrine JM, Cassar N, Bernhardt ES (2018b) Plant and microbial responses to repeated Cu(OH)2 nanopesticide exposures under different fertilization levels in an agro-ecosystem. Front Microbiol 9:1769. https://www.frontiersin.org/articles/10.3389/fmicb.2018.01769/full

Simonin M, Colman BP, Anderson SM, King RS, Ruis MT, Avellan A, Bergemann CM, Perrotta BG, Geitner NK, Ho M, Barrera B, De la Unrine JM, Lowry GV, Richardson CJ, Wiesner MR, Bernhardt ES (2018c) Engineered nanoparticles interact with nutrients to intensify eutrophication in a wetland ecosystem experiment. Ecol Appl 28:1435–1449

Szukics U, Abell GCJ, Hödl V, Mitter B, Sessitsch A, Hackl E, Zechmeister-Boltenstern S (2010) Nitrifiers and denitrifiers respond rapidly to changed moisture and increasing temperature in a pristine forest soil. FEMS Microbiol Ecol 72:395–406

Thiour-Mauprivez C, Martin-Laurent F, Calvayrac C, Barthelmebs L (2019) Effects of herbicide on non-target microorganisms: towards a new class of biomarkers? Sci Total Environ 684:314–325

Tian H, Kah M, Kariman K (2019) Are nanoparticles a threat to Mycorrhizal and Rhizobial symbioses? a critical review. Front Microbiol 10:1660. https://www.frontiersin.org/articles/10.3389/fmicb.2019.01660/full. Accessed 22 Jan 2020

Totsche KU, Amelung W, Gerzabek MH, Guggenberger G, Klumpp E, Knief C, Lehndorff E, Mikutta R, Peth S, Prechtel A, Ray N, Kögel-Knabner I (2018) Microaggregates in soils. J Plant Nutr Soil Sci 181:104–136

Van Bruggen AHC, He MM, Shin K, Mai V, Jeong KC, Finckh MR, Morris JG (2018) Environmental and health effects of the herbicide glyphosate. Sci Total Environ 616–617:255–268

VandeVoort AR, Arai Y (2012) Effect of silver nanoparticles on soil denitrification kinetics. Ind Biotechnol 8:358–364

VandeVoort AR, Skipper H, Arai Y (2014) Macroscopic assessment of nanosilver toxicity to soil denitrification kinetics. J Environ Qual 43:1424–1430

Vicré M, Santaella C, Blanchet S, Gateau A, Driouich A (2005) Root border-like cells of arabidopsis. microscopical characterization and role in the interaction with rhizobacteria. Plant Physiol 138:998–1008

Vitali F, Raio A, Sebastiani F, Cherubini P, Cavalieri D, Cocozza C (2019) Environmental pollution effects on plant microbiota: the case study of poplar bacterial-fungal response to silver nanoparticles. Appl Microbiol Biotechnol 103:8215–8227

Von Uexküll HR, Mutert E (1995) Global extent, development and economic impact of acid soils. Plant Soil 171:1–15

Watteau F, Villemin G (2018) Soil microstructures examined through transmission electron microscopy reveal soil-microorganisms interactions. Front Environ Sci 6:1–10. https://www.frontiersin.org/articles/10.3389/fenvs.2018.00106/full. Accessed 16 Jan 2020

Whittaker RH (1960) Vegetation of the Siskiyou mountains, Oregon and California. Ecol Monogr 30:279–338

Wilpiszeski RL, Aufrecht JA, Retterer ST, Sullivan MB, Graham DE, Pierce EM, Zablocki OD, Palumbo AV, Elias DA (2019) Soil aggregate microbial communities: towards understanding microbiome interactions at biologically relevant scales. Appl Environ Microbiol 85(14):pii: e00324-19

Wolińska A, Stępniewska Z, Pytlak A (2015) The effect of environmental factors on total soil DNA content and dehydrogenase activity. Arch Biol Sci 67(2):493–501

Young IM, Crawford JW, Nunan N, Otten W, Spiers A (2008) Chapter 4 microbial distribution in soils: physics and scaling. In: Advances in agronomy. Academic, San Diego, pp 81–121. http://www.sciencedirect.com/science/article/pii/S0065211308006044. Accessed 23 Jan 2020

Zhai Y, Hunting ER, Wouters M, Peijnenburg WJGM, Vijver MG (2016) Silver nanoparticles, ions, and shape governing soil microbial functional diversity: nano shapes micro. Front Microbiol 7:1123. https://www.frontiersin.org/articles/10.3389/fmicb.2016.01123/full. Accessed 22 Jan 2020

Zhang X, Xu Z, Wu M, Qian X, Lin D, Zhang H, Tang J, Zeng T, Yao W, Filser J, Li L, Sharma VK (2019) Potential environmental risks of nanopesticides: application of Cu(OH)2 nanopesticides to soil mitigates the degradation of neonicotinoid thiacloprid. Environ Int 129:42–50

Zhu Y-G, Johnson TA, Su J-Q, Qiao M, Guo G-X, Stedtfeld RD, Hashsham SA, Tiedje JM (2013) Diverse and abundant antibiotic resistance genes in Chinese swine farms. Proc Natl Acad Sci 110:3435–3440

Zumft WG (1997) Cell biology and molecular basis of denitrification. Microbiol Mol Biol Rev 61:533–616

# Bioactivity of Nanoformulated Synthetic and Natural Insecticides and Their Impact on Environment

**Josef Jampílek, Katarína Kráľová, and Peter Fedor**

*"This chapter is sincerely dedicated to my father, PhMr. Josef Jampílek, a long-time and passionate beekeeper, on the occasion of his 81st birthday, and also to his beloved bees being an important part of the joyful content of his life."*

**Abstract** Insects represent the most diverse group of organisms on our planet with approximately one million described species. While some of them have beneficial effects in ecosystem services through plant pollination and natural pest control, there are numerous quarantine insect pests causing considerable damage to crop production and storage. Consequently, in crop pest management, the application of effective insecticides is extremely needed, and at selection of appropriate active compounds, the effects of insecticides or their residues on non-target organisms should be considered. The application of synthetic insecticides could result in the resistance of the target insect against the applied insecticide. Therefore, recently, a great attention has been devoted to insecticide formulations using active compounds of natural origin that are less toxic than conventional synthetic insecticides, exert the effects exclusively on the target insect and closely related organisms, are very effective in very small doses, are characterized with rapid decomposition, and, due to short exposure, practically do not contribute to environmental pollution. Using a nanotechnology approach, insecticide formulations with the enhanced bioavailability of active ingredients enabling their targeted delivery, controlled release, protection against degradation, and higher effectiveness could be prepared. In this manner, the overuse of

J. Jampílek (✉)
Department of Analytical Chemistry, Faculty of Natural Sciences, Comenius University, Bratislava, Slovakia

K. Kráľová
Institute of Chemistry, Faculty of Natural Sciences, Comenius University, Bratislava, Slovakia

P. Fedor
Department of Environmental Ecology, Faculty of Natural Sciences, Comenius University, Bratislava, Slovakia



L. F. Fraceto et al. (eds.), *Nanopesticides*,
https://doi.org/10.1007/978-3-030-44873-8_7

these toxic compounds could be avoided resulting in the reduced contamination of the environment and representing an economically favorable solution. This chapter gives a comprehensive overview of recent findings related to the bioactivity of nanoformulations of synthetic and natural insecticides against harmful insects causing severe damage to economically important crops or deteriorating stored food products. The impact of nanoinsecticides on the environment, including potential deleterious effects on non-target organisms, is discussed as well.

**Keywords** Nanoparticles · Nanomaterials · Nanoformulations · Insecticides · Organophosphates · Carbamates · Neonicotinoids · Pyrethroids · Disruptors · Avermectins · Metals · Metalloids · Essential oils · Nanoecotoxicology

# 1 Introduction

Insects represent the most diverse group of organisms on Earth with approximately one million of less than two million of all described species, including some 15,000 new descriptions annually. Based on modern estimation methods, several more million species, varying around the average of 5.5 million in total, are supposed to be discovered and described in future (Stork 2018). However, a lot of other professional estimates refer even to higher global biodiversity (May 2010).

For more than 350 million years, insects have developed into one of the most evolutionary successful groups with perfect adaptations for life in various environments. Strong ecological interactions of insects with other organisms, including humans, underline their fundamental and essential role in ecosystems, providing a wide variety of ecosystem services, including biocontrol, decomposition, carbon cycles, pedogenesis, and pollination. Based on the growing evidence on the beneficial role of insects in food production, it could be noted that insect pollination is to some degree necessary for 75% of the crop species used for food (Klein et al. 2007; Bommarco et al. 2013), and crops depending on pollinators represent 35% of global crop production volume (Potts et al. 2016). The overwhelming majority of pollinator species are wild, including >20,000 species of Hymenoptera, Diptera, Lepidoptera, and other insect orders; moreover, there are several domesticated arthropods, with western honey bee (*Apis mellifera*) being the best known species (Potts et al. 2016). A wide variety of predaceous insects provide the important regulation of numerous pest species (Maas et al. 2013). The economic benefits of natural biological coffee pest control have been estimated at US$75–310 $ha^{-1}$ $year^{-1}$ (the plantation benefit corresponds to Costa Rica's average annual income) (Karp et al. 2013; Schowalter et al. 2018). Insects are essential components of ecological systems providing many supporting services, including primary production control, nutrient balance (Belovsky and Slade 2018), organic matter cycles, and nitrogen reduction.

While some of them have beneficial effects on ecosystem services, there are numerous quarantine insect pests causing considerable damage to crop production, storage, or even human health. FAO (2014) has defined pest as "any species, strain, or biotype of plant, animal, or pathogenic agent that is injurious to plants and plant products, materials, or environments and includes vectors of parasites or pathogens of humans and animal diseases and animals causing public health nuisance." According to the WHO report (2017), insect vectors of human diseases, particularly Diptera, Hemiptera, and Anoplura, refer to millions of deaths annually, especially in developing countries. Malaria as the most important mosquito-borne infectious disease in tropical and subtropical regions represents a risk for some 100 countries. The WHO (WHO 2015) estimates that in 2015, there were 214 million new cases of malaria resulting in 438,000 deaths.

Approximately a fifth of the world's global production is believed to be destroyed and damaged by insect pests annually. Farmland, particularly monocultures, represent specific, human supported and usually large size ecosystems with reduced ecological stability and weakened mechanisms of resistance or resilience. With high concentration of nutritious value, they offer suitable conditions for infesting insects (Sallam 2013). Generally, food plants are damaged by more than 10,000 species of insects (Dhaliwal et al. 2007). In some cases, the reduction of yield caused by insect pests grows up to 60–70% (Singhand and Gandhi 2012).

Aridization and global warming models can make crops vulnerable to pest infestation, which should spread to higher latitudes, and many species of limited colds can increase their geographical scope (Thomson et al. 2010) with numerous examples in temperate and Boreal climate territories (e.g., Battisti et al. 2005; Masarovič et al. 2014, 2017; Fedor et al. 2018).

Over the last century climate of the planet has generally warmed up by ca 0.6 °C (Walther et al. 2002), what in decisive manner exerted influence on the distribution and ecological dynamics of many species, including insects (e.g., Walther et al. 2002; Deka et al. 2011). Progressive spread of exotic pests represents a severe problem in Europe adversely affecting not only natural ecosystems but also urban and farmland areas and is associated with environmental, ecological, and even economic consequences, which are also strengthened with the actual climate change and globalization of biological commodity trade (Lodge et al. 2006; Varga and Fedor 2008; Hulme 2009). In fact, up to €12.5 billion a year has been spent on controlling biological invasion in the European Union for over 20 years (Kettunen et al. 2008).

On the other hand, a warmer climate will change at least two agriculturally relevant insect pest characteristics: an increase of metabolic rate of individual insects (and thus, consequentially, of food consumption) and changes in pest insect species richness (growth or decline) in accordance with specific ecological conditions (Dillon et al. 2010; Deutsch et al. 2018).

Pest control has been recently based on a wide variety of pesticides, for example, organophosphates, neonicotinoids, pyrethroids, agents of the group of chitin synthesis inhibitors, avermectins, and botanical insecticides (bioinsecticides). Although each type of insecticides has its specific beneficial and side-effect particularities, insect pests are capable to adapt to new environments and situations, e.g., overcoming the effect of toxic materials or bypassing the natural or artificial resistance of plants, which further confounds the problem (Roush and McKenzie 1987; Sallam 2013; Santos et al. 1990). Therefore, newly, nanoparticles (NPs) of graphene oxide and metals or metal oxides as insecticides with innovative or synergistic effects have been used.

Recently, nanotechnology has become one of the crucial technologies. Nanoscaled materials change their physical and chemical properties (Borm et al. 2006; Čitaković 2019; Dolez 2015; Jeevanandam et al. 2018; Mott 2019), and thus, old materials are innovated and used practically in all fields of human activity. Nanotechnologies can be successfully utilized also in agriculture, where they can considerably contribute to the sustainable intensification of agricultural production. They can be used to fabricate nanoformulations of herbicides, insecticides, fungicides, or fertilizers, and on the other hand, they can be applied as various sensors for monitoring plant growth, infections, etc. Thus, various agrochemicals constitute the most significant field of nanotechnology utilization (Coles and Frewer 2013; Chen et al. 2014; Jampílek and Kráľová 2015, 2017a, 2018a, 2019; Pérez-de-Luque and Hermosín 2013; Prasad et al. 2014; Raliya et al. 2013).

According to the adopted Recommendation on the definition of a nanomaterial by the European Commission (2011) the term "nanomaterial" means "a natural, incidental (e.g., as a result of abrasion/erosion or burning) or manufactured material containing particles, in an unbound state or as an aggregate or as an agglomerate and where, for 50% or more of the particles in the number size distribution, one or more external dimensions is in the size range 1–100 nm. In specific cases and where warranted by concerns for the environment, health, safety, or competitiveness the number size distribution threshold of 50% may be replaced by a threshold between 1 and 50%. By derogation from the above, fullerenes, graphene flakes, and single wall carbon nanotubes with one or more external dimensions below 1 nm should be considered as nanomaterials" (Buzea et al. 2007; European Commission 2011).

As mentioned above, the use of nanotechnologies in agriculture allows the re-use of various old and/or toxic insecticides in the form of nanoformulations and can ensure the enhanced bioavailability of active ingredients, allow their targeted delivery, controlled release, protection against degradation, and higher effectiveness compared to bulk preparations. Lower doses of a nanoformulated insecticide ensuring the required effect could considerably lower the costs. Moreover, in many cases the emergence of resistance against the applied insecticide can be suppressed by the synergistic effect of nanosized and/or other ingredients of the formulation (Amenta et al. 2015; Jampílek and Kráľová 2019; Makarenko and Makarenko 2019).

This chapter give a comprehensive overview of recent findings related to the bioactivity of nanoformulations of synthetic and natural insecticides against harmful insects causing severe damage to economically important crops or deteriorating stored food products. The impact of nanoinsecticides on the environment, including potential deleterious effects on non-target organisms, is discussed as well.

## 2 Synthetic Insecticides

Synthetic, i.e., man-made chemical insecticides play predominant role in control strategies to prevent destruction of crops and infestation of stored food and food products by insect pests. However, synthetic insecticides could exhibit unwarranted toxicity and even though lethal effects on non-target organisms and target insects could develop physiological resistance against these compounds. Their accumulation/persistence in environmental matrices has adverse effect on the environment (Ragaei and Sabry 2014). To mitigate these negative effects formulations that are able to target the pests specifically and do not contribute to environment pollution are required. From these aspects, "gut busters" (i.e., the encapsulated products) that break open only, when they come in contact with the alkaline environment, such as insect intestines, are favorable (Prasad et al. 2014). Nanoformulations could ensure controlled release of encapsulated insecticides and lower dose of active component is necessary to exhibit the desirable insecticidal effects than at applying bulk form of the insecticide. To the major benefits of nanosized insecticide formulations belong their easier application, improved targeting of pest species, higher efficacy, lower doses, and higher environmental safety (Huang et al. 2018; Jampílek et al. 2019; Jampílek and Kráľová 2018b, 2019; Rakhimol et al. 2020; Slattery et al. 2019; Walker et al. 2018; Zhao et al. 2018a). Methods related to fabrication of nanoinsecticide formulations applied in insect's pest control were comprehensively summarized by Sabry and Ragaei (2018). Nanocarriers used for organic insecticides and inorganic insecticidal NPs are presented in Fig. 1. Insecticide Resistance Action Committee classified the insecticides according their mode of action on insects with neurological site of action (targeting nerves and muscles) (Table 1) and those with other targets (Table 2) (IRAC 2019). Structures of some frequently used synthetic insecticides are presented in Figs. 2, 3, 4, 5, and 6.

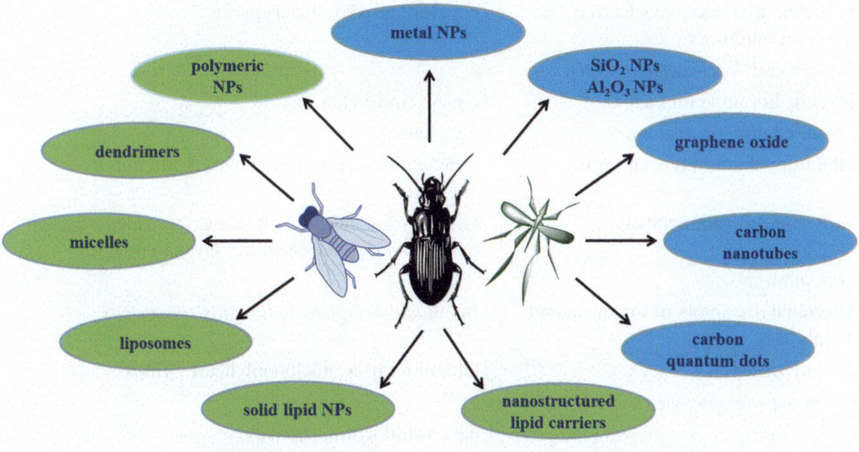


**Fig. 1** Nanocarriers used for organic insecticides and inorganic insecticidal NPs

**Table 1** Groups of insecticides targeting nerves and muscles (IRAC 2019)

| | |
|---|---|
| Acetylcholinesterase (AChE) inhibitors | Carbamates (e.g., methomyl, thiodicarb), Organophosphates (e.g., chlorpyrifos) |
| Nicotinic ACh receptor competitive modulators | Neonicotinoids (e.g., acetamiprid, thiacloprid, thiamethoxam) |
| Nicotinic ACh receptor allosteric modulators | Spinosyns (e.g., spinosad, spinetoram) |
| | GS-omega/kappa HXTX-HV1A peptide |
| Nicotinic ACh receptor blockers | e.g., bensultap, cartap |
| Sodium channel modulators | Pyrethrins (pyrethrin I and II, cinerin I and II, jasmolin I and II), Pyrethroids, (e.g., cypermethrin, cyhalothrin) |
| Voltage dependent sodium channel blockers | Oxadiazines (e.g., indoxacarb), Semicarbazones (e.g., metaflumizone) |
| Glutamate-gated chloride channel allosteric modulators | Avermectins, Milbemycins (e.g., abamectin, emamectin benzoate, lepimectin) |
| GABA-gated chloride channel blockers | Cyclodiene organochlorines (e.g., endosulfan), Phenylpyrazoles (e.g., fipronil) |
| GABA-gated chloride channel allosteric modulators | Meta-diamides (e.g., broflanilide, fluxametamide, isocycloseram) |
| Ryanodine receptor modulators | Diamides (e.g., chlorantraniliprole, cyantraniliprole, cyclaniliprole, flubendiamide, tetraniliprole) |
| Chordotonal organ TRPV channel modulators | e.g., afidopyropen |
| Chordotonal organ modulators | e.g., flonicamid |
| Octopamine receptor agonists | |

**Table 2** Groups of insecticides targeting other than neurological sites (IRAC 2019)

| | |
|---|---|
| *Respiration targets* | |
| Uncouplers of oxidative phosphorylation (proton gradient disruptors) | e.g., chlorfenapyr |
| Mitochondrial complex electron transport inhibitors | e.g., tolfenpyrad, fluacrypyrim |
| *Growth and development targets* | |
| Juvenile hormone mimics | e.g., pyriproxyfen |
| Juvenile hormone analogues | e.g., fenoxycarb |
| Inhibitors of chitin biosynthesis | Benzoylureas (e.g., flufenoxuron, lufenuron, novaluron) |
| Ecdysone receptor agonists | Diacylhydrazines (e.g., methoxyfenozide, tebufenozide, halofenozide, chromafenozide) |
| *Midgut targets* | |
| Microbial disruptors of insect midgut membranes | *Bacillus thuringiensis, Bacillus sphaericus* |
| Baculoviruses | Granuloviruses, nucleopolyhedroviruses |
| *Unknown, non-specific (multi-site) inhibitors* | |
| | e.g., azadirachtin, pyridalyl |

## 2.1 Organophosphate Insecticides

Organophosphate insecticides (OPIs), see Fig. 2, act primarily by phosphorylation of the acetylcholinesterase enzyme (AChE) at nerve endings resulting in loss of available AChE (Fukuto 1990), which is crucial for normal control of nerve impulse transmission from nerve fibers to smooth and skeletal muscle cells, secretory cells and autonomic ganglia, and within the central nervous system (CNS). The loss of AChE function results in muscle contraction, muscle twitching, depressed motor function, and respiratory paralysis (Roberts and Routt-Reigart 2013).

Chlorpyrifos (CPF; *O,O*-diethyl *O*-(3,5,6-trichloropyridin-2-yl)phosphorothioate) is a broad-spectrum, chlorinated OPI used also as acaricide and nematicide. Because they are toxic to non-target organisms (including mammals), it is desirable to prepare stable CPF formulations with controlled release that will selectively target pests and their entry in the environmental matrices will be suppressed. CPF-loaded NPs of an amphiphilic copolymer of chitosan (CS) with polylactic acid (PLA) and 1,2-dipalmitoyl-*sn*-glycero-3-phosphoethanolamine with particle sizes in the range 100–300 nm fabricated by varying the copolymer/CPF mass ratio showed sustained release profiles. Increasing mass ratio of copolymer to CPF resulted in decreased size of NPs, loading content, and encapsulation efficiency (EE) (Zhang et al. 2013). The silicone CPF microcapsules fabricated with polysiloxane sodium carboxylate/gelatin/sodium carboxymethyl cellulose (CMC) complex by coacervation method with average diameter ca 3.5 μm (shell thickness being of 285 nm) showing EE of 50.8% exhibited sustainable CPF release and high spreadability on the rice blades contributing to enhanced residual amount of CPF micro-

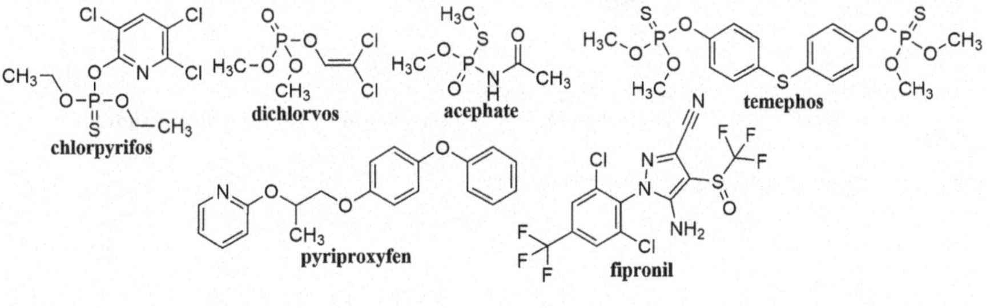


**Fig. 2** Structures of selected frequently used organophosphates

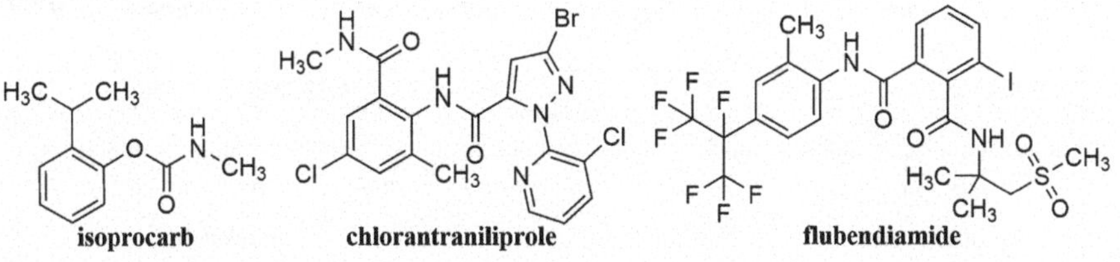


**Fig. 3** Structures of selected used carbamate and diamide insecticides

capsules on blades resulting in improved utilization of the insecticide (Dai et al. 2017). Loading of CPF into the crosslinked network structure of methyl methacrylate-hydroxyethyl methacrylate-methacrylic acid ternary random copolymer notably improved the heat resistance of CPF and the formulation showed sustained release controlled by Fick diffusion mechanism (Chen et al. 2017a). CPF-loaded poly(butyl acrylate-*co*-styrene)/ethylene glycol dimethacrylate (EGDMA) microcapsules with particle sizes 88.36–101.8 nm prepared by emulsion polymerization, in which cross-linking with 0.5–2.5% EGMDA considerably enhanced the extent of sustainable release and a diffusion controlled process of CPF release from microcapsules was observed at monomer ratio 1:2, 0.5% EGDMA or 5 g CPF were prepared by Wang et al. (2015a). A sustained release system responding to pH and ions able to supply CPF consisting of CPF supported on Cu(II) Schiff base mesoporous $SiO_2$ encapsulated in alginate sodium was fabricated by Chen et al. (2016). At pH ≤7 the release rate of insecticide from this formulation decreased with pH increasing, although the rate of CPF release under weak alkaline conditions was slightly higher than under weak acidic conditions. CPF-loaded $SiO_2$ NPs fabricated by sol-gel technique applied at a dose 0.01 g/m$^2$ when evaluated as slurries on Petri dishes caused 100% mortality of adults of *Rhyzopertha dominica* F. and *Tribolium confusum* Jacquelin du Val even after 6 h exposure at 7-d post-treatment time, *T. confusum* being more susceptible than *R. dominica* (Satehi et al. 2018). Porous hydrogel spheres consisting of CPF-nanonetwork-structured polydopamine-modified attapulgite-calcium alginate hydrogel were able to protect CPF from degradation under UV light, they exhibited controlled release of the insecticide and strong pH-responsiveness, which

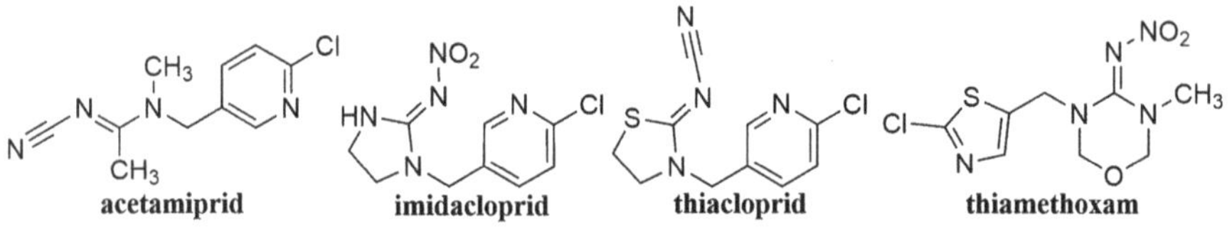


**Fig. 4** Structures of selected used neonicotinoids

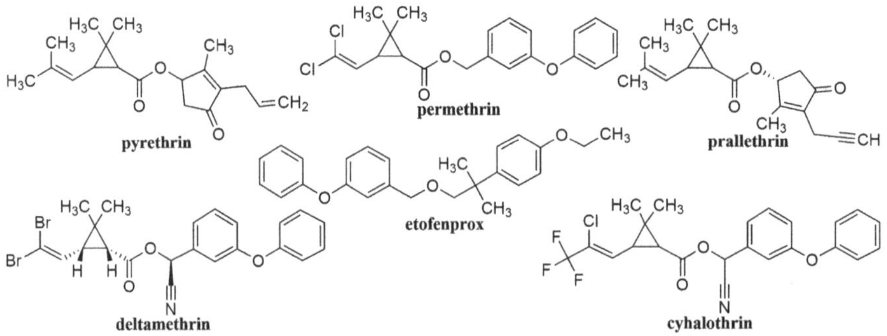


**Fig. 5** Structures of selected pyrethroids

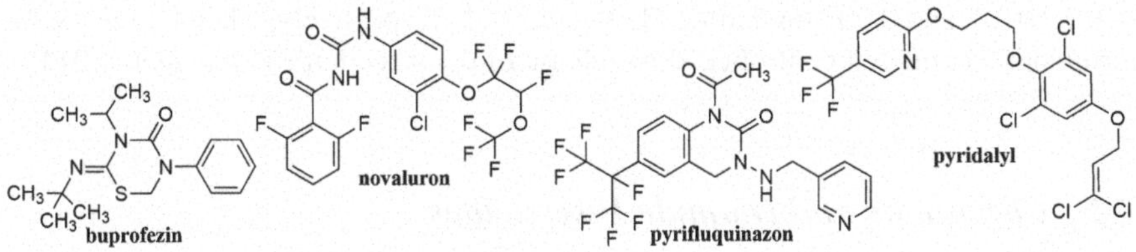


**Fig. 6** Structures of other used synthetic insecticides

was reflected in the collapse of these hydrogel spheres at pH 8.5, whereby the formed small particles possessing nanonetworks structure contributed to higher CPF efficiency against grubs (Xiang et al. 2018).

Starch-AgNPs encapsulated CPF and dichlorvos (DCV; 2,2-dichloroethenyl dimethyl phosphate) with particle size ca 23–35 nm and EE of 95% and 98% for DCV and CPF, respectively, exhibited slower pesticide release than formulation without AgNPs, whereby the AgNPs could exhibit also antimicrobial activity (Ihegwuagu et al. 2016).

A 15 cm band of Inesfly IGR FITO®, a paint containing CPF and pyriproxyfen (2-{[1-(4-phenoxyphenoxy)propan-2-yl]oxy}pyridine), an insecticide that mimics a natural hormone in insects and disrupts their growth, in a microencapsulated formulation showing slow release of pesticides, was painted around citrus trunks at the beginning of the season in two citrus orchards, in which the dominated ant communities were *Lasius grandis* or *Linepithema humile*, respectively. A single application of this paint showed high efficiency and the ants were excluded from canopies throughout the season (Juan-Blasco et al. 2011).

The formulation consisting of the microencapsulated mixture of CPF and fipronil (5-amino-1-[2,6-dichloro-4-(trifluoromethyl)phenyl]-4-[(trifluoromethyl) sulfinyl]-1*H*-pyrazole-3-carbonitrile) tested in peanut fields was found to be considerably protected against decomposition in the environment and treatment of seeds with a single dose of this formulation efficiently controlled white grubs during the whole growing season. In the soil and peanut roots, it was able to maintain 13.6-fold higher concentration of CPF and ca 2.2-fold higher concentration of fipronil compared to conventional formulation; however, the level of residual fipronil in some kernel samples reached the statutory maximum residue limit set by the European Union suggesting risk of the multiple reapplication of this microscaled formulation (Yang et al. 2014).

Nanoformulation of acephate (*O,S*-dimethyl acetylphosphoramidothioate) encapsulated in PEG was found to be biosafe when tested on murine model (Pradhan et al. 2013). On the other hand, bulk acephate was found to induce shortening of the developmental time and early emergence in a non-target insect *Drosophila melanogaster* (Rajak et al. 2013).

The toxicity of the phenyl organothiophosphate insecticide temephos (*O,O,O′,O′*-tetramethyl *O,O′*-(sulfanediyldibenzene-4,1-diyl) bis(phosphorothioate)) nanoencapsulated in PEG against *Culex quinquefasciatus* was reflected in $LC_{50}$ values of

0.013, 0.010, and 0.003 mg/l after 24, 48, and 72 h, respectively, whereby the nanoformulation exhibited controlled slow release of the insecticide (Bhan et al. 2014).

## 2.2 Carbamate and Diamide Insecticides

Carbamate insecticides show mechanism of action like that of OPIs, i.e., they inhibit AChE (Fukuto 1990). Mode of action of diamide insecticides (e.g., flubendiamide, chlorantraniliprole, see Fig. 3) consists in unselective activating the insect ryanodine receptor (Roberts and Routt-Reigart 2013; Troczka et al. 2017).

The nanocomposite prepared by intercalation of isoprocarb (2-(propan-2-yl)phenyl methylcarbamate) into zinc layered hydroxide showed mesoporous-type material characteristics, lower pore size compared to the pristine host, layered zinc layered hydroxide sodium dodecyl sulfate, and better thermal stability compared to the pristine isoprocarb suggesting that its application will be environment-friendly (Muda et al. 2019).

Microcapsule formulations fabricated using a solid in oil in water (S/O/W) double-emulsion method combined with premix membrane emulsion that were loaded with diamide insecticide chlorantraniliprole (CLAP; 3-bromo-*N*-[4-chloro-2-methyl-6-(methylcarbamoyl)phenyl]-1-(3-chloropyridin-2-yl)-1*H*-pyrazole-5-carboxamide) exhibited prolonged sustained release, optimal regulation of which could be obtained by tuning the surface porosity and size of the microcapsules. The insecticidal efficiency of formulations consisting of such porous microcapsules against *Plutella xylostella* exceeded that of the commercial formulation (Liu et al. 2018). By anchoring mechanically interlocked molecules using α-cyclodextrin onto the surface pore rims of hollow mesoporous $SiO_2$ and loaded them with CLAP, enzyme-sensitive controlled release formulations were prepared, in which introduction of external α-amylase could accelerate the insecticide release. The mortality of *P. xylostella* larvae fed with this nanoformulation estimated after 14 days was pronouncedly higher than that observed with the commercial preparation Coragen®, confirming remarkable persistence of prepared nanoformulation (Kaziem et al. 2017). An adhesive organic-inorganic hybrid prepared using hollow mesoporous silica (HMS) as an inlayer material and poly(diacetone acrylamide) as an outer layer with incorporated CLAP showed controlled and sustained release at least 25 days and stronger adhesive property on rice leaves than HMS suggesting that such formulation could be used for photosensitive pesticides applied via foliar spraying (Gao et al. 2018a). Solid nanodispersions of CLAP were fabricated by high pressure homogenization combined with lyophilization with particle sizes <75 nm and the nanoformulation containing 2.5% of insecticide showed average particle size of 29 nm, 97.32% suspensibility in water, and wetting time of 13 s, respectively. The solid nanodispersions reached 1.5- and 3-fold higher retention on the rice leaf than the commercial aqueous suspension concentrate and pure water and their toxicity to diamondback moths were 3.3- and 2.8-fold higher than that of technical and aqueous suspension concentrate, respectively (Cui et al. 2016).

## 2.3 Neonicotinoids

Neonicotinoids (neonics) are a class of insecticides affecting the CNS of insects by strong binding to nicotinic acetycholine receptors in the CNS, resulting in overstimulation of their nerve cells, paralysis, and death. These insecticides are persistent in the environment but they are less toxic to humans than OPIs, carbamates, organochlorides, and pyrethroids, and therefore they are the most frequently used insecticides in the world (Ensley 2018; Roberts and Routt-Reigart 2013). The levels of neonicotinoids estimated in surface waters in Canada and globally could have adverse impact on the aquatic invertebrates (Anderson et al. 2015). To neonicotinoid insecticides belong, for example, imidacloprid (IMI), thiamethoxam (TMX), acetamiprid (ACP), see Fig. 4.

CS (0.1% w/v)-coated liposomes co-encapsulating the IMI ((2*E*)-1-[(6-chloropyridin-3-yl)methyl]-*N*-nitroimidazolidin-2-imine) and pyrethroid insecticide cyhalothrin (see below) with particle size 69 nm and zeta potential of +31 mV, showing EE of 51 and 96% for IMI and cyhalothrin, respectively, exhibited improved insecticidal effects against *Myzus persicae* Sulzer as well as duration of residual activity compared to the effect of individual insecticides and their mixture (Moradi et al. 2019). Submicron particles of amphiphilic CS-*co*-(D,L-lactide) copolymers loaded with IMI prepared by nanoprecipitation and the emulsion/solvent evaporation method showed a sustained insecticide release process, whereby reduction of particle size and IMI loading content was observed with increasing mass ratio of copolymer to IMI (Li et al. 2012). Functional nano-dispensers consisting of IMI encapsulated in poly(D,L-lactide-*co*-glycolide (PLGA) microparticles (MPs; 5–10 μm) fabricated using the solvent-evaporation method exhibited comparable mortality of Asian citrus psyllids (*Diaphorina citri*) as the commercial formulation at a 200-fold lower dose (Meyer et al. 2015). Morphology and size of the nanoscale IMI fabricated using encapsulation of insecticide in the ABA triblock linear dendritic copolymers composed of poly(citric acid) (PCA) as A block and poly(ethylene glycol) (PEG) as B block in the presence of different solvents depended on the used solvent and enabled to prepare particles with mean size of 10–20 nm as well as fiber-like, globular, and tubular particles with sizes from 10 nm to several mm. This formulation was characterized with slower release rate of insecticide at pH 10 corresponding to the pH of *Glyphodes pyloalis* gut compared to neutral pH indicating that its action will be selective and controllable. Moreover, lower dose of IMI was sufficient to achieve the required effect compared to application of bulk insecticide, indicating reduced environmental risk (Memarizadeh et al. 2014). The release of IMI that was loaded in MCM-48 type mesoporous $SiO_2$ NPs showing high surface area was found to be controlled over 48 h and the formulation showed efficient activity against termites in a laboratory experiment (Popat et al. 2012). The time needed for the release of 50% of IMI, which was encapsulated in a composite gel composed of carboxymethyl CS and bentonite, was prolonged to 24 h and in leaching experiments through a soil layer lower amount of insecticide available for leaching due to applied nanocarrier were estimated resulting in lower environmental risk (Li et al. 2012). Nanoaerosol particles from IMI with sizes of 7–300 nm applied at a dose of 2.7 ± 0.1 ng/cm$^3$ showed $T_{50}$ of 88 ± 14 min (at 22 °C) and 36 ± 2 min (at 33 °C), respectively, related to knockdown in half of the *D. melanogaster* insects. Based on the estimated fly knockdowns that were two orders of magnitude lower for the inhaled doses compared to oral doses containing IMI-sucrose mixture it could be suggested that application of such insecticidal nanoaerosol formulations could be very efficient in greenhouses and other closed environments (Morozov and Kanev 2015).

TMX ((4*E*)-3-[(2-chloro-1,3-thiazol-5-yl)methyl]-5-methyl-*N*-nitro-1,3,5-oxadiazinan-4-imine) encapsulated in hydrogel composites fabricated from CMC crosslinked with citric acid in the presence of bentonite exhibited an immediate burst release. The insecticide formulation was found to be able to control insects having pH > 7 in their guts because the observed TMX release rate was higher at pH > 7 compared to neutral pH (Sarkar and Singh 2017). Fast release in a solution with high pH compared to acidic pH was reported also for pH-triggered release formulations of boron and TMX prepared using boric acid cross-linked CMC hydrogels (Sarkar and Singh 2019). The half amount of TMX encapsulated in amphiphilic copolymers fabricated from PEGs and various aliphatic and aromatic diacids able to self-assemble into nanomicellar aggregates in aqueous media was released within 3.56–6.07 days, this time being longer than that from the commercial formulation (Sarkar et al. 2012). Liu et al. (2015) designed water-soluble nanoscaled cationic dendrimers containing hydrophobic dendritic polyesters and peripheral amines which were able to effectively deliver TMX into the live cells, strongly increasing its cytotoxicity.

The EE of ACP ((1*E*)-*N*-[(6-chloropyridin-3-yl)methyl]-*N*′-cyano-*N*-methylethanimidamide) loaded in the micelles of amphiphilic alginate was found to increase from 55 to 96%, respectively, due to the rise in the concentration of $Na^+$ ions from 0.01 M to 0.3 M and a decrease in pH from 5.3 to 2.0 was also reflected in increased EE (55–80%). This formulation exhibited controlled release of ACP (Tang et al. 2018). Sustained release of the insecticide is desirable due to its adverse effects on non-target organisms. For example, ACP exhibited toxicity to zebrafish embryos resulting in high morality and teratogenic effects at concentration >263 mg/l, and malformations such as bent spine (Ma et al. 2019). Self-assembled NPs of cholesteryl-grafted sodium alginate derivatives (CSAD) with particle sizes ca 100 nm effectively encapsulated ACP, whereby at using CSAD with lower molecular weight, ACP release corresponded to Fickian transport (Zhao et al. 2018b). Spherical alginate-CS nanocapsules encapsulating ACP showing EE of 62% exhibited controlled release in vitro in a wide pH range (4–10), maximum release being observed at pH 10, the lowest one at pH 4 (Kumar et al. 2015). ACP immobilized into the layers of montmorillonite modified with cetyltrimethylammonium bromide exhibited a slow and sustained release suggesting that such formulation is suitable to reduce environmental pollution (Yan et al. 2016).

Lorenz et al. (2017a) exposed fourth instar larvae of *Chironomus riparius* to thiacloprid (TCP; {(2*Z*)-3-[(6-chloropyridin-3-yl)methyl]-1,3-thiazolidin-2-ylidene} cyanamide) and to nanoscale zeolites, eventually to zeolite agglomerates composed of primary NPs of 50 nm solely as well as to mixtures of both compounds. While the tested zeolites did not showed toxicity when applied to the insect, they were able to reduce acute toxicity of TCP due to limited bioavailability caused by the sorption of the insecticide on zeolite. Exposure of *Chironomus riparius* to mixtures of TCP and $Al_2O_3$ NPs also resulted in pronounced reduction of the mortality of fourth instar larvae compared to TCP, the effect being more effective with increasing $Al_2O_3$ NPs concentration. However, the presence of $Al_2O_3$ NPs was not able to prevent entirely the mortality of larvae exposed to combine treatment with TCP applied at a dose >0.5 μg TCP/l (larvae showed severe convulsions), although the mortality was delayed (Lorenz et al. 2017b).

As ex vivo application to repel and even kill mosquitoes or flies zirconyl hydrogenphosphate nanocontainers loaded with cypermethrin (cyano(3-phenoxyphenyl)methyl 3-(2,2-dichloroethenyl)-2,2-dimethylcyclopropanecarboxylate) were reported (Rein et al. 2019).

As mentioned above, for neonicotinoids selective binding to insect nicotinic acetylcholine receptors is characteristic. In a structure-activity study of the mode of [H-3]imidacloprid displacement in *Myzus persicae* and *Aphis craccivora* Kayser et al. (2004) used neonicotinoids applied in practice and some newly synthesized analogues and found that the direct competitors (acetamiprid, nitenpyram, thiacloprid, and nithiazine) share the binding site with IMI, whereas non-competitive compounds (TMX, *N*-methyl analogues of IMI and clothianidin) bind to a different site or in a different mode.

## 2.4 Pyrethroid Insecticides

The insecticidal actions of pyrethroids (see Fig. 5) are connected with their ability to bind to and disrupt voltage-gated sodium channels of insect nerves. When the pyrethroid insecticide keeps the channels in their open state, the nerves cannot repolarize, and the axonal membrane remains permanently depolarized, which results in the paralysis of the organism (Soderlund 2012).

Pyrethrins encapsulated in temperature-responsive mixed micelles prepared by a cooperative assembly of poly[2-(2-methoxyethoxy)ethyl methacrylate-*co*-octadecyl methacrylate] and monomethoxyPEG-PLGA in water were protected from degradation with UV light at 26 °C, whereby at increasing of the temperature from 13 to 26 °C a phase transition process from solution state to turbid state was observed. The mixed micelles showed improved larvicidal activity against *Culex pipiens pallens* at 26 °C than at 14 °C or 18 °C and after 24-h exposure at 26 °C they were also more toxic to *C. p. pallens* larvae than the commercial pyrethrin formulation, although at 14 °C their toxicity was lower. The longer-lasting larvicidal activity of this mixed micelle formulation under natural conditions in comparison to the established pyrethrin formulation was observed as well (Zhang et al. 2019). Nanoformulation of natural pyrethrins in water-in-oil microemulsions (MEs) based on non toxic biocompatible materials exhibited enhanced insecticidal activity of insecticides against *Aphis gossypii* (Hemiptera: Aphididae) than two commercial suspension concentrates of natural pyrethrins. On the other hand, nanoformulated preparations did not show toxicity against L3 larvae and four-instar nymphs of the predators *Coccinella septempunctata* L. (Coleoptera: Coccinellidae) and *Macrolophus pygmaeus* Rambur (Hemiptera: Miridae), respectively (Papanikolaou et al. 2018). Also, the insecticidal activity of natural pyrethrin encapsulated in nano-emulsions (NEs) containing globular oil droplets of 36–37 nm in diameter and oil droplets with diameters >150 nm dispersed in the aqueous phase, which was evaluated in laboratory bioassays using target insect *Aphis gossypii* Glover in eggplant, exceeded that of the commercial pyrethrin formulation (Kalaitzaki et al. 2015).

Nanoscaled permethrin (PMT; 3-phenoxybenzyl 3-(2,2-dichloroethenyl)-2,2-dimethylcyclopropanecarboxylate) fabricated by solvent evaporation from an oil-in-water (O/W) volatile ME was ca threefold more effective against *Aedes aegypti* compared to microparticluar formulation of the insecticide (the 24 h $LC_{50}$ of 0.0063 vs. 0.0199 mg/l). Moreover, treatment of maize, cucumber, and tomato seeds with nano-PMT did not adversely affect root length and germination percentage suggesting that the formulation represents a safe alternative of the insecticide for the use in agriculture (Kumar et al. 2013). PMT NEs with the average droplet diameter of $12.4 \pm 1.13$ nm and zeta potential of $-20.4 \pm 0.56$ mV showed $LC_{50}$ values of 0.038 and 0.047 mg/l and 0.049 and 0.063 mg/l against larvae and pupae of *Culex tritaeniorhynchus* and *Ae. aegypti*, respectively, and were found to be nontoxic against non-target organisms (*Closterium* alga, chickpea zebrafish) (Mishra et al. 2019).

The biogenic volatile organic compounds of the home insecticide containing prallethrin ((1*S*)-2-methyl-4-oxo-3-(prop-2-yn-1-yl)cyclopent-2-en-1-yl 2,2-dimethyl-3-(2-methylprop-1-en-1-yl)cyclopropanecarboxylate), α-pinene, cymene, D-limonene, α-terpinene, and α-thujone were reported to be able to initiate secondary aerosol formation under ozone exposure (Bae et al. 2012).

Using nanosized CS (MW 30,000; 01%) carrier controlled release formulation of etofenprox (1-{[2-(4-ethoxyphenyl)-2-methylpropoxy]methyl}-3-phenoxybenzene) with polygonal shaped particles and sizes ≤800 nm showing activity against *Spodoptera litura* was prepared by Hwang et al. (2011).

In deltamethrin (DM; (*S*)-cyano(3-phenoxyphenyl)methyl (1*R*,3*R*)-3-(2,2-dibromoethenyl)-2,2-dimethylcyclopropanecarboxylate)–loaded CS-coated beeswax solid lipid NPs (SLNPs) fabricated using a combination of hot homogenization and sonication and showing 95% EE, the insecticide was efficiently protected against photodegradation and thus such nanoformulation could be applied to improve the effect of the insecticide in the field (Nguyen et al. 2012a, b). Using grafting esterification of dodecanoic acid onto *Nitraria* seed meal substrates Bai et al. (2019) designed a pesticide carrier with distinct hydrophobic surface and irregular holes loaded with DM (loading capacity ca 1068 mg $g^{-1}$) that showed controlled release of the insecticide. The AgNPs-DM core-shell conjugate, in which a 15 nm AgNPs core was surrounded by DM, caused mortality of mosquitoes in a 24 h bioassay at a dose of 0.9 mM. It could be noted that Ag was estimated in the hemolymph of mosquitoes treated with the conjugate (Sooresh et al. 2011). Balaji et al. (2017) prepared a hydrodispersive nanoscaled colloidal form of DM with droplet sizes 35–40 nm from its parental form (PDM), which was hydroimmisicible and exhibited higher efficacy on adult mosquito and larval population of *Cx. tritaeniorhynchus* and *Cx. quinquefasciatus*, even at concentrations lower than PDM, whereby the nanoscale DM exhibited lower toxicity to non-target organisms than PDM suggesting precise targeting of mosquitoes.

Combination of nanoencapsulated DM (protected from esterase-induced enzymatic degradation) with indoxacarb (IDX), an oxadiazine pesticide, enhanced insecticidal activity of IDX against cockroach *Periplaneta americana*, which could result in reducing doses of IDX. Namely, following a rise in intracellular $Ca^{2+}$ levels in insect neurons through the reverse $Na^{+}/Ca^{2+}$ exchanger caused by deltamethrin, the voltage-gated sodium channels showed higher sensitivity to lower concentration of the toxic decarbomethoxylated metabolite of IDX (Caballero et al. 2019).

Cyhalothrin (CHT; cyano(3-phenoxyphenyl)methyl 3-[(1Z)-2-chloro-3,3,3-trifluoroprop-1-en-1-yl]-2,2-dimethylcyclopropanecarboxylate) is an effective pyrethroid insecticide. λ-CHT is a mixture of CHT isomers, while γ-CHT is a single, the most insecticidally active stereoisomer of CHT. γ-CHT-loaded SLMPs with particle diameters 0.3–100 μm exhibited comparable activity against both *Dysdercus cingulatus* nymphs and *Spodoptera littoralis* larvae than the traditional emulsifiable concentrate formulation (Frederiksen et al. 2003). λ-CHT-loaded PLA NPs fabricated by a solvent evaporation method with particle sizes <200 nm and EE >90% showed higher foliage adhesion than the commercial insecticide preparation because of a low surface tension and a low contact angle reflected in improved pesticide utilization (Shen et al. 2018). λ-CHT-loaded PLA microcapsules fabricated via premix membrane emulsification showed prolonged controlled sustained release of insecticide and the activity of such microcapsules (0.68 μm), which were characterized with good UV and thermal stability, against *P. xylostella*, was comparable with that of a commercial microcapsule preparation (Liu et al. 2016). Using ME template with octyl-grafted alginate-amide derivative nanocapsule formulation with λ-CHT showing mean particle size of 25.78 nm and EE of 99.95% and exhibiting restrained release of insecticide in methanol was designed by Hu et al. (2013). A pH-responsive emulsions stabilized by alginate-grafted anisotropic $SiO_2$ applied for the controlled release of λ-CHT were found to be more stable in a pH range from 2.0 to 6.2 due to polymer chain interactions resulting in the creation of a 3-D network. On the other hand, in the pH range from 6.2 to 8.0 the increased emulsion stability was connected with the increasing particle charge. An increase of emulsion pH from 3.0 to 8 led to reduction of cumulative drug release from the formulation from 99.7% to 13.5% (Chen et al. 2017b). CHT-loaded ultrafine particles of poly(2-hydroxyethyl methacrylate)-*co*-PLA, which could be prepared by both nanoprecipitation method and emulsion/solvent evaporation method, were characterized by efficacious dispersity in water and sustained release behavior (Fan et al. 2013). λ-CHT nanosuspension prepared using a melt emulsification (alkylphenol formaldehyde resin polyoxyethylene ether was used as emulsifier) method with average particle size of 12.0 ± 0.1 nm and a polydispersity index (PDI) of 0.279 ± 0.135, that showed improved wettability, stability, and bioavailability compared to conventional suspension concentrates, was reported to be suitable for a broad application in agricultural production systems (Wang et al. 2019a). λ-CHT-loaded biodegradable castor-oil based polyurethane NEs with uniform spheres showing diameters <80 nm, EE of 85%, and insecticide loading capacity ca 40 wt% exhibited sustained and controlled release property. Low surface tension and larger chain mobility of the system as well as H-bond interactions between the polyurethane and foliar surface resulted in significantly improved foliage adhesion compared to the commercial formulations of λ-CHT (Qin et al. 2017). Liu and Guo (2019) designed biodegradable poly(butylene succinate) microspheres with encapsulated λ-CHT that were prepared by the solvent evaporation induced phase separation method. These microspheres demonstrated high-loading capacity and EE as well as long release time. Benzoyl lignin nanospheres fabricated by the reverse solvent method with encapsulated λ-CHT showing diameters of 90–100 nm, in which the benzoyl lignin tended to aggregate on the surface of nanospheres providing them negative charge and the hydrophobic insecticide moved toward the interior of the nanospheres, were prepared by Zhou et al. (2018). λ-CHT loaded polydopamine microcapsules showing good physicochemical stability and sustained release properties exhibited improved bioactivity and long-term efficiency against *Musca domestica* compared to the commercial formulation (Zou et al. 2018).

## 2.5 Chitin Synthesis Inhibitors and Insect Growth Regulators

Insecticides of the group of chitin synthesis inhibitors (Fig. 6) disturb the processes of chitin formation through the preimaginal stages, metamorphosis, and the reproductive development of insects (Dolzhenko and Dolzhenko 2017). By encapsulation of microcrystals of chitin synthesis inhibitor buprofezin ((2*Z*)-2-(*tert*-butylimino)-5-phenyl-3-(propan-2-yl)-1,3,5-thiadiazinan-4-one) with CS and sodium alginate through layer-by-layer (LbL) self-assembly, particles with mean diameter of 1.5 μm and EE of 67.2% showing prolonged release time were designed by Zhang et al. (2011).

Novaluron (NOV; *N*-({3-chloro-4-[1,1,2-trifluoro-2-(trifluoromethoxy)ethoxy] phenyl}carbamoyl)-2,6-difluorobenzamide) insecticide acts as insect growth regulator which disrupts the normal growth and development of immature insects and kills slowly the insects over a period lasting few days (Novaluron 2001). Using direct conversion of O/W MEs with droplet size of 6 nm containing NOV and volatile solvents nanosized powders were prepared. After redispersion NOV particles consisted of NPs aggregates (30–100 nm) reaching a size of 200 nm and showing in vivo toxicity against *S. littoralis* larvae comparable with that of commercial formulation (Elek et al. 2010). Hydrophobic nanoprecipitates formed by NOV or diflubenzuron and β-cyclodextrin inclusion compounds showed higher efficiency against *Ae. aegypti* larvae than free benzoylphenylureas (Bittencourt et al. 2019).

## 2.6 *Unclassified Insecticides*

Pyrifluquinazon (1-acetyl-6-(1,1,1,2,3,3,3-heptafluoropropan-2-yl)-3-[(pyridin-3-ylmethyl)amino]-3,4-dihydroquinazolin-2(1*H*)-one, see Fig. 6) is a new insecticide that interferes with chordotonal receptor neuron function that alters insect behavior by stopping feeding in a short time and the insects starve to death. Nanosized pyrifluquinazon formulated with CS (0.3%; MW 3000) was found to be potent against *M. persicae* at 14 days after exposure, whereby the reaction time was reduced from 14 to 30 days in treated aphids (Kang et al. 2012).

Pyridalyl (2-(3-{2,6-dichloro-4-[(3,3-dichloroprop-2-en-1-yl)oxy]phenoxy} propoxy)-5-(trifluoromethyl)pyridine, see Fig. 6) nanosuspension with sodium alginate with mean micelle size of ca 138 nm, pyridalyl size <100 nm, and zeta potential of $-20 \pm 1$ mV showed insecticidal activity against larvae of *Helicoverpa armigera* with $LC_{50}$ values of 40 μg/ml and in bioassay using leaf dip method it showed 2.26- and 6.25-fold higher effectiveness against *H. armigera* as stomach poison compared to the technical product and commercial preparation, respectively (Saini et al. 2014).

# 3 Macrocyclic Lactone Insecticides

The avermectins, milbemycins, and spinosyns belong to macrocyclic lactones (mostly mixtures of very close complex compounds/derivatives are used) that comprise several classes of chemicals derived from cultures of soil micro-organisms. Avermectins (see Fig. 7) are neurotoxic metabolic products of the bacterium *Streptomyces avermitilis*. Ivermectin (IVM), the most widely used avermectin (AVM) obtained through selective, catalytic hydrogenation of the *cis*-22,23-double bond of the avermectins B1a and B1b, is usually used to control the ecto- and endoparasites (mites and nematodes) of livestock and antifilarial chemotherapy in humans (Lumaret et al. 2012). Arena et al. (1995) reported that the nematocidal effects of avermectins and milbemycins on *Caenorhabditis elegans* are caused by an interaction with a common receptor molecule, glutamate-gated chloride channels. Biochemical mode of action, biological activity, and agricultural importance of avermectins were analyzed by Jansson and Dybas (1998).

AVM/castor oil-based polyurethane NEs designed by Zhang et al. (2018) with particle size <50 nm and EE of >85% showed improved foliar insecticide retention and considerably lower photolysis rate of AVM than pure insecticide, the release of AVM from these NEs being controlled by both diffusion and matrix erosion. Acetylated lignin and benzoylated lignin were used to fabricate nanospheres encapsulating AVM exhibiting superb controlled release properties compared to control group and were able to retain 67.6% and 77.0% of the insecticide after 50 h UV irradiation, while the retention rate of control group reached only 27%. However, it could be mentioned that the higher acylation degree was reflected in reduced retention rate of the insecticide approx, by 15–20% (Zhou et al. 2019).

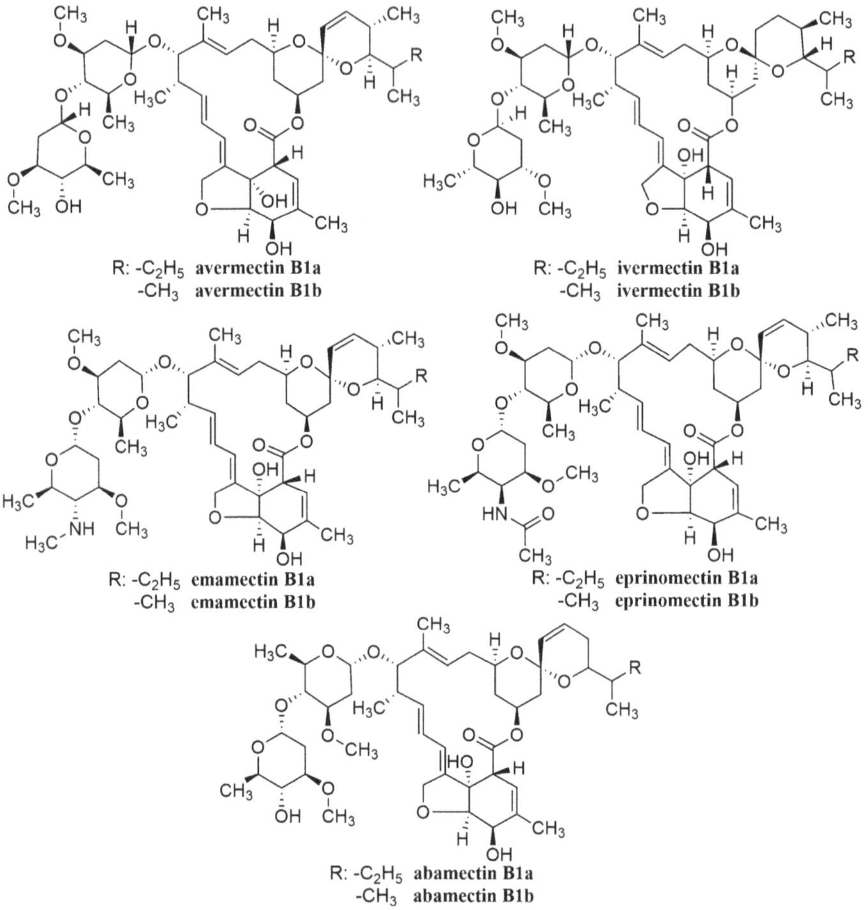


**Fig. 7** Structures of selected avermectines

IVM-loaded CS-alginate NPs of 155 nm and EE of 75.67% exhibited sustained release and their microfilaricidal activity against human lymphatic filariid, *Brugia malayi*, in rodent host following subcutaneous administration of a single dose of 200 µg/kg body weight exceeded that of free IVM applied at a twofold higher dose (Ali et al. 2013). Lipid nanostructured carrier systems for IVM and methoprene showing potential to be used in veterinary applications were developed by Cola et al. (2015, 2016). Solid dispersion of IVM in a lipid matrix (hydrogenated castor oil) exhibiting sustained release, which was evaluated against the ear mange mite, *Notoedres muris* (Astigmata: Sarcoptidae), in rabbits, showed improved bioavailability compared with pure insecticide and was found to provide longer persistence against *N. muris* rabbit's ear mites than a commercial IVM injection (Lu et al. 2017). IVM-loaded lipid nanocapsules with average diameter of 55 nm applied at a dose 0.11 and 0.28%, respectively, showed knockdown $T_{50}$ mortality values for

*Pediculus humanus capitis* De Geer (Anoplura: Pediculidae) 5 and 3 h, respectively, suggesting potential use of such nanoformulation in clinical practice (Ullio-Gamboa et al. 2017). After subcutaneous injection in a rat model the IVM-loaded lipid nanocapsules showed higher systemic disposition (1367 ng h/ml) compared to a commercial preparation (1193 ng h/ml), although considerable differences in the biodistribution pattern were not observed (Gamboa et al. 2016).

Emamectin benzoate (EMB) loaded in ethyl cellulose nanocapsules (219.93 ± 3.89 nm; zeta potential of −26.43 mV), $SiO_2$ NPs (142.77 ± 3.43 nm; zeta potential of −41.0 mV) and MCM-48 particles (119.73 ± 20.28, −36.5 mV) showed insecticidal effect against the third instar larvae of *P. xylostella* with $LC_{50}/L_{90}$ values estimated after 24 h of 0.32/1.67, 7.44/89.03, and 34.79/359.51 mg/l compared to 24.83/311.32 mg/l for pure insecticide. Higher inhibitory activity of nanoformulations containing $SiO_2$ NPs and MCM-48 could be connected with smaller sizes of NPs and higher surface area and it could be assumed that they could better penetrate in the larval body than the active ingredient alone. Moreover, the tested carriers improved the photostability of the entrapped insecticide (Shoaib et al. 2018a). Slow-release microspheres fabricated by the microemulsion polymerization method using polyvinylalcohol (PVA) as stabilizer and polyoxyethylene castor oil as surfactant, which were loaded with EMB showed superb anti-photolysis performance, stability, controlled release properties, and good leaf distribution suggesting that such nanoformulation could improve insecticide efficiency by prolonging its control effect (Wang et al. 2017). The EMB $SiO_2$-epichlorohydrin-CMC microcapsules exhibiting superb cellulase stimuli-responsive properties and a sustained insecticidal efficacy against *Myzus persicae* were designed by Guo et al. (2015).

Hydrophobic nanoprecipitates of inclusion complexes of eprinomectin with β-cyclodextrin (β-CD) improved its larval toxicity against *Ae. aegypti*, while reduced its human cytotoxicity. On the other hand, similar effect due to encapsulation into β-CD was not observed with IVM (Moreira et al. 2018).

Poly(glycidyl methacrylate-*co*-acrylic acid) grafted hollow mesoporous $SiO_2$ composite loaded with abamectin showed pH-dependent release of insecticide, high adhesion on rice leaves and showed higher toxicity than abamectin emulsifiable concentrate in controlling *Cnaphalocrocis medinalis* (Guenee) larvae, a noxious rice pest, during cultivated periods. Moreover, the formulation exhibited long-term efficacy and practically did not affect the growth of rice seedlings (Gao et al. 2019).

## 4 Botanical Insecticides

Botanical insecticides or bioinsecticides are naturally occurring or derived materials from living organisms used to control harmful insects. Over 17,000 plant species produce essential oils (EOs) playing a key role in plant signaling processes (Campolo et al. 2018). Plant extracts and EOs belong to frequently used insecticides as they are less toxic, less persistent, and could be degraded more rapidly than synthetic insecticides, and therefore they are environment-friendly. Moreover, botanical insecticides are safe to humans and non-target organisms (Roberts and Routt-Reigart 2013; Pavela 2016; Hikal et al. 2017; Campos et al. 2016). Boulogne et al. (2012) in an overview paid attention to chemicals of plant origin and species showing insecticidal activity reported that 656 plant species have pronounced insecticidal activities and 17 species of plant families *Lamiaceae* and *Apiaceae* were particularly effective against leaf-cutting ants. The insecticidal effects of EOs on various insect species were discussed in many papers (e.g., Al-Ahmadi 2019; Campos et al. 2016; de Oliveira et al. 2014; Dougoud et al. 2019; Duke et al. 2010; Hikal et al. 2017; Mossa et al. 2018a; Pavela 2016; White and Johnson 2012) and mechanism of action of secondary metabolites of plant origin showing insecticidal activity was overviewed by Rattan (2010). EOs are frequently utilized for the control of preharvest and postharvest phytophagous insects. Structures of selected botanical insecticides are illustrated in Fig. 8.

EOs, i.e., volatile secondary metabolites of many higher plants possessing repellent, insecticidal, or growth-inhibiting activities against a variety of insects could cause neurotoxic effects to insects, whereby they exhibit several mechanisms of action, including mainly inhibition of AChE blockage of GABA-gated chloride channels, eventually they can act as octopamine receptor agonists (Poopathi et al. 2016; Regnault-Roger et al. 2012; Tripathi et al. 2009). Integrity of EOs is greatly affected by light, temperature, and oxygen availability (Turek and Stintzing 2013)

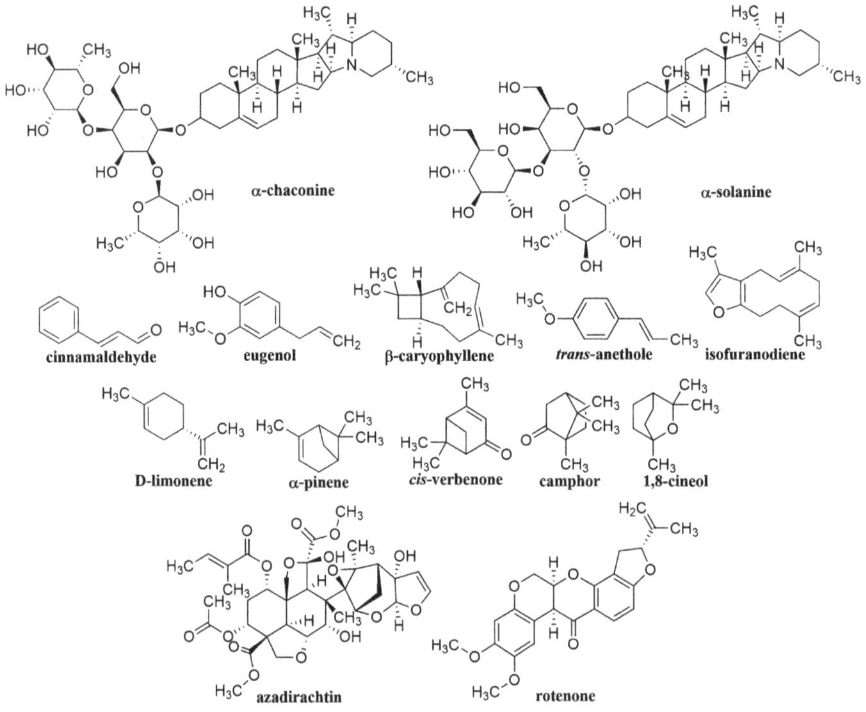


**Fig. 8** Structures of selected botanical insecticides

and therefore it is favorable to protect the active ingredient by encapsulation. Application of powders and extracts of *Azadirachta indica*, *Zanthoxylum zanthoxyloides*, *Anacardium occidentale*, and *Moringa oleifera* against *Sitophilus oryzae* (L), *Oryzaephilus mercator* (Faur), and *Ryzopertha dominica* (Fabr.) with entomocidal activity resulted in the inability of the insects to feed on the paddy coated with these bioinsecticides and therefore starvation; disruption of the respiratory activities of insects caused the asphyxiation and death and potential blockage of the insect spiracles and consecutive suffocation (Ileke and Ogungbite 2014).

Plant extracts containing secondary metabolites such as aliphatic agents, acetogenonins, sterols, alkamides, alkaloids, sesquiterpenes, triterpenes, coumarins, anthraquinones, xanthones, and flavonoids were reported to exhibit neurotoxic effects on mosquitoes, inhibit the detoxificant enzymes and larval development, and/or cause midgut damages (Pavela et al. 2019a). Some EO, plant extracts or their constituents could be considered as a possible alternative to mitigate the harmful effects of synthetic insecticides on beneficial insect species such as pollinators (e.g., Santos et al. 2018; Seixas et al. 2018) or predators of harmful insects (Zandi-Sohani et al. 2018; Thanigaivel et al. 2018; Chellappandian et al. 2018; Gupta et al. 2017). Botanical insecticides having adverse effects only on target insects but not destroying beneficial natural enemies could provide food free from residues (Hikal et al. 2017). On the other hand, botanical insecticides are characterized with short shelf life, photosensitivity, and volatilization that possess limits to their large scale use in plant protection (Campos et al. 2016). Therefore, their application in nanoscale formulations could pronouncedly contribute to mitigate these disabilities (de Oliveira et al. 2014).

Plant-derived compounds used against beetles-pests of stored crops and food, including extracts of *Solanaceae* or *Asteraceae* plants, EOs of *Artemisia absinthium* or *Citrus* spp. as well as some compounds like α-chaconine or α-solanine, and their mode of action were overviewed by Spochacz et al. (2018).

Isofuranodiene, the dominant volatile compound in the EO from *Smyrnium olusatrum* (*Apiaceae*), encapsulated in MEs showed considerable mortality of larvae over time and a pronounced reduction in adult emergence of *Cx. quinquefasciatus* and only a little impact on non-target organisms, aquatic microcrustacean *Daphnia magna* and the earthworm *Eisenia fetida* (Pavela et al. 2019b). As eco-friendly insecticide formulations against *Cx. quinquefasciatus* were reported also MEs with encapsulated EOs of *Pimpinella anisum*, *Trachyspennum cutuni*, and *Crithmum maritimum* showing toxic effects against larvae ($LC_{50}$ values of 1.45–4.01 ml/l), high larval mortality, and low ratio of hatched adults after short-term exposure to sublethal concentrations but low or no mortality to *D. magna* and *E. fetida* (Pavela et al. 2019c). A contact + ingestion assay using wheat grains confirmed that NE prepared using the EO from *Pimpinella anisum* L. (*Apiaceae*) containing 81.2% of (*E*)-anethole with mean droplet size of 198.9 nm and zeta potential of −25.4 ± 4.47 mV was found to be toxic to *Tribolium castaneum* Herbst (Coleoptera: Tenebrionidae) and strongly impacted also its progeny. Pronounced reduction in beetle progeny number was observed with increasing NE concentrations at prolonged exposure, whereby treatment with 10% NE caused 70.85%

reduction in progeny production. It could be noted that the major EO constituent of *P. anisum* EO, (*E*)-anethole, could penetrate throughout the cuticle resulting in irritations and strong damage to various body parts of the insect (Hashem et al. 2018).

*Melissa officinalis* L. EO having as major components geranial, neral, and β-caryophyllene was reported to be an effective insecticide against *T. castaneum* Herbst causing in a contact bioassay 100% mortality of both larvae and adult insects following 24 and 48 h exposure. CS NPs loaded with this EO with average particle size of 362 nm exhibited superb fumigant toxicity as well as repellent activity against *T. castaneum* exceeding that of pure EO or unloaded CS NPs. The ingestion and penetration of encapsulated EO caused strong harm to the midgut region of the insect such as loosening and thinning of epithelial cells, with vacuolated nuclei and modified shapes that triggered feeding deterrence action resulting in the interruption of further feeding. The toxic effects of the nanoformulation were connected mainly with oxidative stress, while treated insects did not show any significant alteration in AChE activity. To the increased efficiency of encapsulated lemon balm EO contributed also the nanoscale size of the formulation enabling easier passive cellular absorption of active ingredients (Upadhyay et al. 2019).

Nanogels of myristic acid-CS loaded with EO obtained from cumin, *Cuminum cyminum* L., were found to be more toxic to *Sitophilus granarius* L. and *T. confusum* Jacquelin du Val. than the free EO and while after 12 days any insecticidal activity of the pure EO was detected, nanogel EO formulation lost only ca 60 and 15% of its activity at application against *S. granarius* and *T. confusum*, respectively (Ziaee et al. 2014).

The ME consisting of carvacrol and methyl salicylate showed efficient insecticidal activity against thrips *Anaphothrips obscurus* in laboratory and field trials. This ME applied at a dose of 600.0 g A.I $hm^{-2}$ was able to control approx. 89.17% of thrips in peppers, and 82.59% in broad bean on the 7th day post application suggesting synergistic action of active constituents and thus a potential to be used as biopesticide (Lu et al. 2020).

Larvicidal activity of *Rosmarinus officinalis* L. EO NEs against *Ae. aegypti*, in which the final concentration of EO was 250 ppm, was reflected in mortality levels of 80 ± 10% and 90 ± 10%, respectively, observed 24 and 48 h after treatment (Duarte et al. 2015). *R. officinalis*-loaded spherical polycaprolactone nanocapsules with an average size of 145 ± 15 nm, zeta potential of −11.0 ± 0.5 mV, and a 78.20 ± 0.93% EE showed improved fumigant and contact toxicity against *T. castaneum* compared to pure EO due to the increased surface area and controlled release of the active ingredients (including α-pinene, 1,8-cineol, camphor, and *cis*-verbenone). In a fumigation toxicity test a 24/72 h exposure of *T. castaneum* to the encapsulated EO at 27.76 μl/l air resulted in 96.6/100% of killed insects compared to 71.6/83.3% at application of the pure oil and 100% mortality was observed also after 72 h exposure of insects at 19.12 μl/l air to *R. officinalis* nanocapsules. Similar results were obtained in contact toxicity test as well (Khoobdel et al. 2017).

*Allium cepa* EO NEs with droplet size 93.4 nm fabricated by ultrasonic emulsification for 35 min showed strong acaricidal activity against the two eriophyid olive mites *Aceria oleae* Nalepa ($LC_{50}$: 298.225 μg/ml) and *Tegolophus hassani* (Keifer) ($LC_{50}$: 309.634 μg/ml) (Mossa et al. 2018b).

Investigation of larvicidal and insecticidal effect of *Cinnamomum zeylanicum* EO applied as pure oil or in nanostructured form against *Alphitobius diaperinus* showed that mortality in larva and adult forms of the insect succeed after treatment with 5 and 10% EO, while for the killing of the insect in both phases of *A. diaperinus* life cycle treatments with NE containing 1% EO or with nanocapsules containing 5% EO were sufficient. Moreover, by encapsulation of EO considerable reduction of its adverse effects on springtails survival and reproduction was achieved (Volpato et al. 2016).

*Pogostemon cablin* EO (containing as main components sesquiterpene hydrocarbons) and its nanoformulation fabricated using polyoxyethylene, ethanol, and water showed superb insecticidal activity and irritability to the leaf-cutting ants: *Atta opaciceps* (Borgmeier, 1939), *Atta sexdens* (Linnaeus, 1758), and *Atta sexdens rubropilosa* (Forel, 1908), whereby concentrations needed to kill 50% of workers were in the range 1.06–2.10 μl/l and ants were dying within 42 h. Moreover, the reduced displacement and velocity speed of workers of *A. opaciceps* and *A. s. rubropilosa* was observed in arenas totally treated with the essential oil of *P. cablin* and its nanoformulation, and in the bioassays with choices, three tested ant species walked less and at a greater speed on the treated side of arena (Rocha et al. 2018).

NEs of *Baccharis reticularia* DC. EO containing as main constituent D-limonene (25.7%) with average droplet sizes ca 90 nm that were applied against *Ae. aegypti* showed larvicidal activity and 48 h after treatment $LC_{50}$ values of 118.94 g/ml and 81.19 μg/ml, respectively, were observed. The mechanism of action of this EO was connected with AChE inhibition and treatment with D-limonene NE resulted in morphological alterations of mosquito larvae (Botas et al. 2017).

Solid lipid NPs of *Melaleuca alternifolia* EO were reported to exhibit both repellent and insecticide action against subterranean termites (*Coptotermes gestroi*) (Clerici et al. 2018).

Campolo et al. (2017) tested the insecticidal activity of the citrus peel EO in form of emulsions or encapsulated in PEG NPs against the invasive tomato pest *Tuta absoluta* (Lepidoptera: Gelechiidae). Eggs were found to be less sensitive than larvae to formulations containing EO, which could be explained with the fact that it is sometimes difficult to reach them with insecticide and the structure of the eggs protecting the developing embryos may interfere with insecticide penetration. The treatment with EO emulsions resulted in stronger contact toxicity to eggs and larvae reflected in higher mortality rate compared to treatment with EO NPs, which could be caused by significantly lower ratio of EO contained in the NPs (ca 10%) compared to pure EO and therefore the EO concentration coming in contact with eggs and larvae was not sufficient to initiate the required biological response. On the other hand, ingestion of EO NPs by larvae exhibited more detrimental impact than the respective EO emulsions.

Neem (*Azadirachta indica* A. Juss) products exhibit behavioral, physiological, and biological effect on insects and were reported to control more than 300 insect species (Nagaraj 2009). Among more than 200 active compounds isolated from neem, the tetranortriterpenoid azadirachtin (AZA), the crucial component of neem oil, shows insecticidal activity acting as an antifeedant, repellent, and repugnant agent and it could prevent oviposition and interrupt sperm production in male's insects resulting in the sterility. The bioinsecticide AZA could be used to replace synthetic toxic insecticides (Chaudhary et al. 2017; Morgan 2009). Low-energy emulsification method was applied to prepare environmentally benign NE formulations of neem EO with particle diameters 208–507 nm for the control of adult *S. oryzae* (Coleoptera: Curculionidae) and *T. castaneum*. *S. oryzae* adults were found to be more susceptible than *T. castaneum* adults to NE formulations. Using food impregnation method a 100% mortality of *S. oryzae* adults was observed following 24 h exposure to the neem EO NE prepared with polysorbate surfactant and 2.0 ml/kg AZA (Choupanian and Omar 2018). After 2 days of exposure to 1% AZA NE the contact toxicity resulted in 85–100% and 74–100% mortality *of S. oryzae* and *T. castaneum* adults, respectively (Choupanian et al. 2017). da Costa et al. (2014) tested the effects of nanoformulated neem products in powder (NC), soluble powder prepared with neem oil (SP), and neem oil emulsifiable concentrate (EC) on the bean weevil, *Zabrotes subfasciatus*, and found that the highest mortality of the insect caused treatment with neem oil ECs containing 1000, 2000, and 4000 ppm of AZA applied at a dose 0.3% (w/v) and the EC formulations also caused reduction of the total number of oviposited eggs. On the other hand, the greatest UV stability was estimated for NC, while SP was found to release AZA more rapidly than the preparation fabricated using biopolymers. Microcapsules of sugarcane bagasse lignin loaded with organic extracts of neem showed increased thermal and photostability of ca 40% compared with control samples and were able to cause 100% mortality of *Spodoptera frugiperda* and *Diatraea saccharalis* insects in shorter time than the controls (Costa et al. 2017). Nanoformulation of neem bark extract crosslinked with polycarboxylic acids loaded on the biogenic $SiO_2$ NPs derived from *Equisetum arvense* showed slow-release properties (60–75% released after 30 days), amelioration in the neem extract stability, and free radical scavenging activity and was able to kill the major workers of *Acromyrmex crassispinus* ant species (Mattos et al. 2017). Maize leaves treated with nanoformulations of encapsulated neem fabricated using poly(ε-caprolactone), poly(β-hydroxybutyrate), or poly(methyl methacrylate) were offered to first instar larvae of *S. frugiperda* during 10 days. It was found that although treatment with some nanopreparations resulted in insect mortality and sublethal effects up to 3 and 7 days after spraying, respectively, the effect of commercial neem oil was higher. On the other hand, all treatments showed phagodeterrence at 1 day after spraying, although this was lost over time (Giongo et al. 2016). Nanoformulations of neem oil showing controlled release of the insecticide were also formulated by some other researchers (e.g., Jerobin et al. 2012; Feng and Peng 2012; Kumar et al. 2010; Sittipummongkol and Pechyen 2018; Mattos et al. 2017).

Forim et al. (2013) prepared NPs and MPs loaded with extracts of *A. indica*, with EE of about 100%, release profile of which based on swelling and relaxation of the polymer or polymer erosion, causing 100% mortality of *P. xylostella* larvae.

Rotenone is a naturally occurring compound with insecticidal activity that interrupts mitochondrial complex I of the electron transport chain and also elicits mitochondrial dysfunction. Rotenone could impair neuronal polarization in cultured hippocampal neurons and cause the inhibition of axonogenesis, which could be connected with its effect on microtubule dynamics, the actin cytoskeleton and their regulatory pathways, small RhoGTPase RhoA being especially affected (Bisbal and Sanchez 2019). As suitable carrier for rotenone *N*-deoxycholic acid-*O*-glycol CS was reported; up to 41 h longer release of the insecticide from the micelles of modified CS was observed compared to free rotenone (Yusoff and Kamari 2018). Controlled release properties showed also deoxycholic acid carboxymethyl CS micelles (67.5–83.3 nm) (Aljafree and Kamari 2018) and oleoyl-carboxymethyl CS micelles (35.5–66.4 nm) with encapsulated rotenone (Kamari et al. 2016).

# 5 Natural Minerals

Diatomaceous earth (DE) is nearly pure amorphous $SiO_2$, fabricated of fossilized diatoms showing insecticidal activity. It can absorb epicuticular lipids and fatty acids causing desiccation in arthropods (Shah and Khan 2014). Particles of insecticidal DE with diameters <10 μm, pH <8.5, marginal number of clay particles, and <1% crystalline $SiO_2$ could be easily picked up by rough bodied insects, and the cuticule damaged by hydrocarbon absorption and abrasion will be permeable to water and the insects die from desiccation (Korunic 1998). The toxicity of $SiO_2$ and $Al_2O_3$ NPs to insects is also connected with their binding to the insect cuticle, and following physical sorption of waxes and lipids causes dehydration of the organism (Benelli 2018a). The insecticidal activity of several porous materials including diatomaceous earth and zeolites against pharaoh ant (*Monomorium pharaonis*) depended predominantly on macroporous surface area and Brunauer–Emmett–Teller (BET) specific surface area, and the removal of the protective epicuticular hydrocarbons resulted in the mortality of insects (Van den Noortgate et al. 2018).

On the other hand, nanoscale zeolites could reduce the bioavailability of synthetic insecticides due to their sorption properties that results in reduced toxicity of the insecticide (Lorenz et al. 2017a). Exposure of eggs of tomato leafminer *T. absoluta* to zeolites showed adverse effect on the development process reflected in the weakening the first instar larvae and increased mortality (De Smedt et al. 2016). In crop protection zeolites are predominantly used as carriers of different active ingredients in slow-release applications (De Smedt et al. 2015).

Using DE Debnath et al. (2010) prepared $Al_2O_3$ NPs and amorphous $SiO_2$ NPs showing strong activity against mustard aphid (*Lipaphis pseudobrassicae*); however, $Al_2O_3$ NPs adversely affected the growth of mustard crops. On the other hand, the insecticidal activity of $TiO_2$ NPs against mustard aphid was only moderate. Bioactivity of DE against various insects such as the subterranean termite *Reticulitermes chinensis* Snyder (Isoptera: Rhinotermitidae) (Gao et al. 2018b), storage pests *Liposcelis paeta*, *Cryptolestes ferrugineus*, *R. dominica*, and

*T. castaneum* (Saeed et al. 2018), lesser mealworm (*Alphitobius diaperinus* Panzer, 1797 [Coleoptera: Tenebrionidae]) (Oliveira et al. 2017) or *Acanthoscelides obtectus* (Say) on chickpeas (*Cicer arientum* L.) (Alkani et al. 2019) and insecticidal potential of zeolites against *S. oryzae* and *Oryzaephilus surinamensis* (Eroglu et al. 2019) or *Acanthoscelides obtectus* (Floros et al. 2018) was reported as well.

Entomotoxicity of amorphous hydrophilic, hydrophobic, and lipophilic $SiO_2$ NPs (15–30 nm) against rice weevil *S. oryzae* exceeded that of bulk $SiO_2$ particles (>1 μm) causing >90% mortality of insects (Debnath et al. 2011). The pulse seeds of *Cajanus cajan*, *Macrotyloma uniflorum*, *Vigna mungo*, *Vigna radiata*, *C. arietinum*, and *Vigna unguiculata* treated with $SiO_2$ NPs were found to be protected to a great extent against the infestation of stored pulse beetle, *Callosobruchus maculatus*, which was reflected in strong reduction in oviposition, adult emergence, and seed damage potential, whereby the treatment did not affect the soil microflora. Complete suppression of insect growth was observed in the treated seeds of pigeon pea (Arumugam et al. 2016). In a laboratory experiment Shoaib et al. (2018b) investigated the entomotoxic effects of $SiO_2$ NPs in form of dust on larvae of *P. xylostella* and at application of a dose of 1 mg/cm$^2$ up to 58% and 85% mortality was observed at 24 and 72 h after treatment and larvae died due to desiccation, body wall abrasion, and spiracle blockage. Seven days after treatment with $SiO_2$ NPs Hala and Elsamahy (2016) estimated $LC_{50}$ values related to the mean lethal concentrations as 316.9, 115.63, and 112.4, 83.0 ppm, for the carmine spider mite, *Tetranychus cinnabarinus* (Boisduval) and the two spotted spider mite, *Tetranychus urticae* (Koch) adult females and eggs, respectively. In predatory species of these insects the $SiO_2$ NPs caused significantly higher mortality in spider mite destroyer, *Stethorus punctillum* (Weise) (97.5%) than in minute pirate bug *Orius insidiosus* (Say) (32.5%) or predatory mite, *Phytoseiulus persimilis* (Athias-Henriot) (35%). Investigation of the impact of $SiO_2$ NPs on third larval instar of the oriental armyworm, *Mythimna separata* (Walker) (Lepidoptera: Noctuidae) showed that at soil treatment and foliar treatment of wheat plants feeding inhibition rate reaching 37.16 and 43.91%, respectively, reduced relative growth rate, caused prolongation of larval stage period by ca 4 days and spraying with $SiO_2$ NPs resulted in 67.69% mortality of larvae (Mousa et al. 2014). In an in vivo study Khandelwal et al. (2015) observed that on 8th day after feeding of *Helicoverpa armigera* with $SiO_2$ nanospheres and rods containing immobilized *Capsicum annuum* proteinase inhibitor (CanPI-13) the insect body mass was reduced by ca 40%. At pH 10 simulating gut milieu of the insect, 56% of the bioactive peptide were released. $SiO_2$ NPs Aerosil® and Nanosav exhibited strong toxicity against *R. dominica* F. and *T. confusum* Jacquelin du Val., *R. dominica* being the more susceptible insect, whereby $SiO_2$ NPs applied on wheat and peeled barley (50–300 mg/kg) were found to be more effective in wheat grain (Ziaee and Ganji 2016).

Hydrophobic $SiO_2$ NPs applied at concentration 112.5 ppm showed strong toxicity also on several mosquitos' species. The larvicidal activity of $SiO_2$ NPs decreased in the order: *Anopheles stephensi* > *Ae. aegypti* > *Cx. quinquefasciatus*, while their pupicidal effect decreased in the order of *A. stephensi* > *Cx. quinquefasciatus* > *Ae. aegypti* (Barik et al. 2012). Analysis of the effect of $SiO_2$ NPs (bare $SiO_2$ particles of 14 nm, 380 nm, and 1430 nm and amine-modified particles of 131 nm and 448 nm, respectively) on the viability of *S. frugiperda* cells (Sf9 cell line) showed that the 14 nm NPs exhibited the highest toxicity, while lower concentrations of positive charged NPs (0.12 or 0.6 mg/ml) stimulated the proliferation of the cells (Santo-Orihuela et al. 2016). $SiO_2$ NPs functionalized with mercaptopropyl-triethoxysilane and hexamethyldisilazane were reported as effective insecticide against *S. litura* larvae (Debnath et al. 2012).

$SiO_2$ NPs and ZnO NPs, which were tested against the newly hatched larvae of the pink bollworm, *Pectinophora gossypiella* (Saunders) (Lepidoptera: Gelechiidae), pronouncedly disrupted the transaminases and carbohydrate enzymes, total lipids, and proteins and were reported to be suitable for controlling this insect (Derbalah et al. 2014).

Investigation of insecticidal impact of nano- to microsized $\alpha$-$Al_2O_3$ powders against *Acanthoscelides obtectus* (Say) (Coleoptera: Chrysomelidae: Bruchinae) showed that survival and progeny number of the insect was reduced with increasing of surface area, pore volume, and diameter, and a decrease in particle size, whereby the adverse effect was more pronounced against males ($LC_{50}$ = 330.4 ppm) than for females ($LC_{50}$ = 409.6 ppm) (Lazarevic et al. 2018). Nanostructured $Al_2O_3$ showed higher toxicity against workers of *Acromyrmex lobicornis* than DE ($LC_{50}$ of 0.14 mg/g vs. 0.36 mg/g) and also more effective attaching to the cuticle of exposed insects. Moreover, the ants were repelled by $Al_2O_3$ NPs neither in laboratory nor in field conditions (Buteler et al. 2018). Under laboratory conditions treatments with nanostructured $Al_2O_3$ (250 and 500 ppm) reduced grain weight loss and frass production in wheat infested by *S. oryzae* and caused progeny (F1) suppression more effectively than DiatomiD®, and Protect-It® (commercial diatomaceous earth) (Lopez-Garcia et al. 2018).

Insecticidal effects of nanostructured $Al_2O_3$ dusts fabricated using a modified glycine-nitrate combustion process was higher on *S. oryzae* than on *R. dominica*, it depended on particle size, particle morphology, and surface area and minimizing particle size and maximizing surface area were found to belong to crucial factors affecting insecticidal effectiveness (Buteler et al. 2015).

## 6 Carbon-Based NPs

The ingesting of water-soluble nanocarbons at a dose of 3 mg/l blocked the growth of the mosquito from the larval stage to adulthood and larvae perished after 4 weeks (Saxena et al. 2013). Investigation of the insecticidal activity of 11 different carbon materials against the pharaoh ant (*Monomorium pharaonis*) showed the shortest median survival time, 25 min, for treatment with activated carbon powder, which was ca fourfold lower than that observed with diatomaceous earth, whereby determined insecticidal activity of activated carbon predominantly depended on the particle size (Van den Noortgate et al. 2019). Structures of carbon-based nanoinsecticides are illustrated in Fig. 9. The details of preparations of carbon-based nanomaterials,

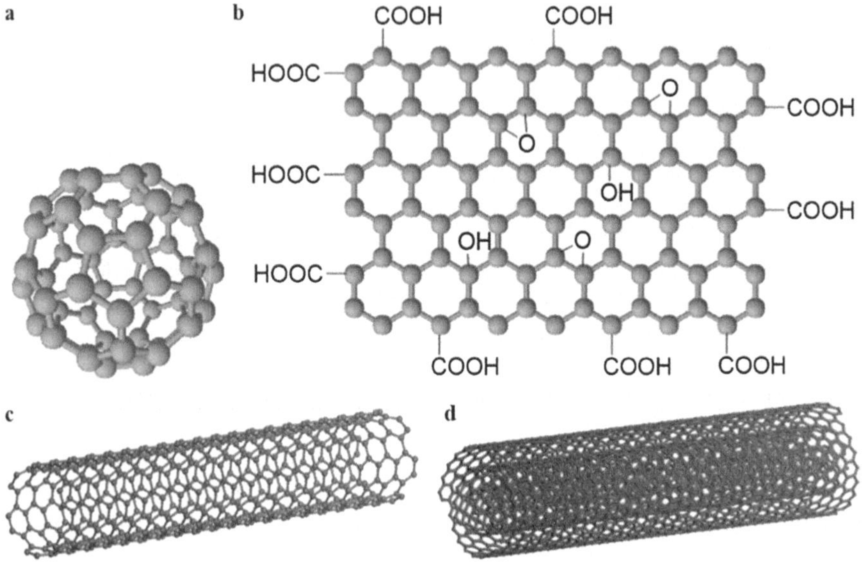


**Fig. 9** Carbon-based nanoinsecticides: $C_{60}$ fullerene (**a**), graphene oxide (**b**), single-walled nanotubes, (**c**) and multi-walled nanotubes (**d**)

their physico-chemical and biological properties were reviewed by Plachá and Jampílek (2019).

Graphene quantum dots (GQDs) showed toxic effects on *A. stephensi* with $LC_{50}$ values ranging from 0157 (larva I) to 6.323 ppm (pupa) and post-treatment with GQDs increased the predation efficiency of non-target organisms *Gambusia affinis*, *Anax immaculifrons*, and *Hoplobatrachus tigerinus* (Murugan et al. 2017).

Sediment-associated fullerenes ($nC_{60}$) were reported to have adverse effect on the growth and development of the sediment-dwelling invertebrate *Chironomus riparius* larvae. The small agglomerates of $nC_{60}$ observed at doses 0.0025–20 mg/kg decreased the body length, at a dose of 0.5 mg/kg delayed emergence rate was estimated, while larger agglomerates occurring at high $nC_{60}$ dose (80 mg/kg) were not toxic (Waissi-Leinonen et al. 2015).

Graphene oxide (GO) NPs strongly affect insect antioxidant and detoxifying enzymes causing oxidative stress and cell death (Benelli 2018a). Oxidative stress reflected in increased enzymatic activity of catalase and glutathione peroxidases and total antioxidant capacity levels were observed in *Acheta domesticus* (L.) crickets after injection of GO NPs into insect hemolymph (Dziewięcka et al. 2016). GO was found to show the synergistic activity with insecticides β-cyfluthrin, monosultap, and imidacloprid on lepidoptera insect Asian corn borer (*Ostrinia furnacalis*) resulting in the 2.1-, 1.51-, and 1.83-fold activity enhancement compared to individual insecticides. The synergistic mechanism could be connected with physical damage to the cement layer of insects leading to dramatic water loss in the insects and with improved penetration of insecticides through the disrupted cement layer (Wang et al. 2019b).

Exposure of carbon black and single-walled carbon nanotubes (SWNTs) in dry form to *D. melanogaster* adults resulted in their strong adhesion to fly surfaces, outperformed natural grooming mechanisms, and deteriorated locomotor function and mortality, while fullerene and multi-walled carbon nanotube (MWNT) arrays showing weak adhesion did not affect locomotor function or survivorship (Liu et al. 2009).

Carbon nanomaterials, such as oxidized MWCNTs and GO, in the diet of *S. frugiperda* larvae (fed from egg hatching to pupation) exhibited adverse impact on the reproductive parameters and the digestive and metabolic efficiency of the insect, especially GO applied at a dose of 1 mg/g caused considerable reduction of the fecundity and fertility of *S. frugiperda* and attenuated efficiency of food conversion into biomass and digestibility (Martins et al. 2019).

MWNCTs and carboxylated MWCNTs did not exhibit toxic effects against the infective juveniles of entomopathogenic nematodes *Steinernema feltiae* (Owinema, Namasys, Nemaplus) and *Heterorhabditis bacteriophora* (Namatop), however they limited the activity of these species (Kuzniar et al. 2011). Injected MWCNTs were found to be incorporated into cells in early *D. melanogaster* embryos, they remained cytoplasmic and were excluded from the nucleus and a rise in cell death of ectodermal but not of neural stem cells suggested stem cell-specific vulnerability to MWCNT exposure (Liu et al. 2014).

## 7 Metal Nanoparticles

The toxicity of metal NPs to living organisms is caused not only by the chemical properties of respective metals and the release of toxic metal ions from NPs but the additional stress occurring due to the surface, nanoscaled size, and shape of these NPs also significantly contributes to their toxicity (Masarovičová and Kráľová 2013; Masarovičová et al. 2014; Kráľová et al. 2019). Nanoscale metal particles can reduce membrane permeability by binding to sulfur and phosphorus atoms in proteins and nucleic acids resulting in organelle and enzyme denaturation followed by cell death (Benelli 2018a; Jampílek and Kráľová 2015, 2017a, b; Pisárčik et al. 2016, 2017, 2018). For example, AgNPs could permeate cell membranes, which results in higher levels of intracellular $Ag^+$ resulting in cytotoxic and genotoxic effects, they induce oxidative stress with consecutive local depletion of glutathione and other antioxidants (Jampílek and Kráľová 2015, 2017a, b; Pisárčik et al. 2016, 2017, 2018). AgNPs could reduce AChE activity, modify the expression of key insect genes, adversely affect protein synthesis and gonadotropin release, which result in developmental damages and reproductive malfunction (Benelli 2018a). Using in vivo model of *D. melanogaster* Alaraby et al. (2019) showed that AgNPs crossed the intestinal barriers and produced primary DNA damage by inducing oxidative stress, even though the effect of $AgNO_3$ was stronger than that of AgNPs.

Ingestion of AgNPs in *D. melanogaster* during adult stage adversely affected egg laying capability, impaired growth of ovary and resulted in reduced survival of larvae suggesting deleterious impact of AgNPs to the reproductive health and survival of the insect. Moreover, trans-generational effect of AgNPs was also observed without feeding progeny (Raj et al. 2017). Thus, AgNPs are considered as a very effective nanoweapon against mosquitoes, because their use is connected with low risk of developing resistance in long-term usage (Muthukumaran et al. 2015; Rouhani et al. 2012; Shanmugasundaram and Balagurunathan 2015; Singh and Mishra 2014; Soni and Prakash 2014; Sutthanont et al. 2019).

Investigation of the impact of AgNPs on *Bombyx mori* using omics technologies showed that feeding of the insect with higher concentrations of AgNPs resulted in downregulation of the expression of digestive enzymes. Consequently, the silkworm tissue was damaged and the AgNPs-induced oxidative stress adversely affected the silkworm digestive system. Destroyed basal lamina and columnar cells were estimated following treatment with 400 mg/l AgNPs (Chen et al. 2019).

Nanosized Ag crystals with *Bauhinia acuminata* phytochemicals as capping agents showing mean particle size of 25 nm exhibited larvicidal activity against *A. stephensi*, *Ae. aegypti*, and *Cx. quinquefasciatus* with $LC_{50}$ values of 24.59, 27.19, and 30.19 μg/ml, respectively, pronouncedly exceeding that of pure *B. acuminata* aqueous leaf extract (204.07, 226.02, and 249.24 μg/ml, respectively) (Alharbi et al. 2018). AgNPs fabricated using entomopathogenic fungus *Beauveria bassiana* were reported to show biological efficiency against mustard aphid (*Lipaphis erysimi* Kalt.) (Kamil et al. 2017). Spherical poly-dispersed Ag nanocomposites (NC) fabricated using the aqueous stem extract of *Achyranthes aspera* and $AgNO_3$ showing mean size of 1–30 nm were recommended as environment-friendly alternative to synthetic insecticide formulations for mosquito control. The larvicidal activity of these NCs against early fourth instars of *Ae. aegypti* depended on the used $AgNO_3$ concentration and the $LC_{50}$ values estimated in 48/72 h bioassays were 1.113/0.610 μg/ml and 0.420/0.407 μg/ml for application of 3 and 4 mM of NC (Sharma et al. 2019).

The AgNPs prepared using *Curcuma zedoaria* EO, which were investigated against larvae of insecticide-sensitive and insecticide-resistant strains of *Cx. quinquefasciatus*, were able to cause 100% larval mortality within 24 h of exposure and the estimated $LC_{50}/LC_{99}$ values against the sensitive strain were 0.57/8.54 ppm and 0.64/8.88 ppm against the resistant strain. On the other hand, using EO alone at similar conditions, the determined $LC_{50}/LC_{99}$ values were 36.32/85.11 ppm against the susceptible, and 37.29/76.79 against the resistant strain, respectively (Sutthanont et al. 2019). Eco-fabricated AgNPs using *Carmona retusa* (Vahl) Masam leaf extract exhibited efficient larvicidal activity against *A. stephensi*, *Ae. aegypti*, and *Cx. quinquefasciatus* with $LC_{50}$ values of 116.681 ppm, 198.766 ppm, and 83.553 ppm, respectively (Rajkumar et al. 2018). AgNPs biosynthesized using seaweed *Sargassum polycystum* with mean particle sizes 20–88 nm exhibited effective larvicidal activity against *Ae. aegypti* and moderate toxicity against *Cx. quinquefasciatus* (90 and 80% mortality after 72 h exposure), while their impact on *A. stephensi* and *Cx. tritaeniorhynchus* larvae was less pronounced (Vinoth et al. 2019). Based on the

$LC_{50}$ value of 5.93 mg/l related to the larvicidal activity of AgNPs biosynthesized using *Garcinia mangostana* bark extract against fourth instar larvae of *Ae. aegypti* it could be concluded that nanoscale silver particles could penetrate the insect cuticle and pass into individual cells and thus interfere with molting and some other physiological processes (Karthiga et al. 2018).

Treatment of *A. aegypti* mosquito larvae with green fabricated AgNPs using *Schinus molle* extract was found to be ca 16.4-fold more effective than application of plant crude extract ($LC_{50}$ values of 13.894 vs. 228.345 ppm) (Hamed et al. 2018). Similar results were obtained at treatment of *Ae. aegypti* larvae with AgNPs prepared using *Chrysanthemum* extract when the insecticidal activity of AgNPs was 17.9-fold higher than that of the plant extract ($LC_{50}$ values of 12.754 ppm vs. 228.345 ppm) (Ghramh et al. 2018). AgNPs green synthesized using the leaf extract of the orchid *Zeuxine gracilis* showed effective insecticidal activity against the larvae of *A. stephensi*, *Ae. Aegypti*, and *Cx. quinquefasciatus* with $LC_{50}$ values of 8.48, 10.39, and 13.21 A µg/ml, respectively (Kovendan et al. 2018). AgNPs (35–55 nm) fabricated using stearic acid from *Catharanthus roseus* leaf extract applied at a concentration of 200 ppm showed high antifeedant and larvicidal activities (87.13% and 93.77%, respectively) against *Earias vittella*, whereby the corresponding $LC_{50}$ values were 45.46 and 25.12 ppm, respectively. These AgNPs also exhibited acute toxicity against *Cx. quinquefasciatus* and *Ae. aegypti* ($LC_{50}$ < 40 ppm) (Pavunraj et al. 2017).

The larvicidal activities of AgNPs fabricated using *Habenaria plantaginea* leaf extract against *A. stephensi*, *Ae. aegypti*, *Cx. quinquefasciatus*, *Anopheles subpictus*, *Aedes albopictus*, and *Cx. tritaeniorhynchus* expressed by $LC_{50}$ values were 12.23, 13.38, 14.78, 14.37, 15.39, 16.89 µg/ml, respectively, and were considerably lower than those observed with *H. plantaginea* extract (102.51, 111.99, 123.47, 123.96, 136.56, 149.42 µg/ml, respectively). Both *H. plantaginea* extract and AgNPs showed only minor toxicity to *Anisops bouvieri*, *Diplonychus indicus*, *Poecilia reticulata*, and *Gambusia affinis*, natural enemies predating mosquito larvae and pupae, with $LC_{50}$ values in the range from 831.82 to 36,212.67 µg/ml (Aarthi et al. 2018). AgNPs fabricated using *Suaeda maritima* extract exhibited larvicidal and pupicidal activity against *Ae. aegypti* showing $LC_{50}$ values in the range from 8.668 (larva I) to 17.975 ppm (pupa), while $LC_{50}$ values against *S. litura* ranged from 20.937 (larva I) to 46.896 ppm (pupa). Exposure to 20 ppm of AgNPs or 250 ppm of *S. maritima* extract was able to reduce egg hatchability by 100% (Suresh et al. 2018). Morejon et al. (2018) reported $LC_{50}/LC_{90}$ values related to insecticidal activity of AgNPs fabricated using leaf extracts of *Ambrosia arborescens* against third instar larvae of *Ae. aegypti* as 1844.61/6043.95 ppm.

The AgNPs (22.5–66.2 nm) fabricated using alkaloids of *Peganum harmala* L. seeds exhibited considerable insecticidal and growth-inhibiting activities against khapra beetle, *Trogoderma granarium* (Everts) (Coleoptera: Dermestidae) exceeding that of pure alkaloids. Considerable decline in the normal growth and development of *T. granarium* was observed following feeding the second instar larvae with grains treated with sublethal concentrations of AgNPs and adverse impact was also reflected in a high portion of malformed larvae and pupae, a prolonged life span of pupae, and in notable drop in adult emergence (Almadiy et al. 2018).

In an experiment, in which AgNPs prepared using the extract of jujube *Ziziphus* sp. were applied to whitefly infested Al-Mustakbal eggplant hybrid grown in a greenhouse, it was shown that exposure to 3000 ppm AgNPs resulted in 100% reduction of population density of *Bemisia tabaci* nymphs after 1, 3, 7 days and in 80% reduction 21 days after treatment (Al Shammari et al. 2018).

Benelli et al. (2018) investigated the ovicidal, larvicidal, and adulticidal toxicity of AgNPs (40.2–70.4 nm) prepared using *Acacia caesia* leaf extract against three mosquito vectors. The larvicidal activity of AgNPs expressed by $LC_{50}$ values decreased as follows: *A. subpictus* (10.33 μg/ml) > *Ae. albopictus* (11.32 μg/ml) > *Cx. tritaeniorhynchus* (12.35 μg/ml), and $LD_{50}$ values estimated in adulticidal assays showed a similar rank and they were 18.66, 20.94, and 22.63 μg/ml; complete inhibition of egg hatchability on three tested vectors was observed at 60, 75, and 90 μg/ml, respectively. The AgNPs were found to show moderate toxicity against non-target aquatic biocontrol agents *A. bouvieri*, *D. indicus*, and *Gambusia affinis*, with $LC_{50}$ values ranging from 684 to 2245 μg/ml.

AgNPs prepared using aqueous extract of *Cassia fistula* fruit pulp exhibited insecticidal activity against *Cx. pipiens pallens* with $LC_{50}$ values ranging from 1.1 mg/L (I instar larva) to 19.0 mg/L (pupae) and *Ae. albopictus* with $LC_{50}$ values ranging from 8.3 mg/L (I instar) to 17.3 mg/L (pupae). At higher doses of AgNPs the internal toxic effects of tiny particles inside cuticle could cause the mortality of larvae and pupae because of absorption of high quantity of AgNPs by larval body. Moreover, binding of AgNPs to sulfur in proteins and phosphorus in nucleotides of DNA results in the denaturation of some organelles and enzymes. The exposure of the larvae of both tested insects to AgNPs notably reduced the total protein level and disturbed the protein metabolism in the larvae suggesting direct toxic effect of AgNPs on the protein synthetic machinery of the larvae. Treatment with AgNPs significantly reduced the level of AChE activity as well (Fouad et al. 2018).

Ingestion of AgNPs by *D. melanogaster* larvae reduced the diversity of the gut microbiota causing an increase in the predominance of *Lactobacillus brevis* and a reduction in *Acetobacter* compared to control and insects treated with AgMPs, while delayed development, shortened adult longevity, and decreased sperm competition were observed in medium containing CuNPs and CuMPs as well (Han et al. 2014). Ag and Ag-Zn NPs applied against the oleander aphid, *Aphis nerii* Boyer de Fonscolombe, showed $LC_{50}$ values of 424.67 and 539.46 mg/ml, respectively, highest insect mortality being observed at 700 mg/ml (Rouhani et al. 2012). Ibrahim and Ali (2018) observed developmental and physiological changes in the larvae and pupae of *S. littoralis* (Lepidoptera: Noctuidae) induced by sublethal concentrations of AgNPs (50–60 nm) and ZnO NPs (10–30 nm). Late second instar larvae of *S. littoralis* treated with NPs dipped castor leaves (10 mg/ml) for 6 days had reduced weight gain and pupal weight compared to control. ZnO NPs ingestion was found to affect the digestive and immunological physiology and the development of the insect, which was reflected in reduced levels of proteins, lipids, and carbohydrates and a considerable enhancement of the activities of some enzymes, including catalase and superoxide dismutase; ZnO NPs also extended larval period. Exposure to AgNPs increased plasmatocytes and their impact on the contents of protein, lipids, and carbohydrates was lower than that of ZnO NPs.

Ingestion of AuNPs (15 and 30 nm) contained in food at a dose of 87.44 μg/g reduced ootheca viability of *Blattella germanica* females and decreased the number of hatched nymphs by 32.8% compared to control. Exposure to AuNPs also decreased the number of nymphs that molted to second and third instars by 35.8% and reduced life span (Small et al. 2016). The larvicidal activity of AuNPs biosynthesized using extract of the *Turbinaria ornata* (Turner) J. Agardh 1848 seaweed against fourth instar larvae of *A. stephensi* expressed by $LC_{50}/LC_{90}$ values was 12.79/78.70 μg/ml exceeding that of pure seaweed extract (37.77/159.55 μg/ml). The application of green synthesized metal NPs against mosquitoes is favorable because long-term use of synthetic insecticides could cause insect resistance to these chemicals, adversely affect non-target aquatic organisms, and disturb the microbial community of the soil (Deepak et al. 2018), even though the possible toxicity of residual metal ions in the aquatic ecosystems could be considered (Benelli et al. 2017). The insecticidal activities of AuNPs on selected insect species, including *Ae. aegypti*, *A. stephensi*, and *Cx. quinquefasciatus*, were overviewed by Benelli (2018b). A review paper focused on the toxicity of AgNPs on insects such as *Bombyx mori* was presented by Pandiarajan and Krishnan (2017).

CuNPs (50–100 nm) prepared using *Aegle marmelos* Correa aqueous leaf extract showed improved larvicidal activity against *A. stephensi* ($LC_{50}$ of 500.06 ppm) than the crude leaf extracts or the insecticide temephos (Angajala et al. 2014). Toxic effects of CuNPs on hematophagous (blood feeding) larvae of *A. subpictus* Grassi, *Cx. quinquefasciatus*, and *Rhipicephalus (Boophilus) microplus* were estimated by Ramyadevi et al. (2011). CuNPs green synthesized using the whole cell biomass of *Fusarium proliferatum* exhibited larvicidal activity against *A. stephensi*, *Ae. aegypti*, and *Cx. quinquefasciatus* mosquitoes with $LC_{50}$ values of 39.25 μg/ml, 81.34 μg/ml, and 21.84 μg/ml (Kalaimurugan et al. 2019).

The impact of chemically fabricated iron-based NPs against *Cx. quinquefasciatus* I instar larvae and pupae was expressed by $LC_{50}$ values that varied in the range from 20.9 (larvae) to 43.7 ppm (pupae) for treatment with Fe(0) NPs and from 4.5 (I instar larvae) to 22.1 ppm (pupae) for the treatment with $Fe_2O_3$ NPs. It could be noted that a single exposure to sublethal doses of both NPs magnified the predation efficiency of the guppy fish, *Poecilia reticulata* (Murugan et al. 2018a). Core-shell nanohybrid fabricated using surface active maghemite NPs as a core having chlorin-e6 photosensitizer as the shell showing high photocidal activity on *Ae. aegypti* larvae could represent a safe alternative to conventional insecticides (Magro et al. 2019). FeS NPs synthesized using *Artemisia herba-alba* leaves extract as reducing and stabilizing agent of the size ca 40 nm showed insecticidal activity against the green peach aphid showing $LC_{50}$ values of 251 and 302 ppm, respectively, against the early and late nymphal instars of the insect (Asoufi et al. 2018a). The biosynthesized FeNPs of 40 nm also exhibited pronounced impact on the green peach aphid longevity and fecundity for three generation (Asoufi et al. 2018b).

ZnO NPs green synthesized using *Pongamia pinnata* leaf extract with mean particle size of 21.3 nm and zeta potential of −12.45 mV reduced the fecundity (eggs laid) and hatchability of *Callosobruchus maculatus*, pronouncedly delayed the larval, pupal, and total development period of the treated insect, reduced the activities of some important enzymes, and caused 100% mortality at a dose 25 μg/ml. Moreover, in treated insects reduced activities of the midgut α-amylase, cysteine protease, β-glucosidase, glutathione *S*-transferase, and lipase were observed (Malaikozhundan and Vinodhini 2018). ZnO NPs biosynthesized using *Ulva lactuca* seaweed extract applied at a dose of 50 μg/ml were reported to cause 100% mortality of *Ae. aegypti* fourth instar larvae within 24 h (Ishwarya et al. 2018). Brown macroalga *Sargassum wightii* Greville ex J. Agardh extract was used to fabricate ZnO NPs showing insecticidal activity against *A. stephensi* with $LC_{50}$ value ranging from 4.330 (larva I) to 7.430 ppm (pupa) and against *Helicoverpa armigera* Hubner with $LC_{50}$ ranging from 12.278 (larva I) to 20.798 ppm (pupa). Moreover, the ZnO NPs greatly reduced longevity and fecundity of both insects as well as food consumption of *H. armigera* individuals. On the other hand, the predation efficiency of non-target guppy *Poecilia reticulata* against I and II instar larvae of *A. stephensi* showed a ca 1.3-fold increase in a ZnO NPs-contaminated environment (Murugan et al. 2018c). Ingestion of ZnO NPs caused considerable toxicity in F1 progenies of *D. melanogaster* and caused reduction in the egg-to-adult viability of the flies, which was associated with the induction of reactive oxygen species (ROS) by metal NPs (Ng et al. 2017). Increased ROS level in the *D. melanogaster* testis due to exposure to AgNPs resulted in the reduced number of germline stem cells compared to control by stimulating premature differentiation of these cells (Ong et al. 2016). *Bacillus thuringiensis* coated ZnO NPs with average particle size of 20 nm and zeta potential of −12.7 mV also decreased the fecundity (eggs laid) and hatchability of *Callosobruchus maculatus* and treatment with 25 μg/ml caused 100% mortality of the insect. The corresponding $LC_{50}$ value of 10.71 μg/ml and decreased activities of midgut α-amylase, cysteine protease, α-glucosidase, and glutathione *S*-transferase (GST) in treated insects were observed as well (Malaikozhundan et al. 2017).

CuO, ZnO, and MgOH NPs green synthesized using aqueous extracts of *Punica granatum* peels, *Olea europaea* leaves, and *Chamaemelum nobile* flowers with particle sizes ranging from 5 nm to 80 nm caused mortality of *Myzus persicae* Sulzer (Homoptera: Aphididae), MgOH NPs being the most efficient (Ghidan et al. 2018). $SiO_2$, $TiO_2$, and ZnO NPs-assisted controlled release of methyl eugenol (synthetic insect attractant) from lure dispensers was found to be maximal at $10^{-5}$ dilution in the temperature range 30–35 °C, whereby the highest number of fruit fly catches for up to 12 weeks was observed with $TiO_2$ NPs (Dharanivasan et al. 2017).

Sunderland and McNeil (2017) investigated the effectiveness of nanosized $TiO_2$ desiccant to protect wool carpets and other fabric made proteinaceous fibers against *Anthrenocerus australis* and *Tineola bisselliella* and found that it was more difficult to reduce the feeding of *T. bisselliella* on carpet, than on fabric. In silkworms fed with $TiO_2$ NPs promoted 20-hydroxyecdysone biosynthesis, shortened developmental progression, and reduced duration of molting was observed (Li et al. 2014). In contrast, administration of $TiO_2$ NPs in diet pronouncedly increased the body size of

*B. mori* and upregulated the insulin/ecdysteroid signaling genes (Shi et al. 2017) and increased cocoon mass, cocoon shell mass, and the ratio of cocoon shell (Li et al. 2016). Pretreatment with $TiO_2$ NPs mitigated the phoxim-induced midgut injury and reduced oxidative stress in the midgut of *B. mori*, which was reflected in increased body weight and survival (Wang et al. 2015b). In silkworms treated with $TiO_2$ NPs at 30 °C the expression of antioxidant genes was stimulated resulting in reduced oxidative stress suggesting that $TiO_2$ NPs could mitigate the high-temperature induced oxidative stress to the insect (Li et al. 2018). Cytotoxic effects on midgut was observed in the third instar larvae of *D. melanogaster* fed by $TiO_2$ NPs (0.08 to 1.60 mg/ml). The primary DNA damage observed following $TiO_2$ NPs exposure in *D. melanogaster* evaluated using the comet assay was explained as to be associated with specific physicochemical properties of $TiO_2$ NPs (Carmona et al. 2015).

Positively charged $CeO_2$ NPs had no effect on the growth of the third instar larvae of *D. melanogaster*, while the negatively charged ones were found to delay the growth of larvae by ca 7 days (Parimi et al. 2019). Treatment with 0.250 mg/l of $CeO_2$ NPs mycosynthesized using *Aspergillus niger* culture filtrate caused 100% mortality on first instar of *Ae. aegypti* after 24 h exposure (Gopinath et al. 2015).

Bismuth oxyiodide (BiOI) nanoflakes synthesized using the hydrothermal method exhibited insecticidal activity against *A. stephensi* showing $LC_{50}$ values of 2.263 ppm (larva I), 3.414 ppm (larva II), 4.956 ppm (larva III), 6.983 ppm (larva IV), and 8.605 ppm (pupae) (Murugan et al. 2018b).

## 8 Insecticide Contamination and Non-target Organisms

As mentioned above, the widespread use of insecticides is one of the reasons of environmental pollution, especially waters and soils, which has negative effects on non-target organisms including, inter alia, beneficial insects (pollinators or harmful insect predators). This has the effect of disturbing the ecosystem balance and is associated with a strong decline in the number of insect species (Künast et al. 2013; Sánchez-Bayo 2012; Sánchez-Bayo and Wyckhuys 2019; Zacharia 2011).

Although the application of nanoformulated insecticides generally results in lower amounts of toxic chemicals entering the environment, in some cases even these lower concentrations are sufficient to cause harm to non-target organisms. Since the number of papers dealing with the effects of nanoinsecticides on non-target organisms is far from the number that focuses on the impact of their bulk form, a few examples of the effects of selected frequently used toxic insecticides are given below.

Although novel generation of insecticides show improved human and environmental safety profiles compared to older insecticide generations, it is necessary to perform comprehensive risk assessments considering effects of insecticides on non-target species (Guedes et al. 2016). Predictive ecotoxicology based on systematic, strict characterization of physiological mechanisms of action enabling more impressive extrapolation of in vitro to in vivo toxicity and in silico ecotoxicology will allow to assess the impact of untested chemicals on environmental organisms (Ashauer and Jager 2018).

Increasing pesticide contamination has adverse impact on regional aquatic biodiversity, causing ca 30% reduction of macroinvertebrate family richness at pesticide levels corresponding to legally accepted regulatory threshold levels. Thus, insecticides applied in agriculture endanger surface waters globally, because measured insecticide concentrations often exceeded the regulatory threshold levels for either surface water or sediments, which was observed mainly for newer-generation insecticides such as pyrethroids also in countries with stringent environmental regulations. Consequently, it is indispensable to improve the current pesticide regulations and agricultural pesticide application practices in global scale (Stehle and Schulz 2015).

The use of agricultural land is considered as a principal contributor of pesticides in streams. Szoecs et al. (2017) reported that regulatory acceptable concentrations were exceeded in 26% of streams, and the highest increases were observed for neonicotinoid insecticides.

Based on a meta-analysis of 32 important insecticides and their degradation products in United States surface waters in the period 1962–2017 Wolfram et al. (2018) was found that about half of the measured insecticide concentrations exceeded their regulatory threshold levels, whereby the overload decreased in the following order: pyrethroids > organophosphates ~ carbamates > organochlorines, and the persistence of neonicotinoids in surface waters contributes to higher risk for biodiversity and endangered species (Wolfram et al. 2018).

In juveniles of the teleost *Prochilodus lineatus* an exposure to 5–500 ng/l λ-CHT using a commercial formulation containing this pyrethroid insecticide as an active ingredient caused specific modifications in biotransformation enzymes, and oxidative stress, hematological adjustments, osmoregulatory disorders, and DNA damage were observed as well. Decreased AChE activity in the muscles of fish at all tested concentrations, and decreases in $Ca^{2+}$ and $Mg^{2+}$ gill ATPases resulting in hypocalcemia were estimated, while at a dose 500 ng/l the activity of $Na^+/K^+$ ATPase increased (Vieira and Martinez 2018). Permethrin was reported to induce oxidative stress and neurotoxic effects connected with drastic depletion of AChE activity in the freshwater beetle *Laccophilus minutus* belonging to predatory aquatic beetles, that represent important components of the aquatic food webs (Touaylia et al. 2019). The encapsulated γ-CHT showing average hydrodynamic diameter of 449 nm was found to be more toxic to a freshwater macroinvertebrate *Ceriodaphnia dubia* than free insecticide or its encapsulated form with hydrodynamic diameter of 758 nm, which was reflected in $EC_{50}$ values of 0.18, 0.57, and 0.65 μg/l, respectively, estimated in an acute immobilization test. The results showed that the properties of insecticidal formulations such as particle size could pronouncedly influence the effects on non-target organisms as well (Slattery et al. 2019). The fate of pyrethroid insecticide bifenthrin, an endocrine disrupting compound, in the environment and biological systems and the toxic effects of the chiral parent compound bifenthrin as well as its main metabolites, including sublethal toxic effects on various non-target organisms were summarized by Yang et al. (2018). Bifenthrin and cypermethrin could be considered as the crucial contributors to toxicity in benthic invertebrates and their simulated hazard quotients for sediment-associated pyrethroids to benthic organisms ranged from 10.5 ± 31.1 (bifenthrin) to 41.7 ± 204 (cypermethrin) (Li et al. 2017). Rogers et al. (2016) found that bifenthrin caused trophic cascade and modified insect emergence in mesocosms. Reduced larval macroinvertebrate abundance, richness, and biomass were observed with $EC_{50}$ values ranging from 197.6 to 233.5 ng bifenthrin/g organic carbon and a decrease of macroinvertebrate scrapers resulted in increased periphyton abundance. Exposure of freshwater mussels *Unio ravoisieri* to permethrin concentration of 50 μg/l during 7 days resulted in pronounced increase in catalase activity, while at treatment with 100 μg/l catalase activity decreased. Glutathione *S*-transferase activity and malondialdehyde levels were increased with increasing permethrin concentration suggesting oxidative stress; 51 and 89% inhibition of AChE activity of mussels was observed at exposure to 50 and 100 μg/l of the insecticide (Khazri et al. 2017). Wieczorek et al. (2018) investigated structural and functional effects of a short-term pyrethroid pulse exposure on invertebrates in outdoor stream mesocosms using etofenprox insecticide. They found that a 6 h pulse exposure to 5.3 μg/l etofenprox was able to cause negative effects up to 100% at the individual and population level and resulted in community structure alterations. Moreover, ca. 2 order lower etofenprox concentration (0.04 μg/l) decreased the abundance of the mayfly *Cloeon simile* by 66% and the feeding rate of *Asellus aquaticus* by 44%. Molecular mechanisms of pyrethroid biotransformation and endocrine toxicity of different pyrethroid types to fishes were discussed by Brander et al. (2016).

The impact of the neonicotinoid IMI alone and in the mixture with PEG-600 on Japanese quails was investigated by Rawi et al. (2019), and it was found that the $LD_{50}$ value related to mortality 24 h after oral administration was 15.98 mg/kg for the mixture of the insecticide with PEG, while for IMI alone it was 17.02 mg/kg. A single dose of IMI or IMI + PEG with concentrations corresponding to a quarter of $LD_{50}$ value strongly affected the activity of plasma AChE and brain monoamines transmitters, the maximal inhibition being observed 72 h after exposure, whereby PEG-adjuvant contributed to enhanced toxic effect. While a mixture of IMI + PEG more strongly affected dopamine alterations, treatment with IMI alone was more effective in inducing changes in serotonin (5-HT). Moreover, treatment with IMI, applied alone as well as in combination with PEG, resulted in neural congestion, neuronal degeneration, pyknosis, and perivascular cuffing with glial cells. Acute contact toxicity of IMI to *Apis mellifera* is caused by much faster and more readily penetration of bee cuticule resulting in its higher steady-state internal body concentrations compared to TCP and ACP (Zaworra et al. 2019). Investigation of the response of estuarine invertebrates to IMI following field applications in Willapa Bay using principal response curve (PRC) analysis showed negative impact of insecticide application only on five assemblages of mollusks and one assemblage of crustaceans, which could be connected with low concentrations of insecticide and short period of exposure, eventually with low toxicological susceptibility to IMI for many taxa (Booth et al. 2019). The effect of the IMI, TCP, and clothianidin on the individual immunocompetence of *Apis mellifera* L. could impair disease resistance capacity (Brandt et al. 2016). The risks of neonicotinoids for pollinators such as honey bees, bumble bees, and solitary bees were summarized in a comprehensive overview by Blacquiere et al. (2012). Raby et al. (2018) estimated and compared the acute (48- or 96-h) toxicity of 6 neonicotinoids (ACP, clothianidin, dinotefuran, IMI, TCP, and TMX) to aquatic invertebrates of 10 aquatic arthropod orders. The most susceptible invertebrates were insects from the orders Ephemeroptera (*Neocloeon triangulifer*) and Diptera (*Chironomus dilut*us), while the least sensitivity showed cladocerans (*Daphnia magna*, *Ceriodaphnia dubia*). Considering fifth percentile hazard concentrations the tested neonicotinoids except IMI did not represent hazard in terms of acute toxicity to aquatic communities in Ontario freshwater streams. Pollinators and aquatic insects were found to be extremely sensitive to the treatments with neonicotinoids, chronic sublethal effects being more prevalent than acute toxicity (Hladik et al. 2018). Neonicotinoids clothianidin and TMX were found to have negative impact on the colonization of invertebrate populations in aquatic microcosms at field-realistic levels, TMX being more toxic. Adverse effects of both insecticides on populations of *Chironomids* (Diptera) and Ostracoda were estimated, while clothianidin at the tested doses 0–15 ppb doses showed any unfavorable effect on Culicidae. Reduction of the invertebrates populations in ephemeral ponds observed at realistic concentrations of neonicotinoids could affect food chain as well (Basley and Goulson 2018). In the full life-cycle toxicity tests using *Chironomus dilutus* the toxicities of 3 neonicotinoids: IMI, clothianidin, and TMX were compared. The estimated 14/40 d median lethal concentrations were 1.52/0.39 g/l (IMI), 2.41/0.28 g/l (clothianidin), and 23.60/4.13 g/l (TMX), respectively. Based on population-relevant endpoints and toxic equivalency factors was found that IMI and clothianidin exhibited similar chronic toxicity to *C. dilutus*, while to achieve comparable effects ca tenfold higher TMX concentrations were necessary, which could be connected with readily degradation of TMX compared to clothianidin under field conditions (Cavallaro et al. 2017). Comparative mammalian hazards of neonicotinoid insecticides among exposure durations with initial thresholds of toxicological concern derived for rat, dog, mouse, and rabbit under comparative experimental scenarios were presented by Wang et al. (2019c).

Based on the hypothesis that improved pesticide tolerance is connected with the generalized stressor response in organisms and could be induced as a response to sublethal exposure to natural and anthropogenic stressors, Jones et al. (2017) exposed larval wood frogs (*Lithobates sylvaticus*) to carbaryl (0.5 or 1.0 mg/l) and predator cue (*Anax* spp.) and using time-to-death assays studied their tolerance to a lethal carbaryl concentration. Estimated carbaryl tolerance observed in tadpoles exposed to concentration 0.5 mg/l and also in tadpoles exposed to predator cues confirmed the above mentioned hypothesis of the researchers. Exposure of honey bees to wettable powder of carbamate insecticide carbaryl under semi-field condition showed toxic effect of the insecticide on bees and affected their gut microbial community (Nogrado et al. 2019). Carbaryl encapsulated in waxy microspheres (15.8–19.8 μm) showing controlled release and exhibiting lower vertical mobility compared to the vertical mobility of the technical-grade product were reported to represent lower potential risk for contamination of groundwater (Quaglia et al. 2001).

Comparison of the toxicity of formulations based on chitin synthesis inhibitor diflubenzuron (DFB) or on OPI temephos, usually used as a larvicides to control *Ae. aegypti*, against freshwater fishes *Oreochromis niloticus* and *Hyphessobrycon eques* showed that DFB still induced mortality and tissue damage in fishes and its formulation was able to reduce body weight of *H. eques* at concentrations 272-fold lower than its $LC_{50}$. DFB caused edemas and aneurisms on gills, and hepatocyte hypertrophy and vascular congestion of the liver in *O. niloticus*. On the other hand, pyknotic nuclei, which may result in irreversible necrosis, were induced also by TMP-based formulation (Abe et al. 2019). *Hyalella curvispina* (Amphipoda) was found to be pronouncedly sensitive to CPF (more than some other crustacean species), which was reflected in 48 h $LC_{50}$ value of 0.38 ± 0.04 g/l and because in stream waters CPF concentrations exceeding that of 48 h $LC_{50}$ value were estimated, adverse effects of insecticide on this organism could be demonstrated (Solis et al. 2019). Ecotoxicological assessment of the impact of synthetic insecticides, CPF, cypermethrin, and their combination with a bioinsecticide, azadirachtin, showed that 45 days post application the microbial community structure of the insecticide-treated soil resembled in only 70% to control rhizospheric soil; however, the effects of biopesticide were comparable with those of synthetic insecticides (Walvekar et al. 2017). Tran et al. (2019) investigated the effects of treatment with CPF combined with warming within and across generations on antipredator behavior of *Cx. pipiens* larvae. Stronger reduction of diving time was observed at 20 °C compared to 24 °C, except in the offspring whose parents had been exposed to 24 °C. However, at combined exposure to insecticide and warming, reduction of escape diving time was observed within each generation and, thus, the larvae become more susceptible to predation. Organophosphate pesticide malathion strongly reduced the abundance of total zooplankton, cyclopoid copepods, copepod nauplii, and *Ceriodaphnia*, while increased the abundance of rotifers suggesting that contamination of aquatic ecosystems with this insecticide could affect the abundance and composition of zooplankton communities (Smith et al. 2018).

## 9 Conclusion

Insecticides are increasingly used in agriculture in order to achieve higher crop yield and to kill harmful insects such as mosquitoes, which cause dangerous diseases such as malaria, threatening human population. Since 2000, the average annual number of human deaths caused by the mosquitoes were approx. two million. Numerous quarantine insect pests cause pronounced damages to the production of economically important crops as well as during their storage, and could adversely affect also human health. Moreover, climate change support an expansion of exotic insects that are needed to be effectively destroyed using selective insecticides that target these specific pest species. The overuse of insecticides frequently results in the development of the resistance of insects to synthetic insecticides. For example, the resistance of mosquitoes to pyrethroids and DDT caused by a single genetic mutation is already spreading in mosquito populations. On the other hand, the overuse of insecticides has adverse impact on the environment, the overall state of important agricultural crops and animals, and, consequently, human health. Although modern insecticides are much more gentle, their overuse can result in water and soil infestations leading to a decline or even eradication of beneficial insects such as pollinators and in adverse impact on non-target organisms, causing sometimes also health complications of the human population. The globally estimated dramatic decline of insect species due to anthropogenic activities requires immediate precautions, including the rational use of insecticides. In general, natural (bio)insecticides are less toxic (including toxicity to mammalians), less persistent, and could be degraded more rapidly than synthetic insecticides, and their use is connected with a low risk of developing resistance in long-term usage; therefore, they are preferred over synthetic agents. Using insecticide nanoformulations characterized with targeted distribution, controlled release, increased efficacy, and thus lower doses of the active ingredient, the environmental and health risks of insecticides could be significantly reduced. Despite the significant innovations and advances achieved in the fabrication of nanoscaled insecticides, it is important not to forget that nanoparticles themselves are toxic to living organisms because their nanodimension enables them to interact with native macromolecules. Therefore, it is important to know in detail possible toxic impacts of nanoinsecticides not only on target organisms, but above all on non-target organisms as well as their behavior in water and soil (persistence or degradation to more toxic metabolites), so that the innovative helper does not become a "nightmare" of the entire Earth's ecosystems.

**Acknowledgments** This study was supported by the Slovak Research and Development Agency (APVV-17-0373 and APVV-17-0318) and by project VEGA 1/0286/20.

## References

Aarthi C, Govindarajan M, Rajaraman P, Alharbi NS, Kadaikunnan S, Khaled JM, Mothana RA, Siddiqui NA, Benelli G (2018) Eco-friendly and cost-effective Ag nanocrystals fabricated using the leaf extract of *Habenaria plantaginea*: toxicity on six mosquito vectors and four non-target species. Environ Sci Pollut Res 25(11):10317–10327. https://doi.org/10.1007/s11356-017-9203-2

Abe FR, Machado AA, Coleone AC, da Cruz C, Machado-Neto JG (2019) Toxicity of diflubenzuron and temephos on freshwater fishes: ecotoxicological assays with *Oreochromis niloticus* and *Hyphessobrycon eques*. Water Air Soil Pollut 230(3):77. https://doi.org/10.1007/s11270-019-4128-7

Al Shammari HI, AL-Khazraji HI, Falih SK (2018) The effectivity of silver nanoparticles prepared by jujube *Ziziphus* sp. extract against whitefly *Bemisia tabaci* nymphs. Res J Pharm Biol Chem Sci 9(6):551–558

Al-Ahmadi MS (2019) Pesticides, anthropogenic activities, and the health of our environment safety. In: Larramendy M, Soloneski S (eds) Pesticides - use and misuse and their impact in the environment. InTech, Rijeka, pp 1–24. https://doi.org/10.5772/intechopen.84161

Alaraby M, Romero S, Hernandez A, Marcos R (2019) Toxic and genotoxic effects of silver nanoparticles in *Drosophila*. Environ Mol Mutagen 60(3):277–285. https://doi.org/10.1002/em.22262

Alharbi NS, Govindarajan M, Kadaikunnan S, Khaled JM, Almanaa TN, Alyahya SA, Al-Anbr MN, Gopinath K, Sudha A (2018) Nanosilver crystals capped with *Bauhinia acuminata* phytochemicals as new antimicrobials and mosquito larvicides. J Trace Elem Med Biol 50:146–153. https://doi.org/10.1016/j.jtemb.2018.06.016

Ali M, Afzal M, Verma M, Misra-Bhattacharya S, Ahmad FJ, Dinda AK (2013) Improved antifilarial activity of ivermectin in chitosan-alginate nanoparticles against human lymphatic filarial parasite, *Brugia malayi* nematode. Parasitol Res 112(8):2933–2943. https://doi.org/10.1007/s00436-013-3466-4

Aljafree NFA, Kamari A (2018) Synthesis, characterisation and potential application of deoxycholic acid carboxymethyl chitosan as a carrier agent for rotenone. J Polym Res 25(6):133. https://doi.org/10.1007/s10965-018-1530-6

Alkani M, Erturk S, Firat TA, Ciftci E (2019) Study of insecticidal and behavioral effects and some characteristic of native diatomaceous earth against the bean weevil, *Acanthoscelides obtectus* (Coleoptera: Chrysomelidae). Fresenius Environ Bull 28(4):2916–2922

Almadiy AA, Nenaah GE, Shawer DM (2018) Facile synthesis of silver nanoparticles using harmala alkaloids and their insecticidal and growth inhibitory activities against the khapra beetle. J Pest Sci 91(2):727–737. https://doi.org/10.1007/s10340-017-0924-2

Amenta V, Aschberger K, Arena M, Bouwmeester H, Botelho Moniz F, Brandhoff P, Gottardo S, Marvin HJP, Mech A, Quiros Pesudo L, Rauscher H, Schoonjans R, Vettori MV, Weigel S, Peters RJB (2015) Regulatory aspects of nanotechnology in the Agri/feed/food sector in EU and non-EU countries. Regul Toxicol Pharmacol 73(1):463–476. https://doi.org/10.1016/j.yrtph.2015.06.016

Anderson JC, Dubetz C, Palace VP (2015) Neonicotinoids in the Canadian aquatic environment: a literature review on current use products with a focus on fate, exposure, and biological effects. Sci Total Environ 505:409–422. https://doi.org/10.1016/j.scitotenv.2014.09.090

Angajala G, Pavan P, Subashini R (2014) One-step biofabrication of copper nanoparticles from *Aegle marmelos* Correa aqueous leaf extract and evaluation of its anti-inflammatory and mosquito larvicidal efficacy. RSC Adv 4(93):51459–51470. https://doi.org/10.1039/c4ra10003d

Arena JP, Liu KK, Paress PS, Frazier EG, Cully DF, Mrozik H, Schaeffer JM (1995) The mechanism of action of avermectins in *Caenorhabditis elegans*: correlation between activation of glutamate-sensitive chloride current, membrane binding, and biological activity. J Parasitol 81(2):286–294

Arumugam G, Velayutham V, Shanmugavel S, Sundaram J (2016) Efficacy of nanostructured silica as a stored pulse protector against the infestation of bruchid beetle, *Callosobruchus maculatus* (Coleoptera: Bruchidae). Appl Nanosci 6(3):445–450. https://doi.org/10.1007/s13204-015-0446-2

Ashauer R, Jager T (2018) Physiological modes of action across species and toxicants: the key to predictive ecotoxicology. Environ Sci Processes Impacts 20(1):48–57. https://doi.org/10.1039/c7em00328e

Asoufi HM, Al-Antary TM, Awwad AM (2018a) Biosynthesis and characterization of iron sulfide (FeS) nanoparticles and evaluation their aphicidal activity on the green peach aphid *Myzus persicae* (Homoptera: Aphididae). Fresenius Environ Bull 27(11):7767–7775

Asoufi HM, Al-Antary TM, Awwad AM (2018b) Effect of biosynthesized iron sulfide (FeS) nanoparticles on the green peach aphid *Myzus persicae* (Homoptera: Aphididae) longevity and fecundity. Fresenius Environ Bull 27(11):7817–7821

Bae GN, Vu TP, Kim SH, Lee SB, Shim SG, Sohn JR (2012) Secondary aerosol pollution by a home insecticide under ozone exposure. Indoor Environ Technol 9(2):97–112

Bai B, Xu XH, Hai JJ, Hu N, Wang HL, Suo YR (2019) Lauric acid-modified Nitraria seed meal composite as green carrier material for pesticide controlled release. J Chem 2019:5376452. https://doi.org/10.1155/2019/5376452

Balaji APB, Sastry TP, Manigandan S, Mukherjee A, Chandrasekaran N (2017) Environmental benignity of a pesticide in soft colloidal hydrodispersive nanometric form with improved toxic precision towards the target organisms than non-target organisms. Sci Total Environ 579:190–201. https://doi.org/10.1016/j.scitotenv.2016.10.240

Barik TK, Kamaraju R, Gowswami A (2012) Silica nanoparticle: a potential new insecticide for mosquito vector control. Parasitol Res 111:1075–1083. https://doi.org/10.1007/s00436-012-2934-6

Basley K, Goulson D (2018) Neonicotinoids thiamethoxam and clothianidin adversely affect the colonisation of invertebrate populations in aquatic microcosms. Environ Sci Pollut Res 25(10):9593–9599. https://doi.org/10.1007/s11356-017-1125-5

Battisti A, Stastny M, Netherer S, Robinet C, Schopf A, Roques A, Larsson S (2005) Expansion of geographic range in the pine processionary moth caused by increased winter temperatures. Ecol Appl 15:2084–2096. https://doi.org/10.1890/04-1903

Belovsky G, Slade J (2018) Grasshoppers affect grassland ecosystem functioning: spatial and temporal variation. Basic Appl Ecol 26:24–34. https://doi.org/10.1016/j.baae.2017.09.003

Benelli G (2018a) Mode of action of nanoparticles against insects. Environ Sci Pollut Res 25(13):12329–12341. https://doi.org/10.1007/s11356-018-1850-4

Benelli G (2018b) Gold nanoparticles - against parasites and insect vectors. Acta Trop 178:73–80. https://doi.org/10.1016/j.actatropica.2017.10.021

Benelli G, Caselli A, Canale A (2017) Nanoparticles for mosquito control: challenges and constraints. J King Saud Univ Sci 29(4):424–435. https://doi.org/10.1016/j.jksus.2016.08.006

Benelli G, Kadaikunnan S, Alharbi NS, Govindarajan M (2018) Biophysical characterization of *Acacia caesia*-fabricated silver nanoparticles: effectiveness on mosquito vectors of public health relevance and impact on non-target aquatic biocontrol agents. Environ Sci Pollut Res 25(11):10228–10242. https://doi.org/10.1007/s11356-017-8482-y

Bhan S, Mohan L, Srivastava CN (2014) Relative larvicidal potentiality of nano-encapsulated Temephos and Imidacloprid against *Culex quinquefasciatus*. J Asia Pac Entomol 17(4):787–791. https://doi.org/10.1016/j.aspen.2014.07.006

Bisbal M, Sanchez M (2019) Neurotoxicity of the pesticide rotenone on neuronal polarization: a mechanistic approach. Neural Regen Res 14(5):762–766. https://doi.org/10.4103/1673-5374.249847

Bittencourt VCE, Moreira AMD, da Silva JG, Gomides AFD, Velloso-Rodrigues C, Kelmann RG, Mendonca LM, Lula IS, Denadai AML (2019) Hydrophobic nanoprecipitates formed by benzoylphenylureas and β-cyclodextrin inclusion compounds: synthesis, characterization and toxicity against *Aedes aegypti* larvae. Heliyon 5(7):e02013. https://doi.org/10.1016/j.heliyon.2019.e02013

Blacquiere T, Smagghe G, van Gestel CAM, Mommaerts V (2012) Neonicotinoids in bees: a review on concentrations, side-effects and risk assessment. Ecotoxicology 21(4):973–992. https://doi.org/10.1007/s10646-012-0863-x

Bommarco R, Kleijn D, Potts SG (2013) Ecological intensification: harnessing ecosystem services for food security. Trends Ecol Evol 28(4):230–238. https://doi.org/10.1016/j.tree.2012.10.012

Booth SR, Patten K, New L (2019) Response of estuarine benthic invertebrates to field applications of insecticide. Estuar Coast Shelf Sci 218:86–94. https://doi.org/10.1016/j.ecss.2018.11.025

Borm PJA, Robbins D, Haubold S, Kuhlbusch T, Fissan H, Donaldson K, Schins R, Stone V, Kreyling W, Lademann J, Krutmann J, Warheit D, Oberdoster E (2006) The potential risks of nanomaterials: a review carried out for ECETOC. Part Fibre Toxicol 3:11. https://doi.org/10.1186/1743-8977-3-11

Botas GD, Cruz RAS, de Almeida FB, Duarte JL, Araujo RS, Souto RNP, Ferreira R, Carvalho JCT, Santos MG, Rocha L, Pereira VPL, Fernandes CP (2017) *Baccharis reticularia* DC. and limonene nanoemulsions: promising larvicidal agents for *Aedes aegypti* (Diptera: Culicidae) control. Molecules 22(11):1990. https://doi.org/10.3390/molecules22111990

Boulogne I, Petit P, Ozier-Lafontaine H, Desfontaines L, Loranger-Merciris G (2012) Insecticidal and antifungal chemicals produced by plants: a review. Environ Chem Lett 10(4):325–347. https://doi.org/10.1007/s10311-012-0359-1

Brander SM, Gabler MK, Fowler NL, Connon RE, Schlenk D (2016) Pyrethroid pesticides as endocrine disruptors: molecular mechanisms in vertebrates with a focus on fishes. Environ Sci Technol 50(17):8977–8992. https://doi.org/10.1021/acs.est.6b02253

Brandt A, Gorenflo A, Siede R, Meixner M, Buechler R (2016) The neonicotinoids thiacloprid, imidacloprid, and clothianidin affect the immunocompetence of honey bees (*Apis mellifera* L.). J Insect Physiol 86:40–47. https://doi.org/10.1016/j.jinsphys.2016.01.001

Buteler M, Sofie SW, Weaver DK, Driscoll D, Muretta J, Stadler T (2015) Development of nanoalumina dust as insecticide against *Sitophilus oryzae* and *Rhyzopertha dominica*. Int J Pest Manag 61(1):80–89. https://doi.org/10.1080/09670874.2014.1001008

Buteler M, Garcia GL, Stadler T (2018) Potential of nanostructured alumina for leaf-cutting ants *Acromyrmex lobicornis* (Hymenoptera: Formicidae) management. Austral Entomol 57(3):292–296. https://doi.org/10.1111/aen.12277

Buzea C, Pacheco I, Robbie K (2007) Nanomaterials and nanoparticles: sources and toxicity. Biointerphases 2:MR17–MR71. https://doi.org/10.1116/1.2815690

Caballero JP, Murillo L, List O, Bastiat G, Flochlay-Sigognault A, Guerino F, Lefrancois C, Lautram N, Lapied B, Apaire-Marchais V (2019) Nanoencapsulated deltamethrin as synergistic agent potentiates insecticide effect of indoxacarb through an unusual neuronal calcium-dependent mechanism. Pestic Biochem Physiol 157:1–12. https://doi.org/10.1016/j.pestbp.2019.03.014

Campolo O, Cherif A, Ricupero M, Siscaro G, Grissa-Lebdi K, Russo A, Cucci LM, Di Pietro P, Satriano C, Desneux N, Biondi A, Zappala L, Palmeri V (2017) Citrus peel essential oil nanoformulations to control the tomato borer, *Tuta absoluta*: chemical properties and biological activity. Sci Rep 7:13036. https://doi.org/10.1038/s41598-017-13413-0

Campolo O, Giunti G, Russo A, Palmeri V, Zappala L (2018) Essential oils in stored product insect pest control. J Food Qual 2017:6906105. https://doi.org/10.1155/2018/6906105

Campos EVR, de Oliveira JL, Pascoli M, de Lima R, Fraceto LF (2016) Neem oil and crop protection: from now to the future. Front Plant Sci 7:1494. https://doi.org/10.3389/fpls.2016.01494

Carmona ER, Escobar B, Vales G, Marcos R (2015) Genotoxic testing of titanium dioxide anatase nanoparticles using the wing-spot test and the comet assay in *Drosophila*. Mutat Res Genet Toxicol Environ Mutagen 778:12–21. https://doi.org/10.1016/j.mrgentox.2014.12.004

Cavallaro MC, Morrissey CA, Headley JV, Peru KM, Liber K (2017) Comparative chronic toxicity of imidacloprid, clothianidin, and thiamethoxam to *Chironomus dilutus* and estimation of toxic equivalency factors. Environ Toxicol Chem 36(2):372–382. https://doi.org/10.1002/etc.3536

Chaudhary S, Kanwar RK, Sehgal A, Cahill DM, Barrow CJ, Sehgal R, Kanwar JR (2017) Progress on *Azadirachta indica* based biopesticides in replacing synthetic toxic pesticides. Front Plant Sci 8:610. https://doi.org/10.3389/fpls.2017.00610

Chellappandian M, Thanigaivel A, Vasantha-Srinivasan P, Edwin ES, Ponsankar A, Selin-Rani S, Kalaivani K, Senthil-Nathan S, Benelli G (2018) Toxicological effects of *Sphaeranthus indicus* Linn. (Asteraceae) leaf essential oil against human disease vectors, *Culex quinquefasciatus* Say and *Aedes aegypti* Linn., and impacts on a beneficial mosquito predator. Environ Sci Pollut Res 25(11):10294–10306. https://doi.org/10.1007/s11356-017-8952-2

Chen H, Seiber JN, Hotze M (2014) ACS select on nanotechnology in food and agriculture: a perspective on implications and applications. J Agric Food Chem 62:1209–1212. https://doi.org/10.1021/jf5002588

Chen HY, Lin YS, Zhou HJ, Zhou XH, Gong S, Xu H (2016) Highly efficient alginate sodium encapsulated chlorpyrifos/copper(II) Schiff base mesoporous silica sustained release system with pH and ion response for pesticide delivery. RSC Adv 6(115):114714–114721. https://doi.org/10.1039/c6ra23836j

Chen HY, Lin YS, Xu H, Cheng DM, Gong S, Liu QH, Zhou XH, Zhou HJ (2017a) Preparation of sustained-release chlorpyrifos particles via the emulsification coacervation method and their sustained-release performance. J Macromol Sci A 54(2):91–96. https://doi.org/10.1080/10601325.2017.1261621

Chen K, Yu GB, He FR, Zhou QF, Xiao DC, Li JC, Feng YH (2017b) A pH-responsive emulsion stabilized by alginate-grafted anisotropic silica and its application in the controlled release of λ-cyhalothrin. Carbohydr Polym 176:203–213. https://doi.org/10.1016/j.carbpol.2017.07.046

Chen L, Meng X, Gu J, Fan WQ, Abdili N, Peprah FA, Wang NN, Zhu FF, Lu P, Ma SS, Chen K (2019) Silver nanoparticle toxicity in silkworms: omics technologies for a mechanistic understanding. Ecotoxicol Environ Saf 172:388–395. https://doi.org/10.1016/j.ecoenv.2019.01.055

Choupanian M, Omar D (2018) Formulation and physicochemical characterization of neem oil nanoemulsions for control of *Sitophilus oryzae* (L., 1763) (Coleoptera: Curculionidae) and *Tribolium castaneum* (Herbst, 1797) (Coleoptera: Tenebrionidae). Turk Entomol Derg-TU 42(2):127–139. https://doi.org/10.16970/entoted.398541

Choupanian M, Omar D, Basri M, Asib N (2017) Preparation and characterization of neem oil nanoemulsion formulations against *Sitophilus oryzae* and *Tribolium castaneum* adults. J Pest Sci 42(3–4):158–165. https://doi.org/10.1584/jpestics.D17-032

Čitaković NM (2019) Physical properties of nanomaterials. Mil Tech Cour 67(1):159–171. https://doi.org/10.5937/vojtehg67-18251

Clerici DJ, de Souza ME, Quatrin PM, Santos RVC, Costa-Leonardo AM, Ourique AF, Raffin RP, Marques CT, Dietrich CRRD, Gundel SD, Klein B, Wagner R (2018) Effects of nanostructured essential oils against subterranean termites (*Coptotermes gestroi*). J Appl Entomol 142(4):406–412. https://doi.org/10.1111/jen.12494

Cola DF, da Silva CMG, Fraceto LF (2015) Lipid nanoparticles as carrier systems for Ivermectin and Methoprene aiming veterinary applications. Toxicol Lett 238(2):S201–S201. https://doi.org/10.1016/j.toxlet.2015.08.602

Cola DF, Pasquoto T, Guilger M, de Lima R, da Silva CMG, Fraceto LF (2016) Sistemas carreadores lipídicos nanoestruturados para ivermectina e metopreno visando controle de parasitas. Quimica Nova 39(9):1034–1043. https://doi.org/10.21577/0100-4042.20160123

Coles D, Frewer LJ (2013) Nanotechnology applied to European food production: a review of ethical and regulatory issues. Trends Food Sci Technol 34:32–43. https://doi.org/10.1016/j.tifs.2013.08.006

Costa ES, Perlatti B, Silva EMD, Matos AP, Silva MFGFD, Fernandes JB, Zuin VG, Silva CMP, Forim MR (2017) Use of lignins from sugarcane bagasse for assembling microparticles loaded with *Azadirachta indica* extracts for use as neem-based organic insecticides. J Brazil Chem Soc 28(1):126–135. https://doi.org/10.5935/0103-5053.20160155

Cui B, Feng L, Wang CX, Yang DS, Yu ML, Zeng ZH, Wang Y, Sun CJ, Zhao X, Cui HX (2016) Stability and biological activity evaluation of chlorantraniliprole solid nanodispersions prepared by high pressure homogenization. PLoS One 11(8):e0160877. https://doi.org/10.1371/journal.pone.0160877

da Costa JT, Forim MR, Costa ES, de Souza JR, Mondego JM, Boica AL (2014) Effects of different formulations of neem oil-based products on control *Zabrotes subfasciatus* (Boheman, 1833) (Coleoptera: Bruchidae) on beans. J Stored Prod Res 56:49–53. https://doi.org/10.1016/j.jspr.2013.10.004

Dai RY, You SY, Lu LM, Liu Q, Li ZX, Wei L, Huang XG, Yang ZY (2017) High blades spreadability of chlorpyrifos microcapsules prepared with polysiloxane sodium carboxylate/sodium carboxymethylcellulose/gelatin via complex coacervation. Colloids Surf A Physicochem Eng Asp 530:13–19. https://doi.org/10.1016/j.colsurfa.2017.06.057

de Oliveira JL, Campos EVR, Bakshi M, Abhilash PC, Fraceto LF (2014) Application of nanotechnology for the encapsulation of botanical insecticides for sustainable agriculture: prospects and promises. Biotechnol Adv 32(8):1550–1561. https://doi.org/10.1016/j.biotechadv.2014.10.010

De Smedt C, Someus E, Spanoghe P (2015) Potential and actual uses of zeolites in crop protection. Pest Manag Sci 71(10):1355–1367. https://doi.org/10.1002/ps.3999

De Smedt C, Van Damme V, De Clercq P, Spanoghe P (2016) Insecticide effect of zeolites on the tomato leafminer *Tuta absoluta* (Lepidoptera: Gelechiidae). Insects 7(4):72. https://doi.org/10.3390/insects7040072

Debnath N, Das S, Brahmachary RL, Chandra R, Sudan S, Goswami A (2010) Entomotoxicity assay of silica, zinc oxide, titanium dioxide, aluminium oxide nanoparticles on *Lipaphis pseudobrassicae*. ICANN 2009. In: Giri PK, Goswami DK, Perumal A, Chattopadhyay A (eds) Book Series: AIP Conference Proceedings 1276:307. https://doi.org/10.1063/1.3504316

Debnath N, Das S, Seth D, Chandra R, Bhattacharya SC, Goswami A (2011) Entomotoxic effect of silica nanoparticles against *Sitophilus oryzae* (L.). J Pest Sci 84(1):99–105. https://doi.org/10.1007/s10340-010-0332-3

Debnath N, Mitra S, Das S, Goswami A (2012) Synthesis of surface functionalized silica nanoparticles and their use as entomotoxic nanocides. Powder Technol 221:252–256. https://doi.org/10.1016/j.powtec.2012.01.009

Deepak P, Sowmiya R, Balasubramani G, Aiswarya D, Arul D, Josebin MPD, Perumal P (2018) Mosquito-larvicidal efficacy of gold nanoparticles synthesized from the seaweed, *Turbinaria ornata* (Turner) J.Agardh 1848. Particul Sci Technol 36(8):974–980. https://doi.org/10.1080/02726351.2017.1331286

Deka S, Barthakur S, Pandey R (2011) Potential effects of climate change on insects pest dynamics. In: Singh MP, Khetarpal S, Pandey R, Kumar P (eds) Climate change: impacts and adaptations in crop plants. Today and Tomorrow's Printers and Publishers, New Delhi, pp 301–312

Derbalah AS, Khidr AA, Moustafa HZ, Taman A (2014) Laboratory evaluation of some non-conventional pest control agents against the pink bollworm *Pectinophora gossypiella* (Saunders). Egypt J Biol Pest Control 24(2):363–368

Deutsch CA, Tewksbury JJ, Tigchelaar M, Battisti DS, Merrill SC, Huey RB, Naylor RL (2018) Increase in crop losses to insect pests in a warming climate. Science 361(6405):916–919. https://doi.org/10.1126/science.aat3466

Dhaliwal GS, Dhawan AK, Singh R (2007) Biodiversity and ecological agriculture: issues and perspectives. Indian J Ecol 34(2):100–109. https://doi.org/10.5367/000000008784648870

Dharanivasan G, Sithanantham S, Kannan M, Chitra S, Kathiravan K, Janarthanan S (2017) Metal oxide nanoparticles assisted controlled release of synthetic insect attractant for effective and sustainable trapping of fruit flies. J Clust Sci 28(4):2167–2183. https://doi.org/10.1007/s10876-017-1215-z

Dillon ME, Wang G, Huey RB (2010) Global metabolic impacts of recent climate warming. Nature 467:704–706. https://doi.org/10.1038/nature09407

Dolez PI (2015) Nanoengineering: global approaches to health and safety issues. Elsevier, Amsterdam

Dolzhenko VI, Dolzhenko TV (2017) Insecticides based on insect chitin synthesis inhibitors. Russ Agric Sci 43(3):238–243. https://doi.org/10.3103/S1068367417030065

Dougoud J, Toepfer S, Bateman M, Jenner WH (2019) Efficacy of homemade botanical insecticides based on traditional knowledge. A review. Agron Sustain Dev 39:37. https://doi.org/10.1007/s13593-019-0583-1

Duarte JL, Amado JRR, Oliveira AEMFM, Cruz RAS, Ferreira AM, Souto RNP, Falcao DQ, Carvalho JCT, Fernandes CP (2015) Evaluation of larvicidal activity of a nanoemulsion of *Rosmarinus officinalis* essential oil. Rev Bras Farmacogn 25(2):189–192. https://doi.org/10.1016/j.bjp.2015.02.2010

Duke SO, Cantrell CL, Meepagala KM, Wedge DE, Tabanca N, Schrader KK (2010) Natural toxins for use in pest management. Toxins 2(8):1943–1962. https://doi.org/10.3390/toxins2081943

Dziewięcka M, Karpeta-Kaczmarek J, Augustyniak M, Majchrzycki Ł, Augustyniak-Jabłokow MA (2016) Evaluation of in vivo graphene oxide toxicity for *Acheta domesticus* in relation to nanomaterial purity and time passed from the exposure. J Hazard Mater 305:30–40. https://doi.org/10.1016/j.jhazmat.2015.11.021

Elek N, Hoffman R, Raviv U, Resh R, Ishaaya I, Magdassi S (2010) Novaluron nanoparticles: formation and potential use in controlling agricultural insect pests. Colloids Surf A Physicochem Eng Asp 372(1–3):66–72. https://doi.org/10.1016/j.colsurfa.2010.09.034

Ensley SM (2018) Neonicotinoids. In: Gupta RC (ed) Veterinary toxicology basic and clinical principles, 3rd edn. Academic Press & Elsevier, pp 521–524

Eroglu N, Sakka MK, Emekci M, Athanassiou CG (2019) Effects of zeolite formulations on the mortality and progeny production of *Sitophilus oryzae* and *Oryzaephilus surinamensis* at different temperature and relative humidity levels. J Stored Prod Res 81:40–45. https://doi.org/10.1016/j.jspr.2018.11.004

European Commission (2011) Recommendation on the definition of a nanomaterial (2011/696/EU). http://ec.europa.eu/environment/chemicals/nanotech/faq/definition_en.htm

Fan TF, Wu XM, Wu Y (2013) Preparation and characterization of cyhalothrin-loaded poly(2-hydroxyethyl methacrylate)-co-polylactide (PHEMA-co-PLA) ultrafine particles. J Appl Polym Sci 129(4):1861–1867. https://doi.org/10.1002/app.38892

FAO (2014) The international code of conduct on pesticide management, Food and Agriculture Organization of the United Nations 2014. http://www.fao.org/agriculture/crops/thematic-sitemap/theme/pests/code/en/

Fedor P, Sigmund J, Zvaríková M, Masarovič R, Štefánik M, Krumpál M, Litavský J, Prokop P (2018) The most northern record of the alien composite thrips *Microcephalothrips abdominalis* in Europe. Plant Protect Sci 54(1):43–47. https://doi.org/10.17221/42/2017-PPS

Feng BH, Peng LF (2012) Synthesis and characterization of carboxymethyl chitosan carrying ricinoleic functions as an emulsifier for azadirachtin. Carbohydr Polym 88(2):576–582. https://doi.org/10.1016/j.carbpol.2012.01.002

Floros GD, Kokkari AI, Kouloussis NA, Kantiranis NA, Damos P, Filippidis AA, Koveos DS (2018) Evaluation of the natural zeolite lethal effects on adults of the bean weevil under different temperatures and relative humidity regimes. J Econ Entomol 111(1):482–490. https://doi.org/10.1093/jee/tox305

Forim MR, Costa ES, da Silva GFMF, Fernandes JB, Mondego JM, Boica AL (2013) Development of a new method to prepare nano−/microparticles loaded with extracts of *Azadirachta indica*, their characterization and use in controlling *Plutella xylostella*. J Agric Food Chem 61:9131–9139. https://doi.org/10.1021/jf403187y

Fouad H, Li HJ, Hosni D, Wei JQ, Abbas G, Ga'al H, Jianchu M (2018) Controlling *Aedes albopictus* and *Culex pipiens pallens* using silver nanoparticles synthesized from aqueous extract of *Cassia fistula* fruit pulp and its mode of action. Artif Cells Nanomed Biotechnol 46(3):558–567. https://doi.org/10.1080/21691401.2017.1329739

Frederiksen HK, Kristenson HG, Pedersen M (2003) Solid lipid microparticle formulations of the pyrethroid gamma-cyhalothrin-incompatibility of the lipid and the pyrethroid and biological properties of the formulations. J Control Release 86(2–3):243–252. https://doi.org/10.1016/S0168-3659(02)00406-6

Fukuto TR (1990) Mechanism of action of organophosphorus and carbamate insecticides. Environ Health Perspect 87:245–254. https://doi.org/10.1289/ehp.9087245

Gamboa GVU, Palma SD, Lifschitz A, Ballent M, Lanusse C, Passirani C, Benoit JP, Allemandi DA (2016) Ivermectin-loaded lipid nanocapsules: toward the development of a new antiparasitic delivery system for veterinary applications. Parsitol Res 115(5):1945–1953. https://doi.org/10.1007/s00436-016-4937-1

Gao YH, Kaziem AE, Zhang YH, Xiao YN, He S, Li JH (2018a) A hollow mesoporous silica and poly(diacetone acrylamide) composite with sustained-release and adhesion properties. Microporous Mesoporous Mater 255:15–22. https://doi.org/10.1016/j.micromeso.2017.07.025

Gao YY, Yu SX, Li JJ, Sun PD, Xiong M, Lei CL, Zhang ZB, Huang QY (2018b) Bioactivity of diatomaceous earth against the subterranean termite *Reticulitermes chinensis* Snyder (Isoptera: Rhinotermitidae). Environ Sci Pollut Res 25(28):28102–28108. https://doi.org/10.1007/s11356-018-2718-3

Gao YH, Zhang YH, He S, Xiao YN, Qin XY, Zhang Y, Li DL, Ma HJ, You H, Li JH (2019) Fabrication of a hollow mesoporous silica hybrid to improve the targeting of a pesticide. Chem Eng J 364:361–369. https://doi.org/10.1016/j.cej.2019.01.105

Ghidan AY, Al-Antary TM, Awwad AM, Ghidan OY, Araj SEA, Ateyyat MA (2018) Comparison of different green synthesized nanomaterials on green peach aphid as aphicidal potential. Fresenius Environ Bull 27(10):7009–7016

Ghramh HA, Al-Ghamdi KM, Mahyoub JA, Ibrahim EH (2018) Chrysanthemum extract and extract prepared silver nanoparticles as biocides to control *Aedes aegypti* (L.), the vector of dengue fever. J Asia Pac Entomol 21(1):205–210. https://doi.org/10.1016/j.aspen.2017.12.001

Giongo AMM, Vendramim JD, Forim MR (2016) Evaluation of neem-based nanoformulations as alternative to control fall armyworm. Cienc Agrotec 40(1):26–36. https://doi.org/10.1590/S1413-70542016000100002

Gopinath K, Karthika V, Sundaravadivelan C, Gowri S, Arumugam A (2015) Mycogenesis of cerium oxide nanoparticles using *Aspergillus niger* culture filtrate and their applications for antibacterial and larvicidal activities. J Nanostructure Chem 5(3):295–303. https://doi.org/10.1007/s40097-015-0161-2

Guedes RNC, Smagghe G, Stark JD, Desneux N (2016) Pesticide-induced stress in arthropod pests for optimized integrated pest management programs. Annu Rev Entomol 61:43–62. https://doi.org/10.1146/annurev-ento-010715-023646

Guo MC, Zhang WB, Ding GL, Guo D, Zhu JL, Wang BT, Punyapitak D, Cao YS (2015) Preparation and characterization of enzyme-responsive emamectin benzoate microcapsules based on a copolymer matrix of silica-epichlorohydrin-carboxymethylcellulose. RSC Adv 5(113):93170–93179. https://doi.org/10.1039/c5ra17901g

Gupta G, Agarwal U, Kaur H, Kumar NR, Gupta P (2017) Aphicidal effects of terpenoids present in *Citrus limon* on *Macrosiphum roseiformis* and two generalist insect predators. Asia Pac Entomol 20(4):1087–1095. https://doi.org/10.1016/j.aspcn.2017.08.007

Hala HA, Elsamahy MFM (2016) Relative toxicity of silica nanoparticles to two tetranychids and three associated predators. Egypt J Biol Pest Control 26(2):283–286

Hamed AG, Jazem AM, Sadeq KA (2018) Synthesis of silver nanoparticles from *Schinus molle* extract and its larvicidal activity against *Aedes aegypti* (L.), the vector of dengue fever in kingdom of Saudi Arabia. Res J Biotechnol 13(5):63–70

Han X, Geller B, Moniz K, Das P, Chippindale AK, Walker VK (2014) Monitoring the developmental impact of copper and silver nanoparticle exposure in *Drosophila* and their microbiomes. Sci Total Environ 487:822–829. https://doi.org/10.1016/j.scitotenv.2013.12.129

Hashem AS, Awadalla SS, Zayed GM, Maggi F, Benelli G (2018) *Pimpinella anisum* essential oil nanoemulsions against *Tribolium castaneum*-insecticidal activity and mode of action. Environ Sci Pollut Res 25(19):18802–18812. https://doi.org/10.1007/s11356-018-2068-1

Hikal WM, Baeshen RS, Said-Al Ahl HAH (2017) Botanical insecticide as simple extractives for pest control. Cogent Biol 3(1):1404274. https://doi.org/10.1080/23312025.2017.1404274

Hladik ML, Main AR, Goulson D (2018) Environmental risks and challenges associated with neonicotinoid insecticides. Environ Sci Technol 52(6):3329–3335. https://doi.org/10.1021/acs.est.7b06388

Hu WT, Li JC, Hou HW, Yan HQ, Feng YH, Mi X, Lin Q (2013) Preparation and characterization of hydrophobic alginate derivative nanocapsules entrapping λ-cyhalothrin, Part A. Asian J Chem 25(17):9904–9908. https://doi.org/10.14233/ajchem.2013.15593

Huang B, Chen F, Shen Y, Qian K, Wang Y, Sun C, Zhao X, Cui B, Gao F, Zeng Z, Cui H (2018) Advances in targeted pesticides with environmentally responsive controlled release by nanotechnology. Nano 8:102. https://doi.org/10.3390/nano8020102

Hulme PE (2009) Trade, transport and trouble: managing invasive species pathways in an era of globalization. J Appl Ecol 46:10–18. https://doi.org/10.1111/j.1365-2664.2008.01600.x

Hwang IC, Kim TH, Bang SH, Kim KS, Kwon HR, Seo MJ, Youn YN, Park HJ, Aoki CY, Yu YM (2011) Insecticidal effect of controlled release formulations of etofenprox based on nano-bio technique. J Fac Agric Kyushu Univ 56(1):33–40

Ibrahim AMA, Ali AM (2018) Silver and zinc oxide nanoparticles induce developmental and physiological changes in the larval and pupal stages of *Spodoptera littoralis* (Lepidoptera: Noctuidae). J Asia Pac Entomol 21(4):1373–1378. https://doi.org/10.1016/j.aspen.2018.10.018

Ihegwuagu NE, Sha'Ato R, Tor-Anyiin TA, Nnamonu LA, Buekes P, Sone B, Maaza M (2016) Facile formulation of starch-silver-nanoparticle encapsulated dichlorvos and chlorpyrifos for enhanced insecticide delivery. New J Chem 40(2):1777–1784. https://doi.org/10.1039/c5nj01831e

Ileke KD, Ogungbite OC (2014) Entomocidal activity of powders and extracts of four medicinal plants against *Sitophilus oryzae* (L), *Oryzaephilus mercator* (Faur) and *Ryzopertha dominica* (Fabr.). Jordan J Biol Sci 7:57–62

IRAC (2019) Insecticide Resistance Action Committee. https://www.irac-online.org/

Ishwarya R, Vaseeharan B, Kalyani S, Banumathi B, Govindarajan M, Alharbi NS, Kadaikunnan S, Al-anbr MN, Khaled JM, Benelli G (2018) Facile green synthesis of zinc oxide nanoparticles using *Ulva lactuca* seaweed extract and evaluation of their photocatalytic, antibiofilm and insecticidal activity. J Photochem Photobiol B 178:249–258. https://doi.org/10.1016/j.jphotobiol.2017.11.006

Jampílek J, Kráľová K (2015) Application of nanotechnology in agriculture and food industry, its prospects and risks. Ecol Chem Eng S 22(3):321–361. https://doi.org/10.1515/eces-2015-0018

Jampílek J, Kráľová K (2017a) Nanopesticides: preparation, targeting and controlled release. In: Grumezescu AM (ed) Nanotechnology in agri-food industry – new pesticides and soil sensors. Elsevier, London, pp 81–127

Jampílek J, Kráľová K (2017b) Nano-antimicrobials: activity, benefits and weaknesses. In: Ficai A, Grumezescu AM (eds) Nanostructures in therapeutic medicine – nanostructures for antimicrobial therapy. Elsevier, Amsterdam, pp 23–54

Jampílek J, Kráľová K (2018a) Nanomaterials applicable in food protection. In: Rai RV, Bai JA (eds) Nanotechnology applications in food industry. Taylor & Francis Group, Boca Raton, pp 75–96

Jampílek J, Kráľová K (2018b) Benefits and potential risks of nanotechnology applications in crop protection. In: Abd-Elsalam K, Prasad R (eds) Nanobiotechnology applications in plant protection. Springer, Cham, pp 189–246

Jampílek J, Kráľová K (2019) Nano-biopesticides in agriculture: state of art and future opportunities. In: Koul O (ed) Nano-biopesticides today and future perspectives. Academic Press & Elsevier, Amsterdam, pp 397–447

Jampílek J, Kráľová K, Campos EVR, Fraceto LF (2019) Bio-based nanoemulsion formulations applicable in agriculture, medicine and food industry. In: Prasad R, Kumar V, Kumar M, Choudhary DK (eds) Nanobiotechnology in bioformulations. Springer, Cham, pp 33–84

Jansson RK, Dybas RA (1998) Avermectins: biochemical mode of action, biological activity and agricultural importance. In: Ishaaya I, Degheele D (eds) Insecticides with novel modes of action. Applied agriculture. Springer, Berlin, pp 152–170. https://doi.org/10.1007/978-3-662-03565-8_9

Jeevanandam J, Barhoum A, Chan YS, Dufresne A, Danquah MK (2018) Review on nanoparticles and nanostructured materials: history, sources, toxicity and regulations. Beilstein J Nanotechnol 9:1050–1074. https://doi.org/10.3762/bjnano.9.98

Jerobin J, Sureshkumar RS, Anjali CH, Mukherjee A, Chandrasekaran N (2012) Biodegradable polymer based encapsulation of neem oil nanoemulsion for controlled release of Aza-A. Carbohydr Polym 90(4):1750–1756. https://doi.org/10.1016/j.carbpol.2012.07.064

Jones DK, Hintz WD, Schuler MS, Yates EK, Mattes BM, Relyea RA (2017) Inducible tolerance to agrochemicals was paved by evolutionary responses to predators. Environ Sci Technol 51(23):13913–13919. https://doi.org/10.1021/acs.est.7b03816

Juan-Blasco M, Tena A, Vanaclocha P, Cambra M, Urbaneja A, Monzo C (2011) Efficacy of a micro-encapsulated formulation compared with a sticky barrier for excluding ants from citrus canopies. J Appl Entomol 135(6):467–472. https://doi.org/10.1111/j.1439-0418.2010.01572.x

Kalaimurugan D, Sivasankar P, Lavanya K, Shivakumar MS, Venkatesan S (2019) Antibacterial and larvicidal activity of *Fusarium proliferatum* (YNS2) whole cell biomass mediated copper nanoparticles. J Clust Sci 30(4):1071–1080. https://doi.org/10.1007/s10876-019-01568-x

Kalaitzaki A, Papanikolaou NE, Karamaouna F, Dourtoglou V, Xenakis A, Papadimitriou V (2015) Biocompatible colloidal dispersions as potential formulations of natural pyrethrins: a structural and efficacy study. Langmuir 31(21):5722–5730. https://doi.org/10.1021/acs.langmuir.5b00246

Kamari A, Aljafree NFA, Yusoff SNM (2016) Oleoyl-carboxymethyl chitosan as a new carrier agent for the rotenone pesticide. Environ Chem Lett 14(3):417–422. https://doi.org/10.1007/s10311-016-0550-x

Kamil D, Prameeladevi T, Ganesh S, Prabhakaran N, Nareshkumar R, Thomas SP (2017) Green synthesis of silver nanoparticles by entomopathogenic fungus *Beauveria bassiana* and their bioefficacy against mustard aphid (*Lipaphis erysimi* Kalt.). Indian J Exp Biol 55(8):555–561

Kang MA, Seo MJ, Hwang IC, Jang C, Park HJ, Yu YM, Youn YN (2012) Insecticidal activity and feeding behavior of the green peach aphid, *Myzus persicae*, after treatment with nano types of pyrifluquinazon. J Asia Pac Entomol 15(4):533–541. https://doi.org/10.1016/j.aspen.2012.05.015

Karp DS, Mendenhall CD, Figueroa Sandí R, Chaumont N, Ehrlich PR, Hadly EA, Daily GC (2013) Forest bolsters bird abundance, pest control and coffee yield. Ecol Lett 16:1339–1347. https://doi.org/10.1111/ele.12173

Karthiga P, Rajeshkumar S, Annadurai G (2018) Mechanism of larvicidal activity of antimicrobial silver nanoparticles synthesized using *Garcinia mangostana* bark extract. J Clust Sci 29(6):1233–1241. https://doi.org/10.1007/s10876-018-1441-z

Kayser H, Lee C, Decock A, Baur M, Haettenschwiler J, Maienfisch P (2004) Comparative analysis of neonicotinoid binding to insect membranes: I. A structure-activity study of the mode of [H-3]imidacloprid displacement in *Myzus persicae* and *Aphis craccivora*. Pest Manag Sci 60(10):945–958. https://doi.org/10.1002/ps.919

Kaziem AE, Gao YH, He S, Li JH (2017) Synthesis and insecticidal activity of enzyme-triggered functionalized hollow mesoporous silica for controlled release. J Agric Food Chem 65(36):7854–7864. https://doi.org/10.1021/acs.jafc.7b02560

Kettunen M, Genovesi P, Gollasch S, Pagad S, Starfinger U, ten Brink P, Shine C (2008) Technical support to EU strategy on invasive species (IAS) – assessment of the impacts of IAS in Europe and the EU (final module report for the European Commission). Belgium, Institute for European Environmental Policy (IEEP): 44. https://ec.europa.eu/environment/nature/invasivealien/docs/Kettunen2009_IAS_Task%201.pdf

Khandelwal N, Doke DS, Khandare JJ, Jawale PV, Biradar AV, Giri AP (2015) Bio-physical evaluation and in vivo delivery of plant proteinase inhibitor immobilized on silica nanospheres. Colloids Surf B Biointerfaces 130:84–92. https://doi.org/10.1016/j.colsurfb.2015.03.060

Khazri A, Sellami B, Hanachi A, Dellali M, Eljarrat E, Beyrem H, Mahmoudi E (2017) Neurotoxicity and oxidative stress induced by permethrin in gills of the freshwater mussel *Unio ravoisieri*. Chem Ecol 33(1):88–101. https://doi.org/10.1080/02757540.2016.1248948

Khoobdel M, Ahsaei SM, Farzaneh M (2017) Insecticidal activity of polycaprolactone nanocapsules loaded with *Rosmarinus officinalis* essential oil in *Tribolium castaneum* (Herbst). Entomol Res 47(3):175–184. https://doi.org/10.1111/1748-5967.12212

Klein AM, Vaissiere BE, Cane JH, Steffan-Dewenter I, Cunningham SA, Kremen C, Tscharntke TD (2007) Importance of pollinators in changing landscapes for world crops. Proc R Soc B Biol Sci 274:303–313. https://doi.org/10.1098/rspb.2006.3721

Korunic Z (1998) Diatomaceous earths, a group of natural insecticides. J Stored Prod Res 34(2–3):87–97. https://doi.org/10.1016/S0022-474X(97)00039-8

Kovendan K, Chandramohan B, Govindarajan M, Jebanesan A, Kamalakannan S, Vincent S, Benelli G (2018) *Zeuxine gracilis*-fabricated silver nanoparticles against dengue, malaria and filariasis mosquito vectors. J Clust Sci 29(2):345–357. https://doi.org/10.1007/s10876-018-1331-4

Kráľová K, Masarovičová E, Jampílek J (2019) Plant responses to stress induced by toxic metals and their nanoforms. In: Pessarakli M (ed) Handbook of plant and crop stress, 4th edn. CRC, Taylor and Francis, Boca Raton, pp 479–522

Kumar J, Shakil NA, Singh MK, Singh MK, Pandey A, Pandey RP (2010) Development of controlled release formulations of azadirachtin-A employing poly(ethylene glycol) based amphiphilic copolymers. J Environ Sci Health B 45(4):310–314. https://doi.org/10.1080/03601231003704457

Kumar RSS, Shiny PJ, Anjali CH, Jerobin J, Goshen KM, Magdassi S, Mukherjee A, Chandrasekaran N (2013) Distinctive effects of nano-sized permethrin in the environment. Environ Sci Pollut Res 20(4):2593–2602. https://doi.org/10.1007/s11356-012-1161-0

Kumar S, Chauhan N, Gopal M, Kumar R, Dilbaghi N (2015) Development and evaluation of alginate-chitosan nanocapsules for controlled release of acetamiprid. Int J Biol Macromol 81:631–637. https://doi.org/10.1016/j.ijbiomac.2015.08.062

Künast C, de Graeff R, Whitmore G (2013) Pesticides and biodiversity agricultural productivity and biodiversity conservation. European Landowners' Organization & European Crop Protection Association. https://www.ecpa.eu/sites/default/files/7584%2BBiodiversity_V04_b%C3%A0t.pdf

Kuzniar T, Ropek D, Lemek T (2011) Impact of multi-walled carbon nanotubes on viability and pathogenicity of entomopathogenic nematodes. Ecol Chem Eng A 18(5–6):757–762

Lazarevic J, Radojkovic A, Kostic I, Krnjajic S, Mitrovic J, Kostic MB, Novakovic T, Brankovic Z, Brankovic G (2018) Insecticidal impact of alumina powders against *Acanthoscelides obtectus* (Say). J Stored Prod Res 77:45–54. https://doi.org/10.1016/j.jspr.2018.02.006

Li JF, Yao J, Li YM, Shao Y (2012) Controlled release and retarded leaching of pesticides by encapsulating in carboxymethyl chitosan/bentonite composite gel. J Environ Sci Health B 47(8):795–803. https://doi.org/10.1080/03601234.2012.676421

Li F, Gu Z, Wang B, Xie Y, Ma L, Xu K, Ni M, Zhang H, Shen W, Li B (2014) Effects of the biosynthesis and signaling pathway of ecdysterone on silkworm (*Bombyx mori*) following exposure to titanium dioxide nanoparticles. J Chem Ecol 40:913–922. https://doi.org/10.1007/s10886-014-0487-0

Li YY, Ni M, Li FC, Zhang H, Xu KZ, Zhao XM, Tian JH, Hu JS, Wang BB, Shen WD, Li B (2016) Effects of $TiO_2$ NPs on silkworm growth and feed efficiency. Biol Trace Elem Res 169(2):382–386. https://doi.org/10.1007/s12011-015-0413-5

Li HZ, Cheng F, Wei YL, Lydy M, You J (2017) Global occurrence of pyrethroid insecticides in sediment and the associated toxicological effects on benthic invertebrates: an overview. J Hazard Mater 324(Part B):258–271. https://doi.org/10.1016/j.jhazmat.2016.10.056

Li JX, Xue B, Cheng XY, Hu JH, Hu JS, Tian JH, Li FC, Yu XH, Li B (2018) $TiO_2$ NPs alleviates high-temperature induced oxidative stress in silkworms. J Econ Entomol 111(2):879–884. https://doi.org/10.1093/jee/toy002

Liu YK, Guo BH (2019) Preparation and characterisation of poly(butylene succinate) microcarriers containing pesticide. Micro Nano Lett 14(1):81–85. https://doi.org/10.1049/mnl.2018.5060

Liu XY, Vinson D, Abt D, Hurt RH, Rand DM (2009) Differential toxicity of carbon nanomaterials in *Drosophila*: larval dietary uptake is benign, but adult exposure causes locomotor impairment and mortality. Environ Sci Technol 43:6357–6363. https://doi.org/10.1021/es901079z

Liu B, Campo EM, Bossing T (2014) *Drosophila* embryos as model to assess cellular and developmental toxicity of multi-walled carbon nanotubes (MWCNT) in living organisms. PLoS One 9(2):e88681. https://doi.org/10.1371/journal.pone.0088681

Liu XX, He BC, Xu ZJ, Yin MZ, Yang WT, Zhang HJ, Cao JJ, Shen J (2015) A functionalized fluorescent dendrimer as a pesticide nanocarrier: application in pest control. Nanoscale 7(2):445–449. https://doi.org/10.1039/c4nr05733c

Liu BX, Wang Y, Yang F, Wang X, Shen H, Cui HX, Wu DC (2016) Construction of a controlled-release delivery system for pesticides using biodegradable PLA-based microcapsules. Colloids Surf B Biointerfaces 144:38–45. https://doi.org/10.1016/j.colsurfb.2016.03.084

Liu BX, Wang Y, Yang F, Cui HX, Wu DC (2018) Development of a chlorantraniliprole microcapsule formulation with a high loading content and controlled-release property. J Agric Food Chem 66(26):6561–6568. https://doi.org/10.1021/acs.jafc.7b01295

Lodge DM, Williams S, MacIsaac HJ, Hayes KR, Leung B, Reichard S, Mack RN, Moyle PB, Smith M, Andow DA, Carlton JT, McMichael A (2006) Biological invasions: recommendations for US policy and management. Ecol Appl 16:2035–2054. https://doi.org/10.1890/1051-0761(2006)016[2035:birfup]2.0.co;2

Lopez-Garcia GP, Buteler M, Stadler T (2018) Testing the insecticidal activity of nanostructured alumina on *Sitophilus oryzae* (L.) (Coleoptera: Curculionidae) under laboratory conditions using galvanized steel containers. Insects 9(3):87. https://doi.org/10.3390/insects9030087

Lorenz CS, Wicht AJ, Guluzada L, Crone B, Karst U, Lee HJ, Triebskorn R, Haderlein SB, Huhn C, Koehler HR (2017a) Nano-sized zeolites as modulators of thiacloprid toxicity on *Chironomus riparius*. Peer J 5:e3525. https://doi.org/10.7717/peerj.3525

Lorenz CS, Wicht AJ, Guluzada L, Luo LL, Jaeger L, Crone B, Karst U, Triebskorn R, Liang Y, Anwander R, Haderlein SB, Huhn C, Koehler HR (2017b) Nano-sized $Al_2O_3$ reduces acute toxic effects of thiacloprid on the non-biting midge *Chironomus riparius*. PLoS One 12(5):e0176356. https://doi.org/10.1371/journal.pone.0176356

Lu MM, Xiong D, Sun WW, Yu T, Hu ZX, Ding JF, Cai YP, Yang SZ, Pan BL (2017) Sustained release ivermectin-loaded solid lipid dispersion for subcutaneous delivery: in vitro and in vivo evaluation. Drug Deliv 24(1):622–631. https://doi.org/10.1080/10717544.2017.1284945

Lu XP, Liu JH, Weng H, Ma ZQ, Zhang X (2020) Efficacy of binary combinations between methyl salicylate and carvacrol against thrips *Anaphothrips obscurus*: laboratory and field trials. Pest Manag Sci. 76(2):589–596. https://doi.org/10.1002/ps.5551

Lumaret JP, Errouissi F, Floate K, Römbke J, Wardhaugh K (2012) A review on the toxicity and non-target effects of macrocyclic lactones in terrestrial and aquatic environments. Curr Pharm Biotechnol 13(6):1004–1060. https://doi.org/10.2174/138920112800399257

Ma X, Li HZ, Xiong JJ, Mehler WT, You J (2019) Developmental toxicity of a neonicotinoid insecticide, acetamiprid to zebrafish embryos. J Agric Food Chem 67(9):2429–2436. https://doi.org/10.1021/acs.jafc.8b05373

Maas B, Clough Y, Tscharntke T (2013) Bats and birds increase crop yield in tropical agroforestry landscapes. Ecol Lett 16:1480–1487. https://doi.org/10.1111/ele.12194

Magro M, Bramuzzo S, Baratella D, Ugolotti J, Zoppellaro G, Chemello G, Olivotto I, Ballarin C, Radaelli G, Arcaro B, De Liguoro M, Coppellotti O, Guidolin L, de Almeida RJ, Bonaiuto E, Zboril R, Vianello F (2019) Self-assembly of chlorin-e6 on $\gamma$-$Fe_2O_3$ nanoparticles: application for larvicidal activity against *Aedes aegypti*. J Photochem Photobiol B 194:21–31. https://doi.org/10.1016/j.jphotobiol.2019.03.004

Makarenko NA, Makarenko VV (2019) Nanotechnologies in crop cultivation: ecotoxicological aspects. Biosyst Divers 27(2):148–155. https://doi.org/10.15421/011920

Malaikozhundan B, Vinodhini J (2018) Nanopesticidal effects of *Pongamia pinnata* leaf extract coated zinc oxide nanoparticle against the pulse beetle, *Callosobruchus maculatus*. Mater Today Commun 14:106–115. https://doi.org/10.1016/j.mtcomm.2017.12.015

Malaikozhundan B, Vaseeharan B, Vijayakumar S, Thangaraj MP (2017) *Bacillus thuringiensis* coated zinc oxide nanoparticle and its biopesticidal effects on the pulse beetle, *Callosobruchus maculatus*. J Photochem Photobiol B 174:306–314. https://doi.org/10.1016/j.jphotobiol.2017.08.014

Martins CHZ, de Sousa M, Fonseca LC, Martinez DST, Alves OL (2019) Biological effects of oxidized carbon nanomaterials (1D versus 2D) on *Spodoptera frugiperda*: material dimensionality influences on the insect development, performance and nutritional physiology. Chemosphere 215:766–774. https://doi.org/10.1016/j.chemosphere.2018.09.178

Masarovič R, Doričová M, Prokop P, Fedor J (2014) "Testing the limits" – an interesting record of the exotic banded greenhouse thrips *Hercinothrips femoralis* (Thysanoptera: Thripidae: Panchaetothripinae) at the high Carpathian mountain altitudes. Biologia 69:1631–1634. https://doi.org/10.2478/s11756-014-0458-5

Masarovič R, Štefánik M, Zvaríková M, Sigmund J, Fedor P (2017) First record of a new alien economically important thrips *Dichromothrips corbetti* (Priesner, 1936) (Thysanoptera: Thripidae) in Slovakia. Plant Protect Sci 53(3):177–180. https://doi.org/10.17221/141/2016-PPS

Masarovičová E, Kráľová K (2013) Metal nanoparticles and plants. Ecol Chem Eng S 20:9–22

Masarovičová E, Kráľová K, Zinjarde SS (2014) Metal nanoparticles in plants. Formation and action. In: Pessarakli M (ed) Handbook of plant and crop physiology, 3rd edn. CRC, Taylor and Francis, Boca Raton, pp 683–731

Mattos BD, Rojas OJ, Magalhaes WLE (2017) Biogenic silica nanoparticles loaded with neem bark extract as green, slow-release biocide, Part 4. J Clean Prod 142:4206–4213. https://doi.org/10.1016/j.jclepro.2016.11.183

May R (2010) Tropical arthropod species, more or less? Science 329(5987):41–42. https://doi.org/10.1126/science.1191058

Memarizadeh N, Ghadamyari M, Adeli M, Talebi K (2014) Preparation, characterization and efficiency of nanoencapsulated imidacloprid under laboratory conditions. Ecotoxicol Environ Saf 107:77–83. https://doi.org/10.1016/j.ecoenv.2014.05.009

Meyer WL, Gurman P, Stelinski LL, Elman NM (2015) Functional nano-dispensers (FNDs) for delivery of insecticides against phytopathogen vectors. Green Chem 17(8):4173–4177. https://doi.org/10.1039/c5gc00717h

Mishra P, Dutta S, Haldar M, Dey P, Kumar D, Mukherjee A, Chandrasekaran N (2019) Enhanced mosquitocidal efficacy of colloidal dispersion of pyrethroid nanometric emulsion with benignity towards non-target species. Ecotoxicol Environ Saf 176:258–269. https://doi.org/10.1016/j.ecoenv.2019.03.096

Moradi FG, Hejazi MJ, Hamishehkar H, Enayati AA (2019) Co-encapsulation of imidacloprid and lambda-cyhalothrin using biocompatible nanocarriers: characterization and application. Ecotoxicol Environ Saf 175:155–163. https://doi.org/10.1016/j.ecoenv.2019.02.092

Moreira AMD, Bittencourt VCE, Costa FLS, de Lima ME, Lopes MTP, Borges WS, Martins GF, Nascimento CS, da Silva JG, Denadai AML, Borges KB (2018) Hydrophobic nanoprecipitates of β-cyclodextrin/avermectins inclusion compounds reveal insecticide activity against *Aedes aegypti* larvae and low toxicity against fibroblasts. J Agric Food Chem 66(28):7275–7285. https://doi.org/10.1021/acsjafc.8001300

Morejon B, Pilaquinga F, Domenech F, Ganchala D, Debut A, Neira M (2018) Larvicidal activity of silver nanoparticles synthesized using extracts of *Ambrosia arborescens* (Asteraceae) to control *Aedes aegypti* L. (Diptera: Culicidae). J Nanotechnol 2018:6917938. https://doi.org/10.1155/2018/6917938

Morgan ED (2009) Azadirachtin, a scientific gold mine. Bioorg Med Chem 17(12):4096–4105. https://doi.org/10.1016/j.bmc.2008.11.081

Morozov VN, Kanev IL (2015) Knockdown of fruit flies by imidacloprid nanoaerosol. Environ Sci Technol 49(20):12483–12489. https://doi.org/10.1021/acs.est.5b03219

Mossa ATH, Mohafrash SMM, Chandrasekaran N (2018a) Safety of natural insecticides: toxic effects on experimental animals. Biomed Res Int 2018:4308054. https://doi.org/10.1155/2018/4308054

Mossa ATH, Afia SI, Mohafrash SMM, Abou-Awad BA (2018b) Formulation and characterization of garlic (*Allium sativum* L.) essential oil nanoemulsion and its acaricidal activity on eriophyid olive mites (Acari: Eriophyidae). Environ Sci Pollut Res 25(11):10526–10537. https://doi.org/10.1007/s11356-017-0752-1

Mott SM (2019) The size of matter: why properties change at the nanoscale. Yale National Initiative, Yale University. https://teachers.yale.edu/curriculum/viewer/initiative_10.05.06_u

Mousa KM, Elsharkawy MM, Khodeir IA, El-Dakhakhni TN, Youssef AE (2014) Growth perturbation, abnormalities and mortality of oriental armyworm *Mythimna separata* (Walker) (Lepidoptera: Noctuidae) caused by silica nanoparticles and *Bacillus thuringiensis* toxin. Egypt J Biol Pest Control 24(2):347–351

Muda Z, Hashim N, Isa IM, Abu Bakar S, Ali NM, Hussein MZ, Mamat M, Sidik SM (2019) Synthesis and characterization of mesoporous zinc layered hydroxide-isoprocarb nanocomposite. J Saudi Chem Soc 23(4):486–493. https://doi.org/10.1016/j.jscs.2018.08.012

Murugan K, Nataraj D, Jaganathan A, Dinesh D, Jayashanthini S, Samidoss CM, Paulpandi M, Panneerselvam C, Subramaniam J, Aziz A, Nicoletti M, Kumar S, Higuchi A, Benelli G (2017) Mosquitoes, *Plasmodium falciparum* and MCF-7 cancer cells: impact on predation of non-target tadpoles, odonate nymphs and mosquito fishes. J Clust Sci 28(1):393–411. https://doi.org/10.1007/s10876-016-1107-7

Murugan K, Dinesh D, Nataraj D, Subramaniam J, Amuthavalli P, Madhavan J, Rajasekar A, Rajan M, Thiruppathi KP, Kumar S, Higuchi A, Nicoletti M, Benelli G (2018a) Iron and iron oxide nanoparticles are highly toxic to *Culex quinquefasciatus* with little non-target effects on larvivorous fishes. Environ Sci Pollut Res 25(11):10504–10514. https://doi.org/10.1007/s11356-017-0313-7

Murugan K, Madhavan J, Samidoss CM, Panneerselvam C, Aziz A, Malathi A, Rajasekar A, Pandiyan A, Kumar S, Alarfaj AA, Higuchi A, Benelli G (2018b) Bismuth oxyiodide nanoflakes showed toxicity against the malaria vector *Anopheles stephensi* and in vivo antiplasmodial activity. J Clust Sci 29(2):337–344. https://doi.org/10.1007/s10876-018-1332-3

Murugan K, Roni M, Panneerselvam C, Aziz A, Suresh U, Rajaganesh R, Aruliah R, Mahyoub JA, Trivedi S, Rehman H, Hatem ANA, Kumar S, Higuchi A, Vaseeharan B, Wei H, Senthil-Nathan S, Canale A, Benelli G (2018c) *Sargassum wightii*-synthesized ZnO nanoparticles reduce the fitness and reproduction of the malaria vector *Anopheles stephensi* and cotton bollworm *Helicoverpa armigera*. Physiol Mol Plant Pathol 101:202–213. https://doi.org/10.1016/j.pmpp.2017.02.004

Muthukumaran U, Govindarajan M, Rajeswary M (2015) Mosquito larvicidal potential of silver nanoparticles synthesized using *Chomelia asiatica* (*Rubiaceae*) against *Anopheles stephensi*, *Aedes aegypti*, and *Culex quinquefasciatus* (Diptera: Culicidae). Parasitol Res 114(3):989–999. https://doi.org/10.1007/s00436-014-4265-2

Nagaraj G (2009) Oilseeds-properties, processing, products and procedures. New India Publishing Agency, New Delhi, p 2017

Ng CT, Yong LQ, Hande MP, Ong CN, Yu LE, Bay BH, Baeg GH (2017) Zinc oxide nanoparticles exhibit cytotoxicity and genotoxicity through oxidative stress responses in human lung fibroblasts and *Drosophila melanogaster*. Int J Nanomedicine 12:1621–1637. https://doi.org/10.2147/IJN.S124403

Nguyen HM, Hwang IC, Park JW, Park HJ (2012a) Photoprotection for deltamethrin using chitosan-coated beeswax solid lipid nanoparticles. Pest Manag Sci 68(7):1062–1068. https://doi.org/10.1002/ps.3268

Nguyen HM, Hwang IC, Park JW, Park HJ (2012b) Enhanced payload and photo-protection for pesticides using nanostructured lipid carriers with corn oil as liquid lipid. J Microencapsul 29(6):596–604. https://doi.org/10.3109/02652048.2012.668960

Nogrado K, Lee S, Chon K, Lee JH (2019) Effect of transient exposure to carbaryl wettable powder on the gut microbial community of honey bees. Appl Biol Chem 62:UNSP 6. https://doi.org/10.1186/s13765-019-0415-7

Novaluron (2001) EPA Pesticide Fact Sheet. https://www3.epa.gov/pesticides/chem_search/reg_actions/registration/fs_PC-124002_24-Sep-01.pdf

Oliveira DGP, Bonini AK, Alves LFA (2017) Field assessments to control the lesser mealworm (Coleoptera: Tenebrionidae) using diatomaceous earth in poultry houses. J Econ Entomol 110(6):2716–2723. https://doi.org/10.1093/jee/tox287

Ong C, Lee QY, Cai Y, Liu XL, Ding J, Yung LYL, Bay BH, Baeg GH (2016) Silver nanoparticles disrupt germline stem cell maintenance in the *Drosophila* testis. Sci Rep 6:20632. https://doi.org/10.1038/srep20632

Pandiarajan J, Krishnan M (2017) Properties, synthesis and toxicity of silver nanoparticles. Environ Chem Lett 15(3):387–397. https://doi.org/10.1007/s10311-017-0624-4

Papanikolaou NE, Kalaitzaki A, Karamaouna F, Michaelakis A, Papadimitriou V, Dourtoglou V, Papachristos DP (2018) Nano-formulation enhances insecticidal activity of natural pyrethrins against *Aphis gossypii* (Hemiptera: Aphididae) and retains their harmless effect to non-target predators. Environ Sci Pollut Res 25(11):10243–10249. https://doi.org/10.1007/s11356-017-8596-2

Parimi D, Sundararajan V, Sadak O, Gunasekaran S, Mohideen SS, Sundaramurthy A (2019) Synthesis of positively and negatively charged $CeO_2$ nanoparticles: investigation of the role of surface charge on growth and development of *Drosophila melanogaster*. ACS OMEGA 4(1):104–113. https://doi.org/10.1021/acsomega.8b02747

Pavela R (2016) History, presence and perspective of using plant extracts as commercial botanical insecticides and farm products for protection against insects – a review. Plant Protect Sci 52(4):229–241. https://doi.org/10.17221/31/2016-PPS

Pavela R, Maggi F, Iannarelli R, Benelli G (2019a) Plant extracts for developing mosquito larvicides: from laboratory to the field, with insights on the modes of action. Acta Trop 193:236–271. https://doi.org/10.1016/j.actatropica.2019.01.019
Pavela R, Pavoni L, Bonacucina G, Cespi M, Kavallieratos NG, Cappellacci L, Petrelli R, Maggi F, Benelli G (2019b) Rationale for developing novel mosquito larvicides based on isofuranodiene microemulsions. J Pest Sci 92(2):909–921. https://doi.org/10.1007/s10340-018-01076-3
Pavela R, Benelli G, Pavoni L, Bonacucina G, Cespi M, Cianfaglione K, Bajalan I, Morshedloo MR, Lupidi G, Romano D, Canale A, Maggi F (2019c) Microemulsions for delivery of Apiaceae essential oils-towards highly effective and eco-friendly mosquito larvicides? Ind Crop Prod 129:631–640. https://doi.org/10.1016/j.indcrop.2018.11.073
Pavunraj M, Baskar K, Duraipandiyan V, Al-Dhabi NA, Rajendran V, Benelli G (2017) Toxicity of Ag nanoparticles synthesized using stearic acid from *Catharanthus roseus* leaf extract against *Earias vittella* and mosquito vectors (*Culex quinquefasciatus* and *Aedes aegypti*). J Clust Sci 28(5):2477–2492. https://doi.org/10.1007/s10876-017-1235-8
Pérez-de-Luque A, Hermosín MC (2013) Nanotechnology and its use in agriculture. In: Bagchi D, Bagchi M, Moriyama H, Shahidi F (eds) Bio-nanotechnology: a revolution in food, biomedical and health sciences. Wiley-Blackwell, West Sussex, pp 299–405
Pisárčik M, Jampílek J, Devínsky F, Drábiková J, Tkacz J, Opravil T (2016) Gemini surfactants with polymethylene spacer: supramolecular structures at solid surface and aggregation in aqueous solution. J Surfactant Deterg 19(3):477–486. https://doi.org/10.1007/s11743-016-1797-8
Pisárčik M, Jampílek J, Lukáč M, Horáková R, Devínsky F, Bukovský M, Kalina M, Tkacz J, Opravil T (2017) Silver nanoparticles stabilised by cationic gemini surfactants with variable spacer length. Molecules 22(10):1794. https://doi.org/10.3390/molecules22101794
Pisárčik M, Lukáč M, Jampílek J, Bilka F, Bilková A, Pašková Ľ, Devínsky F, Horáková R, Opravil T (2018) Silver nanoparticles stabilised with cationic single-chain surfactants. Structure-physical properties-biological activity relationship study. J Mol Liq 272:60–72. https://doi.org/10.1016/j.molliq.2018.09.042
Plachá D, Jampílek J (2019) Graphenic materials for biomedical applications. Nanomaterials 9(12):1758. https://doi.org/10.3390/nano9121758
Poopathi S, Murugan K, Selvakumari J, Mani C, Bala P, de Britto RLJ, Pattnaik S, Goud S, Prabakaran K (2016) Biopesticide production from environmental wastes. JSM Trop Med Res 1:1008. https://www.jscimedcentral.com/TropicalMedicine/tropicalmedicine-1-1008.pdf
Popat A, Liu J, Hu QH, Kennedy M, Peters B, Lu GQ, Qiao SZ (2012) Adsorption and release of biocides with mesoporous silica nanoparticles. Nanoscale 4(3):970–975. https://doi.org/10.1039/c2nr11691j
Potts SG, Imperatriz-Fonseca VL, Ngo HT, Biesmeijer JC, Breeze TD, Dicks LV, Garibaldi LA, Hill R, Settele J, Vanbergen AJ, Aizen MA, Cunningham SA, Eardley C, Freitas BM, Gallai N, Kevan PG, Kovács-Hostyánszki A, Kwapong PK, Li J, Li X, Martins DJ, Nates-Parra G, Pettis JS, Rader R, Viana BF (2016) Summary for policymakers of the assessment report of the intergovernmental science-policy platform on biodiversity and ecosystem services on pollinators, pollination and food production. Secretariat of the Intergovernmental Science-Policy Platform on Biodiversity and Ecosystem Services, Bonn. https://www.actu-environnement.com/media/pdf/news-26331-synthese-ipbes-decideurs-pollinisateurs.pdf
Pradhan S, Roy I, Lodh G, Patra P, Choudhury SR, Samanta A, Goswami A (2013) Entomotoxicity and biosafety assessment of PEGylated acephate nanoparticles: a biologically safe alternative to neurotoxic pesticides. J Environ Sci Health B 48(7):559–569. https://doi.org/10.1080/03601234.2013.774891
Prasad R, Kumar V, Prasad KS (2014) Nanotechnology in sustainable agriculture: present concerns and future aspects. Afr J Biotechnol 13:705–713. https://doi.org/10.5897/AJBX2013.13554
Qin H, Zhang H, Li LX, Zhou XT, Li JP, Kan CY (2017) Preparation and properties of lambda-cyhalothrin/polyurethane drug-loaded nanoemulsions. RSC Adv 7(83):52684–52693. https://doi.org/10.1039/c7ra10640h

Quaglia F, Barbato F, De Rosa G, Granata E, Miro A, La Rotonda MI (2001) Reduction of the environmental impact of pesticides: waxy microspheres encapsulating the insecticide carbaryl. J Agric Food Chem 49(10):4808–4812. https://doi.org/10.1021/jf010418w

Raby M, Nowierski M, Perlov D, Zhao XM, Hao CY, Poirier DG, Sibley PK (2018) Acute toxicity of 6 neonicotinoid insecticides to freshwater invertebrates. Environ Toxicol Chem 37(5):1430–1445. https://doi.org/10.1002/etc.4088

Ragaei M, Sabry AH (2014) Nanotechnology for insect pest control. Int J Sci Environ Technol 3:528–545. https://doi.org/10.1016/j.btre.03.002

Raj A, Shah P, Agrawal N (2017) Dose-dependent effect of silver nanoparticles (AgNPs) on fertility and survival of *Drosophila*: an in-vivo study. PLoS One 12(5):e0178051. https://doi.org/10.1371/journal.pone.0178051

Rajak P, Sahana S, Roy S (2013) Acephate-induced shortening of developmental duration and early adult emergence in a nontarget insect *Drosophila melanogaster* non-target. Toxicol Environ Chem 95(8):1369–1379. https://doi.org/10.1080/02772248.2014.880608

Rajkumar R, Shivakumar MS, Nathan SS, Selvam K (2018) Pharmacological and larvicidal potential of green synthesized silver nanoparticles using *Carmona retusa* (Vahl) Masam leaf extract. J Clust Sci 29(6):1243–1253. https://doi.org/10.1007/s10876-018-1443-x

Rakhimol KR, Sabu T, Volova T, Jayachandran K (2020) Controlled release of pesticides for sustainable agriculture. Springer, Cham

Raliya R, Tarafdar JC, Gulecha K, Choudhary K, Ram R, Mal P, Saran RP (2013) Scope of nanoscience and nanotechnology in agriculture. J Appl Biol Biotechnol 1:41–44. https://doi.org/10.7324/JABB.2013.1307

Ramyadevi J, Jeyasubramanian K, Marikani A, Rajakumar G, Rahuman AA, Santhoshkumar T, Kirthi AV, Jayaseelan C, Marimuthu S (2011) Copper nanoparticles synthesized by polyol process used to control hematophagous parasites. Parasitol Res 109(5):1403–1415. https://doi.org/10.1007/s00436-011-2387-3

Rattan RS (2010) Mechanism of action of insecticidal secondary metabolites of plant origin. Crop Prot 29(9):913–920. https://doi.org/10.1016/j.cropro.2010.05.008

Rawi SM, Al-Logmani AS, Hamza RZ (2019) Neurological alterations induced by formulated imidacloprid toxicity in Japanese quails. Metab Brain Dis 34(2):443–450. https://doi.org/10.1007/s11011-018-0377-1

Regnault-Roger C, Vincent C, Arnason JT (2012) Essential oils in insect control: low-risk products in a high-stakes world. Annu Rev Entomol 57:405–424. https://doi.org/10.1146/annurev-ento-120710-100554

Rein V, Meschkov A, Hagens K, Redinger N, Schepers U, Mehlhorn H, Schaible UE, Feldmann C (2019) Zirconyl hydrogenphosphate nanocontainers for flexible transport and release of lipophilic cytostatics, insecticides, and antibiotics. Adv Funct Mater 29(28):1900543. https://doi.org/10.1002/adfm.201900543

Roberts JR, Routt-Reigart J (2013) Recognition and management of pesticide poisonings, 6th edn. U.S. Environmental Protection Agency, Washington DC. https://www.epa.gov/sites/production/files/2015-01/documents/rmpp_6thed_final_lowresopt.pdf

Rocha AG, Oliveira BMS, Melo CR, Sampaio TS, Blank AF, Lima AD, Nunes RS, Araujo APA, Cristaldo PF, Bacci L (2018) Lethal effect and behavioral responses of leaf-cutting ants to essential oil of *Pogostemon cablin* (Lamiaceae) and its nanoformulation. Neotrop Entomol 47(6):769–779. https://doi.org/10.1007/s13744-018-0615-6

Rogers HA, Schmidt TS, Dabney BL, Hladik ML, Mahler BJ, Van Metres PC (2016) Bifenthrin causes trophic cascade and altered insect emergence in mesocosms: implications for small streams. Environ Sci 50(21):11974–11983. https://doi.org/10.1021/acs.est.6602761

Rouhani M, Samih MA, Kalantari S (2012) Insecticide effect of silver and zinc nanoparticles against *Aphis nerii* Boyer de Fonscolombe (Hemiptera: Aphidae). Chil J Agric Res 72(4):590–594. https://doi.org/10.4067/S0718-58392012000400020

Roush DK, McKenzie JA (1987) Ecological genetics of insecticide and acaricide resistance. Annu Rev Entomol 32:361–380. https://doi.org/10.1146/annurev.en.32.010187.002045

Sabry AH, Ragaei M (2018) Nanotechnology and their applications in insect's pest control. In: Abd-Elsalam KA, Prasad R (eds) Nanobiotechnology applications in plant protection book series: nanotechnology in the life sciences. Springer, Cham, pp 1–28. https://doi.org/10.1007/978-3-319-91161-8_1

Saeed N, Farooq M, Shakeel M, Ashraf M (2018) Effectiveness of an improved form of insecticide-based diatomaceous earth against four stored grain pests on different grain commodities. Environ Sci Pollut Res 25(17):17012–17024. https://doi.org/10.1007/s11356-018-1835-3

Saini P, Gopal M, Kumar R, Srivastava C (2014) Development of pyridalyl nanocapsule suspension for efficient management of tomato fruit and shoot borer (*Helicoverpa armigera*). J Environ Sci Health B 49(5):344–351. https://doi.org/10.1080/03601234.2014.882168

Sallam MN (2013) Insect damage – post harves operations. Food and Agriculture Organization of the United Nations. http://www.fao.org/3/a-av013e.pdf

Sánchez-Bayo F (2012) Ecological impacts of insecticides. In: Khan Perveen F (ed) Insecticides – advances in integrated pest management. InTech, Rijeka, pp 61–90

Sánchez-Bayo F, Wyckhuys KAG (2019) Worldwide decline of the entomofauna: a review of its drivers. Biol Conserv 232:8–27. https://doi.org/10.1016/j.biocon.2019.01.020

Santo-Orihuela PL, Foglia ML, Targovnik AM, Miranda MV, Desimone MF (2016) Nanotoxicological effects of $SiO_2$ nanoparticles on *Spodoptera frugiperda* Sf9 cells. Curr Pharm Biotechnol 17(5):465–470. https://doi.org/10.2174/1389201017051603031656O4

Santos JP, Maia JDG, Cruz I (1990) Damage to germination of seed corn caused by maize weevil (*Sitophilus zeamais*) and angoumois grain moth (*Sitotroga cerealella*). Pesquisa Agropecuaria Brasileira 25(12):1687–1692

Santos ACC, Cristaldo PF, Araujo APA, Melo CR, Lima APS, Santana EDR, de Oliveira BMS, Oliveira JWS, Vieira JS, Blank AF, Bacci L (2018) *Apis mellifera* (Insecta: Hymenoptera) in the target of neonicotinoids: a one-way ticket? Bioinsecticides can be an alternative. Ecotoxicol Environ Saf 163:28–36. https://doi.org/10.1016/j.ecoenv.2018.07.048

Sarkar DJ, Singh A (2017) Base triggered release of insecticide from bentonite reinforced citric acid crosslinked carboxymethyl cellulose hydrogel composites. Carbohydr Polym 156:303–311. https://doi.org/10.1016/j.carbpol.2016.09.045

Sarkar DJ, Singh A (2019) pH-triggered release of boron and thiamethoxam from boric acid crosslinked carboxymethyl cellulose hydrogel based formulations. Polymer Plast Tech Eng 58(1):83–96. https://doi.org/10.1080/03602559.2018.1466165

Sarkar DJ, Kumar J, Shakil NA, Walia S (2012) Release kinetics of controlled release formulations of thiamethoxam employing nano-ranged amphiphilic PEG and diacid based block polymers in soil. J Environ Sci Health A Tox Hazard Subst Environ Eng 47(11):1701–1712. https://doi.org/10.1080/10934529.2012.687294

Satehi AB, Ziaee M, Ashrafi A (2018) Silica nanoparticles: a potential carrier of chlorpyrifos in slurries to control two insect pests of stored products. Entomol Gen 37(1):77–91. https://doi.org/10.1127/entomologia/2017/0406

Saxena M, Sonkar SK, Sarkar S (2013) Water soluble nanocarbons arrest the growth of mosquitoes. RSC Adv 3(44):22504–22508. https://doi.org/10.1039/c3ra44100h

Schowalter TD, Noriega JA, Tscharntke T (2018) Insect effects on ecosystem services—introduction. Basic Appl Ecol 26:1–7. https://doi.org/10.1016/j.baae.2017.09.011

Seixas PTL, Demuner AJ, Alvarenga ES, Barbosa LCA, Marques A, Farias ED, Picanco MC (2018) Bioactivity of essential oils from *Artemisia* against *Diaphania hyalinata* and its selectivity to beneficial insects. Sci Agric 75(6):519–525. https://doi.org/10.1590/1678-992x-2016-0461

Shah MA, Khan AA (2014) Use diatomaceous earth for the management of stored-product pests. Int J Pest Manag 60(2):100–113. https://doi.org/10.1080/09670874.2014.918674

Shanmugasundaram T, Balagurunathan R (2015) Mosquito larvicidal activity of silver nanoparticles synthesised using actinobacterium, *Streptomyces* sp. M25 against *Anopheles subpictus*, *Culex quinquefasciatus* and *Aedes aegypti*. J Parasitol Dis 39(4):677–684. https://doi.org/10.1007/s12639-013-0412-4

Sharma A, Kumar S, Tripathi P (2019) A facile and rapid method for green synthesis of *Achyranthes aspera* stem extract-mediated silver nano-composites with cidal potential against *Aedes aegypti* L. Saudi J Biol Sci 26(4):698–708. https://doi.org/10.1016/j.sjbs.2017.11.001

Shen Y, Zhu HX, Cui JX, Wang AQ, Zhao X, Cui B, Wang Y, Cui HX (2018) Construction of lambda-cyhalothrin nano-delivery system with a high loading content and controlled-release property. Nano 8(12):1016. https://doi.org/10.3390/nano8121016

Shi GF, Zhan PF, Jin WM, Fei JM, Zhao LH (2017) Dietary $TiO_2$ particles modulate expression of hormone-related genes in *Bombyx mori*. Arch Insect Biochem Physiol 95(4):e21397. https://doi.org/10.1002/arch.21397

Shoaib A, Waqas M, Elabasy A, Cheng XL, Zhang QQ, Shi ZH (2018a) Preparation and characterization of emamectin benzoate nanoformulations based on colloidal delivery systems and use in controlling *Plutella xylostella* (L.) (Lepidoptera: Plutellidae). RSC Adv 8(28):15687–15697. https://doi.org/10.1039/c8ra01913d

Shoaib A, Elabasy A, Waqas M, Lin LL, Cheng XL, Zhang QQ, Shi ZH (2018b) Entomotoxic effect of silicon dioxide nanoparticles on *Plutella xylostella* (L.) (Lepidoptera: Plutellidae) under laboratory conditions. Toxicol Environ Chem 100(1):80–91. https://doi.org/10.1080/02772248.2017.1387786

Singh HB, Mishra S (2014) Biosynthesized silver nanoparticles as a nanoweapon against phytopathogens: exploring their scope and potential in agriculture. Appl Microbiol Biotechnol 99(3):1097–1107. https://doi.org/10.1007/s00253-014-6296-0

Singhand A, Gandhi S (2012) Agricultural insect pest: occurrence and infestation level in agricultural fields of Vadodara, Gujarat. Int J Sci Res Publ 2(4):1–5

Sittipummongkol K, Pechyen C (2018) Production, characterization and controlled release studies of biodegradable polymer microcapsules incorporating neem seed oil by spray drying. Food Packag Shelf Life 18:131–139. https://doi.org/10.1016/j.fpsl.2018.09.001

Slattery M, Harper B, Harper S (2019) Pesticide encapsulation at the nanoscale drives changes to the hydrophobic partitioning and toxicity of an active ingredient. Nano 9(1):81. https://doi.org/10.3390/nano9010081

Small T, Ochoa-Zapater MA, Gallelo G, Ribera A, Romero FM, Torreblanca A, Garcera MD (2016) Gold-nanoparticles ingestion disrupts reproduction and development in the German cockroach. Sci Total Environ 565:882–888. https://doi.org/10.1016/j.scitotenv.2016.02.032

Smith GR, Krishnamurthy SVB, Burger AC, Rettig JE (2018) Effects of malathion and nitrate exposure on the zooplankton community in experimental mesocosms. Environ Sci Pollut Res 25(10):9992–9997. https://doi.org/10.1007/s11356-018-1311-0

Soderlund DM (2012) Molecular mechanisms of pyrethroid insecticide neurotoxicity: recent advances. Arch Toxicol 86(2):165–181. https://doi.org/10.1007/s00204-011-0726-x

Solis M, Paracampo A, Bonetto C, Mugni H (2019) Acute toxicity of chlorpyrifos to *Hyalella curvispina*: comparison of species sensitivity and assessment of environmental risk. Environ Processes 6(1):107–117. https://doi.org/10.1007/s40710-019-00352-3

Soni N, Prakash S (2014) Green nanoparticles for mosquito control. Scientific World J 2014:496362. https://doi.org/10.1155/2014/496362

Sooresh A, Kwon F, Tayor R, Pietrantonio P, Pine M, Sayes CM (2011) Surface functionalization of silver nanoparticles: novel applications for insect vector control. ACS Appl Mater Interfaces 3(10):3779–3787. https://doi.org/10.1021/am201167v

Spochacz M, Chowanski S, Walkowiak-Nowicka K, Szymczak M, Adamski Z (2018) Plant-derived substances used against beetles-pests of stored crops and food-and their mode of action: a review. Compr Rev Food Sci Food Saf 17(5):1339–1366. https://doi.org/10.1111/1541-4337.12377

Stehle S, Schulz R (2015) Agricultural insecticides threaten surface waters at the global scale. Proc Natl Acad Sci U S A 112(18):5750–5755. https://doi.org/10.1073/pnas.1500232112

Stork NE (2018) How many species of insects and other terrestrial arthropods are there on earth? Annu Rev Entomol 63:31–45. https://doi.org/10.1146/annurev-ento-020117-043348

Sunderland MR, McNeil SJ (2017) Protecting wool carpets from beetle and moth larvae with nanocidal titanium dioxide desiccant. Clean Technol Environ 19(4):1205–1213. https://doi.org/10.1007/s10098-016-1297-6

Suresh U, Murugan K, Panneerselvam C, Rajaganesh R, Roni M, Aziz A, Al-Aoh HAN, Trivedi S, Rehman H, Kumar S, Higuchi A, Canale A, Benelli G (2018) *Suaeda maritima*-based herbal coils and green nanoparticles as potential biopesticides against the dengue vector *Aedes aegypti* and the tobacco cutworm *Spodoptera litura*. Physiol Mol Plant Pathol 101:225–235. https://doi.org/10.1016/j.pmpp.2017.01.002

Sutthanont N, Attrapadung S, Nuchprayoon S (2019) Larvicidal activity of synthesized silver nanoparticles from *Curcuma zedoaria* essential oil against *Culex quinquefasciatus*. Insects 10(1):27. https://doi.org/10.3390/insects10010027

Szoecs E, Brinke M, Karaoglan B, Schaefer RB (2017) Large scale risks from agricultural pesticides in small streams. Environ Sci Technol 51(13):7378–7385. https://doi.org/10.1021/acs.est.7b00933

Tang YY, Chen K, Li JC, Feng YH, Yu GB, Wang LZ, Zhao XY, Peng Y, Zhang Q (2018) Electrolyte and pH-sensitive amphiphilic alginate: synthesis, self-assembly and controlled release of acetamiprid. RSC Adv 8(56):32193–32199. https://doi.org/10.1039/c8ra05503c

Thanigaivel A, Vasantha-Srinivasan P, Edwin ES, Ponsankar A, Selin-Rani S, Chellappandian M, Kalaivani K, Senthil-Nathan S, Benelli G (2018) Development of an eco-friendly mosquitocidal agent from *Alangium salvifolium* against the dengue vector *Aedes aegypti* and its biosafety on the aquatic predator. Environ Sci Pollut Res 25(11):10340–10352. https://doi.org/10.1007/s11356-017-9102-6

Thomson LJ, MacFayden S, Hoffmann AA (2010) Predicting the effects of climate change on natural enemies of agricultural pests. Biol Control 52:296–306. https://doi.org/10.1016/j.biocontrol.2009.01.022

Touaylia S, Ali M, Abdellhafidh K, Bejaoui M (2019) Permethrin induced oxidative stress and neurotoxicity on the freshwater beetle *Laccophilus minutus*. Chem Ecol 35(5):459–471. https://doi.org/10.1080/02757540.2019.1567718

Tran TT, Janssens L, Dinh KV, Stoks R (2019) An adaptive transgenerational effect of warming but not of pesticide exposure determines how a pesticide and warming interact for antipredator behavior. Environ Pollut 245:307–315. https://doi.org/10.1016/j.envpol.2018.11.022

Tripathi AK, Upadhyay S, Bhuiyan M, Bhattacharya PR (2009) A review on prospects of essential oils as biopesticides in insect-pest management. J Pharmacogn Phytother 1(5):52–63

Troczka BJ, Willianson ML, Field LM, Davies TGE (2017) Rapid selection for resistance to diamide insecticides in *Plutella xylostella* via specific amino acid polymorphisms in the ryanodine receptor. Neurotoxicology 60:224–233. https://doi.org/10.1016/j.neuro.2016.05.012

Turek C, Stintzing FC (2013) Stability of essential oils: a review. Compr Rev Food Sci Food Saf 12(1):40–53. https://doi.org/10.1111/1541-4337.12006

Ullio-Gamboa G, Palma S, Benoit JP, Allemandi D, Picollo MI, Toloza AC (2017) Ivermectin lipid-based nanocarriers as novel formulations against head lice. Parasitol Res 116(8):2111–2117. https://doi.org/10.1007/s00436-017-5510-2

Upadhyay N, Singh VK, Dwivedy AK, Das S, Chaudhari AK, Dubey NK (2019) Assessment of *Melissa officinalis* L. essential oil as an eco-friendly approach against biodeterioration of wheat flour caused by *Tribolium castaneum* Herbst. Environ Sci Pollut Res 26(14):14036–14049. https://doi.org/10.1007/s11356-019-04688-z

Van den Noortgate H, Lagrain B, Sree SP, Kerkhofs S, Wenseleers T, Martens JA (2018) Material properties determining the insecticidal activity of highly divided porous materials on the pharaoh ant (*Monomorium pharaonis*). Pest Manag Sci 74(6):1374–1385. https://doi.org/10.1002/ps.4814

Van den Noortgate H, Sree SP, Ostyn N, Lagrain B, Roeffaers M, Wenseleers T, Martens JA (2019) Material properties determining insecticidal activity of activated carbon on the pharaoh ant (*Monomorium pharaonis*). J Pest Sci 92(2):643–652. https://doi.org/10.1007/s10340-018-1024-7

Varga L, Fedor PJ (2008) First interception of the greenhouse pest *Echinothrips americanus* Morgan, 1913 (Thysanoptera: Thripidae) in Slovakia. Plant Protect Sci 44:155–158. https://doi.org/10.17221/29/2008-PPS

Vieira CED, Martinez CBD (2018) The pyrethroid λ-cyhalothrin induces biochemical, genotoxic, and physiological alterations in the teleost *Prochilodus lineatus*. Chemosphere 210:958–967. https://doi.org/10.1016/j.chemosphere.2018.07.115

Vinoth S, Shankar SG, Gurusaravanan P, Janani B, Devi JK (2019) Anti-larvicidal activity of silver nanoparticles synthesized from *Sargassum polycystum* against mosquito vectors. J Clust Sci 30(1):171–180. https://doi.org/10.1007/s10876-018-1473-4

Volpato A, Baretta D, Zortea T, Campigotto G, Galli GM, Glombowsky P, Santos RCV, Quatrin PM, Ourique AF, Baldissera MD, Stefani LM, Da Silva AS (2016) Larvicidal and insecticidal effect of *Cinnamomum zeylanicum* oil (pure and nanostructured) against mealworm (*Alphitobius diaperinus*) and its possible environmental effects. J Asia Pac Entomol 19(4):1159–1165. https://doi.org/10.1016/j.aspen.2016.10.008

Waissi-Leinonen GC, Nybom I, Pakarinen K, Akkanen J, Leppanen MT, Kukkonen JVK (2015) Fullerenes ($nC_{60}$) affect the growth and development of the sediment-dwelling invertebrate *Chironomus riparius* larvae. Environ Pollut 206:17–23. https://doi.org/10.1016/j.envpol.2015.06.010

Walker GW, Kookana RS, Smith NE, Kah M, Doolette CL, Reeves PT, Lovell W, Anderson DJ, Turney TW, Navarro DA (2018) Ecological risk assessment of nano-enabled pesticides: a perspective on problem formulation. J Agric Food Chem 66(26):6480–6486. https://doi.org/10.1021/acs.jafc.7b02373

Walther GR, Post E, Convey P, Menzel A, Parmesan C, Beebee TJC, Fromentin JM, Heogh-Guldberg O, Bairlein F (2002) Ecological responses to recent climate change. Nature 416:389–395

Walvekar VA, Bajaj S, Singh DK, Sharma S (2017) Ecotoxicological assessment of pesticides and their combination on rhizospheric microbial community structure and function of *Vigna radiata*. Environ Sci Pollut Res 24(20):17175–17186. https://doi.org/10.1007/s11356-017-9284-y

Wang Y, Gao ZD, Shen F, Li Y, Zhang SN, Ren XQ, Hu SW (2015a) Physicochemical characteristics and slow release performances of chlorpyrifos encapsulated by poly(butyl acrylate-co-styrene) with the cross-linker ethylene glycol dimethacrylate. J Agric Food Chem 63(21):5196–5204. https://doi.org/10.1021/acs.jafc.5b01378

Wang L, Su MY, Zhao XY, Hong J, Yu XH, Xu BQ, Sheng L, Liu D, Shen WD, Li B, Hong F (2015b) Nanoparticulate $TiO_2$ protection of midgut damage in the silkworm (*Bombyx mori*) following phoxim exposure. Arch Environ Contam Toxicol 68(3):534–542. https://doi.org/10.1007/s00244-014-0121-8

Wang Y, Wang AQ, Wang CX, Cui B, Sun CJ, Zhao X, Zeng ZH, Shen Y, Gao F, Liu GQ, Cui H (2017) Synthesis and characterization of emamectin-benzoate slow-release microspheres with different surfactants. Sci Rep 7:12761. https://doi.org/10.1038/s41598-017-12724-6

Wang CX, Cui B, Guo L, Wang AQ, Zhao X, Wang Y, Sun CJ, Zeng ZH, Zhi H, Chen HY, Liu GQ, Cui HX (2019a) Fabrication and evaluation of lambda-cyhalothrin nanosuspension by one-step melt emulsification technique. Nano 9(2):145. https://doi.org/10.3390/nano9020145

Wang XP, Xie HC, Wang ZY, He KL, Jing DP (2019b) Graphene oxide as a multifunctional synergist of insecticides against lepidopteran insect. Environ Sci-Nano 6(1):75–84. https://doi.org/10.1039/c8en00902c

Wang Z, Brooks BW, Zeng EY, You J (2019c) Comparative mammalian hazards of neonicotinoid insecticides among exposure durations. Environ Int 125:9–24. https://doi.org/10.1016/j.envint.2019.01.040

White J, Johnson D (2012) Vendors of microbial & botanical insecticides & insect monitoring devices. ENTFACT-124. College of Agriculture, University of Kentucky. https://entomology.ca.uky.edu/ef124

WHO (2015) World Malaria Report 2015. https://www.who.int/malaria/publications/world-malaria-report-2015/report/en/

WHO (2017) Vector-borne diseases. https://www.who.int/news-room/fact-sheets/detail/vector-borne-diseases

Wieczorek MV, Bakanov N, Bilancia D, Szoecs E, Stehle S, Bundschuh M, Schulz R (2018) Structural and functional effects of a short-term pyrethroid pulse exposure on invertebrates in outdoor stream mesocosms. Sci Total Environ 610:810–819. https://doi.org/10.1016/j.scitotenv.2017.08.048

Wolfram J, Stehle S, Bub S, Petschick LL, Schulz R (2018) Meta-analysis of insecticides in United States surface waters: status and future implications. Environ Sci Technol 52(24):14452–14460. https://doi.org/10.1021/acs.est.8b04651

Xiang YB, Zhang GL, Chen CW, Liu B, Cai DQ, Wu ZY (2018) Fabrication of a pH-responsively controlled-release pesticide using an attapulgite-based hydrogel. ACS Sustain Chem Eng 6(1):1192–1201. https://doi.org/10.1021/acssuschemeng.7b03469

Yan HQ, Chen XQ, Feng YH, Xiang F, Li JC, Shi ZF, Wang XH, Lin Q (2016) Modification of montmorillonite by ball-milling method for immobilization and delivery of acetamiprid based on alginate/exfoliated montmorillonite nanocomposite. Polym Bull 73(4):1185–1206. https://doi.org/10.1007/s00289-015-1542-x

Yang DB, Li GX, Yan XJ, Yuan HZ (2014) Controlled release study on microencapsulated mixture of fipronil and chlorpyrifos for the management of white grubs (*Holotrichia parallela*) in peanuts (*Arachis hypogaea* L.). J Agric Food Chem 62(44):10632–10637. https://doi.org/10.1021/jf502537x

Yang Y, Wu NX, Wang CL (2018) Toxicity of the pyrethroid bifenthrin insecticide. Environ Chem Lett 16(4):1377–1391. https://doi.org/10.1007/s10311-018-0765-0

Yusoff SNM, Kamari A (2018) *N*-deoxycholic acid-*O*-glycol chitosan as a potential carrier agent for botanical pesticide rotenone. J Appl Polym Sci 135(47):46855. https://doi.org/10.1002/app.46855

Zacharia JT (2011) Ecological effects of pesticides. In: Stoytcheva M (ed) Pesticides in the modern world – risks and benefits. InTech, Rijeka, pp 129–142

Zandi-Sohani N, Rajabpour A, Yarahmadi F, Ramezani L (2018) Sensitivity of *Bemisia tabaci* (Hemiptera: Aleyrodidae) and the generalist predator *Orius albidipennis* (Hemiptera: Anthocoridae) to vapors of essential oils. J Entomol Sci 53(4):493–502. https://doi.org/10.18474/JES17-113.1

Zaworra M, Koehler H, Schneider J, Lagojda A, Nauen R (2019) Pharmacokinetics of three neonicotinoid insecticides upon contact exposure in the Western honey bee, *Apis mellifera*. Chem Res Toxicol 32(1):35–37. https://doi.org/10.1021/acs.chemrestox.8b00315

Zhang Z, Chi DF, Yu J (2011) Preparation, characterization and properties evaluation of buprofezin nanocomposites obtained by polyelectrolytes assembly. Adv Mater Res 183-185:1677–1681. https://doi.org/10.4028/www.scientific.net/AMR.183-185.1677

Zhang JK, Li M, Fan TF, Xu Q, Wu Y, Chen CY, Huang QL (2013) Construction of novel amphiphilic chitosan copolymer nanoparticles for chlorpyrifos delivery. J Polym Res 20(3):107. https://doi.org/10.1007/s10965-013-0107-7

Zhang H, Qin H, Li LX, Zhou XT, Wang W, Kan CY (2018) Preparation and characterization of controlled-release avermectin/castor oil-based polyurethane nanoemulsions. J Agric Food Chem 66(26):6552–6560. https://doi.org/10.1021/acs.jafc.7b01401

Zhang YF, Chen W, Jing MM, Liu SZ, Feng JT, Wu H, Zhou YW, Zhang X, Ma ZQ (2019) Self-assembled mixed micelle loaded with natural pyrethrins as an intelligent nano-insecticide with a novel temperature-responsive release mode. Chem Eng J 361:1381–1391. https://doi.org/10.1016/j.cej.2018.10.132

Zhao X, Cui H, Wang Y, Sun C, Cui B, Zeng Z (2018a) Development strategies and prospects of nano-based smart pesticide formulation. J Agric Food Chem 66:6504–6512. https://doi.org/10.1021/acs.jafc.7b02004

Zhao XY, Li JC, Feng YH, Yu GB, Zhou QF, He FR, Xiao DC, Chen K, Zhang L (2018b) Self-aggregation behavior of hydrophobic sodium alginate derivatives in aqueous solution and their application in the nanoencapsulation of acetamiprid. Int J Biol Macromol 106:418–424. https://doi.org/10.1016/j.ijbiomac.2017.08.038

Zhou MS, Xiong ZC, Yang DJ, Pang YX, Wang DP, Qiu XQ (2018) Preparation of slow release nanopesticide microspheres from benzoyl lignin. Holzforschung 72(7):599–607. https://doi.org/10.1515/hf-2017-0155

Zhou MS, Wang DP, Yang DJ, Qiu XQ, Li YX (2019) Avermectin loaded nanosphere prepared from acylated alkali lignin showed anti-photolysis property and controlled release performance. Ind Crop Prod 137:453–459. https://doi.org/10.1016/j.indcrop.2019.04.037

Ziaee M, Ganji Z (2016) Insecticidal efficacy of silica nanoparticles against *Rhyzopertha dominica* F. and *Tribolium confusum* Jacquelin du Val. J Plant Prot Res 56(3):250–256. https://doi.org/10.1515/jppr-2016-0037

Ziaee M, Moharramipour S, Mohsenifar A (2014) MA-chitosan nanogel loaded with *Cuminum cyminum* essential oil for efficient management of two stored product beetle pests. J Pest Sci 87(4):91–699. https://doi.org/10.1007/s10340-014-0590-6

Zou AH, Yang Y, Cheng JG, Garamus VM, Li N (2018) Construction and characterization of a novel sustained-release delivery system for hydrophobic pesticides using biodegradable polydopamine-based microcapsules. J Agric Food Chem 66(25):6262–6268. https://doi.org/10.1021/acs.jafc.8b00877

# Environmental Toxicity of Nanopesticides Against Non-Target Organisms: The State of the Art

**Francine Côa, Leandro S. Bortolozzo, Romana Petry, Gabriela H. Da Silva, Carlos H. Z. Martins, Aline M. Z. de Medeiros, Carla M. S. Sabino, Rubens S. Costa, Latif U. Khan, Fabrício S. Delite, and Diego Stéfani T. Martinez**

**Abstract** Nanopesticides is a promising technology for agricultural productivity and innovation. The emergent field of nanopesticides associated with their safety evaluation provides an excellent opportunity for scientists, industry, policy makers, and civil society to interact and share their experiences regarding sustainable agriculture and environmental protection. In this chapter, we present an overview of the state of the art concerning the environmental toxicity of nanopesticides against non-target model organisms, such as soil microorganisms, plants, worms, insects, algae, daphnids, and fishes. Advanced characterization methods are also explored for better understanding the interactions of nanopesticides with complex biological and

---

F. Côa (✉) · A. M. Z. de Medeiros
Brazilian Nanotechnology National Laboratory (LNNano), Brazilian Center for Research in Energy and Materials (CNPEM), Campinas, São Paulo, Brazil

Center for Nuclear Energy in Agriculture (CENA), University of São Paulo (USP), Piracicaba, São Paulo, Brazil
e-mail: francinecoa@usp.br

L. S. Bortolozzo
Brazilian Nanotechnology National Laboratory (LNNano), Brazilian Center for Research in Energy and Materials (CNPEM), Campinas, São Paulo, Brazil

School of Technology, University of Campinas (UNICAMP), Limeira, São Paulo, Brazil

R. Petry
Brazilian Nanotechnology National Laboratory (LNNano), Brazilian Center for Research in Energy and Materials (CNPEM), Campinas, São Paulo, Brazil

Center for Natural and Human Sciences, Federal University of ABC (UFABC), Santo André, São Paulo, Brazil

G. H. Da Silva · L. U. Khan · F. S. Delite
Brazilian Nanotechnology National Laboratory (LNNano), Brazilian Center for Research in Energy and Materials (CNPEM), Campinas, São Paulo, Brazil

C. H. Z. Martins
Laboratory of Solid State Chemistry (LQES), Institute of Chemistry, University of Campinas (UNICAMP), Campinas, São Paulo, Brazil


L. F. Fraceto et al. (eds.), *Nanopesticides*,
https://doi.org/10.1007/978-3-030-44873-8_8

environmental systems. Considering that nanopesticides is a global technology, we highlight the nanoinformatics as an emerging and fundamental approach for international harmonization of nanoecotoxicity protocols, safety-by-design, environmental risk analysis, and management of nanopesticides towards responsible development and regulation.

**Keywords** Nanoecotoxicology · Nanosafety · Nanomaterials · Environment Pesticides

## 1 Nanopesticides: Agricultural, Environmental, and Safety Aspects

Recent advances in applications of engineered nanomaterials (NMs) in agriculture have attracted a research interest to develop novel nanopesticides. Overall, the use of nanopesticides can improve the pesticide efficacy, controlled release, and delivery to the targeted pathogens. Consequently, nanopesticides use can increase the global food production, safety, and security (He et al. 2019). Conventional pesticides have offered several problems regarding their formulations and usages, such as low solubility of active ingredients (AIs) in water, non-selectivity, and uncontrolled release (Kah and Hofmann 2014; Sarlak et al. 2014). It has been reported that about 0.1% of pesticides are delivered to the target, while 99.9% are empty in the surrounding environment (Pimentel 1995; Özkara et al. 2016). This poor delivery efficiency results in several hazardous effects, such as water and soil pollution, increased pest, pathogen resistance, and loss of biodiversity.

C. M. S. Sabino
Brazilian Nanotechnology National Laboratory (LNNano), Brazilian Center for Research in Energy and Materials (CNPEM), Campinas, São Paulo, Brazil

Laboratory of Advanced Functional Materials (LaMFA), Department of Physics, Federal University of Ceará (UFC), Fortaleza, Ceará, Brazil

R. S. Costa
Brazilian Nanotechnology National Laboratory (LNNano), Brazilian Center for Research in Energy and Materials (CNPEM), Campinas, São Paulo, Brazil

Interdisciplinary Laboratory of Advanced Materials, Federal University of Piauí (UFPI), Teresina, Piauí, Brazil

Federal Institute of Education, Science and Technology of Piauí (IFPI), Oeiras, Piauí, Brazil

D. S. T. Martinez (✉)
Brazilian Nanotechnology National Laboratory (LNNano), Brazilian Center for Research in Energy and Materials (CNPEM), Campinas, São Paulo, Brazil

Center for Nuclear Energy in Agriculture (CENA), University of São Paulo (USP), Piracicaba, São Paulo, Brazil

School of Technology, University of Campinas (UNICAMP), Limeira, São Paulo, Brazil
e-mail: diego.martinez@lnnano.cnpem.br

Nanotechnology can undoubtedly play a crucial role in improving this delivery efficiency of pesticides (Kookana et al. 2014). The application of nanotechnology in agriculture has suddenly increased for the past few years. However, the prosperous industrial applications of nanopesticides request considerable research attention.

Nanopesticides can be defined as the agrochemicals that contain engineered nanomaterials as active ingredients either as a whole or part of the respective nanostructures that present biocidal properties. A range of nanopesticides formulation types have been suggested in experimental literature, including nanomicelles, nanocapsules (e.g., with polymers), mesoporous silica, and products containing inorganic nanoparticles (e.g., metals and metal oxides), as illustrated in Fig. 1. These nanoscale products are under extensive research investigations and they show potential to improve the efficiency of traditional pesticides with enhanced environmental safety.

Implementing environmentally friendly practices have been becoming increasingly essential to success in today's nanotechnology businesses. However, the understanding of the environmental impacts of nanopesticides is still an unknown that makes the field of research and development of nanomaterials an inspiring sector to determine the toxicological relationships that these materials can present to the human health and the environment (Adisa et al. 2019).

Ecotoxicology is a multidisciplinary science dedicated to investigating the exposure and adverse effects of chemicals on the organisms and the environment. It is a relatively new study field that emerged following the publication of Silent Spring. In this book, Rachel Carson pointed out, for the first time in history, the undesirable effects of indiscriminate use of pesticides in the 1940s and 1950s. Nanoecotoxicology and nanosafety are emerging scientific disciplines associated with ecotoxicology and responsible for understanding the interactions of nanoscale materials with biological and environmental systems towards responsible innovation, sustainability, safety, and regulation. Currently, there is an increasing number of publications

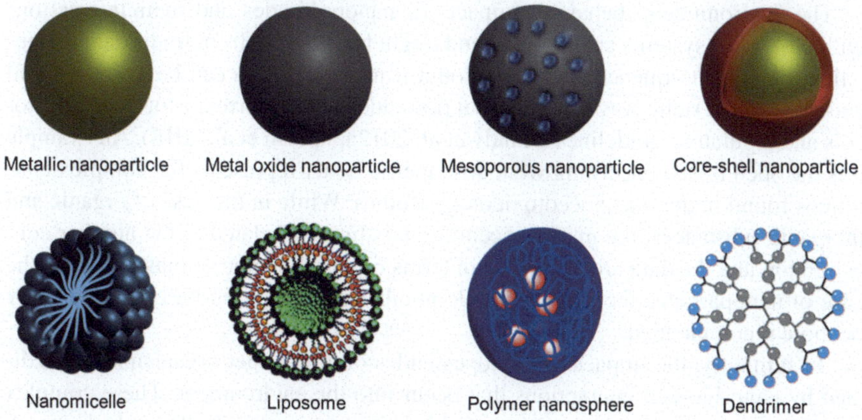


**Fig. 1** Schematic illustration of inorganic and organic nanoparticles used in nanopesticides formulations

reporting the effects of nanoparticles on several biological models, especially focusing on inorganic nanoparticles (e.g. $SiO_2$, AgNPs, CuO, $CeO_3$, etc.) and carbon-based nanomaterials. However, it is missing a consensus about the nanomaterials toxicity and their impacts on earth systems (Klaine et al. 2008; Valsami-Jones and Lynch 2015; Hochella et al. 2019); therefore, it is very important to continue the evaluations and ecotoxicological studies of nanopesticides towards safety applications in agriculture.

## 2 The Importance of Non-Target Organisms in Nanoecotoxicology

The emergency of nanopesticides as a new technology in agriculture field represents an important strategy for controlling several pests. However, their indiscriminate use can trigger an increase of nanopesticides level in the environment, resulting in the exposure of non-target organisms.

Non-target model organisms have potentially contributed to risk assessment of various NPs and to decipher their ecotoxicity mechanisms at nano-bio-ecology interface. The use of these organisms is ethically less controversial than other in vivo invertebrate and vertebrate models and offers several and unique advantages from technical points, such as low maintenance cost and easily handling, short life span and fast offspring turnover, distinct life stages, relatively simple fully sequenced genome, high immunological, structural, and genetic homology with mammals (Milošević et al. 2013; Vecchio 2015; Al Naggar 2016; Horch et al. 2017; Meng et al. 2017). They have a worldwide distribution and are exposed to multiple natural and anthropogenic stressors. Accordingly, researches have adopted an integrative and multidisciplinary approach to investigate the interactions of NPs with the biological systems of these non-target organisms.

The environmental behavior, impacts of nanopesticides and their interactions with biological systems are complex and might be different from traditional formulations. As a consequence, there is a doubt if nanopesticides can be evaluated and classified in the same way as traditional pesticides within current ecotoxicity protocols and regulatory guidelines (Handy et al. 2012a, b; Kah et al. 2018). An example it is the open issue concerning what dose metric better represents the complex conditions found in the nananoecotoxicology studies. While in the case of organic and inorganic substances, the mass concentration is linearly related to the number concentration and the data are expressed in terms of mass/volume or mass/mass, in the case of nanoparticles the use of particle number concentration-based metrics may be more recommended.

In summary, the impacts of nanopesticides on non-target organisms are mediated by nano-bio-eco interactions that occur into the environment. These complex interactions are governed by biotic and abiotic factors, such as biological, chemical, and physical transformations, which modify the physicochemical properties and colloidal behavior of the nanoparticles (He et al. 2018), as demonstrated in Fig. 2.

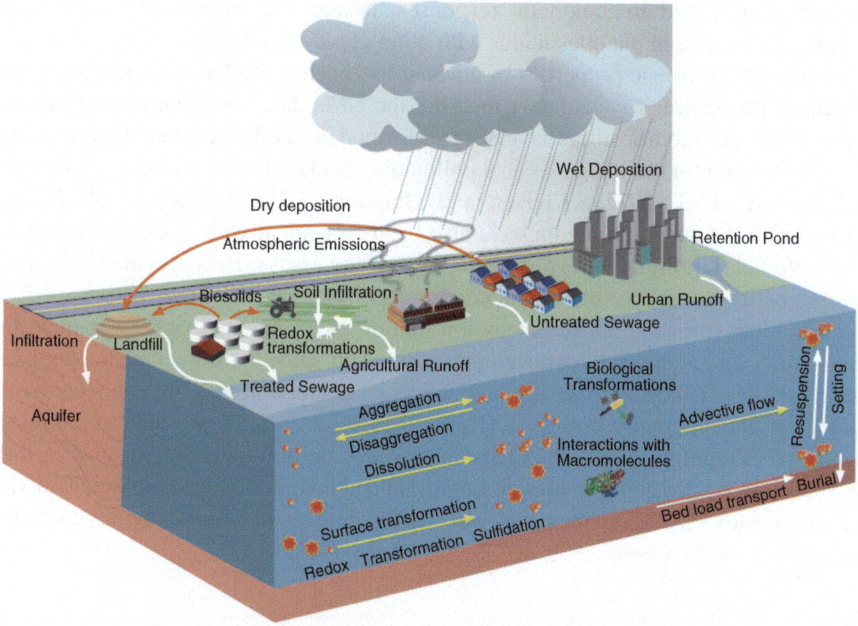


**Fig. 2** Transformations that nanomaterials can suffer into the environment. Reprinted with permission from Dale et al. (2015). Copyright (2019) American Chemical Society

The transformations of NMs involve oxidation-reduction reactions that can be catalyzed by sunlight action (photooxidation and photoreduction), leading to the NMs dissolution and/or sulfidation. In addition, NMs can be physically transformed by homoaggregation and heteroaggregation processes that disturb their colloidal stability, reducing their surface area and reactivity. Thus, these phenomena can induce the NMs sedimentation, extending their persistence in the environment. The surface of NMs can be coated with naturally occurring biomacromolecules and/or geomacromolecules, such as proteins, lipids, and humic substances. This coating, known as biomolecule corona, results in a new biological identity which will govern the nanoparticle-organism interactions (Markiewicz et al. 2018). Furthermore, organic and inorganic ligands (i.e., environmental pre-existing contaminants) can either be attached on the NMs or only combined with them, leading to joint and unexpected toxicity effects. All these transformation processes will impact on the nanopesticides environmental behavior, fate, and toxicity, making it difficult to predict the toxic effects of such substances (Deng et al. 2017).

Since nanopesticides may be transformed into the environment, it is also recommended monitoring the long-term effects of these substances. According to Mancini et al. (2019), this knowledge is crucial for understanding the real impacts on non-target populations and also providing feedback for review of licensing conditions in a post-approval context. This assessment would propitiate meaningful information, not only about the chronic effects of nanopesticides, but also regarding toxicodynamic and toxicokinetic responses, besides the unexpected synergistic interactions between nanopesticides with other classical pollutants.

Due to the intrinsic properties of NMs, it has been a challenge to develop standardized procedures to accurately measure the NMs damage on non-target organisms. Technical modifications have been performed in standard ecotoxicity protocols (i.e., developed for studying classical pollutants) to adapt for the nanomaterials reality (Handy et al. 2012a; Kleiven and Oughton 2015; Petersen et al. 2015). Adaptations of culturing media have also been strategically conducted to improve the colloidal stability of NMs during the ecotoxicity assays (Brinke et al. 2011), and also to better reflect the soil natural conditions (Tyne et al. 2013).

It is clear from the above discussion that an environmental risk assessment is mandatory for nanopesticides development lined up with environmental, health, and safety issues. Likewise, the harmonization of methodologies in nanoecotoxity assays is fundamentally necessary (Kah 2015; Baun et al. 2017). Due to the relevance of evaluating the effects of nanopesticides on non-target organisms, in the next section we are going to present important and commonly used biological models for ecotoxicity assessment, as well as the impact of nanopesticides already reported in the literature.

# 3 Ecotoxicity of Nanopesticides on Non-Target Organisms

## *3.1 Terrestrial Organisms*

### 3.1.1 Soil Microbial Communities

Soil microbial communities composed of archaea, bacteria, and fungi play a crucial role in the soil ecosystem by maintaining its functional integrity. They are responsible for the soil structure formation, decomposition of organic matter, and the nitrogen, carbon, sulfur, and phosphorus biogeochemical cycles. Moreover, the microbial biomass is a food source for other terrestrial animals (Varma and Buscot 2005).

Microorganisms can biodegrade and biotransform pollutants when exposed to environmental contaminants. However, they can be negatively impacted and react reducing the soil nutrients and the availability of organic carbon supplies. For these reasons, soil microbial communities are considered relevant bioindicators to assess perturbations on soil functioning, as the impacts of nanopesticides on terrestrial environment (Holden et al. 2014).

Notably, bacteria is one of the major risk group among soil microbial communities because the metallic composition of some NMs leads to antimicrobial effects (Kim et al. 2007). This characteristic was observed by Sillen et al. (2015) that exposed bacteria and fungi to AgNPs for 75 days and noticed that bacterial communities were more affected than fungi due to the silver dissolution. Xu and Zhang (2018) studied the long-term effects of AgNPs and observed a decrease of microbial community, as well as in the abundance of total bacteria and specific microbes responsible for nitrogen cycling. Samarajeewa et al. (2017) evaluated the effects of AgNPs through multiple endpoints (i.e., heterotrophic plate counting, microbial respiration, organic matter decomposition, soil enzyme activity, biological nitrification, community level physiological profiling, Ion Torrent™ DNA sequencing and denaturing gradient gel electrophoresis). And they reported that AgNPs affected the microbial growth over a concentration range from 49 to 1815 mg Ag $kg^{-1}$. On the other hand, at low concentrations (49 mg $kg^{-1}$), a stimulatory effect (hormesis) in the nitrification process was observed, and the emergence of a silver-tolerant bacteria (*Rhodanobacter* sp.) was reported for the first time in literature.

Monitoring the enzymatic activity is an excellent alternative to evaluate soil and microbiota health since the increase or inhibition of some enzymes can be related to the induction of disturbances in this environment (Utobo and Tewari 2015). Simonin et al. (2016b) applied $TiO_2$ NPs in agricultural soil (silty-clay texture) and by nitrification and denitrification enzymatic activities, they identified alterations on the nitrogen cycle and in the bacterial community structure. In a study performed by Liu et al. (2014), it was demonstrated that enzymatic activity monitoring is a useful strategy to differentiate the effects resulting from single or combined exposure. These authors identified that $Fe_3O_4$ magnetic nanoparticles combined with Diuron cause more inhibition of the metabolic activity of microorganisms than NPs or herbicide themselves. For heavy metals, Jośko et al. (2019) observed that intercellular enzymes were more sensitive than extracellular enzymes. Also, they noticed a stimulatory effect of dehydrogenase activity in three different soils on the first day of nano-ZnO and nano-CuO incubation. However, after 730 days, the enzymatic activities of two soils exposed to NPs were similar in comparison with control, while in another soil, enzymes were inhibited by nano-CuO and stimulated by nano-ZnO.

In fact, the effects of NMs on soil microbiota should be studied case-by-case, since they may be soil-type dependent. Frenk et al. (2013) identified that the effects of $Fe_3O_4$ and CuO NPs on the bacterial community were more significant in soils composed of low organic matter and clay fraction amounts. Besides, they observed that a relevant bacterium group (*Rhizobiales*) responsible for nitrogen fixation was more affected by CuO NPs than $Fe_3O_4$ NPs exposure. Gómez-Sagasti et al. (2019) also reported the soil-dependent toxicity of zero-valent iron (nZVI). While in clay-loam soil no inhibitory effect on microbiota was observed, in a sandy-loam soil, the arylsulfatase activity, bacterial biomass, richness, and diversity of microbiota were negatively impacted. These results suggested that the high clay and organic matter content resulted in the formation of biomolecule corona on nZVI NPs surface that hindered the contact of microorganisms with the nZVI NPs.

Although great efforts have been made to predict the effects of nanopesticides on the terrestrial environment, there are still some issues to be further investigated. Repeated applications of nanopesticides should be considered during the studies, since Simonin et al. (2016a) demonstrated a reduction of archaea and bacteria populations associated with the nitrification and ammonia-oxidizing processes, due to repeated exposure of soil to $TiO_2$ NPs. Other environmentally relevant aspects need to be explored, such as the potential risk of the environmental mixture of nanopesticides with traditional pesticides or classical pollutants. Parada et al. (2019), for instance, reported an increase of atrazine persistence on soil due to its interaction with copper nanoparticles (from 6 to 37 days).

Herein, we demonstrate that the nanopesticides can impact the soil microbial communities, since these materials are susceptible to be accumulated in the topsoil, which is the most active microbial zone of this environment (Simonin et al. 2016a). Therefore, understanding the NPs toxicity on soil microbiome is imperative for maintaining the soil functionality and avoiding indirect effects in the ecosystem.

### 3.1.2 Worms

Worms are indispensable components of terrestrial environment since they transform living and dead organic material in soil organic matter and nutrients. They create pore structures on the ground, which are habitats for other organisms, and facilitate soil aeration, plant roots penetration, and water infiltration. Thus, worms are essential to improve the soil nutrient availability, water-holding capacity, and agricultural productivity (Edwards 2004).

Worms are interesting bioindicators of soil environmental pollution (Jager et al. 2005). They have been applied as model organisms in ecotoxicity studies because the assays with worms present experimental simplicity, efficacy, and reduced cost (Doke and Dhawale 2015; Hunt 2017; Gomes et al. 2019). *Eisenia fetida, Enchytraeus albidus,* and *Enchytraeus crypticus* are some of interesting non-target species within this context; *E. fetida,* for example, is a model employed in ecotoxicity tests based on the principles described by the Organization for Economic Cooperation and Development (OECD). Mohd Firdaus et al. (2018) identified the potential effects of nanoencapsulated bifenthrin systems and compared these with the traditional bifenthrin using *E. fetida* and *Lumbricus terrestris* worms as biological models. In this study, they found that worms exposed to nanoformulations accumulated approximately 50% more bifenthrin than those exposed to traditional bifenthrin. Although nanoencapsulated bifenthrin had quickly absorbed by worms, it was found more bifenthrin in the gut. Then, they concluded that the internalized concentration was lower in the whole organism exposed to nanoencapsulated bifenthrin than that exposed to analytical grade substance. Therefore, this study points out the fact that nanoformulations may increase the efficacy of traditional pesticides by prolonging their lifetime and increasing the oral ingestion. In fact, bifenthrin nanoformulated can show risks to birds and mammals that feed on earthworms, indicating important aspects to be considered during risk evaluation and trophic ecological interactions studies (Mohd Firdaus et al. 2018).

Gomes et al. (2019) investigated the effects of atrazine nanoformulation (nano_ATZ) on *Enchytraeus crypticus*, an invertebrate used as a standard species in toxicological studies (OECD 2016). The results of the nanoformulation were compared with the commercial formulation (Gesaprim®) and atrazine (ATZ). Toxicity endpoints were evaluated through the whole life cycle of *E. crypticus* (i.e., hatching, growth, survival, and reproduction) over a concentration range of 1–400 mg atrazine per kg soil. In terms of hatching, nano_ATZ and ATZ caused similar effects

($EC_{50}$ = 218 mg nano_ATZ per kg and $EC_{50}$ = 208 mg ATZ per kg), with a significant reduction, approximately 40% in relation to control, at 200 mg ATZ per kg. For Gesaprim®, there was more significant variability in response, with the greatest reduction of hatching occurring at 100 mg $kg^{-1}$. Nano_ATZ did not affect adult survival, although there was a reduction in the number of juveniles. In conclusion, this study highlighted that nano_ATZ and pure ATZ were more toxic to *E. crypticus* than the commercial formula Gesaprim® (Gomes et al. 2019).

*Caenorhabditis elegans* model has also been used for studying the potential damage of NMs to non-target organisms (Hunt 2017). It is a transparent free-living nematode found in the soil liquid-phase that presents a short reproductive and life cycle, which enables its use in different types of toxicological assays, including high throughput screening that is more limited in complex animals (Gonzalez-Moragas et al. 2015). The effects of NMs on *C. elegans* have been studied in short and prolonged assays, following oral exposure, topical applications, or microinjection to specific organs (Wu et al. 2019). Several endpoints have been measured, such as lethality, reproduction, fertility, and growth, as well as molecular biomarkers. The primary target organs have been evaluated, since some nanomaterials can enter in the digestive system, from the pharyngeal lumen to the lumen and rectum of the intestine, as well as secondary target organs, which include muscle, neurons, spermatheca, and gonads, passing through the intestinal barrier or the active intestinal cell transport system in *C. elegans*, as demonstrated in Fig. 3 (Wu et al. 2013).

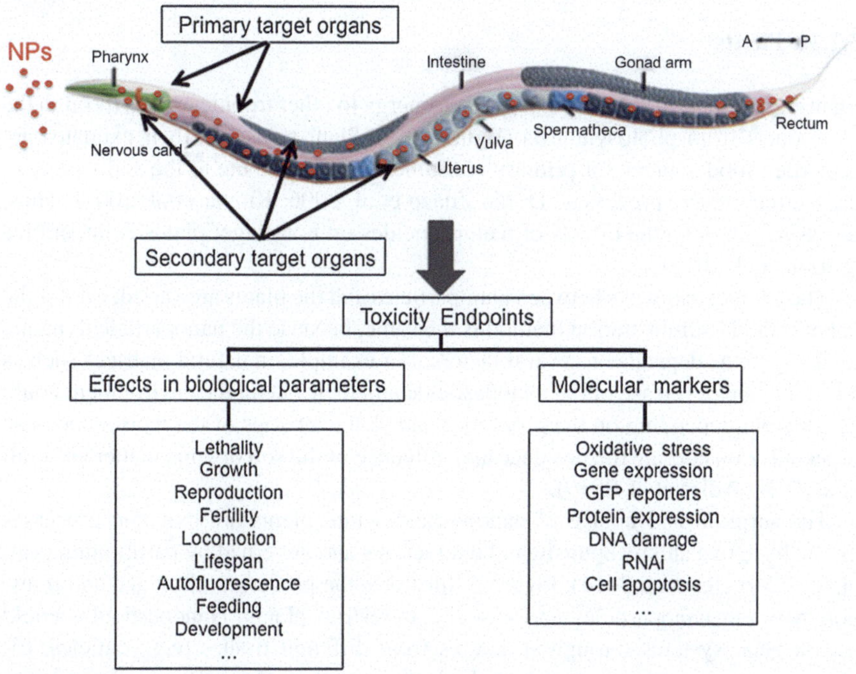


**Fig. 3** Representation of the major organs and endpoints that the *C. elegans* nematode may contribute to the toxicological evaluation of nanopesticides. Reprinted with permission from Wu et al. (2019)

Jaques et al. (2017) evaluated the toxicity of three different nanoparticles associated with conventional pesticides (i.e., atrazine, simazine, and paraquat) on *C. elegans* model. They found that only the NPs and pesticide-associated NPs were toxic at the highest concentrations tested. Toxicity studies of ZnO NPs in association with artificial soil sediment against *C. elegans* showed that these NPs affected the growth and reproduction of the worm, as well as to its locomotive behavior (frequency of body and head curves) and ATP levels. Moreover, significant increases in intracellular reactive oxygen species and lipid peroxidation were induced by long-term exposure to ZnO NPs (Huang et al. 2019).

Silver nanoparticles, which exhibit insecticidal activity against some mosquitoes disease vectors, may also represent risks to *C. elegans* (Kim et al. 2017; Yang et al. 2018). The size distribution of nanoparticles can also impact the toxicity of silver nanoparticles (Bone et al. 2015; Kim et al. 2017; Yang et al. 2018). Furthermore, some reports noticed that media where NPs were tested could influence the NPs toxicity. For instance, Yang et al. (2018) observed that AgNPs effects on the reproduction and neural functioning of *C. elegans* were more pronounced in the highest ionic strength media. Because of this, it was demonstrated that assessing the adverse impact on different media is fundamental to avoid misinterpretations.

Considering the ecological importance of worms, obtaining reliable data from toxicity studies with these organisms is crucial to support the discussion of nanotechnology regulatory structures and their responsible use in the agriculture.

### 3.1.3 Plants

Plants are primary producers that provide energy to other trophic groups through the byproducts from photosynthesis (Wink 1988). Plant roots and their exudates are important food sources for primary consumers and contribute to the soil structure, preventing erosive processes (Durán Zuazo et al. 2006; Kumar et al. 2007). Thus, assessing the potential effects of nanopesticides on non-target plants is imperative (Servin et al. 2015).

The contact pathways between nanoparticles and the plants are varied, even if the nanopesticide adminstration method is the same, because the nanoparticle dynamic on the systems depends on several factors. For example, in natural matrices such as soils, the rate of release of the nanopesticide and their fate in the environment could be different depending on the properties and characteristics of the matrix choosen, especially considering the presence and influence of the soil organic matter (Worrall et al. 2018; Adisa et al. 2019).

The sorption and uptake of nanopesticides into plants are dynamic processes driven by active and receptor-ligand interactions and governed by partitioning principles (Cornelis et al. 2014). Figure 4 illustrates the processes of uptake and transport between nanoparticles and vascular terrestrial plants. Nanopesticides could access plant systems through structures from different tissues (e.g., cuticles, trichrome, stomata, stigma, and hydathodes), as well as through wounds and root junctions (Ruttkay-Nedecky et al. 2017). Nanoparticle-environmental parameters

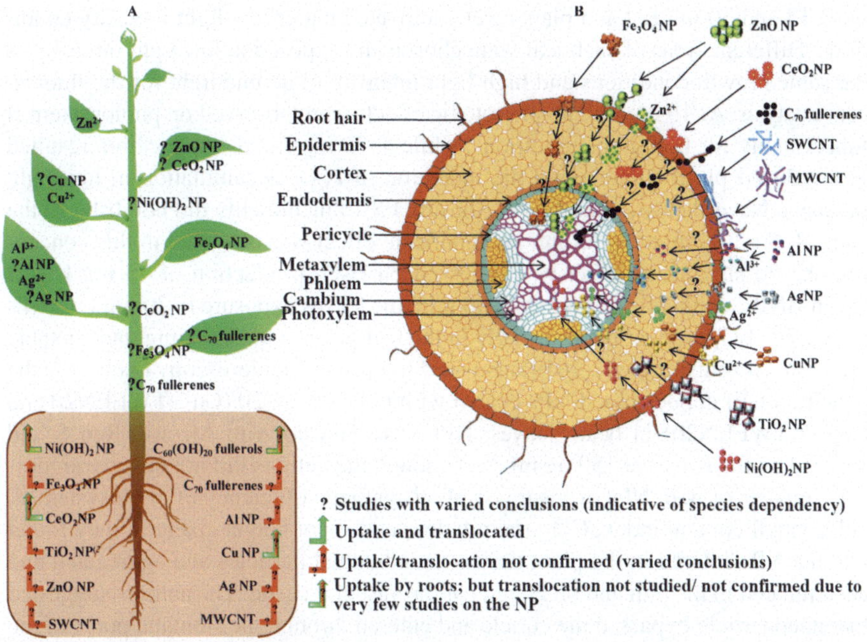


**Fig. 4** Scheme summarizing different nanoparticle and terrestrial vascular plant interactions. Uptake and translocation of nanoparticles in different plant tissues are presented. (**a**) Uptake and translocation of nanoparticles in a plant system; (**b**) transversal cross-section of root adsorption zone and translocation of nanoparticle in the tissues. Reprinted with permission from Rico et al. (2011). Copyright (2019) American Chemical Society

(e.g., aggregation and sedimentation rate, and surface charge) and physiological structures of the biological surface (e.g., membranes composition and surface area) may represent key factors in this interaction (Adisa et al. 2019). Once into the intracellular environment, nanopesticides could be distributed across organs and tissues and the organism response will depend on the cells' potential to react and adjust to this invasion through biological pathways, receiving and sending signals. This response can be analyzed in different hierarchical levels of biological system. In a molecular level, observing the regulation of genes (genomics) or the metabolites produced (metabolomics), in a macromolecular level, through the proteins expression (proteomics), and to organism level, by the morphological and physiological changes (phenotype).

Much of the current literature present metal and metal oxide nanopesticides as tools to control plant pathogens and enhance crop yield. Antonoglou et al. (2018) used chlorophyll fluorescence imaging analysis to access spatiotemporal photosystem II (PSII) efficiency and reactive oxygen species (ROS) formation on *Lycopersicon esculentum* (tomato) exposed to CuZn bimetallic nanoparticles (BNPs) (antifungal) via foliar application. The concentrations of CuZn BNPs used were 15 and 30 mg $L^{-1}$ and plants were cultivated under low light intensity conditions. Different areas of each leaf were chosen and exposed to low light intensity as the same growth conditions and high light intensity of actinic light for the fluorescent analysis. At 15 mg $L^{-1}$, no phytotoxic effects were observed on photosystem II functionality for both light conditions, while at 30 mg $L^{-1}$ the application resulted in a reduced plastoquinone and the induction of $H_2O_2$ accumulation in low light intensity. Nevertheless, after 3 h of exposure PSII functionality did not differ to the control. For high light exposure, no significant effect was observed in this concentration. No significant effect was observed on $H_2O_2$ production of 15 mg $L^{-1}$ of CuZn BNPs, while after 30 min and 90 min, the leaves exposure to 30 mg $L^{-1}$ NPs increased $H_2O_2$ production, especially in the leaf veins, and becoming undetectable after 3 h. Gkanatsiou et al. (2019) also used fluorescence microscopy to observe the penetration of copper-based NPs coated with polysorbate 20 (Cu@TWEEN20 and $Cu_2O$@TWEEN20) in beam leaves. NPs were labeled with Alizarin Red S and sprayed on the leaves at 150 µg $mL^{-1}$. Although the authors did not observe significant impacts of both NPs exposure on photosynthetic efficiency of photosystem II, chlorophyll content index, $CO_2$ assimilation, and shoot height changes, they found that the NPs entered in the plant tissues as individual particles and aggregated and accumulated in the pith and other cells of the vascular tissue. The authors suggested that nanoparticle bypassed the cuticle and entered through the stomatal pores of the leaves and could be transported to other tissues by plant's vascular system, apoplast (through the membranes) and the symplast (cell to cell). Regarding this result, it is important to highlight that the conversion of nanopesticides into fluorescent probes is a critical factor within nanoecotoxicity studies because of the manner that cell plants will "visualize" these compounds and interact with them (i.e., coated with the fluorescent marker), since these processes are mostly ruled by surface-receptors.

Nanoemulsions have also been presented as alternatives to solubilize hydroimmiscible conventional pesticides. Balaji et al. (2017) produced a hydrodispersive nanometric colloidal form of deltamethrin (NDM) used in mosquitoes control and evaluated phase index, mitotic index, and mitotic inhibition by microscopic analysis in *Allium cepa* roots. The onion bulbs were immersed in distilled water for 24 h to germinate and then the roots were immersed in 0.1, 0.5, 1.0, 5.0, and 10.0 mg $L^{-1}$ of NDM and the hydroimmiscible form of deltamethrin at the same conditions for 24 h. The toxicity exhibited by conventional pesticide form was higher than the nanoemulsion. The cells exposed to both types of the pesticides showed reduction of mitotic index due to the blockage of the G1 stage in the cell division. In addition, it was observed chromosomal aberrations, evidenced by laggard chromosomes, sticky chromosomes, anaphase bridges, clumped and disturbed metaphases.

Thereby, as nanopesticides applications are focused in agricultural purposes and plants are important components of the ecosystem and food chain, the knowledge regarding the NPs toxicity on non-target plants is essential to preserve the environment and the human food supply.

### 3.1.4 Insects

Insects represent the most diverse organisms in the history of life (75% of all animal species). They perform ecosystem functions which are of vital importance to both agriculture and biodiversity worldwide, and have a direct and especially economic impact on humans (Noriega et al. 2018; Woodcock et al. 2019). They play a crucial role in nutrient cycling and trophic chain structuring, participating in energy transfers within ecosystems (Schoonhoven et al. 2005). Thus, non-target insects are keystone species to evaluate the potential environmental risks of nanopesticides towards a sustainable agriculture (Lombi et al. 2019).

The literature regarding NPs effects on non-target insects is still an emergent field and has focused on model organisms intensively used in biological research and as environmental pollution bioindicators. The non-target insect models comprise the fruit fly *Drosophila melanogaster* (Diptera: Drosophilidae), the most important organism used in biomedical sciences over the last century (Alaraby et al. 2016); the beneficial honeybees *Apis mellifera* (Hymenoptera: Apidae), a key model essential to global ecology through pollination (Potts et al. 2016); the house crickets *Acheta domesticus* (Orthoptera: Gryllidae), an edible and synanthropic insect (Belluco et al. 2013); and the most representative benthic invertebrate species in freshwater ecosystems, *Chironomus riparius* (Diptera: Chironomidae) (Pedrosa et al. 2017).

*D. melanogaster* is recommended by the European Centre for the Validation of Alternative Methods as an ideal model organism to investigate nanomaterial-mediated toxicity (Mao et al. 2018). Several studies have demonstrated that after the oral exposition, different metal and metal oxide NPs may cross the peritrophic membrane of insect gut and induce the accumulation of ROS, leading to apoptosis and DNA damage (Carmona et al. 2015; Alaraby et al. 2019). This mode of action negatively affects *Drosophila* survival, development, and reproduction, and maybe the next generations (Parimi et al. 2019). Behavioral and phenotypic defects have been observed in larvae and adult fly, indicating that exposure to NPs alters different signaling pathways and influences the expression of biochemical and molecular biomarkers related genes involved in oxidative stress, detoxification of xenobiotics, and antioxidative defense mechanism (Lee et al. 2015; Anand et al. 2017; Sario et al. 2018; Yasinskyi et al. 2019). Drosophila has demonstrated to be a superior model for understanding the mechanistic role of NPs toxicity. Therefore, the enormous knowledge acquired over a century of research in the biomedical field may be extrapolated to understand the interaction of NPs with living systems and to place *Drosophila* as a non-target insect model organism that can translate the potential in vivo toxicity of nanopesticides into effects on human health (Chifiriuc et al. 2016). Approximately 75% of the genes involved in human diseases have related or similar sequences to *D. melanogaster* (Alaraby et al. 2016). Therefore, *D. melanogaster* can substantially contribute as an insect model to further nanopesticides toxicological classification.

The honeybees *A. mellifera* possess a high foraging activity and wide flying range, performing an essential role in biodiversity conservation and global agriculture as pollinators (Requier et al. 2019). However, the potential adverse impact of nanopesticides on honeybees is still sparsely investigated. Only three reports have demonstrated that the acute toxic effects (mortality ratio) of metal and metal oxide NPs on honeybees increased in a dose- and time-dependent manner (Dağlioğlu et al. 2015, 2016; Özkan et al. 2015). Some studies have shown that ZnO, CdO and/or PbO, and $CeO_2$ NPs, exert differential adverse effects on honeybees after NPs chronic exposure. These NPs decreased survival, alter feeding behavior, probably impairing the reproduction, and cause histological and cellular anomalies to honeybees midgut epithelial cells (Milivojević et al. 2015; Al Naggar et al. 2018; Dabour et al. 2019). The expression of genes involved in detoxification of xenobiotics and biochemical biomarkers activities (AChE, catalase, and GST enzymes) appears to be determinant in these processes (Milivojević et al. 2015; Al Naggar et al. 2018). The biological effects depend on NPs type and concentration as well the seasonal bee species and different body compartments analyzed (Kos et al. 2017). In contrast, no chronic effects were observed on *A. mellifera carnica* after exposure to titanium dioxide NPs (Jemec et al. 2016). The short-term exposure of honeybees to ZnO NPs did not affect their mortality rate and GST and AChE activity but altered their feeding behavior (Glavan et al. 2017). These results indicate that distinct compensatory and behavioral mechanisms are involved in honeybees response exposed to these particles at different exposure time. Glycerol Monolaurate nanocapsules were able to increase resistance against a fatal larval bee infection, without toxic effects on honeybees (Lopes et al. 2016). Sublethal doses (1 or 10 ng $mL^{-1}$) of a nanopesticide based on solid lipid nanoparticles loaded with pyrethrum extract was found to be safer for honeybees, and did not cause morphological changes in the digestive cells of bees during acute exposure (Oliveira et al. 2019). Therefore, nanoencapsulation of active biological compounds seems to be an interesting approach to reduce non-target effects on honeybees. These results indicate that further studies are required to determine the impact of NPs on behavior and pollination ecology of honeybees.

The house crickets *A. domesticus* are polyphagic and ground-dwelling insects, which served as a food source for many vertebrates and have attracted considerable attention in research as model organism (Horch et al. 2017). They have emerged as a representative model to study the interaction and trophic transfer of nanomaterials across different levels of biological organization in terrestrial ecosystems. Simulated terrestrial food chains have shown a particle-size dependent accumulation and trophic transfer of $CeO_2$ NPs from zucchini to crickets and wolf spiders (Hawthorne et al. 2014). However, the same size-dependent effect was not observed for $La_2O_3$ trophic transfer and accumulation from lettuce to crickets to mantis (De La Torre Roche et al. 2015). Copper oxide NPs may undergo transformation processes in soil upon weathering that subsequently impact NPs availability in food chains composed of lettuce, house crickets, and lizards (Servin et al. 2017). This model has also been important to understand the influence of soil pH, clay minerals, and soil organic matter on the uptake of silver from different AgNPs in simulated terrestrial systems (Pappas et al. 2017a, b, c, d, 2018). These results indicate that the type of NP and the organisms involved in food chain influence the trophic transfer of NPs in terrestrial ecosystems.

Nanopesticides assessment on non-target aquatic insects is critical to evaluate their potential environmental risks in aquatic ecosystems (Tomilina et al. 2014). In this sense, *C. riparius* is a classical model organism to ecotoxicity evaluation of chemical exposure on benthic invertebrates and water quality bioindicators. *C. riparius* is a sediment-dwelling, detritus-feeding insect, broadly recommended to sediment test toxicity (Oecd 2010). The studies well reported that the main toxicity mechanism of metal and metal oxide NPs, titanium, cerium, and zinc oxide NPs, is related to oxidative stress induced by ROS generation, evidenced at a molecular level (Nair et al. 2013; Walters et al. 2014; Niemuth et al. 2019). However, it is important to highlight that the properties, fate, and toxicity of NPs are promptly modified in aquatic environments, which may differentially affect *C. riparius*, depending on the resulting transformations. In fact, the biological response of *C. riparius* is greatly influenced by several factors, including NPs concentration and aggregation state, coating surface, exposure time, ion dissolution extent, exposure scenario, and ligands-associated sediments (Walters et al. 2014; Weil et al. 2019). The presence of sulfide in sediments affected AgNPs dispersion than AgNPs alone, mitigating the acute and chronic endpoints toxicity to *C. riparius*, whereas dissolved organic matter did not affect NP dispersion and modulated toxicity in the organism (Lee et al. 2016). The surface coatings generally mitigate NPs toxicity when compared to bare-NPs, with the reduced release of ionic silver from coated-AgNPs likely playing an important role (Park et al. 2015). An integrated assessment of $CeO_2$ NPs impacts on an experimental freshwater ecosystem reveals that mesocosm exposure to $CeO_2$ NPs led to different responses depending on the studied organisms (Bour et al. 2016a). No effects were reported on chironomid larvae, despite a significant NP accumulation and trophic transfer to amphibian larvae predator. In a microcosm study, different $CeO_2$ NPs presented different effects on the organisms involved in leaf litter decomposition and differentially affected this process in an aquatic ecosystem (Bour et al. 2016b). The most important impacts of $CeO_2$ were observed with the small, uncoated spheres, which impacted the bacterial communities and teratogenicity on chironomid larvae and decreased litter decomposition. Indeed, the teratogenicity in chironomid larvae can be used as relevant marker of the long-term environmental impacts of NPs (Bour et al. 2016b; Savić-Zdravković et al. 2018).

All these findings pointed out that NPs pose potential ecotoxicological effects in non-target insect models. However, the biological effects observed are dependent on the interplay of various factors and therefore, detailed studies are essential for ensuring their safety in agroecosystem. It can be concluded that is imperative the development of eco-friendly and biocompatible nanomaterials and their ecotoxicity assessment under environmentally relevant conditions.

## 3.2 Aquatic Organisms

### 3.2.1 Algae

Algae are ranked at the base of the aquatic food chain being considered important organisms that maintain the ecological equilibrium in both aquatic and terrestrial environments (Cáceres et al. 2008; Kalia and Gosal 2011). They are sensible to a wide range of chemicals, therefore are excellent bioindicators for environmental water quality and recommended by the United States Environmental Protection Agency and Organisation for Economic Co-operation and Development as aquatic toxicology models (USEPA 1996; OECD 2011). Over the last decade, there is an exponential increase in research regarding the effects of NPs towards this model (Chen et al. 2019a).

There are different mechanisms in which NPs can cause toxicity to algae (Chen et al. 2019a), as described in Fig. 5. The most common is the shading effect that decreased the light absorbed by the organism, disturbing its energy transduction process, inducing oxidative stress, and declining the chlorophyll content in addition to the reduction of photosynthesis and consequently of its growth (Tang et al. 2013; Iswarya et al. 2015). The damages of the shading effect to *Chlorella vulgaris* was

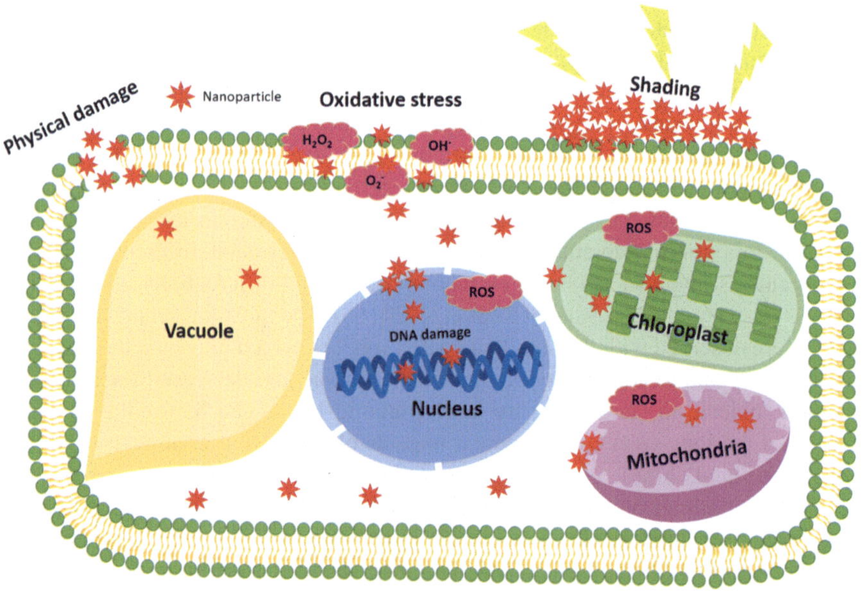


**Fig. 5** NPs toxicity towards algal cell. NPs can physically damage cell and organelles membrane, interacting with DNA and causing DNA damage. The shading effect also contributes to NP toxicity, since it limits the arrival of light in the chloroplast. Moreover, NPs exposure induces ROS production resulting in membrane lipid peroxidation and activation of antioxidative enzymes, such as superoxide dismutase and peroxidase (SOD and POD). The ROS can lead to a decrease in photosynthesis and it can damage the mitochondrial membrane and DNA

demonstrated by Djearamane et al. (2019). They found that the shading effect caused by ZnO NPs exposure led to the reduction of the cells viability, biomass and chlorophyll fluorescence emission. In another example, AgNPs decreased the photosynthetic yield in a dose-dependent manner and significantly altered cell morphology of *Euglena gracilis* (Li et al. 2015). CuO NPs exposure also resulted in alterations of cell morphology, and the NPs were adhered and even penetrated through the cell wall of *Chlorella* sp., influencing the algal growth due to shading effect (Wan et al. 2018).

On the other hand, because of the controllable release properties, nanoformulated pesticides present lower toxicity to aquatic non-target organisms. For example, hexaconazole, i.e., 2-(2,4-Dichlorophenyl)-1-(1H-1,2,4-triazol-1-yl) hexan-2-ol, for example, are less inhibitory to growth and metabolic activity of *Anabaena*, *Nostoc*, *Aulosira*, and *Tolypothrix* algae (Kumar et al. 2016). Also, atrazine-containing nanocapsules were less toxic to *Pseudokirchneriella subcapitata* than the conventional atrazine herbicide (Clemente et al. 2014). Further, the biosafety assessment of the nanoformulated pesticide (Pyrethroid nanometric emulsion) against freshwater algal species of genus *Closterium* showed minimal toxicity at the applied pesticide concentration of 1 mg $L^{-1}$ (Mishra et al. 2019).

In the case of inorganic nanopesticides such as metals (e.g., Ag, Cu), or metal oxides (e.g., $SiO_2$, $TiO_2$, ZnO, $Al_2O_3$), carbon nanotubes, and combinations thereof, the toxicity depends on the type of particle and its proprieties. Metal nanoparticles seem to have higher toxicity potential among inorganic pesticides, for example, $TiO_2$ NP reduced growth rate on *C. vulgaris* under nitrogen limitation and increased lipid peroxidation, glutathione s-transferase, and peroxidase activities (Dauda et al. 2017). Silver nanoparticles also reduce growth rate and pigment contents in *C. vulgaris* and *Dunaliella tertiolecta* (Oukarroum et al. 2012). Further, AgNP can cause the inhibition of the gene expression of superoxide dismutase and peroxidase in *Microcystis aeruginosa* (Qian et al. 2016). However, for metallic nanoparticles, the toxicity usually is lower than for the ionic counterpart. For example, Kleiven et al. (2019) demonstrated that silver nitrate was more toxic for *Raphidocelis subcapitata* than long-rod-shaped silver nanoparticle and spherical silver nanoparticles with 11 and 16 nm of size. The same was observed for $CuSO_4$ that was more toxic than CuO NP to *Chlorella* sp. (Wan et al. 2018).

Although some studies have reported the toxic effects towards non-target organisms, as present above, these substances are being considered more efficient than conventional pesticides, with the possibility to be used at lower rates, reducing application and therefore cause less harmful effects than the conventional ones.

### 3.2.2 Daphnia

*Daphnia* spp., commonly known as water fleas, are planktonic crustaceans belonging to the suborder of Cladocera. They reproduced by either sexual or asexual reproduction. When environmental conditions are good they reproduced by parthenogenesis; however, when conditions are not favorable they reproduced sexually and can produce a haploid egg, named ephippium, that can survive for many decades in hazardous conditions (Siciliano and Gesuele 2013). As filter feeders they play a key position in aquatic food chain as primary consumers and are highly distributed in most aquatic environments (Tatarazako and Oda 2007).

Due to the ability to process a large volume of media, filter feeders, in aquatic environments (e.g., daphnids), may be more vulnerable to nanopesticides than other organisms (Kookana et al. 2014). For this reason, they are commonly used as bioindicators for those materials (Clemente et al. 2014; Son et al. 2015; Slattery et al. 2019). These organisms are extremely sensible to a variety of chemicals. Besides, they are important component of the ecosystem, with a short life cycle, large brood size, easy and cheap to cultivate in the laboratory (OECD 2004, 2012; ASTM 2012; ABNT 2014, 2017). The interaction between nanopesticides and daphnids can be analyzed by different endpoints, such as survival, feeding rate, production of offspring, offspring sex ratio, among others (Siciliano and Gesuele 2013).

The toxicity of nanomaterials (including nanopesticides) depends on their proprieties, for example, size and stability may strongly influence the toxicity of nanopesticides, as observed by Son et al. (2015). They studied the toxicity of nanoencapsulated λ-cyhalothrin and observed a size-dependent effect, since particles with small hydrodynamic diameter were more severely lethal to *Daphnia magna* than particles with large hydrodynamic diameter. Comparing nanoparticles with different colloidal behavior, they found that well dispersed λ-cyhalothrin encapsulated NPs were more lethal (EC50 = 0.063 ug $L^{-1}$) than less stable λ-cyhalothrin encapsulated NPs (EC50= 0.172 ug $L^{-1}$). For *Ceriodaphnia dubia,* the size of nanopesticides also plays a significant role in toxicity. Slattery et al. (2019) observed that, for those organisms, the exposure to smaller encapsulated y-cyhalothrin (EC50 = 0.18 μg $L^{-1}$) caused more effect than larger (EC50 = 0.57 μg $L^{-1}$) and not encapsulated (free) y-cyhalothrin (EC50 = 0.65 μg $L^{-1}$). For metallic nanopesticides, the same trend was observed. ZnO NP with 50 nm, for instance, showed higher toxicity (EC50 = 1.9 mg $L^{-1}$) to *D. magna* in comparison to ZnO NP of 100 nm (3.1 mg $L^{-1}$) (Santo et al. 2014). Besides, for Cu NPs of 25 nm, the lethal concentration was 0,0018 mg $L^{-1}$, while for particles of 40 nm was 2.1 mg $L^{-1}$ and for 80 nm Cu NPs was 1.95 mg $L^{-1}$. Therefore smallest Cu NPs were more toxic to *D. magna* (Rasera et al. 2019).

The type of nanopesticides also plays an important role on toxicity, for example, organic nanopesticides may be more toxic than the pesticide itself, as shown by Slattery et al. (2019). These authors demonstrated that nanosized encapsulated pyrethroids (y-cyhalothrin) were more toxic to *C. dubia* than the conventional y-cyhalothrin. In another study, atrazine-containing nanocapsules were more toxic to *Daphnia similis* than atrazine itself (Clemente et al. 2014). In the case of metal nanoparticles, usually, it has been observed that the ionic form is more toxic than the NPs. For instance, Cu and CuO NPs were less toxic (0.2 and 1.61 mg $L^{-1}$, respectively) to *D. magna* than $CuSO_4$ (0.111 mg $L^{-1}$) (Arratia et al. 2019). The same was also observed for AgNPs, which $AgNO_3$ was more lethal than AgNP (Kim et al. 2011).

A concern about nanoparticles is that they may be transferred through food chain, causing long-term effects. Most studies in this topic have been conducted using algae and daphnids, and the nanoparticles tested are primarily metal oxides (Tangaa et al. 2016). For example, Kalman et al. (2015) studied the bioaccumulation and trophic transfer of silver nanoparticles in the green alga *C. vulgaris* and the crustacean *D. magna*, and observed that AgNPs were assimilated more efficiently than aqueous Ag. Moreover, it was observed that diet is the primary route of uptake for AgNPs in *D. magna,* also even after a depuration period, the NPs were not totally eliminated from daphnids, which may lead to a possible transport along the food chain. The same was observed after feeding *C. dubia* with algae (*Scenedesmus obliquus*) exposed to ZnO NPs of 50 and 100 nm. It was found a significant amount of Zn accumulated even after a 48 h depuration period. In addition, the authors found that size is related with the amount of NPs that were accumulated, as ZnO NPs uptake was more pronounced for 50 nm ($840 \pm 25$ μg $g^{-1}$ dry wt) than for 100 nm ZnO NPs ($650 \pm 10$ μg $g^{-1}$ dry wt) (Bhuvaneshwari et al. 2018).

To our knowledge, there is no work exploring the bioaccumulation of organic nanopesticides on daphnids and there are only few works in the literature focusing on the toxicity and bioaccumulation of nanomaterials for agriculture use (i.e., nanopesticides). Once they are applied directly into the environment, the chances that those materials reach the aquatic environment are high. Therefore, efforts to fully understanding the risks associated with the use of those materials are required.

### 3.2.3 Fishes

In the aquatic environment, among the main species affected by non-intentional exposure, fishes represent a major group. They have a wide geographical distribution with a presence in a broad type of environment. Moreover, they feature several eating habits and, in this way, take part in several trophic levels of the food chain.

Thus, because their ecological importance and due to fishes show fundamental biological mechanisms similar (conservative) to vertebrates, they have been extensively used as a model to evaluate the potential risks of nanopesticides (Sieber et al. 2019). Therefore, the obtained results in ecotoxicological assays can be extrapolated to other vertebrate species. Besides, because fishes are complex organisms, it is possible to analyze the toxicity mechanisms into different levels of the biological organization leading to phenotype modification.

Within the different methodologies recommended to assess the adverse effects of pollutants on fishes, the acute fish embryo toxicity (FET) assay has been largely explored. This assay allows the analysis of the fish embryo survival and embryonic development, with an excellent correlation to the conventional acute fish toxicity (Braunbeck et al. 2014). Also, embryos have been considered as an alternative model because they can be used as a replacement and refinement in the sense of 3Rs (replacement, reduction, refinement) principles for animal experiments.

Since 2013, the FET has been optimized and standardized by OECD guideline TG 236 (OECD 2013) for *Danio rerio* (zebrafish). This species presents several advantages that make it a suitable model. For example, the eggs and larvae show optical clarity which allows the development monitoring. Further, it offers fast development. The major organs of zebrafish embryo are formed during 24 h post-fertilization (hpf), and the larvae are fully developed at 96 hpf (Lee et al. 2017). Moreover, fish cell lines could be used to identify cytotoxic effect after exposure. Overall, in vitro tests have been applied to evaluate impacts on cell viability, pro-inflammatory responses, metabolic activity, mitochondrial function, oxidative stress, among others.

The toxicity of inorganic nanoparticles in the aquatic environment can be originated by the release of metal ions (dissolution) and/or by nanoparticles themselves (Wang et al. 2016). Overall, the sub-lethal effects of metal and metal-oxide nanoparticles on fish are very similar but attenuate than that observed when the fish are exposed to free metal ions (i.e., provided by their respective metal salt). And organ pathologies have been observed in gills, intestine, liver, and brain of fish (Shaw and Handy 2011).

Understanding the dynamic of metal NPs dissolution is fundamental to the accurate interpretation of biological effects observed in the toxicity assays (Skjolding et al. 2016; Baun et al. 2017). The dissolution of metal NPs depends on both intrinsic and extrisinc factors of NPs characteristics. The intrinsic factors refere to NPs properties, such as their surface charge, size, and surface coatings. Whereas the extrinsic factors are related to the composition of the surrounding biological medium, such as the amount of chloride ions and monovalent and divalent cations. The influence of NPs intrinsic characteristics on toxicity has reported in some studies with zebrafish embryos. Overall, smallest-sized particles (with largest surface/volume ratio) led to more significant toxicity effects to the embryos (Walters et al. 2014). Liu et al. (2019), for instance, observed that smaller silver nanoparticles (20 nm) were more toxic than larger silver NPs (100 nm NPs). In addition, they found that citrate-coated AgNPs showed higher toxicity than PVP-coating AgNPs, concluding that the particle sizes and surface coatings could impact on AgNPs dissolution and toxicity. Impacts of medium composition on metal NPs dissolution and toxicity. Impacts of medium composition on metal NPs dissolution and toxicity have also demonstrated. Some studies reported that the chloride ions present in the media may bind to NPs surface leading to the reduction of the AgNPs dissolution to silver ions, which are normally related to the toxicity. This effect was observed in the studies of Groh et al. (2014) and Lee et al. (2018).

The development of nanoformulations from conventional pesticides has been highlighted as a promissing strategy to increase the efficacy of these substances and to reduce their adverse effects to non-target organisms. For example, Mishra et al. (2019) studied the application of a nanoemulsion of permethrin against the mosquitos *Culex tritaeniorhynchus* and *Aedes aegypti.* They found a significant improvement of the permethrin pupicial and larvicidal efficacy when they applied the nanoemulsion instead of the conventional permethrin. In addition, they reported that the effects of nanoformulated permethrin did not cause severe distortions in the zebrafish gills.

The atrazine nanoencapsulated with poly ε-caprolactone (PCL) also showed a reduction of deleterious effects in comparison to conventional herbicide in the fish *Prochilodus lineatus* (de Andrade et al. 2019). They evaluated different biomarkers (i.e., genotoxic, biochemical, and physiological) in various tissues (gills, liver, and blood) and observed that atrazine caused a more significant adverse effect than its nanoencapsulated form.

Samadder et al. (2019) observed a decrease of cypermethrin toxicity on the fish *Oreochromis mossambica* due to cypermethrin nanoencapsulation with pelargonidin, a type of plant pigment. The reduction of deleterious effects were observed in the L6 muscle cell line viability, DNA activity, and in the anti-oxidative enzymes (superoxide dismutase, catalase, and lipid peroxidase). Meredith et al. (2016) also observed the reduction of deleterious effect of pyrethroid in zebrafish embryos (e.g., tremors, malformation, edema, mortality) after its encapsulation inside a polymeric shell. However, they reported that there was not difference between the effects observed in the embryos exposed to nano-sized (~200 nm) or micro-sized (~2200 nm) tested capsules.

Hence, nanoformulations do not change the toxicity of active ingredient but modify the exposure dynamics due to changes at the liberation rate of AI and ions. Consequently, nanoformulations impact directly in their bioavailability and in the uptake pathways of AI on species affected by non-intentional exposure. In this way, it could be acting in the toxicity profile and reducing the harmful effects of conventional pesticides by lower rates of AI in non-target species, such as fishes.

## 4 Nanopesticides Trophic Transfer in Ecosystems

Understanding how nanopesticides are transferred across different trophic levels in food chains plays a central role to evaluate their ecological impacts on ecosystem and the resulting implications for environmental integrity and human health (Tangaa et al. 2016; Vázquez Núñez and de la Rosa-Álvarez 2018). Accordingly, a great effort has been performed concerning the development of integrative tools and methods to clarify the effects of NMs on food chain structuring. Most studies have focused on aquatic environment using simulated food chains under well-controlled experimental conditions containing single species of two or three trophic levels (Lekamge et al. 2018). These food chains are generally composed by representative models organisms of different taxa, including decomposers (microorganisms), producers (algae and plants), primary consumers (protozoa, rotifers, mollusks, and arthropods) and secondary consumers (amphibians, and fishes). These organisms participate in a variety of trophic interactions and perform important ecosystem functions in aquatic ecosystems.

Microbial food chains were pioneer models in this topic and showed the transfer to and biomagnification of quantum dots via trophic interactions between bacteria-ciliate-rotifers (Holbrook et al. 2008) and bacteria-ciliate (Werlin et al. 2011), respectively. The algae-daphnids-predators interaction has been a food chain model intensively explored, demonstrating the trophic transfer primarily of metal oxides NMs in a range of exposure scenarios. This system investigated the transfer and/or bioaccumulation of NMs via trophic interactions in algae-daphnids (McTeer et al. 2014; Chen et al. 2015a; Iswarya et al. 2018), algae-fish (Zhu et al. 2010; Skjolding et al. 2014), algae-bivalve (Renault et al. 2008), algae-amphipod (Jackson et al. 2012), and algae-daphnids-fish (Chae and An 2016). The food chain models insect-fish (Asztemborska et al. 2014), plant-fish (Asztemborska et al. 2018), and worms-fish (Lammel et al. 2019) also demonstrated the transport of NMs via trophic interactions in simulated aquatic ecosystems. Only one report did not observe a probable biomagnification and trophic transfer from algae to daphnia after exposure to chemically and biologically synthesized nano-zero-valent iron materials in the experimental conditions adopted (Bhuvaneshwari et al. 2018).

These studies have been crucial to demonstrate the influence of the physicochemical characteristics of NMs and the surrounding environment on toxicity and mechanisms of NP bioaccumulation and trophic transfer in freshwater simplified food chains. Lekamge et al. (2019) observed that the type of AgNP coating and medium critically influences the degree of aggregation and the behavior of AgNPs, which differentially impact on their toxicity, feeding behavior, bioaccumulation, and trophic transfer from the algae *Raphidocelis subcapitata* to daphnids. The effects of size and crystal structure of nanoparticulate $TiO_2$ ($nTiO_2$) on the algae-daphnids model showed that acute toxicity, bioconcentration, and biomagnification all decreased with increasing exposure concentration, particle size, or rutile content of $nTiO_2$. At a molecular level, the disturbance of distinct metabolic pathways indicates that the presence of algae alleviated the impact of $nTiO_2$ on the metabolism of daphnia (Chen et al. 2019b). The bioaccumulation patterns of 10 and 20-μm silver nanowires (AgNWs) in algae-daphnia-zebrafish food chain indicate that toxic amounts of AgNWs can be transferred from algae to fish through water fleas, and large AgNWs are more toxic to organisms, whereas small AgNWs show a greater tendency to bioaccumulation (Chae and An 2016).

However, although these food chain models have illuminated important issues on the ecotoxicity of NMs in aquatic systems, they do not represent a real environmental scenario or mimic ecological functions (Lead et al. 2018). In this sense, microcosms and mesocosms studies have been emerged as a critical tool to understand how NMs effects on a single trophic level can disrupt ecological process and influence the ecosystem sustainability. They are employed to approximate environmental conditions and simulate natural phenomena (Donia and Carbone 2019). Inside of the laboratory, an integrated assessment of $CeO_2$ NPs impacts on a freshwater ecosystem model reveals that mesocosm exposure to $CeO_2$ NPs led to different responses depending on the studied organisms (Bour et al. 2016a). No effects were observed on litter decomposition or on the associated fungal biomass, but the bacterial communities was changed. Chironomid larvae was not affected, despite a significant NP bioaccumulation and toxicity to *Pleurodeles* larvae via insect predation. These results indicate that the interaction of $CeO_2$ with microorganisms indirectly affect the *Pleurodeles* larvae or the dissolution of NP may have occurred in mesocosms, following changes in Ce speciation, leading to toxic compounds (Bour et al. 2016a). Outside of the laboratory, the food web effects of $TiO_2$ nanoparticles in a freshwater mesocosm experiment showed that environmentally relevant concentrations of $TiO_2$ NP may negatively affect certain parameters and taxa of the freshwater aquatic model (sediment, phytoplankton, zooplankton, macroinvertebrates, macrophytes, and fish). However, the ecosystem sustainability was not affected, since no significative impacts were observed on its trophic state or primary production (Jovanović et al. 2016). The size-based effect of $CeO_2$ nanoparticles was evaluate in outdoor mesocosms over 9 months. The mesocosms simulated a complex environment including a submerged aquatic area, wetland, and dry terrestrial zones (Geitner et al. 2018). The food web investigated was composed of several trophic levels, represented by a range of terrestrial and aquatic plants, algal communities, various invertebrates, and a population of eastern mosquitofish (*Gambusia holbrooki*). It was demonstrate that $CeO_2$ small particles were significantly reduced from Ce(IV) to Ce(III) and accumulated in the upper floc layers of aquatic sediment, whereas larger particles transformed more slowly and settled quickly, making the aquatic sediment a sink for untransformed nanoparticles. Cerium from the small particles was also significantly more bioavailable to aquatic plants, snails, and insects. This work highlights the critical role of trace metals originating from NMs in the overall bioavailable pool of metal for biouptake and trophic transfer, as they have much greater distribution through multiple compartments in this complex ecosystem when compared to their larger counterparts (Geitner et al. 2018). In the real world, the bioaccumulation and trophic transfer of nanoparticulate Ag and Ti, relative to other Ag and Ti species, were investigated in the natural environment of Taihu Lake, China (Xiao et al. 2019). Both nanoparticulate Ag (18.8–41.0 nm) and Ti (46.6–116 nm) were detected in all natural samples. Nanoparticulate Ag showed stronger bioaccumulation than nanoparticulate Ti and also than its other chemical counterparts. Nanoparticulate Ag was biomagnified in the fish web while nanoparticulate Ti was found to be diluted in the aquatic food chain. This study underlines the distinctive bioaccumulation and biomagnification behaviors of nanoparticles as opposed to their other chemical counterparts along natural aquatic food chain. Additionally, call attention to the fact that the great bioaccumulation and biomagnification potential of smaller nanoparticles (nano-Ag) in invertebrates and fishes may pose considerable risks to human health due to consumption of aquatic food products.

The trophic transfer of NMs through terrestrial food chains is still a phenomena poorly explored (Gardea-Torresdey et al. 2014; Kwak and An 2016a). The experimental strategy adopted so far is very similar to that used in aquatic ecosystems. However, no mesocosms experiments were performed to assess the impact of NMs on ecological process in terrestrial ecosystems via trophic interactions. Simulated food chain models demonstrate the transfer and/or bioaccumulation of different NMs via trophic interactions between bacteria-nematode (Kim et al. 2013, 2016), earthworm-bullfrog (Unrine et al. 2012), plant-insect (Judy et al. 2012; Koo et al. 2015; Majumdar et al. 2016), earthworm-springtail (Kwak and An 2016b), yeast-springtail-pillbug (Chae et al. 2016), and plant-snails (Dang et al. 2019). The bacteria-nematode trophic interaction has been important to show that metal oxide NPs can be transferred and accumulated from *Escherichia coli* to *Caenorhabditis elegans* through food chain. The NMs were not bioaccumulated, but negatively impacted germ cell, affecting the development and reproduction of the nematode. Importantly, these effects can be transferred to the next generations (Kim et al. 2013; Luo et al. 2016). Small AgNPs (25 nm) were more easily accumulated in the food chain and exhibited a stronger toxicity to the higher trophic level than larger AgNPs (75 nm), indicating a size-dependently transfer and toxicity (Luo et al. 2016). Most of studies on this topic have used the plant-insect-predators trophic interactions as model. Hawthorne et al. (2014) demonstrated a particle-size dependent accumulation and trophic transfer of $CeO_2$ NPs in a simulated food chain composed by zucchini, crickets and wolf spiders. $CeO_2$ NPs were biomagnified through the food chain and Ce was detected at 5.49 ng $g^{-1}$ in the spider. In contrast, the particle-size dependent effect was not observed for $La_2O_3$ NPs trophic transfer and accumulation from lettuce to crickets to mantis (De La Torre Roche et al. 2015). Servin et al. (2017) found that weathering in soil increases nanoparticle CuO bioaccumulation within a terrestrial food chain composed of lettuce, house crickets, and lizards, highlighting the role of particle transformation and soil conditions in terrestrial trophic transfer of NMs. The transfer of $TiO_2$ NPs in the trophic food chain composed of the plant *Aristolochia debilis* and the swallowtail butterfly *Atrophaneura alcinous* was confirmed in vivo by X-ray analytical microscopy. The results demonstrated that $TiO_2$ NPs were transferred from the plant to larvae and eliminated through the feces. The dissemination of NP-contaminated feces might pose a potential environmental hazard (Kubo-Irie et al. 2016). This is an interesting approach to obtain the biodistribution of NPs in the whole "living" organism without any treatment. Soil-based studies with earthworm and different predators revealed that AuNPs are transferred from soil to *Eisenia fetida* and accumulate in juvenile bullfrogs (*Rana catesbeiana*) through food chain exposure rather than via direct exposure. However, a significative decrease in Au content at each trophic level was observed (Judy et al. 2012). A dose-dependent effect on transfer and toxicity of AgNPs was observed in the food chain soil-*Eisenia Andrei-Lobella sokamensis* (Collembola). High concentrations of AgNPs resulted in juvenile earthworm mortality and increased transfer of AgNPs to Collembola, which subsequently inhibited their locomotion (Kwak and An 2016b). Vijayaraj et al. (2018) combined two microcosms mimicking both terrestrial (soil, soil microbial communities, alfalfa plants, and snails) and aquatic (soil leachates and amphibian larvae) compartments to evaluate the transfer and ecotoxicity of $TiO_2$ NPs within and between terrestrial and aquatic ecosystems. It was demonstrated an upward transfer of Ti in the terrestrial ecosystem from soil to plants and a downward transfer of Ti from soil to the amphibian exposure medium. This study highlighted the potential risks of TiO2 NPs on the environment in different ways: (1) the negative effects on soil bacteria can have impact on soil fertility, ecosystem functions, and crop production; (2) the indirect effect of nitrifiers on crop production, and the uptake of Ti by alfalfa plants, lead to the question of food security; (3) potential bioaccumulation of Ti by snails across the food chain over the years; and (4) runoff from contaminated soils ends up in freshwater ecosystems and can be genotoxic to amphibian larvae.

All these findings significantly extend the knowledge regarding the effects of NMs on food chain trophic structuring and can contribute to improve our understanding of the impact of NMs on energy transfer between living systems. A number of studies demonstrated that NMs have indirect effects on non-target organisms via their transfer to higher trophic levels through food chains and potential accumulation. It was also evidenced that NMs effects on a single trophic level can expand through food chain and disrupt ecosystem functions affecting the environmental sustainability. However, some results are conflicting and inconsistent, as the transfer of NPs at different trophic levels across food chains is highly variable and depends on the investigated species, NP compositions, and exposure conditions. Furthermore, NMs interaction with complex ecosystems is strongly influenced by biotic and abiotic parameters that can transform and alter the physicochemical properties of NMs, modulating fate, transport, bioavailability, and their potential ecotoxicity in the environments.

From the above discussion, it is essential that future researchers focus on the development of reliable microcosm and mesocosms experiments outside of the laboratory using large scales and multiple species, mimicking environmental realistic conditions. Comprehensive research connecting the behavioral and physicochemical changes of NMs throughout their full life cycle in the environment to their long-term effects on trophic interactions and transport across food chain is still a challenge to assess the risks and benefits of nanopesticides use in agriculture.

## 5 Towards Advanced Methods for Nanopesticides Characterization in Biological and Environmental Systems

Understanding and controlling the impacts of nanopesticides on biological and environmental systems are dependent of using integrated characterization techniques at the interface between Materials Sciences and Environmental Sciences. A myriad of advanced methods has been applied for characterizing the nano-bio-eco interactions. Herein, we will describe some selected emerging techniques that are promising for nanopesticides characterization after exposure to biological model organisms and environmental matrices. Of course, there are many others techniques to be considered and this knowledge can be found in more specialized review articles and books (Lin et al. 2014; Gunsolus and Haynes 2016; Ma 2016). In fact, it has already been a challenge in nanotoxicology research using integrated techniques for characterization of nanomaterials towards linking nanomaterial physicochemical properties to biological outcomes in complex and dynamic environmental systems (Qiu et al. 2018).

## 5.1 *Mass Spectrometry-Based Techniques*

Since its invention by Joseph John Thompson in the early 1990s, mass spectrometry (MS) has been shown a key analytical technology for great discoveries over history, for example, demonstrating the isotopes existence. Nowadays, this technique has been extensively explored in several areas, such as for proteomics and metabolomics analyses, detecting pesticide residues and determining the elemental composition of biological samples.

The principle of mass spectrometry involves identifying and quantifying the compounds from their gas-phase ions. During the measurement, the sample is introduced into the system by chromatography or by a direct insertion probe, and then an ionization source produces the gas-phase ions of the analyte. The ions are separated in the mass analyzer according to their mass-to-charge-ratio (*m/z*). Lastly, a detector counts the ions, measures their abundance, and converts them into electrical signals that are processed in the computer resulting in the mass spectrum, which presents the ion relative abundance versus mass-to-charge ratio of analyte (de Hoffmann and Stroobant 2007). Several ionization sources have been developed over the years. Among these, the inductively coupled plasma (ICP) has stood out due to its high capacity for efficient vaporization, atomization, and ionization of samples. As a consequence, combining the high precision of ICP source with the high MS resolution has allowed the determination of major and trace metal elements in a myriad of sample types (Gray 1985). ICP-MS has been applied for evaluating the composition of animal tissues, fertilizers, drinking water, and soils (Yamasaki 2000).

Recently, the advantages of ICP-MS have associated with a particle counting technique, allowing an individual analysis of nanoparticles, in a system called single-particle (sp)ICP-MS (Degueldre and Favarger 2003). Hence, sp-ICP-MS has provided information concerning the size, size distribution, mass concentration, and number concentrations of NMs, in addition to distinguishing the NMs of ionic forms (Laborda et al. 2016). sp-ICP-MS has been employed for characterizing and quantifying NMs in biological matrices at environmentally relevant levels (ng $L^{-1}$), besides supporting the investigation regarding the NMs environmental transformations (i.e., oxidation and reduction reactions, dissolution, and sulfidation). Employing sp-ICP-MS, Bao et al. (2016) observed that AgNPs were accumulated in the apoplast of root tissues, and transported to shoot tissues. Moreover, it was possible identifying that the NPs were biotransformed or/and suffered aggregation after or during internalization on *Arabidopsis* plants. Wojcieszek et al. (2019) verified that radish roots preferentially uptake smaller $CeO_2$ NPs than aggregated. Moreover, the NPs adsorption into the radish tissues was also confirmed by laser ablation ICP-MS, as illustrated in Fig. 6 (LA-ICP-MS).

Coupling ICP-MS with laser ablation has been proved a promising strategy for investigating the uptake and transport of NMs on biological tissues, because it enables mapping the spatial distribution of metal and non-metal compounds from solid samples (Drescher et al. 2012). The advantages of LA-ICP-MS include high sensitivity and spatial resolution, the possibility of obtaining isotopic information,

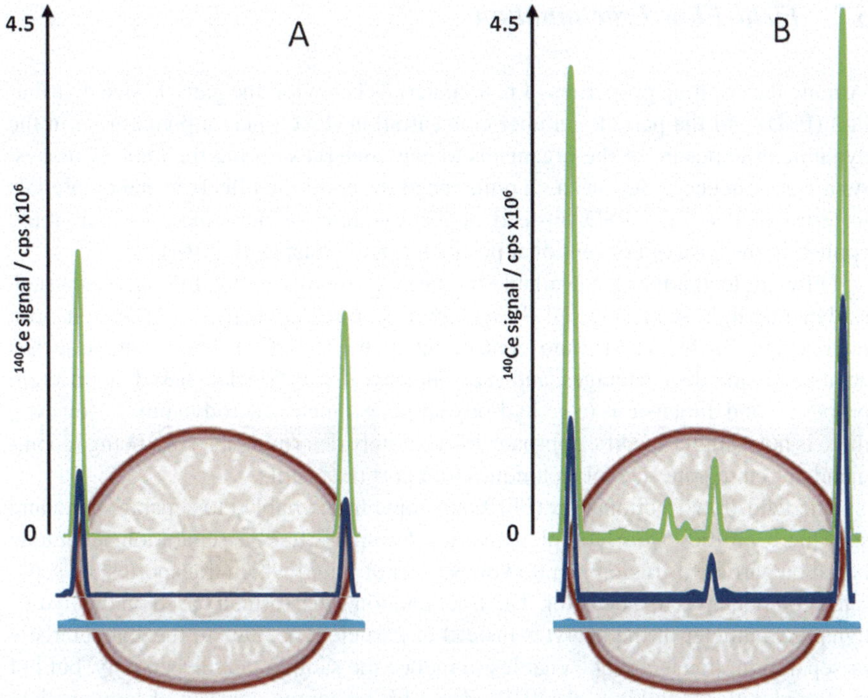


**Fig. 6** LA-ICP-MS analysis of radish roots treated with (**a**) 5 mg $L^{-1}$ $CeO_2$ NPs and (**b**) 50 mg $L^{-1}$ $CeO_2$ NPs. Light blue line: control plant; dark blue line: plants treated for 1 day; green line: plants treated for 2 days. Reprinted with permission from Wojcieszek et al. (2019)

and minimal sample preparation (Limbeck et al. 2015). On the other hand, for sp-ICP-MS analysis, sample preparation is a critical and challenging step. There are no well-established or generic methods to extract NPs from biological matrices, and the commonly applied reagents to digest the tissues (i.e., acid assisted) may modify the properties of interest of NMs (e.g., particle size and composition) (Monikh et al. 2019). Also, the methods generally employed for removal of non-associated NMs from the organisms may be ineffective, resulting in misinterpretations (Petersen et al. 2019). Therefore, the development of suitable and accurate methods is urgently needed for understanding the real risks of nanopesticides.

Mass spectrometry techniques have enabled the investigation of the environmental behavior, transformation, and fate of NMs, as well as their adsorption, distribution, and accumulation on the organisms. Therefore, they have provided the acquisition of multiparametric information that is imperative for identifying the potential risks of NMs and supporting regulatory decisions.

## 5.2 Field-Flow Fractionation

Among the defining properties of nanomaterials behavior, the particle size distribuiton (PSD) and the particle number concentration (PNC) play a pivotal role in the dynamic of exposure of the organisms to nanomaterials during the toxicity assays. As a consequence, it has been recommended to report the effects of nanomaterials in terms of PNC and PSD, instead of mass/volume or mass/mass as commonly applied in the case of conventional pesticides. (Kookana et al. 2014).

Different techniques are available for particle size distribuion measuements such as dynamic light scattering (DLS), nanoparticle tracking analyses (NTA), electron microscopy (SEM, TEM), and centrifugal methods (DCS). Each technique has advantages and disadvantages, and each one provides a PSD data based on different properties and dimensions (e.g., hydrodynamic, geometric, aerodynamic). However, there is not a single specific approach to accurately determine the size, being recommended to integrate several techniques to report the results.

The field-flow fractionation (FFF) technique have enabled to separate nano and sub-micro entities from complex matrices/samples. It is a separation technique based on particle diffusion in a flowing stream of liquid which transportes the sepa-reted componentes to a detector. The fractionation is performed based on the hydrodynamic diameter of the analytes instead of particle type. Due to the characteristic of separation, this technique enables to reduce the sample polydispersivity, but not its complexity (Monikh et al. 2019). The FFF technique consists of laminar flow inside a chamber that is typically a rectangular chamber. Thus, the flow presents different velocity along the channel, with the slowest lamina near to the wall and the fastest lamina flowing in the center. In normal mode, the separation is caused by cross-flow perpendicular to the migration flow. The analytes are towards the bottom wall ("accumulation wall') according to their hydrodynamic sizes. In equilibrium, an exponential analyte concentration profile is built up with the smallest is positionally near the wall, and the larger species are placed at center, where the velocity is higher (Contado 2017).

A variety of techniques can be coupled in FFF, allowing the determination of nanopesticide size. For example, it is possible to coupling FFF online with ICP-MS. It provides the element-based size distribution of the fractionated sample. This technique has been used to separate and characterize metal nanoparticles as AgNPs, AuNPs, $SiO_2$ NPs and quantum dots at relevant concentrations in ecotoxicology studies (Monikh et al. 2019).

The major advantage of FFF is the use of carrier liquid, usually the aqueous buffer used for the dispersion. Also, it is not necessary a stationary phase or gel media since the separation is based on flow dynamics (Moon 2019). However, these methods present some disadvantages as the strong particle-membrane interaction which may provide a low recovery (Petersen et al. 2016). Besides, this characterization technique is not habitual in the pesticide analysis laboratory and represents a significant challenge in a routine procedure in the future.

### 5.3 Hyperspectral-Enhanced Dark Field Microscopy

Hyperspectral-enhanced dark field microscopy (HEDFM) has become a strategic microscopy technique for studying the interface between nanomaterials, biosystems, and the environment. In nanoecotoxicology, this technique is very useful for understanding the nanoparticle cell-interaction, internalization, biodistribution, and nanotoxicity (Roth et al. 2015; Zamora-Perez et al. 2018).

In HEDFM technique, a dark field-based illuminator focuses a highly collimated light at oblique angles on the sample to obtain images with improved contrast and signal-to-noise ratio. This imaging technique utilizes the intrinsic scattering properties of objects such as nanoparticles, organic and inorganic colloids, and other particulate materials, and therefore neither staining nor a contrast agent is required to visualize the sample (Meyer et al. 2010). The microscope uses a cone of light in a "dark field" configuration to excite the samples in the absence of scattering sources; no light at all is transmitted through the microscope objective. However, NPs and other highly light scattering objects, when present, appear as bright features on an otherwise black background (Vallotton et al. 2015). With this configuration, it is possible to record a dark field optical image. A built-in hyperspectral camera produces a hyperspectral image that looks similar to optical image. However, each pixel of the hyperspectral image contains the spectral response for that pixel's spatial area. These hyperspectral images are observed using powerful image analysis software, which can easily adjust how the spectral data is displayed. By modulating the contrast and enhancement settings in the hyperspectral image analysis software, NPs can be instantly visible. This enables easy identification of their spectral response for accurate spectral mapping throughout the sample (Grasseschi et al. 2015).

Hyperspectral microscopy as a complementary imaging technique provides a potential solution to some sample preparations issues, since samples are imaged directly on glass slides, often eliminating steps such as fixation and dehydration (Roth et al. 2015). Dark field microscopy associated with hyperspectral mapping is a novel optical approach with great potential in bio-related studies because it allows the identification and quantitative determination of specific components in biological media (Zamora-Perez et al. 2018).

Exploring HEDFM technique and an inorganic nanopesticide (i.e., AgNPs), it was possible to detect the nanoparticles in the tissues of organisms commonly used in nanoecotoxicology such as bacteria (Eady and Park 2019), *R. subcapitata* (Sekine et al. 2017), plant roots (Gao et al. 2018), *D. magna* (Botha et al. 2016), *C. elegans* (Chatterjee et al. 2017), and *D. rerio* (Botha et al. 2019). However, it is an open challenge exploring this technique to investigate the organic nanopesticides in biological tissues such liposomes and polymeric-based nanoparticles.

HEDFM technique was used to further assess the internal distribution of Au nanoparticle in the daphnia. The technique allowed the visualization of the nanoparticles in the aqueous medium; particles were also detected in the body surface, appendages, mandibles, and gut (Fig. 7).

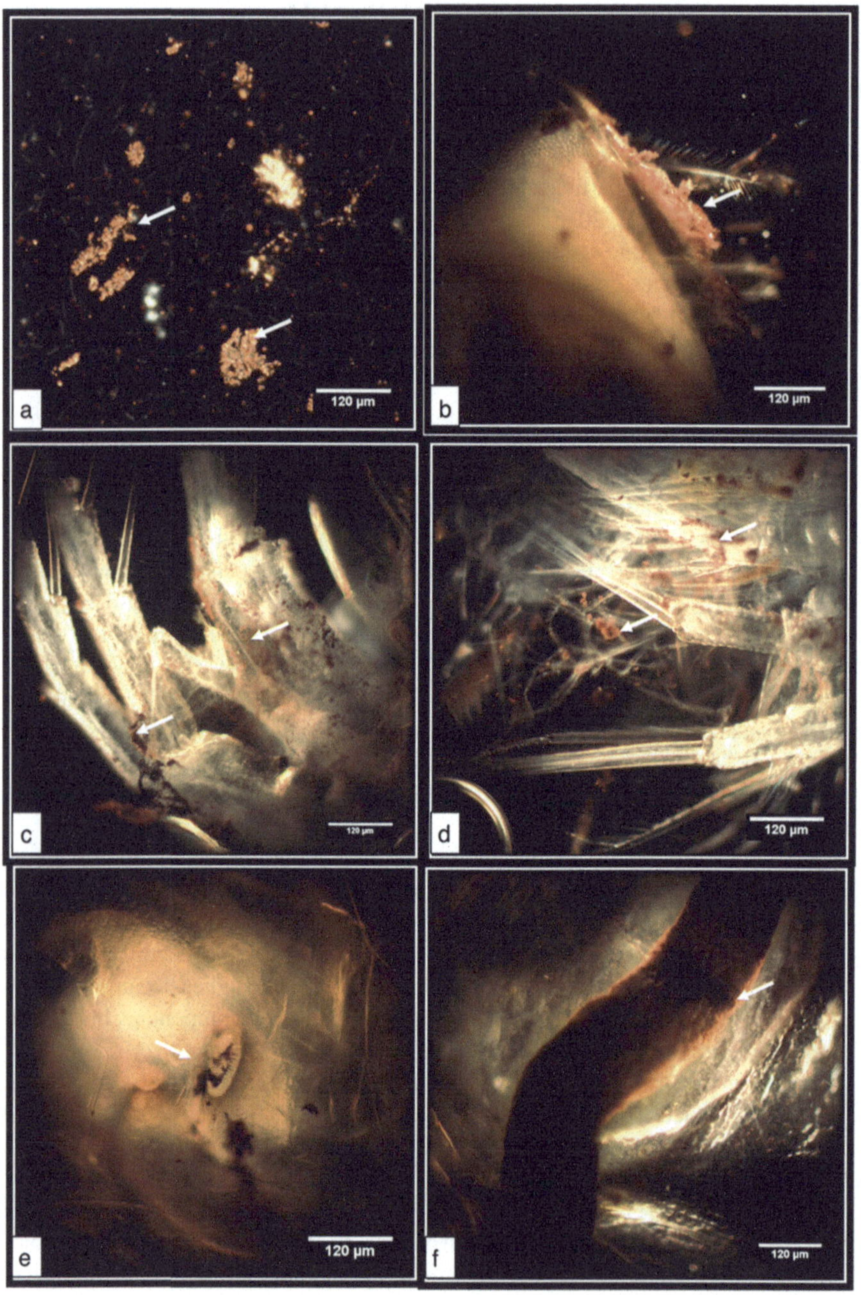


**Fig. 7** Daphnia magna exposed to Au particles in aquatic medium viewed by HEDFM imaging showing (**a**) aquatic medium, (**b**) body surface, (**c**, **d**) appendages, (**e**) mandibles, and (**f**) gut. Reprinted with permission from Botha et al. (2016). Copyright (2019) Clearance Center

Considering terrestrial nanoecotoxicity assessment, *Lumbriculus variegatus* (earthworm) was exposed to AgNPs or silver nitrate ($AgNO_3$), and the uptake was assessed through hyperspectral imaging. In this study, the nanoparticle bioaccumulation was concentration dependent across treatments. Although significant differences were observed in total Ag body burdens of *L. variegatus* of depurated versus undepurated organisms, the results suggest that Ag particles may have similar behavior and uptake patterns regardless of size or composition within sediment and biological systems (Coleman et al. 2013).

Recently, the HEDFM technique has been applied for understanding the interactions of AgNPs in aquatic systems. Exposure of *R. subcapitata* cultures caused size and surface-dependent responses in the algae growth rate. In this study, control samples showed a relatively dark and uniform image of the algal cells, while exposed cells scattered light more intensely and they appeared brighter in the dark field images. Furthermore, the exposed cells appeared to be brighter along the cell walls compared to the control, and there were multiple point scattering commonly associated with silver nanoparticles. The hyperspectral profiles of these spots showed intense scattering at shorter wavelengths, which is typical of AgNPs (Sekine et al. 2017).

HEDFM was also used to monitoring, with a high temporal and spatial resolution, transformations of the AgNPs in wastewaters and biosolids. Reference spectral libraries were first generated from control solutions of silver nanoparticles in deionized water, after HEDFM was used to investigate if silver nanoparticles would be detected in the wastewater matrix. In the absence of added silver nanoparticles, the technique was able to detect nanoparticle aging (chemical transformation) in complex media by measuring the spectral properties of individual nanoparticles in wastewater matrix (Théoret and Wilkinson 2017).

Therefore, HEDFM technique has been proved a pivotal strategy for assessing the ecotoxicity mechanisms at nano-bio-ecology interface and supporting the transformation studies of NPs.

## *5.4 Atomic Force Microscopy*

Atomic force microscopy (AFM) has application in several scientific fields, such as investigating the structure of the biological tissues, cells, and molecules (Harjumäki et al. 2019), analyzing cell adhesion properties (Sagvolden et al. 1999), characterizing materials and surfaces (Côa et al. 2017; Franqui et al. 2019), among others. The versatility of this technique made it important to multidisciplinary areas as nanotoxicology, for example, with AFM is possible to evaluate the morphology of the constituent nanoparticles in pre- and post-incubation conditions in the biological environment, checking the stability and possible morphological changes caused by the medium, such as aggregation, degradation, or adsorption of biomolecules. Nanopesticides have been attracting increasing attention of regulatory institutions and nanotoxicology researches to access their environmental impact and ecotoxicity. They comprise a great variety of materials (e.g., nanoemulsions, nanocarriers with pesticides, and nanoparticles with biological activity) and the proper safety assessment of these materials involves the understanding of the interactions, modifications, and destinies that they are subjected in the environment (Kah et al. 2013).

In this context, Jaques et al. (2017) studied the toxicity of formulations of different nanoparticles loaded with the herbicides atrazine, simazine, and paraquat for the soil nematode, *Caenorhabditis elegans*. The results showed that the formulations increased *C. elegans* lethality, as well as, change the rate of its development, affecting the worm length even at low concentrations. AFM imaging revealed that the morphology of nanopesticides and nanoparticles did not change when incubated in saline medium and the materials remained stable after 30 min of exposure. The morphological stability of nanoparticles observed in AFM images together with other results allowed them to affirm that the effects observed in the biological model studied were caused by the nanomaterial and not necessarily by impact of pesticides, indicating that the formulations need to be improved. On the other hand, Kent and Vikesland (2016) used AFM analysis to evaluate the rate of dissolution of Cu-based NPs (Metallic copper (Cu), cupric oxide/hydroxide ($Cu_{ox}$), and copper sulfide ($Cu_xS$) NPs) fabricated on solid supports in undersaturated solutions concerning copper mineral phases. They found that $Cu_{ox}$NPs dissolve completely in these conditions within a matter of hours, even at neutral to basic pH, the dissolution time of the metallic Cu NPs varied from a few hours to days, whereas $Cu_xS$ NPs showed no significant dissolution over the time scales studied. The results were confirmed in field deployment of Cu-based NP samples in a freshwater stream, which suggested that Cu and $Cu_{ox}$ NPs will be short-lived in the environment (Kent and Vikesland 2016). Roseline et al. (2019) developed AgNPs synthesized using the aqueous extract of four red seaweeds (*Gracilaria corticata, G. edulis, Hypnea musciformis*, and *Hypnoides spyridia)* for application as nanopesticide. They evaluated the antifungal potential of the AgNPs against the species *Ustilaginoidea virens*, also, the antibacterial potential against the species *Xanthomonas axonopodis pv. citri* and *X. oryzae pv. oryzae.* The aqueous extracts of algae used in nanocomposite synthesis have reducing properties that converted $Ag^+$ ions into $Ag^0$, and their composition influenced the size, shape, and stability of the generated nanoparticles. AFM images recorded in liquid state allowed verifying a spherical morphology in nanostructures with average diameters of 37, 54, 53, and 49 nm in *G. corticata, G. edulis, H. musciformis*, and *S. hypnoides*, respectively. In addition, by AFM it was possible to analyze the degree of aggregation of the particles in the medium, which is correlated with the sedimentation of the AgNPs and their absorption in the soil.

The knowledge of how nanomaterials interact with common environmental biomolecules consists of an essential aspect of their safety assessment or even of the effectiveness of their applications, once these molecules can module their properties, interacting with living organisms and changes the NPs environmental fate. In this context, AFM has shown as useful tool for evaluating the distribution and morphology of molecules over materials surface mainly because of the range of mode of analysis and the possibility of operating in environmental conditions, such as immersed in fluids.

## 5.5 *Synchrotron X-Ray Fluorescence Microscopy*

The recent advances in development of next generation synchrotron light sources in certain regions of the world enable the nano-focus beams with efficient photons flux for microscopic screening and imaging of the biological specimen at nanoscale level with high spatial resolution. The available synchrotron X-ray fluorescence microscopy facilities at Petra III Desy Germany, Max IV Sweden, ESRF Grenoble France, Diamond Oxford UK, LNLS-Sirius Campinas Brazil, etc., provide the researchers excellent platforms of elemental mapping and tomography of micro- and nanoscopic objects for wide range of biomedical, environmental, and agricultural applications.

Synchrotron X-ray fluorescence microscopy (SXRF) is a multi-elemental analysis technique with sensitivity in the range of $\mu g.g^{-1}$ (Chen et al. 2015b). These characteristics enable the study of the spatial distribution of small-scale chemical elements and the study of heavy metal trace and nanoparticulate materials, in various scales organisms, different tissues with different physiological and pathological characteristics and even in single cells. Currently, the SXRF technique has been pointed out as an efficient tool for characterizing nanomaterials and evaluating their distribution in cells, plants and living organisms such as *C. elegans*, *Daphnia* (Gao et al. 2008; Jackson et al. 2009; Wang et al. 2010; Castillo-Michel et al. 2017; Lv et al. 2019).

The SXRF technique is extremely powerful for quantitatively determining the elements concentrations in wide range of samples from material science to biological and environmental sciences. Specialized microprobes have the potential to achieve spatial resolutions in the tens of nanometer domain, but most typically operate in the 200 nm to 10 μm range (Kopittke et al. 2018). A typical synchrotron light source consists of a linear particle accelerator (linac), a booster, a storage ring and beamlines or experimental end stations where the photons will meet the sample (usually an X-ray beam). The energy of these photons usually vary from $10^{-3}$ to $10^5$ eV depending on the characteristics of the origin source (Kopittke et al. 2018). The advantage of SXRF technique is the high sensitivity and spatial resolution provided by the availability of more efficient fluorescence detectors and scanning systems since last two decades (Lombi and Susini 2009; Kopittke et al. 2018). The deep penetrating and highly focus beam of X-ray probes, which vary from several micrometers to a millimeter, are suitable for analyzing the agriculture and environmental samples with variety of metal contaminants. The depth of analysis is determined by the sample matrix, the edge energy of the analyzing element, and the efficiency of the output fluorescence X-rays. This analysis is performed on extremely small sample volume, for example, a 1 $\mu m^2$ incident beam illuminated a plant sample of 20–40 μm thickness, which detects the femtograms of the interested elements in absolute manner (Kopittke et al. 2018; van der Ent et al. 2018).

Nanoparticulate systems, after coming in contact with soil, water, and air can inevitably interact with the plants, consequently, influencing the plant physiology and possibly food safety. Nano-phytotoxicity studies related to nanomaterials applications in agriculture have been highlighted in the literature, presenting harmful and beneficial effects for the plants at physiological, biochemical, and genetic levels (Kopittke et al. 2018). Associated with these studies are SRXRF techniques that aid in understanding the absorption, distribution, and accumulation of nanomaterials in thin section of the plant systems (Majumdar et al. 2012). Ma et al. (2017) reported a test of 25 nm $CeO_2$ nanoparticle on hydroponic cucumber plants using split-root system. In this study, they observed through SRXRF and μ-XANES analyses (Fig. 8) the transportation of $CeO_2$ NPs from roots to the shoots through the xylem and also found that the 15% Ce (IV) in NPs were reduced to Ce (III) in the roots of the treated side (TS), but Ce was remained as a Ce(IV) in the blank side (BS). In addition, only the $CeO_2$ NPs were transported back from the shoots to the roots through the phloem. The distribution of $CeO_2$ in the roots of the treated side (TS) are also shown in Fig. 8, which exhibits high concentration of Ce in epidermis of the root and some is being spread in the cortex, as consistent with the TEM images. In addition, the figure shows the μ-XANES spectra of the high intensity points that have been selected on the core-coded μ-XRF map. The study observed that the absorption spectra of Spot 1 and 2 in Fig. 8b are resembled to that of the reference compound Ce (IV), which further confirmed that Ce was translocated from shoots back to roots as $CeO_2$.

Lv et al. (2019) exposed ZnO NPs to a hydroponic maize species (*Zea mays* L. cv. Zhengdan 958) and evaluated the spatial distribution of Zn by SXRF, in which a microwave probe mapping was performed, allowing the direct in situ quantitative visualization of the elemental distribution in maize tissues. The SXRF maps

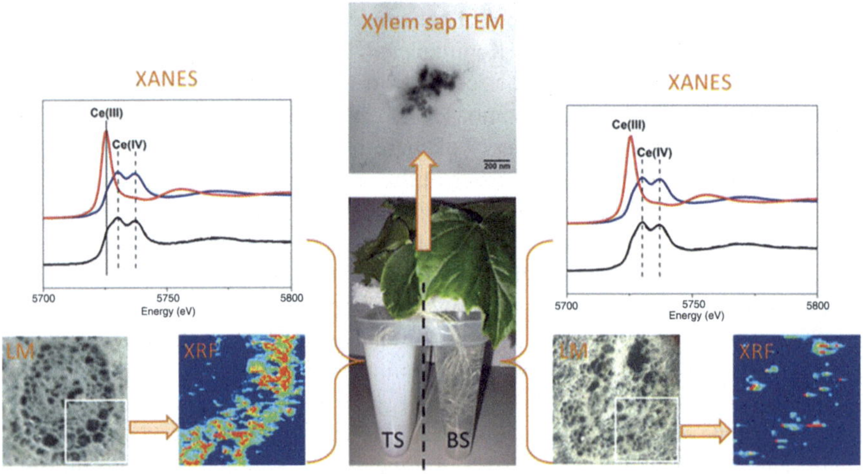


**Fig. 8** SXRF images of Ce in treated side (TS) (**a**) and blank side (BS) (**b**) roots. The study was conducted under treatments of 200 and 2000 mg/L $CeO_2$ NPs. Points 1 and 2 indicated by arrows on panel B are the points where μ-XANES was purchased. (**c**) Ce (III) and Ce (IV) μ-XANES spectra as well as points 1 and 2 selected in image (**b**). (**d**), treated side (TS) (**a**) and blank side (BS) of cucumber plants. Reprinted with permission from (Ma et al. 2017). Copyright (2017) American Chemical Society

exhibited distinct profile in distribution and content of Zn in the leaves and stems of the two treatments. However, more hot spots (representing Zn accumulation) in the root cortex were found for treatment with ZnO NPs than for Zn. This study indicated that ZnO NPs released $Zn^{+2}$ ions, which were absorbed by the roots and accumulated in maize tissues, mainly as inorganic and organic Zn phosphate. At the same time, the authors observed a small fraction of the ZnO NPs adsorbed on root surfaces entered into root cortex due to rapid cell division and elongation of root tips (Lv et al. 2019).

In addition to plants, synchrotron microprobe X-ray fluorescence technique can also be applied to study the toxicological effects of NMs in other organisms, such as *C. elegans* and *D. magna*. Jackson et al. (2009) studied the interaction between Se and Zn quantum dots towards *D. magna* and observed that the biodistribution of these materials appeared to be confined inside the gastrointestinal tract. Additionally, the SXRF images demonstrated that Cu NPs can be accumulated in the intestine and appendages of *D. magna* during acute and chronic toxicity testing (Rasera et al. 2019).

Gao et al. (2008) studied the distribution of copper nanoparticles (Cu NPs) using the in vivo model *C. elegans*. It was possible to observe Cu NPs mainly in the head and a third of the body path from the tail of the small nematode. In account of this, this study indicated that the microbeam SXRF technique could be used to perform investigations on a single tiny living organism instead of fixed or processed samples.

In summary, synchrotron radiation is an advanced light source with a wide range of frequencies, from infrared to the highest energy consumption of X-rays. This light is highly polarized, tunable, and collinear which can be focused on a small area with far more photons than a conventional source. All these properties significantly improve the signal-to-noise ratio, reducing the acquisition time of the results, and improving spatial and temporal resolution. These make Synchrotron radiation an excellent tool for nanopesticides science, environmental interactions, and ecotoxicity studies (Wang et al. 2010; Majumdar et al. 2012; Maret 2018).

## 6 Nanoinformatics

One of the highest goals and challenges of nanoecotoxicology is to elucidate the main relationships between physical-chemical characteristics of the nanomaterials and their biological effects and environmental fate (Nel et al. 2013; Bañares et al. 2017). The variety of size, surface structure, and chemical composition, among other properties, makes it impossible to experimentally evaluate a significant number of descriptors important to predict nanomaterials toxicity and environmental behavior. Moreover, other complicating factors include the transformations which these materials are subject in the biological medium (e.g. degradation, oxidation, and adsorption of biomolecules), as well as, the lack of standard toxicity assays methodologies, which often hamper comparations between different studies, and the development general statements about nanomaterials toxicity, risk and safe usage (Kookana et al. 2014; Fadeel et al. 2018).

The toxicity evaluation and risk assessment of nanopesticides found the whole combination of the challenges listed above. They comprise a great variety of nanoscale components, each with different properties and functions, such as nanoemulsions and dispersions of existing pesticides, nanomaterials as active pesticide agents (e.g., silica, titania, silver, etc.) or as nanocarriers for their delivery (polymer nanoparticles) (Hayles et al. 2017). Furthermore, they are directly applied in the environment, which makes them subject to various processes that interfere in their mode of action, toxicity, and environmental fate, for instance, degradation, leaching, and interactions with different substrates, biomolecules, or even other pesticides (Kookana et al. 2014). Therefore, the toxicology of nanopesticides involves aspects as nano-delivery and transformations in the environment.

In this context, computational approaches as simulation methodologies and Machine Learning (ML) algorithms have been increasingly important to, respectively, elucidate mechanisms of bio-nano interactions and toxicity, and deal with a great volume of data, extracting information to link nanomaterials properties to complex biological responses (Cai et al. 2018).

In silico approaches applied, for example, Molecular Dynamics (MD) or Density Functional Theory (DFT) to get detailed information about the structure, relative stabilities, and physical-chemical properties of molecules and nanomaterials. These techniques use well-established physical concepts of Newtonian mechanics (MD) and quantum mechanics (DFT) to delineate the behavior of atoms and their constituents (nucleus and electrons), respectively (Lee 2016). It is possible to describe mechanisms of, for example, interactions between NMs and pesticides or other organic molecules; intermolecular interactions; behavior of molecules and materials in environments with different conditions. Besides that, quantum chemical methods, which include DFT and others, are capable of finding reaction pathways of important degradation processes, such as photolysis and hydrolysis, from an environmental point of view (Villaverde et al. 2017a).

Despite providing valuable mechanistic knowledge to correlate nanomaterials' physical-chemical characteristics to their biological effects, computational simulations require exceptional computational power and extensive theoretical knowledge, which may become limited factors when dealing big data sets and endpoints, as well as, with complex systems (Winkler 2016; Villaverde et al. 2017b; Zhou et al. 2019). On the other hand, advances in artificial intelligence and machine learning techniques have allowed the development of predicting models from empirical data, finding implicit patterns without complex calculations and with better time efficiency. Various machine learning algorithms are already implemented in different programming languages, such as clustering, regression, decision trees, artificial neural networks, and evolutionary algorithms (Casman and Gernand 2016; Winkler 2016; Yan et al. 2019).

Quantitative Structure-Activity Relationship (QSAR) and Quantitative Structure-Property Relationship (QSPR) tools consist of statistical modeling and machine learning algorithms already implemented for risk assessment of pesticides

(Villaverde et al. 2018). Both tools have the ability of built prediction models, that link physicochemical properties to toxicological/ecotoxicological effects and environmental fate, learning from datasets with measured or calculated parameters of molecular properties and their respective biological (e.g., eco/toxicity) properties (Winkler 2016; Yan et al. 2019). Each one of known and unknown physicochemical or biological properties is termed an endpoint.

QSAR and QSPR methodologies are currently accepted and encouraged by European legislation about pesticides; the Regulation (EC) N° 1907/2006 concerning the Registration, Evaluation, Authorization and Restriction of Chemicals (REACH) considers that information generated using these non-experimental methods should be allowed, besides that, publications of the European Chemicals Agency (ECHA) emphasize the importance of QSARs to avoid animal testing (Villaverde et al. 2019). However, the use of these tools to perform the risk assessment of nanopestices still present limitations. The main issues comprise the need of specific descriptors, the complexity of the biological effects (e.g., (eco)toxicity) and environmental behavior, in addition to the scarce comprehensive empirical and experimental data to fully assess the endpoints, and the wide variety of nanoparticles without a full-accepted criterion classification. Furthermore, due to a large number of factors that may influence the biological activity of nanomaterials, traditional QSAR approach needs algorithm optimizations before application since the modeling effort required increase linearly with the complexity of the endpoints to be modeled (Winkler 2016).

The set of endpoints and the quality of the dataset used for QSAR modeling are fundamental for the accuracy of the predictions (Forest et al. 2019). For nanopesticides, well-defined endpoints may include physicochemical (e.g., surface area and size, diameter, length, shape, composition of the particles, etc.), environmental (e.g., bioaccumulation, degradation, etc.), ecotoxicological (e.g., aquatic toxicity, effects on terrestrial organisms, etc.), or toxicological (e.g., carcinogenicity, mutagenicity, etc.) descriptors; moreover, more recently, other computational chemistry approaches, such as DFT, have been used to calculation and prediction of suitable molecular descriptors, for example, HOMO–LUMO gap energy, heat of formation, total energy, frontier orbitals energy, electrostatic potential, among others (Villaverde et al. 2017a, 2018; Wang et al. 2017).

Finally, considering the issues and challenges of the risk assessment of nanomaterials in different fields (e.g., environmental and human health) and the increase of the importance of allied computational approaches to experimental evaluation of nanotoxicity, international efforts have been undertaken in the area of nanotoxicology to create databases with (1) applications, products, and commercialization, (2) characterization, (3) exposure data, (4) omics data, and (5) human and environmental toxicity of nanomaterials (Bañares et al. 2017; Haase and Klaessig 2018). The purpose of these projects is to enable data exchange and the use of these in computational prediction tools, for example, ML algorithms and QSAR models, thereby allowing the analysis to risk assessment and implementation of the Safe-by-Design concept in the development of new nanomaterials. In this context, it has been stressed the importance of standardization of methodologies and the organization and storage of obtained results to ensure information reliability (Fadeel et al. 2018). Moreover, recently, the Europe and USA are discussing a roadmap cooperation for nanoinformatics development for the next decade, including applications on nanosafety research (Haase and Klaessig 2018). Therefore, it is very important to increase the number of countries associated with this initiative towards open and transnational access of these platforms and nanoinformatics tools in the context of nanopesticides production and safety.

# 7 Final Remarks

Nanopesticides application is a promising alternative for controlling several pests and improving the food production. However, their indiscriminate use can result in undesirable effects on terrestrial and aquatic non-target organisms. Nanopesticides can extend the lifetime of traditional pesticides, increasing their availability, and changing the exposure dynamic in the environment. Some researchers have found that NPs can cross the cell biological barriers leading to structure damages and cell death. There are reports that confirm the potential of nanopesticides to be accumulated and transferred through food chain, causing long-term effects to the ecosystem. Furthermore, nanoparticles can suffer environmental transformations, which can impact on their physicochemical properties, transport, fate, and toxicological profile.

The adoption of a preventive assessment associated with a safe-by-design approach is a most suitable strategy to prevent the risks posed by nanopesticides. Also, the use of biopesticides (e.g., nanoencapsulated neem oil) can be encouraged, since they offer the possibility of controlling several pests at the same time that did not present serious impacts on the environment (Campos et al. 2016).

The uptake pathways, bioaccumulation, and biomagnification are issues to be further investigated. Additionally, environmental transformations and mixture toxicity of nanopesticides with classical pollutants need more attention of the ecotoxicology community. Indeed, it is critical to focus research efforts on more environmental relevant exposure conditions during ecotoxicity assessment of nanopesticides. The biomolecule corona formation effects need to be investigated (Docter et al. 2015; Pulido-Reyes et al. 2017; Baalousha et al. 2018). For example, the humic acid-corona formation on a nanopesticide model (AgNPs) had modulated its toxicity on plant roots as wells as on several biological models commonly used in nanoecotoxicology (Baalousha et al. 2018; Markiewicz et al. 2018). So far, it is a current challenge to develop analytical methods for corona characterization linked to toxicity and safety evaluation protocols for nanopesticides (Chetwynd et al. 2019).

The use of advanced techniques may be necessary to understand the biological and environmental impacts of nanopesticides. Many of these techniques (e.g., electron microscopy, mass spectrometry, atomic force microscopy, field-flow fractionation, etc.) are under harmonization and standardization by international organizations such as OECD and International Organization for Standardization

(ISO). However, no specific guidelines were reported for nanopesticide formulation characterization and ecotoxicity so far. Therefore, it is necessary to support the development of these guidelines for the ecotoxicological assessment of nanopesticides.

Nanoinformatics is an emerging approach in nanotoxicology and environmental nanosafety to make data FAIR (findable, accessible, interoperable, and reusable). The implementation of informatics tools will be very interesting for supporting the nanoecotoxicology community and regulatory agencies during the risk assessment and management of nanopesticides. Besides, nanoinformatics will collaborate to reduce the costs of nanosafety research involving nanopesticides as well as to enhance the quality and confidence of scientific results during regulatory discussions and decisions by governmental agencies and stakeholders involved in this topic. Nanoinformatics therefore will be of pivotal importance to Safe-by-Design on nanopesticide research and development. Finally, considering that nanotechnology is a global technology with many implications in the international commercialization of nano-products and regulatory issues between countries, the implementation of transnational online platforms will be essential for international access to nanosafety data, quality control, and harmonized protocols and methodologies.

## References

ABNT (2017) Associação Brasileira de Normas Técnicas. NBR 13373: Ecotoxicologia Aquática-Toxicidade Crônica-Método de Ensaio com Ceriodaphnia spp (Crustacea-Cladocera)

ABNT (2014) Associação Brasileira de Normas Técnicas. NBR 12713: Ecotoxicologia aquática-Toxicidade aguda-Método de ensaio com Daphnia spp (Cladocera, crustacea)

Adisa IO, Pullagurala VLR, Peralta-Videa JR et al (2019) Recent advances in nano-enabled fertilizers and pesticides: a critical review of mechanisms of action. Environ Sci Nano 6:2002–2030. https://doi.org/10.1039/c9en00265k

Al Naggar Y (2016) Effects of metal and metal oxide nanoparticles on honey bees (Apis millefera L.). J Aquat Pollut Toxicol 1:1

Al Naggar Y, Dabour K, Masry S et al (2018) Sublethal effects of chronic exposure to CdO or PbO nanoparticles or their binary mixture on the honey bee (Apis millefera L.). Environ Sci Pollut Res. https://doi.org/10.1007/s11356-018-3314-2

Alaraby M, Annangi B, Marcos R, Hernández A (2016) Drosophila melanogaster as a suitable in vivo model to determine potential side effects of nanomaterials: a review. J Toxicol Environ Health B Crit Rev 19:65–104. https://doi.org/10.1080/10937404.2016.1166466

Alaraby M, Romero S, Hernández A, Marcos R (2019) Toxic and genotoxic effects of silver nanoparticles in Drosophila. Environ Mol Mutagen 60:277–285. https://doi.org/10.1002/em.22262

Anand AS, Prasad DN, Singh SB, Kohli E (2017) Chronic exposure of zinc oxide nanoparticles causes deviant phenotype in Drosophila melanogaster. J Hazard Mater 327:180–186. https://doi.org/10.1016/j.jhazmat.2016.12.040

de Andrade LL, Santo Pereira ADE, Fraceto LF, dos Reis Martinez CB (2019) Can atrazine loaded nanocapsules reduce the toxic effects of this herbicide on the fi sh Prochilodus lineatus? A multibiomarker approach. Sci Total Environ 663:548–559. https://doi.org/10.1016/j.scitotenv.2019.01.380

Antonoglou O, Moustaka J, Adamakis IDS et al (2018) Nanobrass CuZn nanoparticles as foliar spray nonphytotoxic fungicides. ACS Appl Mater Interfaces 10:4450–4461. https://doi.org/10.1021/acsami.7b17017

Arratia F, Olivares-Ferretti P, García-Rodríguez A et al (2019) Comparative toxic effects of copper-based nanoparticles and their microparticles in *Daphnia magna* by using natural freshwater media. N Z J Mar Freshw Res 53:460–469. https://doi.org/10.1080/00288330.2019.1598447

ASTM (2012) ASTM E1193 - 97(2012) Standard guide for conducting Daphnia magna life-cycle toxicity tests, p 19

Asztemborska M, Jakubiak M, Książyk M et al (2014) Silver nanoparticle accumulation by aquatic organisms – neutron activation as a tool for the environmental fate of nanoparticles tracing. Nukleonika 59:169–173. https://doi.org/10.2478/nuka-2014-0023

Asztemborska M, Jakubiak M, Stęborowski R et al (2018) Titanium dioxide nanoparticle circulation in an aquatic ecosystem. Water Air Soil Pollut 229:208. https://doi.org/10.1007/s11270-018-3852-8

Baalousha M, Afshinnia K, Guo L (2018) Natural organic matter composition determines the molecular nature of silver nanomaterial-NOM corona. Environ Sci Nano 5:868–881. https://doi.org/10.1039/c8en00018b

Balaji APB, Sastry TP, Manigandan S et al (2017) Environmental benignity of a pesticide in soft colloidal hydrodispersive nanometric form with improved toxic precision towards the target organisms than non-target organisms. Sci Total Environ 579:190–201. https://doi.org/10.1016/j.scitotenv.2016.10.240

Bañares MA, Haase A, Tran L et al (2017) CompNanoTox2015: novel perspectives from a European conference on computational nanotoxicology on predictive nanotoxicology. Nanotoxicology 11:839–845. https://doi.org/10.1080/17435390.2017.1371351

Bao D, Oh ZG, Chen Z (2016) Characterization of silver nanoparticles internalized by Arabidopsis plants using single particle ICP-MS analysis. Front Plant Sci 7:32. https://doi.org/10.3389/fpls.2016.00032

Baun A, Sayre P, Steinhäuser KG, Rose J (2017) Regulatory relevant and reliable methods and data for determining the environmental fate of manufactured nanomaterials. NanoImpact 8:1–10. https://doi.org/10.1016/J.IMPACT.2017.06.004

Belluco S, Losasso C, Maggioletti M et al (2013) Edible insects in a food safety and nutritional perspective: a critical review. Compr Rev Food Sci Food Saf 12:296–313. https://doi.org/10.1111/1541-4337.12014

Bhuvaneshwari M, Iswarya V, Vishnu S et al (2018) Dietary transfer of zinc oxide particles from algae (Scenedesmus obliquus) to daphnia (Ceriodaphnia dubia). Environ Res 164:395–404. https://doi.org/10.1016/J.ENVRES.2018.03.015

Bone AJ, Matson CW, Colman BP et al (2015) Silver nanoparticle toxicity to Atlantic killifish (Fundulus heteroclitus) and Caenorhabditis elegans: a comparison of mesocosm, microcosm, and conventional laboratory studies. Environ Toxicol Chem 34:275. https://doi.org/10.1002/etc.2806

Botha TL, Boodhia K, Wepener V (2016) Adsorption, uptake and distribution of gold nanoparticles in Daphnia magna following long term exposure. Aquat Toxicol 170:104–111. https://doi.org/10.1016/j.aquatox.2015.11.022

Botha TL, Brand SJ, Ikenaka Y et al (2019) How toxic is a non-toxic nanomaterial: behaviour as an indicator of effect in Danio rerio exposed to nanogold. Aquat Toxicol 215:105287. https://doi.org/10.1016/j.aquatox.2019.105287

Bour A, Mouchet F, Cadarsi S et al (2016a) Toxicity of CeO2 nanoparticles on a freshwater experimental trophic chain: a study in environmentally relevant conditions through the use of mesocosms. Nanotoxicology 10:245–255. https://doi.org/10.3109/17435390.2015.1053422

Bour A, Mouchet F, Cadarsi S et al (2016b) Impact of CeO2 nanoparticles on the functions of freshwater ecosystems: a microcosm study. Environ Sci Nano 3:830–838. https://doi.org/10.1039/c6en00116e

Braunbeck T, Kais B, Lammer E et al (2014) The fish embryo test (FET): origin, applications, and future. Environ Sci Pollut Res 22:16247. https://doi.org/10.1007/s11356-014-3814-7

Brinke M, Heininger P, Traunspurger W (2011) A semi-fluid gellan gum medium improves nematode toxicity testing. Ecotoxicol Environ Saf 74:1824–1831. https://doi.org/10.1016/j.ecoenv.2011.07.007

Cáceres TP, Megharaj M, Naidu R (2008) Biodegradation of the pesticide fenamiphos by ten different species of green algae and cyanobacteria. Curr Microbiol 57:643–646. https://doi.org/10.1007/s00284-008-9293-7

Cai P, Zhang X, Wang M et al (2018) Combinatorial nano–bio interfaces. ACS Nano 12:5078–5084. https://doi.org/10.1021/acsnano.8b03285

Campos EVR, de Oliveira JL, Pascoli M et al (2016) Neem oil and crop protection: from now to the future. Front Plant Sci 7. https://doi.org/10.3389/fpls.2016.01494

Carmona ER, Escobar B, Vales G, Marcos R (2015) Genotoxic testing of titanium dioxide anatase nanoparticles using the wing-spot test and the comet assay in Drosophila. Mutat Res Genet Toxicol Environ Mutagen 778:12–21. https://doi.org/10.1016/j.mrgentox.2014.12.004

Casman EA, Gernand JM (2016) Seeing the trees for the forest. Nat Nanotechnol 11:405–407. https://doi.org/10.1038/nnano.2016.5

Castillo-Michel HA, Larue C, Pradas del Real AE et al (2017) Practical review on the use of synchrotron based micro- and nano- X-ray fluorescence mapping and X-ray absorption spectroscopy to investigate the interactions between plants and engineered nanomaterials. Plant Physiol Biochem 110:13–32. https://doi.org/10.1016/j.plaphy.2016.07.018

Chae Y, An Y-J (2016) Toxicity and transfer of polyvinylpyrrolidone-coated silver nanowires in an aquatic food chain consisting of algae, water fleas, and zebrafish. Aquat Toxicol 173:94–104. https://doi.org/10.1016/j.aquatox.2016.01.011

Chae Y, Kim SW, An Y-J (2016) In vivo visual evaluation of nanoparticle transfer in a three-species terrestrial food chain. Chemosphere 151:101–107. https://doi.org/10.1016/j.chemosphere.2016.02.075

Chatterjee N, Kim Y, Yang J et al (2017) A systems toxicology approach reveals the Wnt-MAPK crosstalk pathway mediated reproductive failure in Caenorhabditis elegans exposed to graphene oxide (GO) but not to reduced graphene oxide (rGO). Nanotoxicology 11:76–86. https://doi.org/10.1080/17435390.2016.1267273

Chen F, Xiao Z, Yue L et al (2019a) Algae response to engineered nanoparticles: current understanding, mechanisms and implications. Environ Sci Nano 6:1026–1042. https://doi.org/10.1039/C8EN01368C

Chen J, Li H, Han X, Wei X (2015a) Transmission and accumulation of nano-TiO2 in a 2-step food chain (Scenedesmus obliquus to Daphnia magna). Bull Environ Contam Toxicol 95:145–149. https://doi.org/10.1007/s00128-015-1580-y

Chen S, Paunesku T, Yuan Y et al (2015b) The bionanoprobe: synchrotron-based hard X-ray fluorescence microscopy for 2D/3D trace element mapping. Micros Today 23:26–29. https://doi.org/10.1017/s1551929515000401

Chen X, Zhu Y, Yang K et al (2019b) Nanoparticle TiO2 size and rutile content impact bioconcentration and biomagnification from algae to daphnia. Environ Pollut 247:421–430. https://doi.org/10.1016/j.envpol.2019.01.022

Chetwynd AJ, Wheeler KE, Lynch I (2019) Best practice in reporting corona studies: minimum information about nanomaterial biocorona experiments (MINBE). Nano Today 28:100758. https://doi.org/10.1016/j.nantod.2019.06.004

Chifiriuc MC, Ratiu AC, Popa M, Ecovoiu AA (2016) Drosophotoxicology: an emerging research area for assessing nanoparticles interaction with living organisms. Int J Mol Sci 17:1–14. https://doi.org/10.3390/ijms17020036

Clemente Z, Grillo R, Jonsson M et al (2014) Ecotoxicological evaluation of poly (ε-caprolactone) nanocapsules containing triazine herbicides. J Nanosci Nanotechnol 14:4911–4917. https://doi.org/10.1166/jnn.2014.8681

Côa F, Strauss M, Clemente Z et al (2017) Coating carbon nanotubes with humic acid using an eco-friendly mechanochemical method: application for Cu(II) ions removal from water and aquatic ecotoxicity. Sci Total Environ 607–608:1479–1486. https://doi.org/10.1016/j.scitotenv.2017.07.045

Coleman JG, Kennedy AJ, Bednar AJ et al (2013) Comparing the effects of nanosilver size and coating variations on bioavailability, internalization, and elimination, using Lumbriculus variegatus. Environ Toxicol Chem 32:2069–2077. https://doi.org/10.1002/etc.2278

Contado C (2017) Field flow fractionation techniques to explore the nano-world. Anal Bioanal Chem 409:2501. https://doi.org/10.1007/s00216-017-0180-6

Cornelis G, Hund-Rinke K, Kuhlbusch T et al (2014) Fate and bioavailability of engineered nanoparticles in soils: a review. Crit Rev Environ Sci Technol 44:2720–2764. https://doi.org/10.1080/10643389.2013.829767

Dabour K, Al Naggar Y, Masry S et al (2019) Cellular alterations in midgut cells of honey bee workers (Apis millefera L.) exposed to sublethal concentrations of CdO or PbO nanoparticles or their binary mixture. Sci Total Environ 651:1356–1367. https://doi.org/10.1016/j.scitotenv.2018.09.311

Dağlioğlu Y, Kabakçi D, Akdeniz G (2015) Toxicity of nano and non-nano boron particles on Apis mellifera (honey bee). Res J Chem Environ Sci 3:6–13

Dağlioğlu Y, Kabakci D, Akdenİz G, Çelebi MS (2016) Determining the acute toxicity effects of poly(vinylferrocenium) supported platinum nanoparticle (PT/PVF+NPs) on Apis mellifera. Mugla J Sci Technol 2:1–8

Dale AL, Casman EA, Lowry GV et al (2015) Modeling nanomaterial environmental fate in aquatic systems. Environ Sci Technol 49:2587–2593. https://doi.org/10.1021/es505076w

Dang F, Chen Y-Z, Huang Y-N et al (2019) Discerning the sources of silver nanoparticle in a terrestrial food chain by stable isotope tracer technique. Environ Sci Technol 53:3802–3810. https://doi.org/10.1021/acs.est.8b06135

Dauda S, Chia MA, Bako SP (2017) Toxicity of titanium dioxide nanoparticles to Chlorella vulgaris Beyerinck (Beijerinck) 1890 (Trebouxiophyceae, Chlorophyta) under changing nitrogen conditions. Aquat Toxicol 187:108–114. https://doi.org/10.1016/j.aquatox.2017.03.020

De La Torre Roche R, Servin A, Hawthorne J et al (2015) Terrestrial trophic transfer of bulk and nanoparticle La2O3 does not depend on particle size. Environ Sci Technol 49:11866–11874. https://doi.org/10.1021/acs.est.5b02583

Degueldre C, Favarger P-Y (2003) Colloid analysis by single particle inductively coupled plasma-mass spectroscopy: a feasibility study. Colloids Surf A Physicochem Eng Asp 217:137–142. https://doi.org/10.1016/S0927-7757(02)00568-X

Deng R, Lin D, Zhu L et al (2017) Nanoparticle interactions with co-existing contaminants: joint toxicity, bioaccumulation and risk. Nanotoxicology 11:591–612. https://doi.org/10.1080/17435390.2017.1343404

Djearamane S, Wong LS, Lim YM, Lee PF (2019) Short-term cytotoxicity of zinc oxide nanoparticles on Chlorella vulgaris. Sains Malaysiana 48:69–73. https://doi.org/10.17576/jsm-2019-4801-08

Docter D, Westmeier D, Markiewicz M et al (2015) The nanoparticle biomolecule corona: lessons learned – challenge accepted? Chem Soc Rev 44:6094–6121. https://doi.org/10.1039/C5CS00217F

Doke SK, Dhawale SC (2015) Alternatives to animal testing: a review. Saudi Pharm J 23:223–229. https://doi.org/10.1016/j.jsps.2013.11.002

Donia DT, Carbone M (2019) Fate of the nanoparticles in environmental cycles. Int J Environ Sci Technol 16:583–600. https://doi.org/10.1007/s13762-018-1960-z

Drescher D, Giesen C, Traub H et al (2012) Quantitative imaging of gold and silver nanoparticles in single eukaryotic cells by laser ablation ICP-MS. Anal Chem 84:9684–9688. https://doi.org/10.1021/ac302639c

Durán Zuazo VH, Martínez JRF, Pleguezuelo CRR et al (2006) Soil-erosion and runoff prevention by plant covers in a mountainous area (SE Spain): implications for sustainable agriculture. Environmentalist 26:309–319. https://doi.org/10.1007/s10669-006-0160-4

Eady M, Park B (2019) The influence of environmental growth conditions on Salmonella spectra obtained from hyperspectral microscope images. Food Anal Methods 12:2638–2646. https://doi.org/10.1007/s12161-019-01618-0

Edwards CA (2004) Earthworm ecology, 2nd edn. CRC Press, Boca Raton

Fadeel B, Farcal L, Hardy B et al (2018) Advanced tools for the safety assessment of nanomaterials. Nat Nanotechnol 13:537–543. https://doi.org/10.1038/s41565-018-0185-0

Forest V, Hochepied J-F, Pourchez J (2019) Importance of choosing relevant biological end points to predict nanoparticle toxicity with computational approaches for human health risk assessment. Chem Res Toxicol 32:1320–1326. https://doi.org/10.1021/acs.chemrestox.9b00022

Franqui LS, De Farias MA, Portugal RV et al (2019) Interaction of graphene oxide with cell culture medium: evaluating the fetal bovine serum protein corona formation towards in vitro nanotoxicity assessment and nanobiointeractions. Mater Sci Eng C 100:363–377. https://doi.org/10.1016/J.MSEC.2019.02.066

Frenk S, Ben-Moshe T, Dror I et al (2013) Effect of metal oxide nanoparticles on microbial community structure and function in two different soil types. PLoS One 8:e84441. https://doi.org/10.1371/journal.pone.0084441

Gao X, Avellan A, Laughton S et al (2018) CuO nanoparticle dissolution and toxicity to wheat (Triticum aestivum) in Rhizosphere soil. Environ Sci Technol 52:2888–2897. https://doi.org/10.1021/acs.est.7b05816

Gao Y, Liu N, Chen C et al (2008) Mapping technique for biodistribution of elements in a model organism, Caenorhabditis elegans, after exposure to copper nanoparticles with microbeam synchrotron radiation X-ray fluorescence. J Anal At Spectrom 23:1121–1124. https://doi.org/10.1039/b802338g

Gardea-Torresdey JL, Rico CM, White JC (2014) Trophic transfer, transformation, and impact of engineered nanomaterials in terrestrial environments. Environ Sci Technol 48:2526–2540. https://doi.org/10.1021/es4050665

Geitner NK, Cooper JL, Avellan A et al (2018) Size-based differential transport, uptake, and mass distribution of ceria (CeO2) nanoparticles in wetland mesocosms. Environ Sci Technol 52:9768–9776. https://doi.org/10.1021/acs.est.8b02040

Gkanatsiou C, Karamanoli K, Menkissoglu-Spiroudi U, Dendrinou-Samara C (2019) Composition effect of Cu-based nanoparticles on phytopathogenic bacteria. Antibacterial studies and phytotoxicity evaluation. Polyhedron 170:395–403. https://doi.org/10.1016/j.poly.2019.06.002

Glavan G, Milivojević T, Božič J et al (2017) Feeding preference and sub-chronic effects of ZnO nanomaterials in honey bees (Apis mellifera carnica). Arch Environ Contam Toxicol 72:471–480. https://doi.org/10.1007/s00244-017-0385-x

Gomes SIL, Scott-Fordsmand JJ, Campos EVR et al (2019) On the safety of nanoformulations to non-target soil invertebrates-an atrazine case study. Environ Sci Nano 6:1950–1958. https://doi.org/10.1039/c9en00242a

Gómez-Sagasti MT, Epelde L, Anza M et al (2019) The impact of nanoscale zero-valent iron particles on soil microbial communities is soil dependent. J Hazard Mater 364:591–599. https://doi.org/10.1016/j.jhazmat.2018.10.034

Gonzalez-Moragas L, Roig A, Laromaine A (2015) C. elegans as a tool for in vivo nanoparticle assessment. Adv Colloid Interf Sci 219:10–26. https://doi.org/10.1016/j.cis.2015.02.001

Grasseschi D, Lima FS, Nakamura M, Toma HE (2015) Hyperspectral dark-field microscopy of gold nanodisks. Micron 69:15–20. https://doi.org/10.1016/j.micron.2014.10.007

Gray AL (1985) The ICP as an ion source—origins, achievements and prospects. Spectrochim Acta B At Spectrosc 40:1525–1537. https://doi.org/10.1016/0584-8547(85)80176-2

Groh KJ, Dalkvist T, Piccapietra F et al (2014) Critical influence of chloride ions on silver ion-mediated acute toxicity of silver nanoparticles to zebrafish embryos. Nanotoxicology 5390:1–11. https://doi.org/10.3109/17435390.2014.893379

Gunsolus IL, Haynes CL (2016) Analytical aspects of nanotoxicology. Anal Chem 88:451–479. https://doi.org/10.1021/acs.analchem.5b04221

Haase A, Klaessig F (2018) EU US roadmap nanoinformatics 2030. EU Nanosafety Clust 127. https://doi.org/10.5281/zenodo.1486012

Handy RD, Cornelis G, Fernandes T et al (2012a) Ecotoxicity test methods for engineered nanomaterials: practical experiences and recommendations from the bench. Environ Toxicol Chem 31:15–31. https://doi.org/10.1002/etc.706

Handy RD, van den Brink N, Chappell M et al (2012b) Practical considerations for conducting ecotoxicity test methods with manufactured nanomaterials: what have we learnt so far? Ecotoxicology 21:933–972. https://doi.org/10.1007/s10646-012-0862-y

Harjumäki R, Nugroho RWN, Zhang X et al (2019) Quantified forces between HepG2 hepatocarcinoma and WA07 pluripotent stem cells with natural biomaterials correlate with in vitro cell behavior. Sci Rep 9:7354. https://doi.org/10.1038/s41598-019-43669-7

Hawthorne J, De La Torre Roche R, Xing B et al (2014) Particle-size dependent accumulation and trophic transfer of cerium oxide through a terrestrial food chain. Environ Sci Technol 48:13102–13109. https://doi.org/10.1021/es503792f

Hayles J, Johnson L, Worthley C, Losic D (2017) Nanopesticides: a review of current research and perspectives. In: New pesticides and soil sensors. Elsevier, London, pp 193–225

He X, Deng H, Hwang H (2019) The current application of nanotechnology in food and agriculture. J Food Drug Anal 27:1–21. https://doi.org/10.1016/j.jfda.2018.12.002

He X, Fu P, Aker WG, Hwang HM (2018) Toxicity of engineered nanomaterials mediated by nano–bio–eco interactions. J Environ Sci Health C 36:21–42

Hochella MF, Mogk DW, Ranville J et al (2019) Natural, incidental, and engineered nanomaterials and their impacts on the Earth system. Science 363:eaau8299. https://doi.org/10.1126/science.aau8299

de Hoffmann E, Stroobant V (2007) Mass spectrometry: principles and applications, 3rd edn. Wiley, New York

Holbrook RD, Murphy KE, Morrow JB, Cole KD (2008) Trophic transfer of nanoparticles in a simplified invertebrate food web. Nat Nanotechnol 3:352–355. https://doi.org/10.1038/nnano.2008.110

Holden PA, Schimel JP, Godwin HA (2014) Five reasons to use bacteria when assessing manufactured nanomaterial environmental hazards and fates. Curr Opin Biotechnol 27:73–78. https://doi.org/10.1016/j.copbio.2013.11.008

Horch HW, Mito T, Popadi A et al (2017) The cricket as a model organism. Springer, Tokyo

Huang C-W, Li S-W, Liao VH-C (2019) Long-term sediment exposure to ZnO nanoparticles induces oxidative stress in Caenorhabditis elegans. Environ Sci Nano 6:2602–2614. https://doi.org/10.1039/C9EN00039A

Hunt PR (2017) The C. elegans model in toxicity testing. J Appl Toxicol 37:50–59. https://doi.org/10.1002/jat.3357

Iswarya V, Bhuvaneshwari M, Alex SA et al (2015) Combined toxicity of two crystalline phases (anatase and rutile) of Titania nanoparticles towards freshwater microalgae: Chlorella sp. Aquat Toxicol 161:154–169. https://doi.org/10.1016/J.AQUATOX.2015.02.006

Iswarya V, Bhuvaneshwari M, Chandrasekaran N, Mukherjee A (2018) Trophic transfer potential of two different crystalline phases of TiO2 NPs from Chlorella sp. to Ceriodaphnia dubia. Aquat Toxicol 197:89–97. https://doi.org/10.1016/j.aquatox.2018.02.003

Jackson BP, Bugge D, Ranville JF, Chen CY (2012) Bioavailability, toxicity, and bioaccumulation of quantum dot nanoparticles to the amphipod Leptocheirus plumulosus. Environ Sci Technol 46:5550–5556. https://doi.org/10.1021/es202864r

Jackson BP, Pace HE, Lanzirotti A et al (2009) Synchrotron X-ray 2D and 3D elemental imaging of CdSe/ZnS quantum dot nanoparticles in Daphnia magna. Anal Bioanal Chem 394:911–917. https://doi.org/10.1007/s00216-009-2768-y

Jager T, van der Wal L, Fleuren RHLJ et al (2005) Bioaccumulation of organic chemicals in contaminated soils: evaluation of bioassays with earthworms. Environ Sci Technol 39:293–298. https://doi.org/10.1021/es035317o

Jaques TM, Oliveira J, Campos E et al (2017) Safety assessment of nanopesticides using the roundworm Caenorhabditis elegans. Ecotoxicol Environ Saf 139:245–253. https://doi.org/10.1016/j.ecoenv.2017.01.045

Jemec A, Milivojevic T, Drobne D et al (2016) No chronic effects on biochemical biomarkers, feeding and survival of carnolian honeybees (Apis mellifera carnica) after exposure to nano-sized carbon black and titanium dioxide. Acta Biol Slov 59:45–55

Jośko I, Oleszczuk P, Dobrzyńska J et al (2019) Long-term effect of ZnO and CuO nanoparticles on soil microbial community in different types of soil. Geoderma 352:204–212. https://doi.org/10.1016/j.geoderma.2019.06.010

Jovanović B, Milosević D, Piperac MS, Savić A (2016) In situ effects of titanium dioxide nanoparticles on community structure of freshwater benthic macroinvertebrates. Environ Pollut 213:278–282. https://doi.org/10.1016/j.envpol.2016.02.024

Judy JD, Unrine JM, Rao W, Bertsch PM (2012) Bioaccumulation of gold nanomaterials by manduca sexta through dietary uptake of surface contaminated plant tissue. Environ Sci Technol 46:12672–12678. https://doi.org/10.1021/es303333w

Kah M (2015) Nanopesticides and nanofertilizers: emerging contaminants or opportunities for risk mitigation? Front Chem 3:64. https://doi.org/10.3389/fchem.2015.00064

Kah M, Beulke S, Tiede K, Hofmann T (2013) Nanopesticides: state of knowledge, environmental fate, and exposure modeling. Crit Rev Environ Sci Technol 43:1823–1867. https://doi.org/10.1080/10643389.2012.671750

Kah M, Hofmann T (2014) Nanopesticide research: current trends and future priorities. Environ Int 63:224–235. https://doi.org/10.1016/j.envint.2013.11.015

Kah M, Kookana RS, Gogos A, Bucheli TD (2018) A critical evaluation of nanopesticides and nanofertilizers against their conventional analogues. Nat Nanotechnol 13:677–684. https://doi.org/10.1038/s41565-018-0131-1

Kalia A, Gosal SK (2011) Effect of pesticide application on soil microorganisms. Arch Agron Soil Sci 57:569–596. https://doi.org/10.1080/03650341003787582

Kalman J, Paul KB, Khan FR et al (2015) Characterisation of bioaccumulation dynamics of three differently coated silver nanoparticles and aqueous silver in a simple freshwater food chain. Environ Chem 12:662. https://doi.org/10.1071/EN15035

Kent RD, Vikesland PJ (2016) Dissolution and persistence of copper-based nanomaterials in undersaturated solutions with respect to cupric solid phases. Environ Sci Technol 50:6772–6781. https://doi.org/10.1021/acs.est.5b04719

Kim J, Kim S, Lee S (2011) Differentiation of the toxicities of silver nanoparticles and silver ions to the Japanese medaka ( *Oryzias latipes*) and the cladoceran *Daphnia magna*. Nanotoxicology 5:208–214. https://doi.org/10.3109/17435390.2010.508137

Kim JH, Lee SH, Cha YJ et al (2017) C. elegans-on-a-chip for in situ and in vivo Ag nanoparticles' uptake and toxicity assay. Sci Rep 7:40225. https://doi.org/10.1038/srep40225

Kim JS, Kuk E, Yu KN et al (2007) Antimicrobial effects of silver nanoparticles. Nanomed Nanotechnol Biol Med 3:95–101. https://doi.org/10.1016/j.nano.2006.12.001

Kim SW, Kwak JI, An Y-J (2016) Fluorescent approach for visually observing quantum dot uptake in living organisms. Chemosphere 144:1763–1770. https://doi.org/10.1016/j.chemosphere.2015.10.065

Kim SW, Kwak JI, An Y-J (2013) Multigenerational study of gold nanoparticles in Caenorhabditis elegans: transgenerational effect of maternal exposure. Environ Sci Technol 47:5393–5399. https://doi.org/10.1021/es304511z

Klaine SJ, Alvarez PJJ, Batley GE et al (2008) Nanomaterials in the environment: behavior, fate, bioavailability, and effects. Environ Toxicol Chem 27:1825. https://doi.org/10.1897/08-090.1

Kleiven M, Macken A, Oughton DH (2019) Growth inhibition in Raphidocelis subcapita – evidence of nanospecific toxicity of silver nanoparticles. Chemosphere 221:785–792. https://doi.org/10.1016/J.CHEMOSPHERE.2019.01.055

Kleiven M, Oughton D (2015) Standard operating procedure: toxicity test with the nematode caenorhabiditis elegans for the NANoREG core nanomaterials

Koo Y, Wang J, Zhang Q et al (2015) Fluorescence reports intact quantum dot uptake into roots and translocation to leaves of Arabidopsis thaliana and subsequent ingestion by insect herbivores. Environ Sci Technol 49:626–632. https://doi.org/10.1021/es5050562

Kookana RS, Boxall ABA, Reeves PT et al (2014) Nanopesticides: guiding principles for regulatory evaluation of environmental risks. J Agric Food Chem 62:4227–4240. https://doi.org/10.1021/jf500232f

Kopittke PM, Punshon T, Paterson DJ et al (2018) Synchrotron-based X-ray fluorescence microscopy as a technique for imaging of elements in plants. Plant Physiol 178:507–523. https://doi.org/10.1104/pp.18.00759

Kos M, Jemec Kokalj A, Glavan G et al (2017) Cerium(IV) oxide nanoparticles induce sublethal changes in honeybees after chronic exposure. Environ Sci Nano 4:2297–2310. https://doi.org/10.1039/c7en00596b

Kubo-Irie M, Yokoyama M, Shinkai Y et al (2016) The transfer of titanium dioxide nanoparticles from the host plant to butterfly larvae through a food chain. Sci Rep 6:23819. https://doi.org/10.1038/srep23819

Kumar R, Bhatia R, Kukreja K et al (2007) Establishment of Azotobacter on plant roots: chemotactic response, development and analysis of root exudates of cotton (Gossypium hirsutum L.) and wheat (Triticum aestivum L.). J Basic Microbiol 47:436–439. https://doi.org/10.1002/jobm.200610285

Kumar R, Gopal M, Pabbi S et al (2016) Effect of nanohexaconazole on nitrogen fixing blue green algae and bacteria. J Nanosci Nanotechnol 16:643–647. https://doi.org/10.1166/jnn.2016.10901

Kwak JI, An Y-J (2016a) The current state of the art in research on engineered nanomaterials and terrestrial environments: different-scale approaches. Environ Res 151:368–382. https://doi.org/10.1016/j.envres.2016.08.005

Kwak JI, An Y-J (2016b) Trophic transfer of silver nanoparticles from earthworms disrupts the locomotion of springtails (Collembola). J Hazard Mater 315:110–116. https://doi.org/10.1016/j.jhazmat.2016.05.005

Laborda F, Bolea E, Jiménez-Lamana J (2016) Single particle inductively coupled plasma mass spectrometry for the analysis of inorganic engineered nanoparticles in environmental samples. Trends Environ Anal Chem 9:15–23. https://doi.org/10.1016/j.teac.2016.02.001

Lammel T, Thit A, Mouneyrac C et al (2019) Trophic transfer of CuO NPs and dissolved cu from sediment to worms to fish – a proof-of-concept study. Environ Sci Nano 6:1140–1155. https://doi.org/10.1039/C9EN00093C

Lead JR, Batley GE, Alvarez PJJ et al (2018) Nanomaterials in the environment: behavior, fate, bioavailability, and effects-An updated review. Environ Toxicol Chem 37:2029–2063. https://doi.org/10.1002/etc.4147

Lee JG (2016) Computational materials science, second. CRC Press, Boca Ranton

Lee KY, Jang GH, Byun CH et al (2017) Zebrafish models for functional and toxicological screening of nanoscale drug delivery systems: promoting preclinical applications. Biosci Rep 37:1–13. https://doi.org/10.1042/BSR20170199

Lee SH, Lee HY, Lee EJ et al (2015) Effects of carbon nanofiber on physiology of Drosophila. Int J Nanomedicine 10:3687–3697. https://doi.org/10.2147/IJN.S82637

Lee SW, Park SY, Kim Y et al (2016) Effect of sulfidation and dissolved organic matters on toxicity of silver nanoparticles in sediment dwelling organism, Chironomus riparius. Sci Total Environ 553:565–573. https://doi.org/10.1016/j.scitotenv.2016.02.064

Lee W, Kim E, Cho H et al (2018) The relationship between dissolution behavior and the toxicity of silver nanoparticles on Zebrafish embryos in different ionic environments. Nano 8:652. https://doi.org/10.3390/nano8090652

Lekamge S, Ball AS, Shukla R, Nugegoda D (2018) The toxicity of nanoparticles to organisms in freshwater. Springer, Cham, pp 1–80

Lekamge S, Miranda AF, Ball AS et al (2019) The toxicity of coated silver nanoparticles to Daphnia carinata and trophic transfer from alga Raphidocelis subcapitata. PLoS One 14:e0214398. https://doi.org/10.1371/journal.pone.0214398

Li X, Schirmer K, Bernard L et al (2015) Silver nanoparticle toxicity and association with the alga Euglena gracilis. Environ Sci Nano 2:594–602. https://doi.org/10.1039/C5EN00093A

Limbeck A, Galler P, Bonta M et al (2015) Recent advances in quantitative LA-ICP-MS analysis: challenges and solutions in the life sciences and environmental chemistry. Anal Bioanal Chem 407:6593–6617. https://doi.org/10.1007/s00216-015-8858-0

Lin P-C, Lin S, Wang PC, Sridhar R (2014) Techniques for physicochemical characterization of nanomaterials. Biotechnol Adv 32:711–726. https://doi.org/10.1016/j.biotechadv.2013.11.006

Liu H, Wang X, Wu Y et al (2019) Toxicity responses of different organs of zebrafish (Danio rerio) to silver nanoparticles with different particle sizes and surface coatings. Environ Pollut 246:414–422. https://doi.org/10.1016/j.envpol.2018.12.034

Liu W, Yao J, Cai M et al (2014) Synthesis of a novel nanopesticide and its potential toxic effect on soil microbial activity. J Nanopart Res 16:2677. https://doi.org/10.1007/s11051-014-2677-7

Lombi E, Donner E, Dusinska M, Wickson F (2019) A one health approach to managing the applications and implications of nanotechnologies in agriculture. Nat Nanotechnol 14:523–531. https://doi.org/10.1038/s41565-019-0460-8

Lombi E, Susini J (2009) Synchrotron-based techniques for plant and soil science: opportunities, challenges and future perspectives. Plant Soil 320:1–35. https://doi.org/10.1007/s11104-008-9876-x

Lopes LQS, Santos CG, de Almeida VR et al (2016) Evaluation of antimicrobial activity of glycerol monolaurate nanocapsules against American foulbrood disease agent and toxicity on bees. Microb Pathog 97:183–188. https://doi.org/10.1016/j.micpath.2016.05.014

Luo X, Xu S, Yang Y et al (2016) Insights into the ecotoxicity of silver nanoparticles transferred from Escherichia coli to Caenorhabditis elegans. Sci Rep 6:36465. https://doi.org/10.1038/srep36465

Lv J, Christie P, Zhang S (2019) Uptake, translocation, and transformation of metal-based nanoparticles in plants: recent advances and methodological challenges. Environ Sci Nano 6:41–59. https://doi.org/10.1039/C8EN00645H

Ma Y (2016) Characterization of nanomaterials in nanotoxicological analyses. In: Toxicology of nanomaterials. Wiley, Weinheim, pp 1–21

Ma Y, He X, Zhang P et al (2017) Xylem and phloem based transport of CeO2 nanoparticles in hydroponic cucumber plants. Environ Sci Technol 51:5215–5221. https://doi.org/10.1021/acs.est.6b05998

Majumdar S, Peralta-Videa JR, Castillo-Michel H et al (2012) Applications of synchrotron μ-XRF to study the distribution of biologically important elements in different environmental matrices: a review. Anal Chim Acta 755:1–16. https://doi.org/10.1016/j.aca.2012.09.050

Majumdar S, Trujillo-Reyes J, Hernandez-Viezcas JA et al (2016) Cerium biomagnification in a terrestrial food chain: influence of particle size and growth stage. Environ Sci Technol 50:6782–6792. https://doi.org/10.1021/acs.est.5b04784

Mancini F, Woodcock BA, Isaac NJB (2019) Agrochemicals in the wild: identifying links between pesticide use and declines of non-target organisms. Curr Opin Environ Sci Health 11:53. https://doi.org/10.1016/j.coesh.2019.07.003

Mao BH, Chen ZY, Wang YJ, Yan SJ (2018) Silver nanoparticles have lethal and sublethal adverse effects on development and longevity by inducing ROS-mediated stress responses. Sci Rep 8:1–16. https://doi.org/10.1038/s41598-018-20728-z

Maret W (2018) Metallomics: the science of biometals and biometalloids. Springer, Cham, pp 1–20

Markiewicz M, Kumirska J, Lynch I et al (2018) Changing environments and biomolecule coronas: consequences and challenges for the design of environmentally acceptable engineered nanoparticles. Green Chem 20:4133–4168. https://doi.org/10.1039/c8gc01171k

McTeer J, Dean AP, White KN, Pittman JK (2014) Bioaccumulation of silver nanoparticles into Daphnia magna from a freshwater algal diet and the impact of phosphate availability. Nanotoxicology 8:305–316. https://doi.org/10.3109/17435390.2013.778346

Meng X, Zhu F, Chen K (2017) Silkworm: a promising model organism in life science. J Insect Sci 17:97. https://doi.org/10.1093/jisesa/iex064

Meredith AN, Harper B, Harper SL (2016) The influence of size on the toxicity of an encapsulated pesticide : a comparison of micron- and nano-sized capsules. Environ Int 86:68–74. https://doi.org/10.1016/j.envint.2015.10.012

Meyer JN, Lord CA, Yang XY et al (2010) Intracellular uptake and associated toxicity of silver nanoparticles in Caenorhabditis elegans. Aquat Toxicol 100:140–150. https://doi.org/10.1016/j.aquatox.2010.07.016

Milivojević T, Glavan G, Božič J et al (2015) Neurotoxic potential of ingested ZnO nanomaterials on bees. Chemosphere 120:547–554. https://doi.org/10.1016/j.chemosphere.2014.07.054

Milošević D, Simić V, Stojković M et al (2013) Spatio-temporal pattern of the Chironomidae community: toward the use of non-biting midges in bioassessment programs. Aquat Ecol 47:37–55. https://doi.org/10.1007/s10452-012-9423-y

Mishra P, Dutta S, Haldar M et al (2019) Enhanced mosquitocidal efficacy of colloidal dispersion of pyrethroid nanometric emulsion with benignity towards non-target species. Ecotoxicol Environ Saf 176:258–269. https://doi.org/10.1016/J.ECOENV.2019.03.096

Mohd Firdaus MA, Agatz A, Hodson ME et al (2018) Fate, uptake, and distribution of nanoencapsulated pesticides in soil-earthworm systems and implications for environmental risk assessment. Environ Toxicol Chem 37:1420–1429. https://doi.org/10.1002/etc.4094

Monikh FA, Chupani L, Vijver MG et al (2019) Analytical approaches for characterizing and quantifying engineered nanoparticles in biological matrices from an (eco) toxicological perspective: old challenges, new methods and techniques. Sci Total Environ 660:1283–1293. https://doi.org/10.1016/j.scitotenv.2019.01.105

Moon MH (2019) Trends in analytical chemistry flow field-flow fractionation: recent applications for lipidomic and proteomic analysis. Trends Anal Chem 118:19–28. https://doi.org/10.1016/j.trac.2019.05.024

Murugan K, Dinesh D, Nataraj D et al (2018) Iron and iron oxide nanoparticles are highly toxic to Culex quinquefasciatus with little non-target effects on larvivorous fishes. Environ Sci Pollut Res 25:10504–10514. https://doi.org/10.1007/s11356-017-0313-7

Nair PMG, Park SY, Choi J (2013) Evaluation of the effect of silver nanoparticles and silver ions using stress responsive gene expression in Chironomus riparius. Chemosphere 92:592–599. https://doi.org/10.1016/j.chemosphere.2013.03.060

Nel A, Xia T, Meng H et al (2013) Nanomaterial toxicity testing in the 21st century: use of a predictive toxicological approach and high-throughput screening. Acc Chem Res 46:607–621. https://doi.org/10.1021/ar300022h

Niemuth NJ, Curtis BJ, Hang MN et al (2019) Next-generation complex metal oxide nanomaterials negatively impact growth and development in the benthic invertebrate Chironomus riparius upon settling. Environ Sci Technol 53:3860–3870. https://doi.org/10.1021/acs.est.8b06804

Noriega JA, Hortal J, Azcárate FM et al (2018) Research trends in ecosystem services provided by insects. Basic Appl Ecol 26:8–23. https://doi.org/10.1016/j.baae.2017.09.006

OECD (2010) OECD sediment-water chironomid life-cycle toxicity test using spiked water or spiked sediment, New York, p 29. https://doi.org/10.1787/9789264090910-en

OECD (2016) Test no. 220: enchytraeid reproduction test. OECD publishing, Paris. https://doi.org/10.1787/9789264264472-en

OECD (2011) Freshwater alga and cyanobacteria, growth inhibition test. Test guideline 201. OECD Guidelines for the Testing of Chemicals, pp 1–26

OECD (2004) Test no. 202: Daphnia sp. acute immobilisation test. OECD

OECD (2012) Test no. 211: Daphnia magna reproduction test. OECD

OECD (2013) Fish embryo toxicity test. OECD Publishing, Paris, pp 1–22. https://doi.org/10.1787/9789264203709-en

Oliveira CR, Domingues CEC, de Melo NFS et al (2019) Nanopesticide based on botanical insecticide pyrethrum and its potential effects on honeybees. Chemosphere 236:124282. https://doi.org/10.1016/j.chemosphere.2019.07.013

Oukarroum A, Bras S, Perreault F, Popovic R (2012) Inhibitory effects of silver nanoparticles in two green algae, Chlorella vulgaris and Dunaliella tertiolecta. Ecotoxicol Environ Saf 78:80–85. https://doi.org/10.1016/j.ecoenv.2011.11.012

Özkan Y, Irende İ, Akdeniz G et al (2015) Evaluation of the comparative acute toxic effects of TiO2, Ag-TiO2 and ZnO-TiO2 composite nanoparticles on honey bee (Apis mellifera). J Int Environ Appl Sci 10:26

Özkara A, Akyil D, Konuk M (2016) Pesticides, environmental pollution, and health. In: Environmental health risk-hazardous factors to living species. IntechOpen, London

Pappas S, Turaga U, Kumar N et al (2017a) Effect of concentration of silver nanoparticles on the uptake of silver from silver nanoparticles in soil. Int J Environ Agric Res 3:80–90. https://doi.org/10.25125/agriculture-journal-IJOEAR-MAY-2017-12

Pappas SA, Kumar N, Turaga U et al (2018) The influence of soil pH on the uptake of silver nanoparticles in a terrestrial system. J Environ Anal Toxicol 08:1–5. https://doi.org/10.4172/2161-0525.1000545
Pappas SA, Turaga U, Kumar N et al (2017b) The influence of soil organic matter on the uptake of silver nanoparticles in a terrestrial system. Int J Environ Agric Res 3:52–60
Pappas SA, Turaga U, Kumar N et al (2017c) Uptake of silver from polyvinylpyrrolidine coated silver nanoparticles in a uptake of silver from polyvinylpyrrolidine coated silver nanoparticles in a terrestrial system. Int J Environ Agric Res 3:104–112
Pappas SA, Turaga U, Kumar N et al (2017d) The influence of vermiculite on the uptake of silver nanoparticles in a terrestrial system. Int J Environ Agric Res 2:56–64
Parada J, Rubilar O, Diez MC et al (2019) Combined pollution of copper nanoparticles and atrazine in soil: effects on dissipation of the pesticide and on microbiological community profiles. J Hazard Mater 361:228–236. https://doi.org/10.1016/j.jhazmat.2018.08.042
Parimi D, Sundararajan V, Sadak O et al (2019) Synthesis of positively and negatively charged CeO2 nanoparticles: investigation of the role of surface charge on growth and development of Drosophila melanogaster. ACS Omega 4:104–113. https://doi.org/10.1021/acsomega.8b02747
Park SY, Chung J, Colman BP et al (2015) Ecotoxicity of bare and coated silver nanoparticles in the aquatic midge, Chironomus riparius. Environ Toxicol Chem 34:2023–2032. https://doi.org/10.1002/etc.3019
Pedrosa J, Campos D, Cocchiararo B et al (2017) Evolutionary consequences of historical metal contamination for natural populations of Chironomus riparius (Diptera: Chironomidae). Ecotoxicology 26:534–546. https://doi.org/10.1007/s10646-017-1784-5
Petersen EJ, Diamond SA, Kennedy AJ et al (2015) Adapting OECD aquatic toxicity tests for use with manufactured nanomaterials: key issues and consensus recommendations. Environ Sci Technol 49:9532–9547. https://doi.org/10.1021/acs.est.5b00997
Petersen EJ, Flores-Cervantes DX, Bucheli TD et al (2016) Quantification of carbon nanotubes in environmental matrices: current capabilities, case studies, and future prospects. Environ Sci Technol 50:4587–4605. https://doi.org/10.1021/acs.est.5b05647
Petersen EJ, Mortimer M, Burgess RM et al (2019) Strategies for robust and accurate experimental approaches to quantify nanomaterial bioaccumulation across a broad range of organisms. Environ Sci Nano 6:1619–1656. https://doi.org/10.1039/C8EN01378K
Pimentel D (1995) Amounts of pesticides reaching target pests: environmental impacts and ethics. J Agric Environ Ethics 8:17–29. https://doi.org/10.1007/BF02286399
Potts SG, Imperatriz-Fonseca V, Ngo HT et al (2016) Safeguarding pollinators and their values to human well-being. Nature 540:220–229. https://doi.org/10.1038/nature20588
Pulido-Reyes G, Leganes F, Fernández-Piñas F, Rosal R (2017) Bio-nano interface and environment: a critical review. Environ Toxicol Chem 36:3181–3193. https://doi.org/10.1002/etc.3924
Qian H, Zhu K, Lu H et al (2016) Contrasting silver nanoparticle toxicity and detoxification strategies in Microcystis aeruginosa and Chlorella vulgaris: new insights from proteomic and physiological analyses. Sci Total Environ 572:1213–1221. https://doi.org/10.1016/j.scitotenv.2016.08.039
Qiu TA, Clement PL, Haynes CL (2018) Linking nanomaterial properties to biological outcomes: analytical chemistry challenges in nanotoxicology for the next decade. Chem Commun 54:12787–12803. https://doi.org/10.1039/C8CC06473C
Rasera JRS, Sant' Anna Neto A, Rosim Monteiro RT et al (2019) Toxicity, bioaccumulation and biotransformation of Cu oxide nanoparticles in Daphnia magna. Environ Sci Nano 6:2897. https://doi.org/10.1039/c9en00280d
Renault S, Baudrimont M, Mesmer-Dudons N et al (2008) Impacts of gold nanoparticle exposure on two freshwater species: a phytoplanktonic alga (Scenedesmus subspicatus) and a benthic bivalve (Corbicula fluminea). Gold Bull 41:116–126. https://doi.org/10.1007/BF03216589
Requier F, Garnery L, Kohl PL et al (2019) The conservation of native honey bees is crucial. Trends Ecol Evol 34:789–798. https://doi.org/10.1016/j.tree.2019.04.008

Rico CM, Majumdar S, Duarte-Gardea M et al (2011) Interaction of nanoparticles with edible plants and their possible implications in the food chain. J Agric Food Chem 59:3485–3498. https://doi.org/10.1021/jf104517j

Roseline TA, Murugan M, Sudhakar MP, Arunkumar K (2019) Nanopesticidal potential of silver nanocomposites synthesized from the aqueous extracts of red seaweeds. Environ Technol Innov 13:82–93. https://doi.org/10.1016/j.eti.2018.10.005

Roth GA, Tahiliani S, Neu-Baker NM, Brenner SA (2015) Hyperspectral microscopy as an analytical tool for nanomaterials. Wiley Interdiscip Rev Nanomed Nanobiotechnol 7:565–579. https://doi.org/10.1002/wnan.1330

Ruttkay-Nedecky B, Krystofova O, Nejdl L, Adam V (2017) Nanoparticles based on essential metals and their phytotoxicity. J Nanobiotechnol 15:1–19. https://doi.org/10.1186/s12951-017-0268-3

Sagvolden G, Giaever I, Pettersen EO, Feder J (1999) Cell adhesion force microscopy. Proc Natl Acad Sci 96:471–476. https://doi.org/10.1073/pnas.96.2.471

Samadder A, Tarafdar D, Das R (2019) Efficacy of nanoencapsulated pelargonidin in ameliorating pesticide toxicity in fish and L6 cells: modulation of oxidative stress and signalling cascade. Sci Total Environ 671:466–473. https://doi.org/10.1016/j.scitotenv.2019.03.381

Samarajeewa AD, Velicogna JR, Princz JI et al (2017) Effect of silver nano-particles on soil microbial growth, activity and community diversity in a sandy loam soil. Environ Pollut 220:504–513. https://doi.org/10.1016/j.envpol.2016.09.094

Santo N, Fascio U, Torres F et al (2014) Toxic effects and ultrastructural damages to Daphnia magna of two differently sized ZnO nanoparticles: does size matter? Water Res 53:339–350. https://doi.org/10.1016/J.WATRES.2014.01.036

Sario S, Silva AM, Gaivão I (2018) Titanium dioxide nanoparticles: toxicity and genotoxicity in Drosophila melanogaster (SMART eye-spot test and comet assay in neuroblasts). Mutat Res Genet Toxicol Environ Mutagen 831:19–23. https://doi.org/10.1016/j.mrgentox.2018.05.001

Sarlak N, Taherifar A, Salehi F (2014) Synthesis of nanopesticides by encapsulating pesticide nanoparticles using functionalized carbon nanotubes and application of new nanocomposite for plant disease treatment. J Agric Food Chem 62:4833–4838. https://doi.org/10.1021/jf404720d

Savić-Zdravković D, Jovanović B, Đurđević A et al (2018) An environmentally relevant concentration of titanium dioxide (TiO2) nanoparticles induces morphological changes in the mouthparts of Chironomus tentans. Chemosphere 211:489–499. https://doi.org/10.1016/j.chemosphere.2018.07.139

Schoonhoven LM, Loon JJA, Dicke M (2005) Insect-plant biology, 2nd edn. Oxford University Press, New York

Sekine R, Moore KL, Matzke M et al (2017) Complementary imaging of silver nanoparticle interactions with green algae: dark-field microscopy, electron microscopy, and nanoscale secondary ion mass spectrometry. ACS Nano 11:10894–10902. https://doi.org/10.1021/acsnano.7b04556

Servin A, Elmer W, Mukherjee A et al (2015) A review of the use of engineered nanomaterials to suppress plant disease and enhance crop yield. J Nanopart Res 17:92. https://doi.org/10.1007/s11051-015-2907-7

Servin AD, Pagano L, Castillo-Michel H et al (2017) Weathering in soil increases nanoparticle CuO bioaccumulation within a terrestrial food chain. Nanotoxicology 11:1–45. https://doi.org/10.1080/17435390.2016.1277274

Shaw BJ, Handy RD (2011) Physiological effects of nanoparticles on fish: a comparison of nanometals versus metal ions. Environ Int 37:1083–1097. https://doi.org/10.1016/j.envint.2011.03.009

Siciliano A, Gesuele R (2013) How Daphnia (Cladocera) assays may be used as bioindicators of health effects? J Biodivers Endanger Species 01:1–6. https://doi.org/10.4172/2332-2543.s1-005

Sieber S, Grossen P, Bussmann J et al (2019) Zebrafish as a preclinical in vivo screening model for nanomedicines. Adv Drug Deliv Rev. https://doi.org/10.1016/j.addr.2019.01.001

Sillen WMA, Thijs S, Abbamondi GR et al (2015) Effects of silver nanoparticles on soil microorganisms and maize biomass are linked in the rhizosphere. Soil Biol Biochem 91:14–22. https://doi.org/10.1016/j.soilbio.2015.08.019

Simonin M, Martins JMF, Uzu G et al (2016a) Combined study of titanium dioxide nanoparticle transport and toxicity on microbial nitrifying communities under single and repeated exposures in soil columns. Environ Sci Technol 50:10693–10699. https://doi.org/10.1021/acs.est.6b02415

Simonin M, Richaume A, Guyonnet JP et al (2016b) Titanium dioxide nanoparticles strongly impact soil microbial function by affecting archaeal nitrifiers. Sci Rep 6:33643. https://doi.org/10.1038/srep33643

Skjolding LM, Sørensen SN, Hartmann NB et al (2016) Aquatic ecotoxicity testing of nanoparticles-the quest to disclose nanoparticle effects. Angew Chem Int Ed 55:15224–15239. https://doi.org/10.1002/anie.201604964

Skjolding LM, Winther-Nielsen M, Baun A (2014) Trophic transfer of differently functionalized zinc oxide nanoparticles from crustaceans (Daphnia magna) to zebrafish (Danio rerio). Aquat Toxicol 157:101–108. https://doi.org/10.1016/j.aquatox.2014.10.005

Slattery M, Harper B, Harper S (2019) Pesticide encapsulation at the nanoscale drives changes to the hydrophobic partitioning and toxicity of an active ingredient. Nano 9:81. https://doi.org/10.3390/nano9010081

Son J, Hooven LA, Harper B, Harper SL (2015) Effect of pH and ionic strength on exposure and toxicity of encapsulated lambda–cyhalothrin to Daphnia magna. Sci Total Environ 538:683–691. https://doi.org/10.1016/j.scitotenv.2015.08.035

Tang Y, Li S, Qiao J et al (2013) Synergistic effects of nano-sized titanium dioxide and zinc on the photosynthetic capacity and survival of Anabaena sp. Int J Mol Sci 14:14395–14407. https://doi.org/10.3390/ijms140714395

Tangaa SR, Selck H, Winther-Nielsen M, Khan FR (2016) Trophic transfer of metal-based nanoparticles in aquatic environments: a review and recommendations for future research focus. Environ Sci Nano 3:966–981. https://doi.org/10.1039/C5EN00280J

Tatarazako N, Oda S (2007) The water flea Daphnia magna (Crustacea, Cladocera) as a test species for screening and evaluation of chemicals with endocrine disrupting effects on crustaceans. Ecotoxicology 16:197–203. https://doi.org/10.1007/s10646-006-0120-2

Théoret T, Wilkinson KJ (2017) Evaluation of enhanced darkfield microscopy and hyperspectral analysis to analyse the fate of silver nanoparticles in wastewaters. Anal Methods 9:3920–3928. https://doi.org/10.1039/C7AY00615B

Tomilina II, Gremyachikh VA, Grebenyuk LP, Klevleeva TR (2014) The effect of zinc oxide nano- and microparticles and zinc ions on freshwater organisms of different trophic levels. Inland Water Biol 7:88–96. https://doi.org/10.1134/S1995082914010155

Tyne W, Lofts S, Spurgeon DJ et al (2013) A new medium for Caenorhabditis elegans toxicology and nanotoxicology studies designed to better reflect natural soil solution conditions. Environ Toxicol Chem 32:1711–1717. https://doi.org/10.1002/etc.2247

Unrine JM, Shoults-Wilson WA, Zhurbich O et al (2012) Trophic transfer of au nanoparticles from soil along a simulated terrestrial food chain. Environ Sci Technol 46:9753–9760. https://doi.org/10.1021/es3025325

USEPA (1996) Ecological effects test guidelines. OPPTS 850.5400 algal toxicity, tiers I and II. EPA 11

Utobo EB, Tewari L (2015) Soil enzymes as bioindicators of soil ecosystem status. Appl Ecol Environ Res 13:147–169. https://doi.org/10.15666/aeer/1301_147169

Vallotton P, Angel B, Mccall M et al (2015) Imaging nanoparticle-algae interactions in three dimensions using Cytoviva microscopy. J Microsc 257:166–169. https://doi.org/10.1111/jmi.12199

Valsami-Jones E, Lynch I (2015) How safe are nanomaterials? Science 350:388–389. https://doi.org/10.1126/science.aad0768

van der Ent A, Przybyłowicz WJ, de Jonge MD et al (2018) X-ray elemental mapping techniques for elucidating the ecophysiology of hyperaccumulator plants. New Phytol 218:432–452. https://doi.org/10.1111/nph.14810

Varma A, Buscot F (eds) (2005) Microorganisms in soils: roles in genesis and functions. Springer, Berlin

Vázquez Núñez E, de la Rosa-Álvarez G (2018) Environmental behavior of engineered nanomaterials in terrestrial ecosystems: uptake, transformation and trophic transfer. Curr Opin Environ Sci Health 6:42–46. https://doi.org/10.1016/j.coesh.2018.07.011

Vecchio G (2015) A fruit fly in the nanoworld: once again Drosophila contributes to environment and human health. Nanotoxicology 9:135–137. https://doi.org/10.3109/17435390.2014.911985

Vijayaraj V, Liné C, Cadarsi S et al (2018) Transfer and ecotoxicity of titanium dioxide nanoparticles in terrestrial and aquatic ecosystems: a microcosm study. Environ Sci Technol 52:12757–12764. https://doi.org/10.1021/acs.est.8b02970

Villaverde JJ, López-Goti C, Alcamí M et al (2017a) Quantum chemistry in environmental pesticide risk assessment. Pest Manag Sci 73:2199–2202. https://doi.org/10.1002/ps.4641

Villaverde JJ, Sevilla-Morán B, López-Goti C et al (2017b) Computational methodologies for the risk assessment of pesticides in the European Union. J Agric Food Chem 65:2017–2018. https://doi.org/10.1021/acs.jafc.7b00516

Villaverde JJ, Sevilla-Morán B, López-Goti C et al (2018) Considerations of nano-QSAR/QSPR models for nanopesticide risk assessment within the European legislative framework. Sci Total Environ 634:1530–1539. https://doi.org/10.1016/j.scitotenv.2018.04.033

Villaverde JJ, Sevilla-Morán B, López-Goti C et al (2019) Contributions of computer-based chemical modeling technologies on the risk assessment and the environmental fate study of (nano)pesticides. In: Environmental concerns and sustainable development. Springer, Singapore, pp 1–27

Walters CR, Pool EJ, Somerset VS (2014) Ecotoxicity of silver nanomaterials in the aquatic environment: a review of literature and gaps in nano-toxicological research. J Environ Sci Health A 49:1588–1601. https://doi.org/10.1080/10934529.2014.938536

Wan J-K, Chu W-L, Kok Y-Y, Cheong K-W (2018) Assessing the toxicity of copper oxide nanoparticles and copper sulfate in a tropical Chlorella. J Appl Phycol 30:3153–3165. https://doi.org/10.1007/s10811-018-1408-3

Wang B, Wang Z, Feng W et al (2010) New methods for nanotoxicology: synchrotron radiation-based techniques. Anal Bioanal Chem 398:667–676. https://doi.org/10.1007/s00216-010-3752-2

Wang D, Lin Z, Wang T et al (2016) Where does the toxicity of metal oxide nanoparticles come from: the nanoparticles, the ions, or a combination of both? J Hazard Mater 308:328–334. https://doi.org/10.1016/j.jhazmat.2016.01.066

Wang W, Sedykh A, Sun H et al (2017) Predicting nano-bio interactions by integrating nanoparticle libraries and quantitative nanostructure activity relationship modeling. ACS Nano 11:12641–12649. https://doi.org/10.1021/acsnano.7b07093

Weil M, Mackenzie K, Foit K et al (2019) Environmental risk or benefit? Comprehensive risk assessment of groundwater treated with nano Fe 0 -based Carbo-Iron®. Sci Total Environ 677:156–166. https://doi.org/10.1016/j.scitotenv.2019.04.360

Werlin R, Priester JH, Mielke RE et al (2011) Biomagnification of cadmium selenide quantum dots in a simple experimental microbial food chain. Nat Nanotechnol 6:65–71. https://doi.org/10.1038/nnano.2010.251

Wink M (1988) Plant breeding: importance of plant secondary metabolites for protection against pathogens and herbivores. Theor Appl Genet 75:225–233. https://doi.org/10.1007/BF00303957

Winkler DA (2016) Recent advances, and unresolved issues, in the application of computational modelling to the prediction of the biological effects of nanomaterials. Toxicol Appl Pharmacol 299:96–100. https://doi.org/10.1016/j.taap.2015.12.016

Wojcieszek J, Jiménez-Lamana J, Bierła K et al (2019) Uptake, translocation, size characterization and localization of cerium oxide nanoparticles in radish (Raphanus sativus L.). Sci Total Environ 683:284–292. https://doi.org/10.1016/j.scitotenv.2019.05.265

Woodcock BA, Garratt MPD, Powney GD et al (2019) Meta-analysis reveals that pollinator functional diversity and abundance enhance crop pollination and yield. Nat Commun 10:1–10. https://doi.org/10.1038/s41467-019-09393-6

Worrall E, Hamid A, Mody K et al (2018) Nanotechnology for plant disease management. Agronomy 8:285. https://doi.org/10.3390/agronomy8120285

Wu Q, Yin L, Li X et al (2013) Contributions of altered permeability of intestinal barrier and defecation behavior to toxicity formation from graphene oxide in nematode Caenorhabditis elegans. Nanoscale 5:9934. https://doi.org/10.1039/c3nr02084c

Wu T, Xu H, Liang X, Tang M (2019) Caenorhabditis elegans as a complete model organism for biosafety assessments of nanoparticles. Chemosphere 221:708–726. https://doi.org/10.1016/j.chemosphere.2019.01.021

Xiao B, Zhang Y, Wang X et al (2019) Occurrence and trophic transfer of nanoparticulate Ag and Ti in the natural aquatic food web of Taihu Lake, China. Environ Sci Nano 6:3431–3441. https://doi.org/10.1039/C9EN00797K

Xu JY, Zhang H (2018) Long-term effects of silver nanoparticles on the abundance and activity of soil microbiome. J Environ Sci 69:3–4. https://doi.org/10.1016/j.jes.2018.06.009

Yamasaki S (2000) Inductively coupled plasma mass spectrometry in environmental analysis. In: Encyclopedia of analytical chemistry. Wiley, Chichester

Yan X, Sedykh A, Wang W et al (2019) In silico profiling nanoparticles: predictive nanomodeling using universal nanodescriptors and various machine learning approaches. Nanoscale 11:8352–8362. https://doi.org/10.1039/C9NR00844F

Yang Y, Xu G, Xu S et al (2018) Effect of ionic strength on bioaccumulation and toxicity of silver nanoparticles in Caenorhabditis elegans. Ecotoxicol Environ Saf 165:291–298. https://doi.org/10.1016/j.ecoenv.2018.09.008

Yasinskyi Y, Protsenko O, Maistrenko O et al (2019) Reconciling the controversial data on the effects of C 60 fullerene at the organismal and molecular levels using as a model Drosophila melanogaster. Toxicol Lett 310:92–98. https://doi.org/10.1016/j.toxlet.2019.03.006

Zamora-Perez P, Tsoutsi D, Xu R, Rivera Gil P (2018) Hyperspectral-enhanced dark field microscopy for single and collective nanoparticle characterization in biological environments. Materials (Basel) 11:243. https://doi.org/10.3390/ma11020243

Zhou Q, Ouyang S, Ao Z et al (2019) Integrating biolayer interferometry, atomic force microscopy, and density functional theory calculation studies on the affinity between humic acid fractions and graphene oxide. Environ Sci Technol 53:3773–3781. https://doi.org/10.1021/acs.est.8b05232

Zhu X, Wang J, Zhang X et al (2010) Trophic transfer of TiO2 nanoparticles from daphnia to zebrafish in a simplified freshwater food chain. Chemosphere 79:928–933. https://doi.org/10.1016/j.chemosphere.2010.03.022

# Overview of Nanopesticide Environmental Safety Aspects and Regulatory Issues: The Case of Nanoatrazine

**Felícia Pereira de Albuquerque, Ana Cristina Preisler, Leonardo F. Fraceto, Halley Caixeta Oliveira, and Vera Lucia S. S. de Castro**

**Abstract** The use of nanotechnology to create new formulations has shown great potential to reduce the indiscriminate use of pesticides and provide environmentally safer alternatives. Pesticides formulated using controlled release nanosystems are designed to efficiently provide sufficient and targeted quantities of active ingredients to target organisms, thus improving crop yields and reducing environmental contamination with pesticides. However, the possible harmful effects of these nanomaterials on the environment are not yet well understood, highlighting the need for studies assessing the fate and behavior of nanopesticides in the environment. This chapter will discuss the major challenges and advances in the research regarding nanopesticide risk analysis. It will also discuss the difficulty in developing regulations about the commercialization of nanoproducts, due to the underlying specific features of nanomaterials that drive their reactivity and toxicity. Finally, the case of nanoatrazine will be reviewed, providing an example of how the nanoencapsulation can affect herbicide efficiency and influence its toxicity to different non-target organisms.

**Keywords** Atrazine · Ecotoxicology · Nanoherbicide · Nanotechnology · Risk

---

F. P. de Albuquerque · L. F. Fraceto
Institute of Science and Technology of Sorocaba, São Paulo State University (UNESP), Sorocaba, Brazil

A. C. Preisler · H. C. Oliveira (✉)
Department of Animal and Plant Biology, State University of Londrina, Londrina, Brazil
e-mail: halley@uel.br

V. L. S. de Castro (✉)
Embrapa Environment, Jaguariúna, Brazil
e-mail: vera-lucia.castro@embrapa.br



L. F. Fraceto et al. (eds.), *Nanopesticides*,
https://doi.org/10.1007/978-3-030-44873-8_9

## 1 Introduction

Nanotechnology is an interdisciplinary research area with important applications in many fields. Nanomaterials are objects which have a size range from 1 to 100 nm at least in one dimension, thus presenting specific properties of size, shape, porosity, etc. (Gebre and Sendeku 2019). Depending on the area of study, some definitions cover materials that are smaller or larger than the 1–100 nm range but retain the main properties of nanoscale elements. The classification of nanomaterials in complex environmental samples is therefore not trivial since these materials can be composed of substances having many properties (as size, shape, composition) determined by a distribution of values (Miernicki et al. 2019).

Agriculture, food, and natural resources are part of sustainability challenges. However, around 90% of pesticides flow into the environment in the application process and remain in agricultural products as a result of using conventional pesticide formulation and its disadvantages—solvent use, low dispersion, etc. In this scenario, less toxic approaches to pest control should be developed to ensure worker safety, provide healthy food, and provide economic sustainability for farmers. Since there is a number of toxicological issues regarding pesticides (Damalas and Eleftherohorinos 2011; Kim et al. 2017; Bagheri et al. 2019), the improvement of pesticide formulations already used for safer ones could reduce the adverse effects of agriculture and in particular the toxic effects of pesticides. Public interest in reducing the risks to human health and the environment associated with pesticide use may claim for policies that encourage the development of other management alternatives.

In recent years, the use of nanotechnology to create new formulations has shown great potential for improving pesticide efficacy and safety (Zhao et al. 2018). As a result, nanotechnology can improve crop yields and reduce harmful effects on the environment. Nano-based agricultural products have been suggested to offer many benefits compared to conventional products, including better formulation characteristics (stability, permeability, and dispersion of the active pesticide ingredient), ease of application, improved targeting to pest species, higher efficacy, decreased application doses, prevention of premature degradation, and increased environmental safety. Nanopesticides have a much greater surface area than the conventional formulations, which may improve their interaction with target pests at reduced doses (Walker et al. 2018; Chariou et al. 2019). Nanopesticides may also decrease losses of active ingredients by improving absorption and/or bioavailability and, as a consequence, their permeation into target organism tissues, among others (Chhipa 2017). These properties allow lowering the pesticide dose and achieving better pest control with reduced treatment frequency (Kumar et al. 2017).

Despite these advances, continuous evaluation of these novel nano-based pesticides is necessary, having in mind the effect of dose, and impact on non-target organisms (Adisa et al. 2019). Nanoencapsulated formulas utilize the wide variety of available nanomaterials, including polymers, lipids, mesoporous silica, clay, and other materials. In addition to composition, there are other relevant factors in nanopesticide formulations. For example, nanoparticle size determines their colloidal behavior, thereby altering the fate and toxicity of the active ingredient. Thus, the risk profile of nanopesticide products can be modified by disrupting foundational assumptions about chemical behavior (Kah et al. 2013; Nuruzzaman et al. 2016; Slattery et al. 2019).

This chapter will discuss the major challenges regarding the nanopesticide risk analysis and regulatory issues. It will then discuss the case of nanoatrazine, providing an example of how the nanoencapsulation can affect herbicide efficiency and influence its toxicity to different non-target organisms.

## 2 Nanopesticide Risk Analysis and Regulatory Issues

As the possible harmful effects of nanomaterials on the environment are not yet well understood (Deng et al. 2017; Kah et al. 2019), assessing the fate and behavior of nanopesticides during and after their application to the environment is important to determine their potential impact on ecosystems (Kah et al. 2018). Comparisons of nanoformulations with the pure active ingredient and conventional formulations are needed to clarify the mechanisms by which nanomaterials alter the pesticide behavior. The improved bioavailability and solubility of nanopesticides may affect their environmental fate, as well as their toxicokinetic and dynamic behavior once adsorbed by organisms. For example, greater bioavailability could involve a greater effect on non-target organisms. Similarly, less nanopesticide degradation and/or synergistic effects may also result in unexpected toxicity and/or damage to non-target organisms (Villaverde et al. 2018). There is also evidence that nanopesticides and conventional pesticides differ in their environmental behavior, so it is necessary to understand the fate of nanopesticides in detail to ensure they comply with regulatory guidelines and legislation (Chariou et al. 2019). Although commercial formulations may vary with respect to unknown ingredients, adverse-effects comparative studies of their marketing and technical-level formulations appear to be necessary for a more accurate and appropriate risk assessment (Mossa and Abbassy 2012; Puglis and Boone 2011; Mansano et al. 2017). Therefore, a robust toxicological assessment of the potential risks associated with the use of nanopesticides should be performed (Iavicoli et al. 2017; Kah et al. 2018).

Agri-nanotechnologies have the potential to significantly increase the direct release of nanomaterials into the environment and add a significant pathway of exposure to humans through the food chain (Lombi et al. 2019) wherein they would likely undergo transformation during their life cycles (Iavicoli et al. 2017).

Persistence and bioaccumulation of the nanoagrochemicals in higher trophic level species and their translocation through the food chain are not well studied. Like conventional pesticides, they can reach the trophic chain through exposure routes due to nanopesticides residues in soil and crops (Iavicoli et al. 2017). Indirect pathways of exposure due to the leaching, spray-drifting, and runoff of nanopesticides to non-target environments may become significant with its increasing use

(Lombi et al. 2019). Potential adaptations to existing environmental risk assessment tests and procedures for use with nanopesticides have to be discussed, addressing aspects such as analysis and characterization, environmental fate and exposure assessment, uptake by biota, ecotoxicity, and risk assessment of nanopesticides in aquatic and terrestrial ecosystems (Pourzahedi et al. 2018).

The majority of products currently in development consists of nanocarrier systems loaded with a registered active ingredient already in use as insecticides or herbicides. Efficient utilization of nanopesticides will result in efficient management schemes to surpass some problems of chemical pesticides, as premature degradation, plant resistance, etc. For this type of products, it is necessary to establish the durability of the active ingredient–nanocarrier complex upon application in the field for exposure assessment. If the durability of the nanoformulation is short, exposure is likely to be similar to that of conventional pesticide formulations. However, if the nanocarrier–active ingredient complex persists in the environment, a more complex assessment of exposure also considering the nanocarrier properties may be required but there are no standard protocols to measure the durability of the active ingredient–nanocarrier complex (Kah et al. 2014). This measurement is a complex task due to the soil characteristics that influence the nanocarrier active ingredient release rate (Kah et al. 2018). Likewise, to assess the risk of nanopesticides, it will be important to obtain a mechanistic understanding of nanomaterial uptake by plants from soils or leaves, translocation throughout plant vascular system, the dependence of uptake levels on plant species, interactions with soil microorganisms and soil properties, nanomaterial transformations, and transport under different environmental conditions (Henchion et al. 2019).

Given the environmental complexity and variability, the implications of laboratory data on actual environmental systems are not yet adequately established. Thus, there is a great demand for studies that assess the possible non-target organisms toxic and environmental effects. Although the factors leading to nanomaterial trophic transfer are poorly understood, most studies of nanomaterial behavior in the food chain are done with aquatic organisms (Tangaa et al. 2016). Beyond that, harmful substances of nanopesticides remaining in edible plants and animals are likely to enter the human body through the food chain and thus endanger human health (Sun et al. 2019).

Unexpected risk of nanopesticides comes from their possible toxicity to non-target organisms, transportation, bioaccumulation, and from interactions with other pollutants in the environment (Li et al. 2019). Nanomaterials can be transferred from one trophic level to another by feeding since in nature organisms are exposed to multiple contaminants and feed on other organisms previously exposed to one or more xenobiotics in various exposure scenarios (Gardea-Torresdey et al. 2014; Karimi et al. 2018; Marucci et al. 2019). In this way, nanopesticide toxicity at the lower trophic level can also affect organisms at higher levels via the food chain and higher concentrations can be found in organisms belonging to higher trophic levels. Altered communities can in turn alter ecosystem processes such as those that recycle carbon, nutrients, and energy (Werlin et al. 2011; Holden et al. 2013). Considering the persistence of certain nanopesticides, the combined toxicity of nanomaterials and pesticide, and the connectivity of the entire ecosystem, it is possible that nanopesticides could have a significant impact on biodiversity (Sun et al. 2019).

As a result, the lack of information on the fate and behavior in the environment of nanopesticides and their effects on human and animal health may be inhibiting their incorporation into agriculture (Pandey et al. 2018). Despite the advantages of using nanotechnology systems, their development requires caution, since it is necessary to understand their behavior in the environment and their effects on non-target organisms to promote safer product development strategies. Also, it is important to develop nanoformulations that are based on green nanotechnology and that offer low cost, simple procedures, and controlled-release features as an effective tool for sustainable agricultural development (Kim et al. 2018; Camara et al. 2019).

These nanoproducts, to enter the market, must demonstrate the safety of their use for the consumer and the environment. The introduction of nanotechnology in agricultural applications is affected by several factors including technological feasibility, cost effectiveness, regulatory requirements, and consumer acceptance (Lombi et al. 2019; Lai et al. 2018; Henchion et al. 2019). Incipient agricultural commercial nanotechnology development is due to inconsistent national legislative frameworks, limited regulatory guidelines, and a lack of public licensing initiatives. The need for fit regulatory arrangements to support the nanotechnological development in agriculture has been identified as a main priority by a range of international and national organizations (Mitter and Hussey 2019).

As with any other regulated product, applicants seeking market approval for nanopesticides must demonstrate the safe use of these new products without posing undue safety risks to the consumer and the environment. Thus, regulatory frameworks are becoming increasingly important in order to properly address and manage the potential risks of nanotechnology. These milestones should be discussed in collaboration among countries around the world to exchange information and ensure a high level of protection for humans and the environment (Amenta et al. 2015). Therefore, there is a need for the development of better alternative approaches to facilitate the study of the possible damaging effects for responsible nanotechnology development (Deng et al. 2017; Kah et al. 2019). For preventing the negative impact of nanopesticides on humans and the environment, consolidated efforts of scientists, governmental organizations, and business are needed (Makarenko and Makarenko 2019).

Regulation and legislation play a fundamental and key role for the implication of nanotechnology and also serve as official sources for public knowledge and awareness. In order to enable decision-makers to rapidly assess the potential risks that may be imposed by nanomaterials, particularly when confronted by the reality of limited hazard or exposure data, risk assessment tools are needed (Romero-Franco et al. 2017). Otherwise, it is difficult to trace and monitor the distribution of nanomaterials as a result of the complicated nano-bio-eco interactions (He et al. 2018). Then, unfortunately, each type of nanomaterial has its unique properties, leading to a "case-by-case" evaluation regarding regulation and legislation (He et al. 2019; Kah et al. 2019).

In this scenario, ecotoxicity tests are instruments used in environmental risk assessment to answer questions about potential intrinsic hazards in exposure assessments. Toxicological research should be considered as the basis of the identification and characterization of hazards. In the environmental case, the indicators to be used are related to the potential risks associated with the persistence and bioaccumulation of nanomaterial in the environment, soil and water contamination due to its dispersion when applied in agriculture (Miller and Wickson 2015; Villaverde et al. 2018). At present, the main directions of experimental research are studying of translocation/division of nanomaterials in biotic/non-biotic systems, studying migration to biota and humans, revealing and monitoring quantity of entry of nanomaterial into the environment, studying interaction between physical and chemical properties of nanomaterials and nanotoxicity.

Nanomaterial regulatory authorities need systematic and consistent experimental data, along with well-defined experimental protocols. Given that research data on nanomaterials have been obtained using different experimental procedures, current requirements for building reliable datasets should include the use of standard protocols to examine their accuracy and suitability. In nanoecotoxicology, the establishment of possible biomarkers of exposure and effects is the subject of intense discussion. Biomarkers are biological responses to environmental pollutants that can be measured indicating the presence, effects and, in some cases, the degree of contamination. Biomarker analysis is a basic tool for risk assessment and enable short-term biological responses to exposure. It has low analysis costs compared to conventional techniques indicative of pollutant exposure (Ballesteros et al. 2017). In this perspective, the development of indicators can be an ally in the decision-making process for the release of nanomaterials for human consumption or application to the environment (Villaverde et al. 2018).

However, research on the environmental risks of nanopesticides is still in the beginning since nanomaterial risk assessment generally follows the established procedures for conventional chemicals, which consist of four major steps: hazard identification, hazard characterization, exposure assessment, and risk quantification. For nanomaterials, each step involves challenges that add uncertainty to the final calculated risk quotient. The exposure assessment is also hampered by the scarcity of available data on nanomaterial use and of reliable detection methods for model validation (Sørensen et al. 2019). Especially for aquatic systems in the laboratory, monitoring the stability and consistency of nanomaterials during testing is one of the biggest challenges in nanoecotoxicity assessments, as it depends on a greater number of physicochemical parameters than conventional agrochemicals. As a result, data interpretation and comparability require standardization of dispersion methods and knowledge of the dissolution kinetics of nanomaterials.

After obtaining data in the laboratory step, research should be performed on a larger scale. The information obtained may be used by regulatory agencies to assess the potential toxicity and risks associated with nanomaterials throughout the different stages of the product life cycle. Then, nanotechnology governance efforts require more integrative ways of assessing security, including approaches such as life-cycle assessment and multi-criteria decision analysis, as well as improving data sharing and risk communication. Consequently, the study and monitoring of the environmental risks of nanomaterials require models to estimate the flow, fate, and transport thereof, as well as their absorption, bioavailability, and hazard for various organisms. Despite further development-oriented initiatives and strategies in this direction, there are still no specific national and international laws that provide the regulatory requirements and validation of test methodologies (Cerrillo et al. 2017). In silico models have been increasingly used for chemicals hazard characterization—species sensitivity distribution (SSD) and quantitative structure activity relationship (QSAR) models. The environmental risk assessment models comprise both the hazard and exposure assessment of nanomaterial-related risks (Makarenko and Makarenko 2019; Sørensen et al. 2019). However, there is no universal and unique model for all nanosystems and the lack of a consistent framework for the integration of in silico results can lead to uncertainty and even contradictions across models. Consequently, their predictive power should always be validated (Villaverde et al. 2018; Benfenati et al. 2019).

For making decisions as to implement nanotechnologies into agricultural practices we have to assess correctly the direction of changes of nanomaterial properties in the environment and make an objective prognosis of ecotoxicological risks as a result of their use. Due to the difference in ecological risk assessment of nanopesticides from that of conventional pesticides, new parameters are needed to allow an adequate evaluation of the new products (Kah et al. 2018). Little is known about the number of nanomaterials that may enter food webs or induce direct or indirect toxic effects to plants, microbes, or other soil organisms, and thus affect the ecosystem where it was applied for agricultural purposes. These complex relationships may provide variable risk profiles for the exposed populations and require the definition of strategies to define models able to predict possible adverse outcomes (Iavicoli et al. 2017). However, generalizations are hard to make as the impacts depend on the type of nanocarrier, pesticide, and soil, as well as on the time scale (Fojtová et al. 2019). Due to the complexity regarding the quantification of environmental exposures to nanomaterials and the scarcity of toxicity data at the organism level, various alternative approaches have been proposed for assessing the potential environmental impacts of nanomaterials. These approaches have provided insight and criticism into the several elements of the assessment methods (Lai et al. 2018; Romero-Franco et al. 2017).

Therefore, the assessment of the ecological risks of nano-based products will demand the establishment of the proposed use pattern. During the risk analysis and characterization stages of the assessment, the significance of these particular properties of nano-based formulations must be taken into consideration. Application rate should not result in ecosystem levels above acceptable guideline values. For example, the conceptual model for a nano-based pendimethalin formulation has identified some modifications in the environmental fate and behavior of the active ingredient in comparison with the conventional herbicide. Then, to be safely introduced to the market, the nanoformulation should be developed considering the nature of the product being assessed and the products associated with the application scenario (Walker et al. 2018). That being so, the use of nanotechnology in agriculture raises a number of issues for regulators. Knowing the potential risks of nanomaterials is important for their production, marketing, and disposal done in an appropriate and sustainable manner. Thus, nanopesticides should be carefully compared against products currently available on the market. Moreover, the obtained results should be wisely divulgated to stakeholders to prevent unjustified expectations or perceived fears associated with the use of nanomaterials (Kah et al. 2018; Lombi et al. 2019). This procedure would support the responsible development of nanotechnology for agricultural purposes and the introduction of nanopesticides into the market (Jantunen et al. 2018; Oomen et al. 2018; Adisa et al. 2019). Finally, the promotion of information exchange among all involved from manufacturing to disposal, in addition to regulatory agencies, will contribute to achieving safer and more efficient farming practices as environment and human health are intrinsically connected.

It is paramount for policymakers and other stakeholders to understand the public opinion about nanotechnology. The evaluation of consumer concerns and preferences and the elicitation of acceptance with regard to nanotechnology applied to agriculture is crucial for its success, notwithstanding their known benefits to farmers, processors, and consumers (Handford et al. 2015; Henchion et al. 2019). All of these efforts are vital to the development of new, competitive nanoproducts that may increase the agricultural sustainability.

## 3 The Case of Nanoatrazine: Efficiency Toward Target Plants and Adverse Effects on Non-target Organisms

The constant demand for higher yields in agriculture has been accompanied by the increased use of herbicides, as they play an important role to weed control, avoiding the resource competition of these undesirable plants with crops. Atrazine is one of the most commonly used herbicide to control weeds in maize (*Zea mays*), sorghum (*Sorghum bicolor*), and sugarcane (*Saccharum* ssp.) cultures in the Americas and Australia (Yue et al. 2017; Lushchak et al. 2018). However, its widespread and incorrect use has arisen environmental and health concerns, as atrazine has high runoff potential, prolonged soil persistence, moderate adsorption to soil organic matter and clay particles, and high toxicity to living organisms, causing damage well beyond the application area (Singh et al. 2018; Albuquerque et al. 2020). Its toxicity level can be associated with the used concentration, route of contamination, and time of exposure.

In this context, atrazine-loaded nano- and microformulations have been developed with the aim to reduce the adverse effects of this herbicide to the environment (Grillo et al. 2010, 2012; Souza et al. 2012; Pereira et al. 2014; Oliveira et al. 2015a, b; Schnoor et al. 2018; Taverna et al. 2018; Xiao-Ting and Wang 2019). Only part of these studies evaluated the herbicidal activity of the developed formulations against target plants, and they demonstrated that atrazine incorporation into nanoparticles increased its efficacy, thereby decreasing the amount required for weed control (Table 1).

**Table 1** Effects of different nanoatrazine formulations on target plants

| Nanoformulation | Test organism | Route of contamination | Atrazine concentration | Time of exposure | Toxicity[a] | Reference |
|---|---|---|---|---|---|---|
| PCL nanocapsules and nanospheres containing atrazine | *Brassica* sp. | Pre-emergence | 2500 g $ha^{-1}$ | 14 days | Increased the herbicidal activity | Pereira et al. (2014) |
| PCL nanocapsules containing atrazine | *Brassica juncea* | Post-emergence | 200 and 2000 g $ha^{-1}$ | 1, 2, 3, and 7 days | Increased the herbicidal activity (up to ten times), as indicated by more severe symptoms, inhibition of photosystem II activity, and induction of oxidative stress | Oliveira et al. (2015a) |
| Solid lipid nanoparticles containing atrazine and simazine | *Raphanus raphanistrum* | Pre- and post-emergence | 300 and 3000 g $ha^{-1}$ | 7 and 13 days | Increased phytotoxic effects on shoot and roots | Oliveira et al. (2015b) |
| PCL nanocapsules containing atrazine | *Amaranthus viridis Bidens pilosa* | Post-emergence | 200 and 2000 g $ha^{-1}$ | 14 days | Caused a greater decrease in the photosystem II activity, and in root and shoot growth | Sousa et al. (2018) |
| Poly(lactic-co-glycolic acid) nanoparticles containing atrazine | *Solanum tuberosum* | In vitro (microcuttings) | 0.7, 6.3, and 54.0 μg $mL^{-1}$ | 22–30 days | Increased the inhibition of growth parameters | Schnoor et al. (2018) |
| PCL nanocapsules containing atrazine | *Brassica juncea* | Post-emergence | 200 g $ha^{-1}$ | Up to 7 days | Increased the herbicidal activity (up to ten times), leading to chloroplast degradation, due to their adherence to the leaf surface and penetration into the mesophyll through stomata pores | Bombo et al. (2019) |

[a]Compared to the respective non-nano formulation

Among the abovementioned formulations, more advance on the evaluation of biological activity in different types of organisms has been obtained up to date with atrazine-containing poly(ε-caprolactone) (PCL) nanocapsules, synthesized by the method of interfacial deposition of pre-formed polymer (Grillo et al. 2012). PCL is a biodegradable, water-insoluble polymer with slow degradation in aqueous media and low risk in the environment in the bulk form. Atrazine-loaded PCL nanocapsules have a hydrodynamic size of around 260 nm and zeta potential of nearly −30 mV, presenting high suspension stability, no aggregate formation, high encapsulation efficiency, and controlled release profile (Grillo et al. 2012).

This nanoatrazine formulation showed efficient herbicidal activity in post-emergence tests using different target plants (Oliveira et al. 2015a; Sousa et al. 2018). Atrazine nanoencapsulation enhanced the inhibitory effects of the herbicide on photosynthetic and growth parameters of *Brassica juncea*, *Amaranthus viridis,* and *Bidens pilosa.* As a consequence, the treatment with tenfold diluted nanoatrazine was as efficient as conventional atrazine at the full dose to control these plants (Oliveira et al. 2015a; Sousa et al. 2018). In pre-emergent assays with *B. pilosa*, the use of nanoatrazine also allowed the reduction of atrazine dose by tenfold without compromising the herbicidal activity (Preisler et al. 2020). Overall, these results suggested nanoatrazine as a promising alternative for efficient weed control, allowing a reduced dose utilization and diminished input of the herbicide in the environment.

However, an important concern is related to the possible side effects of this nanoherbicide to crops (Table 2). To address this question, maize, an atrazine-resistant crop, has been treated with nanoatrazine both at pre- and post-emergence, resulting in no persistent toxic effects evaluated up to 7 days after the exposure. Nanoatrazine induced alterations on physiological parameters of maize leave only up to 2 days after post-emergent treatment, but these effects were transient and did not affect shoot growth (Oliveira et al. 2015c). In another approach, soybean (*Glycine max*), an atrazine-sensitive crop often cultivated in succession with maize, was sowed in soil that had been previously treated with the nano and conventional atrazine. The residual effects of the herbicide on soybean plants were potentiated by nanoencapsulation when sowing occurred 17 days after soil treatment with the formulations. However, the long-term residual effects of the herbicide on soybean plants (sowing 60 days after soil treatment) did not increase with the use of nanoatrazine compared with the conventional formulation. Considering the use of tenfold lower dose, the residual effect of atrazine on soybean would be even decreased by nanoencapsulation, provided that the safety interval from application to sowing is respected (Preisler et al. 2020).

Several studies have also performed bioassays with different types of non-target organisms to address nanoatrazine ecotoxicity (Table 2). Part of the studies has indicated a lower toxicity of nanoatrazine when compared with the herbicide alone, which would suggest this nanoherbicide as an environmentally viable alternative. For example, the effect of nanoencapsulation in reducing atrazine genotoxicity has been reported for the model plant *Allium cepa*, the neotropical fish *Prochilodus lineatus*, and human cells (Grillo et al. 2012; Clemente et al. 2014; Andrade et al.

**Table 2** Adverse effects of micro and nanoatrazine on non-target organisms

| Nanoformulation | Test organism | Route of contamination | Concentration | Time of exposure | Toxicity[a] | Reference |
|---|---|---|---|---|---|---|
| Polyhydroxybutyrate-co-hydroxyvalerate microspheres containing atrazine | *Lactuca sativa* | Water | 1, 5, and 10 mg $L^{-1}$ | 24 h | Reduced genotoxicity | Grillo et al. (2010) |
| PCL nanocapsules containing atrazine | *Allium cepa* | Water | 1, 10, and 100 mg $L^{-1}$ | 24 h | Reduced geno and cytotoxicity | Grillo et al. (2012) |
| | Human lymphocyte cells | In vitro | 100 mg $mL^{-1}$ | 1 h | | |
| PCL nanocapsules containing atrazine | *Pseudokirchneriella subcapitata* | Water | In a gradient employing a factor of 2.0 | 24, 48, 72, and 96 h | Reduced toxic effect on algal growth | Clemente et al. (2014) |
| | *Daphnia similis* | Water | In a gradient employing a factor of 1.8 | 24 and 48 h | Enhanced toxicity after 48 h, causing immobility | |
| | Human lymphocyte cells | In vitro | 1 mg $mL^{-1}$ | 72 h | Reduced cell damage | |
| PCL nanocapsules and nanospheres containing atrazine | *Zea mays* | Pre-emergence | 2500 g $ha^{-1}$ | 14 days | No effect on plant development | Pereira et al. (2014) |
| | *Allium cepa* | Water | 0.7, 2.1, 6.3, 18, and 54 μg $mL^{-1}$ | Until roots reached a length of 2 cm | Produced fewer chromosomal aberrations at low concentrations, however increased the damage index at high concentrations, although the increases were smaller than those observed for free atrazine | |

(continued)

**Table 2** (continued)

| Nanoformulation | Test organism | Route of contamination | Concentration | Time of exposure | Toxicity[a] | Reference |
|---|---|---|---|---|---|---|
| Solid lipid nanoparticles containing atrazine and simazine | *Zea mays* | Pre- and post-emergence | 300 and 3000 g $ha^{-1}$ | 7 and 13 days | No effect on plant development | Oliveira et al. (2015b) |
| | Rat 3T3 fibroblast cells | In vitro | 15.6, 31.25, and 62.5 μg $mL^{-1}$ | 24 h | Decreased cytotoxicity | |
| PCL nanocapsules containing atrazine | *Zea mays* | Pre- and post-emergence | 200 and 2000 g $ha^{-1}$ | 1, 2, 4, and 8 days | Increased the intensity of the transient inhibition of photosystem II activity, but did not cause persistent effects | Oliveira et al. (2015a) |
| Solid lipid nanoparticles containing atrazine and simazine | *Caenorhabditis elegans* | Water | 0.025, 0.05, 0.1, 0.2, and 0.25 mg $mL^{-1}$ | 48 h | Induction of mortality and decrease in the body length (PCL nanocapsules were the most toxic) | Jacques et al. (2017) |
| PCL nanocapsules containing atrazine | | | 0.1, 0.2, 0.3, 0.4, and 0.5 mg $mL^{-1}$ | | | |
| PCL nanocapsules containing atrazine | *Enchytraeus crypticus* | Soil | 1, 5, 10, 50, 100, and 200 mg $kg^{-1}$ | 48 h, 7, 11, 13, 25, 28, and 46 days | Caused a hatching reduction due to the delayed development, decreasing the number of adults and juveniles, moreover this species did not exhibit avoidance behavior to atrazine nanoformulation | Gomes et al. (2019) |
| PCL nanocapsules containing atrazine | *Prochilodus lineatus* | Water | 2 and 20 μg $L^{-1}$ | 24 and 96 h | Reduced toxicity, as they did not induce alterations in glycemia, antioxidant response, carbonic anhydrase activity, and in the frequency of nuclear erythrocyte abnormalities | Andrade et al. (2019) |

[a]Compared to the respective non-nano formulation

2019). The incorporation into PCL nanocapsules also decreased the toxicity of atrazine to the alga *Pseudokirchneriella subcapitata* (Clemente et al. 2014). In contrast, the same study reported toxicity of PCL nanocapsules (containing or not atrazine) for the microcrustacean *Daphnia similis* after 48 h of exposure. Similarly, Jacques et al. (2017) observed deleterious effects of PCL nanocapsules and solid lipid nanoparticles loaded or not with atrazine on the development of the nematode *Caenorhabditis elegans*.

In a bioassay with *Enchytraeus crypticus*, atrazine-loaded PCL nanocapsules showed less toxicity to the soil invertebrate when compared with the herbicide alone, but higher toxicity when compared with a commercial atrazine formulation (Gomes et al. 2019). This finding requires an evaluation of an exposure concentration reduction versus the higher toxicity of the nanoformulation. In this case, further investigation focusing on specific life stages may clarify the mechanisms of toxicity and contribute to improve the safety of nanopesticides.

Currently the adverse effects of nanoatrazine to non-target organisms have been attributed to some component of the nanoformulation, as the surfactant (Clemente et al. 2014; Jacques et al. 2017). In addition, they might be associated with the form of exposure and the used dose (Clemente et al. 2014; Jacques et al. 2017). It is also interesting to verify the toxicity of the polymer present in nanoformulations, as well as other polymers could be tested for the development of new nanoatrazine formulations, highlighting zein and chitosan. Therefore, new ecotoxicological bioassays are necessary to verify the effects of nanoatrazine on other species to determine the environmental risks of its application.

Another important challenge regarding nanoatrazine and nanoherbicides in general is the understanding of their mechanisms of action and their interactions with plants and non-target organisms. An approach to unveil the uptake, translocation, and accumulation of polymeric nanoparticles in different organisms is the use of fluorescently labeled materials. Recent studies with fluorescent zein nanoparticles (ZNPs) have demonstrated the adhesion and absorption of ZNPs by sugarcane and soybean roots (Nguyen et al. 2016; Prasad et al. 2018). They have also reported that a fraction of ZNPs were translocated via xylem to the leaf as the exposure dose was increased, following the same root pattern, but at smaller amounts (Nguyen et al. 2016; Prasad et al. 2018). Fluorescent PCL nanocapsules have been demonstrated to adhere to the leaf surface and penetrate the mesophyll through the hydathode and stomata pores, which could result in a more efficient atrazine delivery at its site of action (Bombo et al. 2019). Further studies with other methodological strategies are still needed for a better understanding of the interaction of atrazine-loaded PCL nanocapsules with plants.

# 4 Conclusions

The demonstration of real and practical advantages of nanoformulated systems requires the design of scalable processes, risk, toxicity, and life-cycle assessment of the nanopesticides and also simplicity in the regulations about the commercialization of nanoproducts. For this complex task, nanomaterial ecotoxicity assessment needs improvement and appropriateness of current toxicity protocols, which are not designed for nano and environment-dependent transformations, as it is necessary to understand how they behave under different environmental conditions (Kim et al. 2018; Henchion et al. 2019).

On the other hand, since nanomaterial toxicity cannot be generalized due to the underlying nano-specific features that drive nanomaterials reactivity and toxicity, nanosafety has increasingly turned toward the development of safety-by-design approach. This point merits careful consideration to determine the appropriate relationship between gains from added specificity and the difficulty in performing complex models (Henchion et al. 2019). Stakeholders will need to find the right balance between the precautionary principle and the promotion of innovation for solving problems (Kah et al. 2019). Less complex risk assessment models that demand less time and user expertise fit the necessities of stakeholders for these early stages with no substantive resource investments. Tool refinement throughout the next years may change the balance in scoring and assessment among particular tools. A flow-through from research tools to simplified and easily operationalized systems may ultimately deliver the needed balance between rigor and ease (Sørensen et al. 2019).

Risk reduction strategies are part of nanotechnology governance and should be periodically refined. As nanomaterials continue to be developed, the risk assessment and management tools should equally evolve to assist the risk governance of this emerging technology. Such tools differ in input requirements and capacity to provide risk-based or multi-criteria driven outputs. Thus, each individual tool and regulatory strategy is suited for specific stakeholders and specific assessment and governance purposes (Trump et al. 2018; Roig 2018).

The attractiveness and efficacy of the risk governance frameworks for stakeholders would be increased with the development of the frameworks into user-friendly web-based decision-support tools, suitable to guide different stakeholder and public groups categories to fulfill the specific requirements and needs of each phase. Further initiatives in this direction would benefit the implementation of effective risk governance practices and would support safe innovation in all fields of application of nanotechnology (Isigonis et al. 2019).

Since there is still no harmonized basis for nanomaterial risk governance across different sectors (Lombi et al. 2019), this knowledge can be useful for support of policy frameworks, and public funding will be crucial to invest in these technologies evaluation development viewing sustainable farming practices. It is expected that in the near future the controlled release of pesticides by nanotechnologies will turn into key tools to improve cropland productivity and protection with minimum impact to the environment and health.

**Acknowledgements** The authors thank São Paulo Research Foundation (FAPESP, Grant Number 2017/21004-5) and Conselho Nacional de Desenvolvimento Científico e Tecnológico (CNPq, Grant Number 306583/2017-8) for financial support. This study was financed in part by the Coordenação de Aperfeiçoamento de Pessoal de Nível Superior—Brasil (CAPES)—Finance Code 001.

## References

Adisa IO, Pullagural VLR, Peralta-Vide JR et al (2019) Recent advances in nano-enabled fertilizers and pesticides: a critical review of mechanisms of action. Environ Sci Nano 6:2002–2030

Albuquerque FP, Oliveira JL, Moschini-Carlos V, Fraceto LF (2020) An overview of the potential impacts of atrazine in aquatic environments: perspectives for tailored solutions based on nanotechnology. Sci Total Environ 700:1–9

Amenta V, Aschberger K, Arena M et al (2015) Regulatory aspects of nanotechnology in the agri/feed/food sector in EU and non-EU countries. Regul Toxicol Pharmacol 73:463–476

Andrade LL, Pereira AES, Fraceto LF et al (2019) Can atrazine loaded nanocapsules reduce the toxic effects of this herbicide on the fish *Prochilodus lineatus*? A multibiomarker approach. Sci Total Environ 663:548–559

Bagheri A, Bondori A, Allahyari MS et al (2019) Modeling farmers' intention to use pesticides: an expanded version of the theory of planned behavior. J Environ Manag 248:109291

Ballesteros ML, Rivetti NG, Morillo DO et al (2017) Multi-biomarker responses in fish (*Jenynsia multidentata*) to assess the impact of pollution in rivers with mixtures of environmental contaminants. Sci Total Environ 595:711–722

Benfenati E, Chaudhry Q, Gini G et al (2019) Integrating in silico models and read-across methods for predicting toxicity of chemicals: a step-wise strategy. Environ Int 131:105060

Bombo AB, Pereira AES, Lusa MG et al (2019) A mechanistic view of interactions of a nanoherbicide with target organism. J Agric Food Chem 67:4453–4462

Camara MC, Campos EVR, Monteiro RA et al (2019) Development of stimuli-responsive nano-based pesticides: emerging opportunities for agriculture. J Nanobiotechnology 17:100

Cerrillo C, Barandika G, Igartua A et al (2017) Key challenges for nanotechnology: Standardization of ecotoxicity testing. J Environ Sci Health Part C 35(2):104–126

Chariou P, Dogan AB, Welsh AG et al (2019) Soil mobility of synthetic and virus-based model nanopesticides. Nat Nanotechnol 14:712–718

Chhipa H (2017) Nanofertilizers and nanopesticides for agriculture. Environ Chem Lett 15:15–22

Clemente Z, Grillo R, Jonsson M (2014) Ecotoxicological evaluation of poly (epsilon-caprolactone) nanocapsules containing triazine herbicides. J Nanosci Nanotechnol 13:1–7

Damalas CA, Eleftherohorinos IG (2011) Pesticide exposure, safety issues, and risk assessment indicators. Int J Environ Res Public Health 8:1402–1419

Deng R, Lin D, Zhu L et al (2017) Nanoparticle interactions with co-existing contaminants: joint toxicity, bioaccumulation and risk. Nanotoxicology 11:591–612

Fojtová D, Vašíčková J, Grillo R (2019) Nanoformulations can significantly affect pesticide degradation and uptake by earthworms and plants. Environ Chem 16:470–481

Gardea-Torresdey JL, Rico CM, White JC (2014) Trophic transfer, transformation, and impact of engineered nanomaterials in terrestrial environments. Environ Sci Technol 48:2526–2540

Gebre SH, Sendeku MG (2019) New frontiers in the biosynthesis of metal oxide nanoparticles and their environmental applications: an overview. SN Appl Sci 1:928

Gomes SIL, Scott-Fordsmand JJ, Campos EVR et al (2019) On the safety of nanoformulations to non-target soil invertebrates-an atrazine case study. Environ Sci Nano 6(6):1950–1958

Grillo R, Melo NFS, Lourenço LR (2010) Characterization of atrazine-loaded biodegradable poly (hydroxybutyrate-cohydroxyvalerate) microspheres. J Polym Environ 18:26–32

Grillo R, Dos Santos NZP, Maruyama CR et al (2012) Poly(ε-caprolactone) nanocapsules as carrier systems for herbicides: physic-chemical characterization and genotoxicity evaluation. J Hazard Mater 231:1–9
Handford CE, Dean M, Spence M et al (2015) Awareness and attitudes towards the emerging use of nanotechnology in the agri-food sector. Food Control 57:24–34
He X, Fu P, Aker WG et al (2018) Toxicity of engineered nanomaterials mediated by nano-bio-eco interactions. J Environ Sci Health C Environ Carcinog Ecotoxicol Rev 36:21–42
He X, Deng H, Hwang H (2019) The current application of nanotechnology in food and agriculture. J Food Drug Anal 27:1–21
Henchion M, McCarthy M, Dillon EJ et al (2019) Big issues for a small technology: consumer trade-offs in acceptance of nanotechnology in food. Innov Food Sci Emerg Technol 58:102210
Holden PA, Nisbet RM, Lenihan HS et al (2013) Ecological nanotoxicology: integrating nanomaterial hazard considerations across the subcellular, population, community, and ecosystems levels. Acc Chem Res 463:813–822
Iavicoli I, Leso V, Beezhold DH et al (2017) Nanotechnology in agriculture: opportunities, toxicological implications, and occupational risks. Toxicol Appl Pharmacol 329:96–111
Isigonis P, Hristozov D, Benighaus C et al (2019) Risk governance of nanomaterials: review of criteria and tools for risk communication, evaluation, and mitigation. Nano 9:696
Jacques MT, Oliveira JL, Campos EVR (2017) Safety assessment of nanopesticides using the roundworm *Caenorhabditis elegans*. Ecotoxicol Environ Saf 139:245–253
Jantunen APK, Gottardo S, Rasmussen K et al (2018) An inventory of ready-to-use and publicly available tools for the safety assessment of nanomaterials. NanoImpact 12:18–28
Kah M, Beulke S, Tiede K et al (2013) Nanopesticides: state of knowledge, environmental fate, and exposure modeling. Crit Rev Environ Sci Technol 43:1823–1867
Kah M, Machinski P, Koerner P et al (2014) Analysing the fate of nanopesticides in soil and the applicability of regulatory protocols using a polymer-based nanoformulation of atrazine. Environ Sci Pol 21(20):11699–11707
Kah M, Kookana RS, Gogos A et al (2018) A critical evaluation of nanopesticides and nanofertilizers against their conventional analogues. Nat Nanotechnol 13:677–684
Kah M, Tufenkji N, White JC et al (2019) Nano-enabled strategies to enhance crop nutrition and protection. Nat Nanotechnol 14:532–540
Karimi M, Sadeghi R, Kokini J (2018) Human exposure to nanoparticles through trophic transfer and the biosafety concerns that nanoparticle-contaminated foods pose to consumers. Trends Food Sci Technol 75:129–145
Kim KH, Kabir E, Jahan SA (2017) Exposure to pesticides and the associated human health effects. Sci Total Environ 575:525–535
Kim D-Y, Kadam A, Shinde S et al (2018) Recent developments in nanotechnology transforming the agricultural sector: a transition replete with opportunities. J Sci Food Agric 98:849–865
Kumar S, Bhanjana G, Sharma A (2017) Development of nanoformulation approaches for the control of weeds. Sci Total Environ 586:1272–1278
Lai RWS, Yeung KWY, Yung MMN et al (2018) Regulation of engineered nanomaterials: current challenges, insights and future directions. Environ Sci Pollut Res Int 25:3060–3077
Li L, Xu Z, Kah M et al (2019) Nanopesticides: a comprehensive assessment of environmental risk is needed before widespread agricultural application. Environ Sci Technol 53:7923–7924
Lombi E, Donner E, Dusinska M (2019) One health approach to managing the applications and implications of nanotechnologies in agriculture. Nat Nanotechnol 14:523–531
Lushchak VI, Matviishyna TM, Husaka VV et al (2018) Pesticide toxicity: a mechanistic approach. EXCLI J 17:1101–1136
Makarenko NA, Makarenko VV (2019) Nanotechnologies in crop cultivation: ecotoxicological aspects. Biosyst Divers 27(2):148–155
Mansano AS, Moreira RA, Dornfeld HC et al (2017) Effects of diuron and carbofuran and their mixtures on the microalgae *Raphidocelis subcapitata*. Ecotoxicol Environ Saf 142:312–321

Marucci RC, Freitas LM, Santos-Rasera JR et al (2019) Are cerium oxide nanoparticles transferred from plants to the aphid *Myzus persicae* (Hemiptera: Aphididae)? Fla Entomol 102:555–561
Miernicki M, HofmannT EI et al (2019) Legal and practical challenges in classifying nanomaterials according to regulatory definitions. Nat Nanotechnol 14:208–216
Miller G, Wickson F (2015) Risk analysis of nanomaterials: exposing nanotechnology's naked emperor. Rev Policy Res 32:485–512
Mitter N, Hussey K (2019) Moving policy and regulation forward for nanotechnology applications in agriculture. Nat Nanotechnol 14:508–510
Mossa AT, Abbassy M (2012) Adverse haematological and biochemical effects of certain formulated insecticides in male rats. Res J Environ Toxicol 6:160–168
Nguyen MH, Hwang IC, Bui CB et al (2016) Effects of the physical state of nanocarriers on their penetration into the root and upward transportation to the stem of soybean plants using confocal laser scanning microscopy. Crop Prot 87:25–30
Nuruzzaman M, Rahman MM, Liu Y et al (2016) Nanoencapsulation, nano-guard for pesticides: a new window for safe application. J Agric Food Chem 64:1447–1483
Oliveira HC, Stolf MR, Martinez CBR et al (2015a) Nanoencapsulation enhances the post-emergence herbicidal activity of atrazine against mustard plants. PLoS One 10:1–12
Oliveira JL, Campos EVR, Silva CMG et al (2015b) Solid lipid nanoparticles co-loaded with simazine and atrazine: preparation, characterization, and evaluation of herbicidal activity. J Agric Food Chem 63:422–432
Oliveira HC, Stolf-Moreira R, Martinez CBR (2015c) Evaluation of the side effects of poly(epsilon-caprolactone) nanocapsules containing atrazine toward maize plants. Front Chem 3:1–9
Oomen AG, Steinhäuser KG, Bleeker EAJ et al (2018) Risk assessment frameworks for nanomaterials: scope, link to regulations, applicability, and outline for future directions in view of needed increase in efficiency. NanoImpact 9:1–13
Pandey S, Giri K, Kumar R et al (2018) Nanopesticides: opportunities in crop protection and associated environmental risks. Proc Natl Acad Sci India Sect B Biol Sci 88:1287–1308
Pereira AES, Grillo R, Mello NFS et al (2014) Application of poly (epsilon-caprolactone) nanoparticles containing atrazine herbicide as an alternative technique to control weeds and reduce damage to the environment. J Hazard Mater 268:207–215
Pourzahedi L, Pandorf M, Ravikumar D et al (2018) Life cycle considerations of nano-enabled agrochemicals: are today's tools up to the task? Environ Sci Nano 5:1057–1069
Prasad A, Astete CE, Bodoki AE et al (2018) Zein nanoparticles uptake and translocation in hydroponically grown sugar cane plants. J Agric Food Chem 66:6544–6551
Preisler AC, Pereira AE, Campos EV et al (2020) Atrazine nanoencapsulation improves pre-emergence herbicidal activity against *Bidens pilosa* without enhancing long-term residual effect on *Glycine max*. Pest Manag Sci 76:141–149
Puglis HJ, Boone MD (2011) Effects of technical-grade active ingredient vs. commercial formulation of seven pesticides in the presence or absence of UV radiation on survival of green frog tadpoles. Arch Environ Contam Toxicol 60:145–155
Roig A (2018) Nanotechnology governance: from risk regulation to informal platforms. Nano Ethics 12:115–121
Romero-Franco M, Godwin HA, Bilal M et al (2017) Needs and challenges for assessing the environmental impacts of engineered nanomaterials (ENMs). Beilstein J Nanotechnol 8:989–1014
Schnoor B, Elhendawy A, Joseph S et al (2018) Engineering atrazine loaded poly (lactic-co-glycolic acid) nanoparticles to ameliorate environmental challenges. J Agric Food Chem 66:7889–7898
Singh S, Kumar V, Chauhan A et al (2018) Toxicity, degradation and analysis of the herbicide atrazine. Environ Chem Lett 16(1):211–237
Slattery M, Harper B, Harper S (2019) Pesticide encapsulation at the nanoscale drives changes to the hydrophobic partitioning and toxicity of an active ingredient. Nanomaterials 9:81

Sørensen SN, Baun A, Burkard M et al (2019) Evaluating environmental risk assessment models for nanomaterials according to requirements along the product innovation stage-gate process. Environ Sci Nano 6:505–518
Sousa GFM, Gomes DG, Campos EVR et al (2018) Post-emergence herbicidal activity of nanoatrazine against susceptible weeds. Front Environ Sci 6:1–6
Souza PMS, Lobo FA, Rosa AH et al (2012) Desenvolvimento de nanocápsulas de poli-e-caprolactona contendo o herbicida atrazina. Quim Nova 35:132–137
Sun Y, Liang J, Tang L et al (2019) Nano-pesticides: a great challenge for biodiversity? Nano Today 28:100757
Tangaa SR, Selck H, Winther-Nielsen M et al (2016) Trophic transfer of metal-based nanoparticles in aquatic environments: a review and recommendations for future research focus. Environ Sci Nano 3:966–981
Taverna ME, Busatto CA, Lescano MR et al (2018) Microparticles based on ionic and organosolv lignins for the controlled release of atrazine. J Hazard Mater 359:139–147
Trump BD, Hristozov D, Malloy T et al (2018) Risk associated with engineered nanomaterials: different tools for different ways to govern. Nano Today 21:9–13
Villaverde JJ, Sevilla-Morán B, López-Goti C et al (2018) Considerations of nano-QSAR/QSPR models for nanopesticide risk assessment within the European legislative framework. Sci Total Environ 634:1530–1539
Walker GW, Kookana RS, Smith NE et al (2018) Ecological risk assessment of nano-enabled pesticides: a perspective on problem formulation. J Agric Food Chem 66:6480–6486
Werlin R, Priester JH, Mielke RE et al (2011) Biomagnification of cadmium selenide quantum dots in a simple experimental microbial food chain. Nat Nanotechnol 6:65–71
Xiao-Ting C, Wang T (2019) Preparation and characterization of atrazine-loaded biodegradable PLGA nanospheres. J Integr Agric 18:1035–1041
Yue L, Ge C, Feng D et al (2017) Adsorption–desorption behavior of atrazine on agricultural soils in China. J Environ Sci 57:180–189
Zhao X, Cui H, Wang Y et al (2018) Development strategies and prospects of nano-based smart pesticide formulation. J Agric Food Chem 66:6504–6512

# Risk Assessment of Nanofertilizers and Nanopesticides

**Flávia Suelen de Oliveira Pereira, Danielle Araújo Agarrayua, Caroline Brandão Quines, and Daiana Ávila**

**Abstract** The use of nanoagrochemicals is fundamental to modern agriculture and the applications of nanotechnology in the form of nanopesticides or nanofertilizers are growing all over the world. Notwithstanding, risk assessment of these products has been far outpaced by their development. Environmental and human health impacts are general concerns that have been addressed, as reviewed in this chapter. Many studies focus on the level of toxicity testing using organisms such as prokaryotes, plants, and invertebrate and vertebrate animals. The data, however, are still insufficient, as many studies are incomplete, lacking full physical-chemical analysis, adequate controls, or continuity. This impairs the development of regulatory policies and, consequently, the marketing of these products. Risk assessment is needed to ensure the safety of the nanoagrochemicals and the benefits that the environment may attain from their use. This field urges for researchers, funding and organized international collaborative initiatives to thrive.

**Keywords** Nanopesticides · Nanofertilizers · Risk assessment · Toxicity Regulation

## 1 Introduction

Nanotechnology has become one of the most promising tools in modern agriculture. Agricultural and food themes have been focusing on sustainability and protection of food crops for human consumption and for animal feeding (Iavicoli et al. 2017). Considering the increase of world population and the broad search for food, a large

F. S. de O. Pereira · D. A. Agarrayua · C. B. Quines · D. Ávila (✉)
Laboratory of Biochemistry and Toxicology in Caenorhabditis elegans, Graduation Program in Biochemistry, Federal University of Pampa Campus Uruguaiana, Uruguaiana, Brazil
e-mail: daianaavila@unipampa.edu.br



L. F. Fraceto et al. (eds.), *Nanopesticides*,
https://doi.org/10.1007/978-3-030-44873-8_10

scale of application of pesticides in agriculture has occurred, aiming higher production (Salata 2004; Kah et al. 2013). The use of nanoagrochemicals is crucial to modern agriculture and the applications of nanotechnology in the form of nanopesticides or nanofertilizers are growing all over the world (Wang et al. 2016a; Shang et al. 2019; He et al. 2019; Avila et al. 2018).

The fast development of nanotechnology has been facilitating the revolutions of traditional food and agriculture sectors. Numerous novel nanomaterials (NM) have been developed for improving food quality with safer environmental conditions (He et al. 2019). Nanotechnology provides new agrochemical products with new delivery mechanisms to improve productivity while promises to reduce pesticides and fertilizers use. Its applications include: (1) pesticides and fertilizers nanoformulations for crop improvement; (2) nanosensors for crop protection (identification of diseases and residues of agrochemicals); (3) nanodevices for genetic manipulation of plants; and (4) postharvest management (Sekhon 2014). These innovative farming techniques can improve crop yields without damaging soil and water, reducing nitrogen loss, and enhancing nutrients long-term incorporation by soil microorganisms.

On the other hand, there are a high number of nanoformulations potentially employed in agriculture, even though there are uncertainties concerning possible interactions with environment and living organisms. As composition, morphology, chemical properties, application sites become more complex, NMs safety has become a major concern. If nanoparticles (NP) accumulate in the environment or interact with molecules or chemical components of the living organisms, health risks may also occur (Rao 2014). In addition, the dynamic behavior of NMs in the environment and the consequent changes in their physicochemical properties make exposure evaluation challenging. These variables prevent adequate data to suitably support exposure modeling, particularly for those slow/targeted release nanoformulations.

The focus of this review will be outcomes from risk assessment in different models, taking into consideration how these studies are insufficient and how is necessary to establish flows to fully evaluate any NM from the environmental to the human health perspectives, providing basis for regulatory policies.

## 2 Hazards and Assessment Strategies

The development of nanotechnology has far outpaced the studies on occupational and environmental safety for regulatory agencies. Since 2004, the National Institute for Occupational Safety and Health (NIOSH) has established a Nanotechnology Research Center to identify the hazardous implications of nanotechnology for workers health. Moreover, this Center helps companies to plan and manage conducts to protect workers from any identified adverse health effects from working with NMs. However, at the present time, the development of effective risk management for

NMs is still incipient. The same issue has been observed for environmental risk assessment.

For effective risk analysis of any material, a tiered or layered approach focused on main impact points is used (Kookana et al. 2014). Each tier will involve the estimation of a predicted environmental concentration (PEC), which is estimated based on data obtained from analysis from soil, groundwater, and other sources.

The first tier of assessment process involves the toxicity tests using aquatic organisms (such as algae, microcrustaceans, and fish) and soil organisms (as earthworms, springtails, and microbial communities). In the second tier, toxicokinetics/toxicodynamics assays and modified exposure studies (different species and experimental conditions) are performed to broad the assessment. The third tier is based on simple artificial experiments based on natural environments under controlled conditions. The last tier includes the constant biomonitoring using the ecosystem where the product is applied, such as observing birds, bees, etc. (Kookana et al. 2014).

For nanopesticides and nanofertilizers, the guidelines need to be adapted as other variables need to be taken into consideration: particle number concentration, size distribution, zeta potential, shape, coating, and the ratio of "free" and nanoparticle-bound active ingredient (AI). Variables such as the movement of a molecule through the soil to groundwater, phase partitioning, aggregation, and bioconcentration can be obtained by some models and should be included in these analyses (Kookana et al. 2014). The characterization of a nanopesticide at different stages in its environmental life cycle and throughout fate and effect also are necessary for these studies; however, current knowledge does not allow to accurately estimating the exposure and effects of a nanopesticide in specific cases. For instance, if the nanoformulation is more active that the AI, then all the risk assessment studies on that particular pesticide will give a minor contribution to the nanopesticide risk assessment. Studies demonstrate that the average time AI release from nanopesticides is increased about four times in relation to the free pesticide, which totally changes the molecule fate and its impact on the environment (Kah et al. 2018).

Kookana et al. suggest a series of analysis for proper risk assessment of nanopesticides (Kookana et al. 2014): (1) Study of the durability of the product after application; (2) Persistence analysis; (3) Mobility parameters as transport model inputs to estimate leaching or runoff/erosion movement; (4) soils and sediment–water test systems for sorption and degradation studies; (5) Fate and exposure modeling; (6) Uptake into biota. For this last one, it should be taken into consideration the following: (6a) knowledge on how NMs enter the organisms; (6b) understanding whether these NPs can translocate along the different tissues of the organism; (6c) evaluating what adverse biological effects can derive from NM exposure, i.e., its toxicity; (6d) finding the best method to correlate exposure dose and toxicity and (6e) finding the best manner to monitor exposed subjects to detect undesirable effects. These studies need to be performed concomitantly with free AI for suitable comparison. In the future, other variables such as studies using natural sources rather than standard media and assessing the interaction of nanoagrochemicals with other contaminants should be included.

We will next focus on the nanopesticides and nanofertilizers risk assessments at the level of toxicity testing using organisms (prokaryotes, plants, invertebrate and vertebrate animals).

## 3 Nanopesticides Risk Assessment: Toxicity Testing

The application of nanotechnology has been well stablished in several areas such as medicine, engineered materials and cosmetics. The principle of the applicability of this tool is to modify the materials or compounds to a nanometric scale with greater stability and target properties according to the needs of the industry and/ or population. In agriculture, the need to reduce the pesticide use has emerged to minimize their effects on the ecosystem (Fig. 1).

For instance, surfactants improve pesticides efficacy and dispersion by increasing their solubility in water, thus facilitating river contaminations and consequently reflecting risks for organisms and also human health. Therefore, the development of nanopesticides containing the AI in its core may be a great breakthrough in the agricultural field by reducing the amount of surfactants as well as the quantities of pesticides applied to the crops, facilitating their dispersal and maintaining the pesticide properties after application (Kah 2015).

It is expected that these formulations present less environmental and human health problems, in relation to AI only. The rapid development and utilization of nanotechnology resulted in a high activity in this field in the scientific community. In the last decade, approximately 60 papers involving nanopesticides were published (Avila et al. 2018). To evaluate the environmental and health impacts, ani-

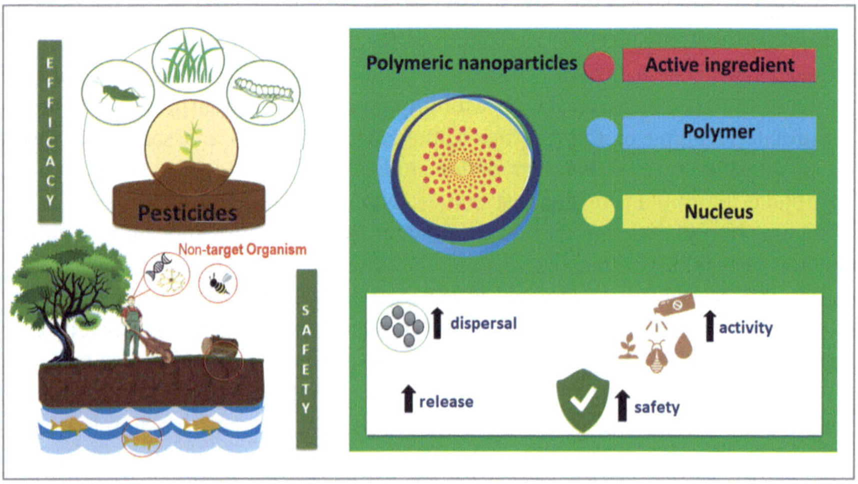


**Fig. 1** Nanopesticides use in agriculture: advantages vs environmental risks

mals, plants, insects, water, and soils, for example, are used to determine nanopesticides toxicity and safety. Remarkably, different materials are used as carriers of the AI, with unknown biological effects (Kah 2015). Therefore, it is necessary to evaluate their safety and their risks. It is important to consider in these tests that the dispersion of the pesticides in the environment affects the area and also non-target organisms, as the residues can deposit and accumulate due to the lipophilic characteristics of these particles (Nicolopoulou-Stamati et al. 2016). These nanoformulations can be formed through polymers (organic) or metals (inorganic), according to the desired application. Inorganic nanoparticles (NP) are related with better release and performance. However, the formulation type can vectorize and facilitate AI incorporation and improve its biological function (Anselmo and Mitragotri 2015).

For example, when silver NP-chitosan encapsulated paraquat was applied to the soil for a large period as 1 month, it maintained the herbicide activity and did not affect soil macro and micronutrients and soil microflora (Namasivayam and Aruna 2014). When phytotoxicity was investigated in the non-target *Vigna mungo* (green gram), no changes in plant growth were observed, whereas free paraquat affected all growth parameters. This demonstrates how NP can reduce the collateral damage caused by pesticides application; however, no further risk assessments for this NP has been published to date (Namasivayam and Aruna 2014).

Copper(II) hydroxide $Cu(OH)_2$ nanopesticides have been extensively used; however, current state of knowledge is not adequate for reliably assessing its environmental risk. Studies with crop plants evaluating genetic, metabolic, and physiological responses depicted alterations in antioxidant defense mechanisms of spinach (*Spinacia oleracea*) and lettuce (*Lactuca sativa*) leaves (Simonin et al. 2018; Zhao et al. 2016; Zhao et al. 2017). A study with the microcrustacean Daphnia magna exposed to Kocide 3000® revealed that genes involved in detoxification and reproduction have changed their expression, depending on the time of exposure to $Cu(OH)_2$ NPs (Aksakal and Arslan 2019). Zhang et al. (2019) demonstrated that $Cu(OH)_2$ NP was able to induce alterations in soil microbiota, which impaired the degradation of neonicotinoid thiacloprid (Zhang et al. 2019). When one pesticide or any contaminant deposits in the soil, it may consequently affect all the micro- and macrocosms who reside in the environment, which brings hazards to all living organisms (Shang et al. 2019).

Aiming to reduce potential toxicity, organic NP come as a promising alternative. Solid lipid NP (SLNs), for instance, provide slow content release and may consequently reduce the toxicological impacts (Naseri et al. 2015). SLN-loaded γ-cyhalothrin presented less toxicity to fish (*Brachydanio rerio*), and thus a potential applicability for the agrochemical agents in the nanotechnology without harm to the environment (Shen et al. 2018). Atrazine and simazine-loaded SLNs demonstrated great stability and slow release, with low cytotoxicity in fibroblasts cells (20% less than the commercial formulation) observed by MTT assay. When the same nanopesticides were tested to non-target organism (Zea mays), SLNs did not cause alterations in plant growth, whereas herbicidal activity against *Raphanus raphanistrum* was ten times stronger than the free pesticides (de Oliveira et al.

2015). On the other hand, unloaded SLNs demonstrated in vivo toxicity, affecting the survival and body length of *C. elegans* nematodes. As the effects were the same in loaded and unloaded NPs, authors attributed the toxicity to SLN formulation and not to the pesticides (Jacques et al. 2017).

The insecticide emamectin benzoate loaded into SLN was tested against pests (moths and aphid) that are responsible for damaging crops. This formulation has demonstrated better toxicity against these insects when compared to conventional pesticide formulations. Furthermore, the formulation has shown good stability, great dispersibility, and interesting biological activity (Shen et al. 2018). However, this formulation lacks further toxicity assessments.

Nanoformulations coated with organic polymers are of particular interest considering the greater biocompatibility with different organisms and, therefore, putatively lower ecotoxicological impact. Imaging studies have demonstrated that NPs can be accumulated in the intestine of soil organisms, as earthworms (Gomes et al. 2019; Jacques et al. 2017). However, it is expected that the accumulation and the toxic effects of nanopesticides are lower than commercial nanoformulations. Studies investigating the parameters of distribution and bioavailability of nanoencapsulated pesticides (bifenthrin) in soil-earthworms (*Eisenia fetida* and *Lumbricus terrestris*), verified that the nanoformulations were accumulated in the worms guts; however, the elimination was better than the free bifenthrin (Mohd Anuar et al. 2018).

One of the most successful examples of risk assessment is the use of poly-ε-caprolactone (PCL)-loaded atrazine. Developed to improve the herbicide activity of atrazine NPs depicted better efficacy towards mustard plants (*Brassica juncea)* than free atrazine form (Pereira et al. 2014). Notably, plant growth parameters were unaffected in non-target plant *Zea mays*, which is an indicator of vectoring pesticide (de Oliveira et al. 2015). Risk assessment assays were conducted as genotoxicity (using *Allium cepa* and human cells), cytotoxicity (with human cells), and ecotoxicological tests (using the alga *Pseudokirchneriella subcapitata* and the worm *Enchytraeus crypticus*), which have indicated a reduced toxicity of PCL nanocapsules containing atrazine towards non-target organisms, as compared to the free herbicide (Clemente et al. 2014; Grillo et al. 2012; Gomes et al. 2019). Using a more complex and vertebrate organism, Andrade et al. demonstrated that nanoencapsulation of atrazine attenuated the biochemical alterations caused by the herbicide to the fish *Prochilodus lineatus* (Andrade et al. 2019*)*. The PCL-loaded atrazine set of studies is a great example of how risk assessment should flow: from the simpler to more complex organisms.

The use of natural-derived polymers is an eco-friendly trend in nanoagrochemicals development. Some examples are chitosan, derived from chitin and zein, derived from corn (*Zea mays*), which have been used to produce biopolymers that are biocompatible, nontoxic, and biodegradable, characteristics that are in accordance with reducing NMs hazards. The literature brings incipient studies of botanical pesticides using zein as biopolymer. For instance, eugenol, geraniol, R-citronellal, and cinnamaldehyde-loaded zein NPs demonstrated improved efficacy against pests such as *Tetranychus urticae*, *Chrysodeixis* includes in comparison to emulsified compounds. Notably, the encapsulation with zein NPs reduced the usual cytotoxicity of the compounds, as observed in vitro in two cell lines (V79-4 and 3 T3) (de Oliveira et al. 2019; Oliveira et al. 2018). Pascoli et al. analyzed the safety of neem oil-loaded zein NPs using *Allium cepa* and *C. elegans*. The study demonstrated that these NPs reduced genotoxicity and also did not affect soil microbiota (Pascoli et al. 2019). Notably, these loaded and unloaded zein NPs did not cause any toxicity to the worms, unlike synthetic-based NPs (Jacques et al. 2017). Studies with rodents testing other Zein NPs confirmed its low toxicity (Penalva et al. 2017; Jain et al. 2018), being considered safe by Food and Drug Administration (FDA). On the other hand, chitosan coated zein NPs induced anxiety-like behavior and memory impairment in exposed rats (Lima et al. 2019). Authors attributed these effects to chitosan, since it has been previously demonstrated that this polymer can alter serotonin levels in zebrafish (*Danio rerio*) (Ozel et al. 2011). Zein data are very promising and open great perspectives towards safe and agricultural practices.

## 4 Nanofertilizers Risk Assessment: Toxicity Testing

Nanofertilizers refer to NMs and nanoenabled bulk materials used as fertilizers. These NP can influence metabolic activities of the plant by delivering required nutrients, thus promoting plant growth, development and productivity in a more efficient manner than regular fertilizers (Raliya et al. 2018). In the context of sustainable agriculture, employing nanofertilizers is a promising approach to increase agronomic yields with environmental safety (Liu and Lal 2015). Depending on the role of the NM and the nutrients in use, nanofertilizers can be separated into four different categories: NMs made of macronutrients or micronutrients, and NMs acting as carriers for macronutrients or micronutrients (Ghormade et al. 2011). Their fundamental properties, such as size, surface area, crystal phase, and surface coating, can control nutrient dissolution and reactivity and also material behavior during application. In addition, they prevent the losses of up to 50% of the nitrogen applied which could be lost by volatilization or leaching (Ghormade et al. 2011; Raliya et al. 2018). Nutrients can be encapsulated by NMs, coated with a thin protective nanoscale polymeric film, or delivered as nano-emulsions or NPs (DeRosa et al. 2010).

NM-enhanced fertilizers may increase plant-uptake efficiency of nutrients and/or reduce the adverse impacts of conventional fertilizer application (Liu and Lal 2015). Substituting traditional methods of fertilizer applications for nanofertilizers is a manner to release nutrients into the soil progressively and in a targeted and controlled manner, thus preventing impacts as water and soil pollution (Sekhon 2014). For instance, nutrient-augmented-zeolites confer a high specific surface area and a high selection towards plant macronutrients, which may be slowly released for specific plant uptake, thus reducing nutrient loss and environmental risks and improving their efficacy (Zulfiqar et al. 2019; Kamaluddin 2010).

Nanotoxicity is a great issue for these products. It is a fact that there are naturally occurring NP in ecosystems and plants possibly have mechanisms of protection against these NPs. In addition, plants do not need high amounts of macro- and micronutrients, therefore any levels above the required would cause phytotoxicity. Phytotoxicity can be caused by the easier dissolution of NPs in comparison to larger and bulk solids (Mayo et al. 1999) which can directly enter plant cells through the sieve-like cell wall structures, delivering high ion levels to the cytoplasm and causing cellular damage. In other words, the indiscriminate use of NP as fertilizers may have potential risks to the environment and therefore, their interactions with aquatic and terrestrial environments have been the subject of investigation by several groups (Kookana et al. 2014; Zulfiqar et al. 2019; Pradhan and Mailapalli 2017).

The toxicity of different hydroxyapatite (HAp) NP was assessed against algae *Pseudokirchneriella subcapitata*, one of the basis of the aquatic trophic chain (Pereira et al. 2017). High concentrations of coprecipitated HAp samples significantly decreased cell density and caused morphological changes on the algal cells surface. Nano-hydroxyapatite (NHAp) has been used to remediate Cd contamination in aqueous and soil environments (Jin et al. 2016; Wang et al. 2016a). However, the complex NHAp-Cd causes severe phytotoxicity to rice seedlings, which showed growth retardation due to oxidative stress caused by Cd (Huang et al. 2017). This same complex has been evaluated in *Daphnia magna* and it has shown to cause increased lipid peroxidation, catalase and peroxidase activities, thus indicating oxidative stress in this specie (Gao et al. 2018). However, further studies assessing the safety of ingested NHAp in more complex animals are still needed.

NHAp can be easily taken up by cells and have been described to cause cell death due to oxidative stress and by triggering inflammatory responses (Rao et al. 2019). Depending on its shape, this NM can induce apoptosis and even cause endothelial and epithelial damage (Santos et al. 2018; Rao et al. 2019). The needle form of NHAp is not even recommended for use in cosmetics (Scientific Committee of Consumer Safety and Bernauer 2018).

The application of ZnONPs enhanced root elongation of germinated radish (*Raphanus sativus*) and rape (*Brassica napus*) in comparison to control. Similarly, metallic ZnNPs improved growth of ryegrass (*Lolium perenne*) seedlings (Lin and Xing 2007). On the other hand, supra-optimal concentrations of ZnNPs resulted in inhibitory or toxic effects on these seedlings. Lee et al. reported the phytotoxicity of ZnONPs based on toxicity indicators (seed germination, root elongation, and number of leaves) on the development of *Arabidopsis thaliana* (thale cress) (Lee et al. 2010). These NPs have also inhibited *Anabaena sp.* growth with generation of reactive oxygen species (ROS) and cellular lipid peroxidation (Tang et al. 2015). Although required by plants at low concentrations, high concentrations of Zn are toxic (Eisler 1993) and can bioaccumulate throughout the food chain, causing damage to other organisms (Lahive et al. 2015).

ZnNPs caused toxicity in a concentration dependent manner to aquatic organisms as *Artemia salina*, *D. magna*, and *Mytilus edulis* (Danabas et al. 2019; Wu et al. 2020). In earthworms (*Xiphinema vuittenezi, C. elegans, and Eisenia fetida*), ZnNP are better uptake than ionic Zn, but also more toxic (Savoly et al. 2016; Heggelund et al. 2014; O'Donnell et al. 2017). Studies show increased mortality, decreased reproduction, and even increasing the number of apoptotic cells. A study using intranasal ZnONPs exposure in rats demonstrated the persistence of these NMs in liver and that their accumulation led to alteration in the expression of antioxidant and metabolic genes in this organ (Guo et al. 2019).

Studies with CuNP showed that, at low concentrations, they stimulated the photosynthesis rate of waterweed (*Elodea densa*) (Nekrasova et al. 2011). Soil adjusted with metallic CuNPs significantly increased lettuce (*Lactuca sativa*) seedling growth (Shah and Belozerova 2009). On the other hand, higher concentrations of metallic CuNPs exhibited toxic effects on seedling growth of mung bean (*Vigna radiata*), wheat (*Triticum aestivum*), yellow squash (*Cucurbita pepo* subsp. *ovifera*), chickpea (*Cicer arietinum*), and soybean (*Glycine max*) (Lee et al. 2008; Adhikari et al. 2012; Musante and White 2012). Root elongation was severely inhibited by NPs exposure when compared with control, whereas the last two plants did not even germinate. In many cases, root necrosis has occurred. CuNP exposure in hydroponically grown lettuce (*Lactuca sativa*) and alfalfa (*Medicago sativa*) reduced the size of both species, modified the nutrient content, and altered enzymatic activity (Hong et al. 2015).

Although essential to most living organisms, supra-optimal concentrations of Cu can be very toxic, and this is no different for CuNPs. CuONPs affect photosynthesis in several algal species; however this toxicity was reduced in a more realistic approach (natural water) (Joonas et al. 2019). These NPs caused dopaminergic neurodegeneration in nematode *C. elegans* (Mashock et al. 2016). CuNPs were very toxic to three fish species rainbow trout (*Oncorhynchus mykiss*), fathead minnow (*Pimephales promelas*), and zebrafish (*D. rerio*), which lowest-observed-effect concentration (LOEC) were below the values of CuNPs predicted to be present in the environment. In addition, at higher temperatures (26 °C), these NPs aggregate, which can increase toxicity to aquatic organisms (Song et al. 2015).

A very elegant study analyzed the bioavailability and toxicity of CuONPs. Lettuce were cultivated in the presence of these NPs, then digested and cultured with intestinal cell culture. The liquid produced by the intestinal cells (which were absorbed by apical cells and exported by the basolateral ones) was collected and added to human liver cells, mimicking the ingestion and metabolization of a contaminated plant and the outcomes. NPs caused intestinal cell damage; however, the levels that "reached" liver cells did not cause necrosis; however, there was increase in ROS levels (Li et al. 2020). The literature brings that CuNPs are basically toxic to liver and kidneys by causing oxidative stress, apoptosis, and genotoxicity; however, the investigations used doses that may not be realistic considering environmental or work exposure (Hejazy et al. 2018).

Titanium is not a required nutrient; however, many have suggested it as a suitable nutrient. $TiO_2$NPs is one of the most studied NP for seed germination, plant growth, and pest management (Liu and Lal 2015). $TiO_2$NPs have been tested on various crops, such as lettuce (Wang et al. 2016b), spinach (Zahra et al. 2015; Hong et al. 2005), duckweed (*Lemna minor*) (Linglan et al. 2008), tomato (*Solanum lycopersicum*) (Song et al. 2012), wheat (*Triticum aestivum*) (Raliya and Biswas 2015), watermelon (*Citrullus lanatus*) (Feizi et al. 2012), beans (*Phaseolus vulgaris*) (Wang et al. 2013), and millets (*Panicum miliaceum*) (Raliya and Biswas 2015;

Tarafdar et al. 2013). These studies conclude that $TiO_2$NP increases plant biomass/yield, chlorophyll content, photosynthetic activity, nutrient contents, and germination rate.

A genotoxicity study with *Allium cepa* (onion) has shown that $TiO_2$NPs have high potential to interact with DNA and cause damage in root meristem cells after 18 h of exposure (Demir et al. 2014). Different concentrations of NP-$TiO_2$ in *Mentha piperita* (peppermint) exhibited reduced gemination percentage (Samadi et al. 2014), which has been also observed in tobacco (*Nicotiana tabacum*), besides exhibiting genotoxic effects (Ghosh et al. 2010). $TiO_2$NPs human health risks were assessed by using different modeling approaches based on literature data. The study concluded that, when considering toxicokinetic information, human health risk can be expected in liver and spleen, ovaries and testes, but mainly towards the reproductive organs (Heringa et al. 2016).

Nano-carbon fertilizers can enhance water and nutrient absorption by plants by enhancing N, P, and K uptake into the plant and it has been suggested that the combined application of N and these NPs could increase the yield and quality of crops (Wu et al. 2010). At the same time, carbon nanotubes can cause damage in plants by forming ROS in plant tissues, consequently leading to cell death. Reduced germination and growth in *Brassica juncea* and *Phaseolus mungo* were observed following exposure to multi-walled carbon nanotubes (Ghodake et al. 2010). Another study has shown that after 20 days of exposure to graphene, there was a significant growth inhibition and reduced biomass in cabbage (*Brassica oleracea*), red tomato, spinach, and lettuce in relation to untreated plants. The reduced number and size of leaves, increased ROS production and cell death of the graphene-treated plants occurred in a dose-dependent manner (Begum et al. 2011). Safety assessment of nano-carbon materials has been growing, but usually inhalation exposure is assessed. In addition, the toxicity depends on the characteristics of the NMs such as length, shape, and concentration, to name a few. Aggregation is a particular characteristic that can lead to chronic issues since these materials can attach to organs and induce inflamatory responses (Francis and Devasena 2018).

A study comparing CuONP, $TiO_2$NP, ZnONP, and carbon nanoparticles and multiwalled carbon nanotubes capacity to cause citotoxicity and DNA damage demonstrated that CuONP has higher toxicity among them all (Karlsson et al. 2008). It is important to emphasize that although many studies have been conducted with nanofertilizers, the risk assessments have not been satisfactorily finalized and, therefore, safe concentrations/doses cannot be indicated.

## 5 Regulation

Although nanotechnology is considered as one of the key advances in agriculture, it has to be incorporated with caution (Shang et al. 2019). The two key considerations of potential risks of nanoagrochemicals application on human are low environmental impacts (water resources, and residues on food products) and worker's safety and secondly, concerning ethical, legal, and social impacts of nanotechnology (Wang et al. 2016a; Shang et al. 2019).

In this context the Global Horizon Scanning Project (GHSP) is an innovative initiative that aims to identify important global environmental quality research needs. Recently, this project has identified 20 key research questions about environment, pollution, and contamination from Latin America (Furley et al. 2018), 40 priority researches to North America (Fairbrother et al. 2019), and 22 priorities to Europe (Van den Brink et al. 2018). These questions covered primary questions about which chemicals we should be mostly concerned regarding their impacts on the planet and sustainability of chemicals through nanotechnology (Furley et al. 2018; Fairbrother et al. 2019; Van den Brink et al. 2018).

Beyond that, there are major concerns about the possible stigmatization of nanoagrochemicals (Shang et al. 2019). In fact, the European Crop Protection Association explains: "*the combination of nanotechnology, food and pesticides has a high potential of public concern. The crop protection industry is afraid of the possibility of a scenario comparable with the rejection of genetically modified organisms*" (Parisi et al. 2014). Peoples apprehensions are focused on NP used as ingredients and additives to food and are related to the presence of pesticides (Kah 2015). Then, the suspicion towards these products and increasing regulations play a role in the nanoagrochemical industry (Parisi et al. 2014).

In an attempt to coordinate discussions, international governmental and nongovernmental organizations have joined scientific, industrial, and regulatory spheres (Kah 2015). These sectors have been debating the future of nanoagrochemicals and new directions towards the reduction in their impact on human and environmental health (Shang et al. 2019). Industry and scientist need to share the necessary data and product information for regulators (Kah 2015). Indeed, some industries like Monsanto, Syngenta, and BASF are developing pesticide-loaded NP and can contribute to regulatory process (Kah et al. 2018).

In Europe, regulation 1107/2009 for marketing authorization of plant protection products requires the novel characteristics of a nanopesticide or nanofertilizer to be considered in the assessment process. It should be considered the interactions between the pesticide AI, safener (chemical substances used in combination with pesticides to make them "safer" and targeted), synergists, and co-formulants during the assessment (Kookana et al. 2014). In 2014, a workshop about nanopesticides was held with the objective to facilitate a harmonized risk assessment of NMs in different research groups. Changes in test methodology, research priorities, and recommendations were proposed in order to facilitate the development of regulatory flows for nanopesticides (Kookana et al. 2014).

In agreement, nanoagrochemicals development is strongly influenced by the regulatory system that controls their entry into the market (Raliya et al. 2018). In the USA, the National Nanotechnology Initiative (NNI) coordinated academic and government research activities on the environmental effects of nanotechnologies (Fairbrother et al. 2019). Indeed, United States Food and Drug Administration (U.S. FDA) has already delivered guidelines for the use of NMs, like nanofertilizers in animal feed (FDA U 2015). The assembled knowledge indicates that food products containing NP available in the market are probably safe to ingest, but this still needs to be more further investigated (Dimkpa and Bindraban 2018; Chaudhry and Castle 2011). Indeed, before nanoagrochemicals in general can be used there is a need for better understanding their modes of function according to standardized assessment approaches, which should be developed to ensure safe use of such agrochemicals (Raliya et al. 2018; Dimkpa and Bindraban 2018).

## 6 Concluding Remarks

Agriculture is in critical need of innovation to reach the increasing and worldwide demand for food, while reducing its impact on the environment. The accountable application of nanotechnologies can turn agriculture more sustainable (Kah et al. 2018). To that end it is imperative that this field receives significant investments for the development of new products and for complete and adequate risk assessment.

Nanopesticides are a great alternative to reduce agrochemical use and their consequences to the environment. Numerous studies report promising results; however, most of the findings are incomplete, i.e., were not investigated in different animal models to provide a full risk assessment panel. The major current concerns regarding nanoagrochemicals are scaling up the synthesis, manufacturing, and designing of lower-cost, precise, safe, and sustainable NMs. In a future perspective, the development of NPs that can act as both fertilizers and pesticides with sustained release and stability for plant protection management, with elucidated nano-biointeractions, transport, and fate in the plant and environment, is desired.

Currently, risk assessments did not provide satisfactory data to predict safe doses/concentrations of NPs in food products and in the environment. Extrapolation of the obtained data is also a limitation, since studies on in vivo gastrointestinal absorption, distribution, and bioavailability are still insufficient. These challenges make the assessment risk a quite interesting area to be explored. The investigations are not well standardized and there is low connection between researchers that study the same subject. In order to understand the human and environmental impacts caused by engineered NMs, additional international collaborative initiatives must be created among leading institutes/countries to develop common strategies and goals to the development of products that contribute to agrotechnology with low risks to environment and human health.

**Acknowledgements** DSA is recipient of CNPq researcher scholarship and supported by grants CNPq (Universal Grants) 453963/2014-5, FAPERGS/PqG # 18/2551-0000434-0, CNPq/FAPERGS/DECIT/SCTIE-MS/PRONEM #16/2551-0000248-7, FAPESP (Grant Number 2017/21004-5) and PROPPI/UNIPAMPA. FSOP and CBQ are recipients of scholarships from Coordenação de Aperfeiçoamento de Pessoal de Nível Superior—Brasil (CAPES)—Finance Code 001.

## References


Adhikari T, Kundu S, Biswas AK, Tarafdar JC, Rao AS (2012) Effect of copper oxide nano particle on seed germination of selected crops. J Agric Sci Technol A 2(6A):815

Aksakal FI, Arslan H (2019) Detoxification and reproductive system-related gene expression following exposure to Cu(OH)2 nanopesticide in water flea (Daphnia magna Straus 1820). Environ Sci Pollut Res Int 27:6103. https://doi.org/10.1007/s11356-019-07414-x

Andrade LL, do Espirito Santo Pereira A, Fernandes Fraceto L, Bueno Dos Reis Martinez C (2019) Can atrazine loaded nanocapsules reduce the toxic effects of this herbicide on the fish Prochilodus lineatus? A multibiomarker approach. Sci Total Environ 663:548–559. https://doi.org/10.1016/j.scitotenv.2019.01.380

Anselmo AC, Mitragotri S (2015) A review of clinical translation of inorganic nanoparticles. AAPS J 17(5):1041–1054. https://doi.org/10.1208/s12248-015-9780-2

Avila DS, GARCIA SC, ARBO MD, NARDI J, JACQUES MT (2018) Use of nanomaterials in agriculture: potential benefits and challenges. In: Singh HB, Mishra S, Fraceto LF, de Lima R (eds) Emerging trends in agri-nanotechnology: fundamental and applied aspects, vol 1. CABI, Egham

Begum P, Ikhtiari R, Fugetsu B (2011) Graphene phytotoxicity in the seedling stage of cabbage, tomato, red spinach, and lettuce. Carbon 49(12):3907–3919

Chaudhry Q, Castle L (2011) Food applications of nanotechnologies: an overview of opportunities and challenges for developing countries. Trends Food Sci Technol 22(11):595–603

Clemente Z, Grillo R, Jonsson M, Santos NZ, Feitosa LO, Lima R, Fraceto LF (2014) Ecotoxicological evaluation of poly(epsilon-caprolactone) nanocapsules containing triazine herbicides. J Nanosci Nanotechnol 14(7):4911–4917. https://doi.org/10.1166/jnn.2014.8681

Danabas D, Ates M, Ertit Tastan B, Cicek Cimen IC, Unal I, Aksu O, Kutlu B (2019) Effects of Zn and ZnO nanoparticles on Artemia salina and Daphnia magna organisms: toxicity, accumulation and elimination. Sci Total Environ 711:134869. https://doi.org/10.1016/j.scitotenv.2019.134869

de Oliveira JL, Campos EV, Goncalves da Silva CM, Pasquoto T, Lima R, Fraceto LF (2015) Solid lipid nanoparticles co-loaded with simazine and atrazine: preparation, characterization, and evaluation of herbicidal activity. J Agric Food Chem 63(2):422–432. https://doi.org/10.1021/jf5059045

de Oliveira JL, Campos EVR, Germano-Costa T, Lima R, Vechia JFD, Soares ST, de Andrade DJ, Goncalves KC, do Nascimento J, Polanczyk RA, Fraceto LF (2019) Association of zein nanoparticles with botanical compounds for effective pest control systems. Pest Manag Sci 75(7):1855–1865. https://doi.org/10.1002/ps.5338

Demir E, Kaya N, Kaya B (2014) Genotoxic effects of zinc oxide and titanium dioxide nanoparticles on root meristem cells of Allium cepa by comet assay. Turk J Biol 38(1):31–39

DeRosa MC, Monreal C, Schnitzer M, Walsh R, Sultan Y (2010) Nanotechnology in fertilizers. Nat Nanotechnol 5(2):91. https://doi.org/10.1038/nnano.2010.2

Dimkpa CO, Bindraban PS (2018) Nanofertilizers: new products for the industry? J Agric Food Chem 66(26):6462–6473. https://doi.org/10.1021/acs.jafc.7b02150

Eisler R (1993) Zinc hazards to fish, wildlife, and invertebrates: a synoptic review. U.S. Department of the Interior, Fish and Wildlife Service, Laurel

Fairbrother A, Muir D, Solomon KR, Ankley GT, Rudd MA, Boxall ABA, Apell JN, Armbrust KL, Blalock BJ, Bowman SR, Campbell LM, Cobb GP, Connors KA, Dreier DA, Evans MS, Henry CJ, Hoke RA, Houde M, Klaine SJ, Klaper RD, Kullik SA, Lanno RP, Meyer C, Ottinger MA, Oziolor E, Petersen EJ, Poynton HC, Rice PJ, Rodriguez-Fuentes G, Samel A, Shaw JR, Steevens JA, Verslycke TA, Vidal-Dorsch DE, Weir SM, Wilson P, Brooks BW (2019) Toward sustainable environmental quality: priority research questions for North America. Environ Toxicol Chem 38(8):1606–1624. https://doi.org/10.1002/etc.4502

FDA U (2015, August) Guidance for industry: use of nanomaterials in food for animals. FDA, Silver Spring

Feizi H, Moghaddam PR, Shahtahmassebi N, Fotovat A (2012) Impact of bulk and nanosized titanium dioxide (TiO 2) on wheat seed germination and seedling growth. Biol Trace Elem Res 146(1):101–106

Francis AP, Devasena T (2018) Toxicity of carbon nanotubes: a review. Toxicol Ind Health 34(3):200–210. https://doi.org/10.1177/0748233717747472

Furley TH, Brodeur J, Silva de Assis HC, Carriquiriborde P, Chagas KR, Corrales J, Denadai M, Fuchs J, Mascarenhas R, Miglioranza KS, Miguez Carames DM, Navas JM, Nugegoda D, Planes E, Rodriguez-Jorquera IA, Orozco-Medina M, Boxall AB, Rudd MA, Brooks BW (2018) Toward sustainable environmental quality: identifying priority research questions for Latin America. Integr Environ Assess Manag 14(3):344–357. https://doi.org/10.1002/ieam.2023

Gao M, Zhang Z, Lv M, Song W, Lv Y (2018) Toxic effects of nanomaterial-adsorbed cadmium on Daphnia magna. Ecotoxicol Environ Saf 148:261–268. https://doi.org/10.1016/j.ecoenv.2017.10.038

Ghodake G, Seo YD, Park D, Lee DS (2010) Phytotoxicity of carbon nanotubes assessed by Brassica juncea and Phaseolus mungo. J Nanoelectron Optoelectron 5(2):157–160

Ghormade V, Deshpande MV, Paknikar KM (2011) Perspectives for nano-biotechnology enabled protection and nutrition of plants. Biotechnol Adv 29(6):792–803. https://doi.org/10.1016/j.biotechadv.2011.06.007

Ghosh M, Bandyopadhyay M, Mukherjee A (2010) Genotoxicity of titanium dioxide (TiO2) nanoparticles at two trophic levels: plant and human lymphocytes. Chemosphere 81(10): 1253–1262

Gomes S, Scott-Fordsmand J, Campos E, Grillo R, Fraceto L, Amorim M (2019) On the safety of nanoformulations to non-target soil invertebrates – an atrazine case study. Environ Sci Nano 6:1950. https://doi.org/10.1039/C9EN00242A

Grillo R, dos Santos NZ, Maruyama CR, Rosa AH, de Lima R, Fraceto LF (2012) Poly(epsilon-caprolactone)nanocapsules as carrier systems for herbicides: physico-chemical characterization and genotoxicity evaluation. J Hazard Mater 231–232:1–9. https://doi.org/10.1016/j.jhazmat.2012.06.019

Guo Z, Luo Y, Zhang P, Chetwynd AJ, Qunhui Xie H, Abdolahpur Monikh F, Tao W, Xie C, Liu Y, Xu L, Zhang Z, Valsami-Jones E, Lynch I, Zhao B (2019) Deciphering the particle specific effects on metabolism in rat liver and plasma from ZnO nanoparticles versus ionic Zn exposure. Environ Int 136:105437. https://doi.org/10.1016/j.envint.2019.105437

He X, Deng H, Hwang HM (2019) The current application of nanotechnology in food and agriculture. J Food Drug Anal 27(1):1–21. https://doi.org/10.1016/j.jfda.2018.12.002

Heggelund LR, Diez-Ortiz M, Lofts S, Lahive E, Jurkschat K, Wojnarowicz J, Cedergreen N, Spurgeon D, Svendsen C (2014) Soil pH effects on the comparative toxicity of dissolved zinc, non-nano and nano ZnO to the earthworm Eisenia fetida. Nanotoxicology 8(5):559–572. https://doi.org/10.3109/17435390.2013.809808

Hejazy M, Koohi MK, Pour ABM, Najafi D (2018) Toxicity of manufactured copper nanoparticles - a review. Nanomed Res J 3(1):1–9. https://doi.org/10.22034/NMRJ.2018.01.001

Heringa MB, Geraets L, van Eijkeren JC, Vandebriel RJ, de Jong WH, Oomen AG (2016) Risk assessment of titanium dioxide nanoparticles via oral exposure, including toxicokinetic considerations. Nanotoxicology 10(10):1515–1525. https://doi.org/10.1080/17435390.2016.1238113

Hong F, Yang F, Liu C, Gao Q, Wan Z, Gu F, Wu C, Ma Z, Zhou J, Yang P (2005) Influences of nano-TiO 2 on the chloroplast aging of spinach under light. Biol Trace Elem Res 104(3):249–260

Hong J, Rico CM, Zhao L, Adeleye AS, Keller AA, Peralta-Videa JR, Gardea-Torresdey JL (2015) Toxic effects of copper-based nanoparticles or compounds to lettuce (Lactuca sativa) and alfalfa (Medicago sativa). Environ Sci: Processes Impacts 17(1):177–185

Huang Y, Qiu W, Yu Z, Song Z (2017) Toxic effect of cadmium adsorbed by different sizes of nano-hydroxyapatite on the growth of rice seedlings. Environ Toxicol Pharmacol 52:1–7

Iavicoli I, Leso V, Beezhold DH, Shvedova AA (2017) Nanotechnology in agriculture: opportunities, toxicological implications, and occupational risks. Toxicol Appl Pharmacol 329:96–111. https://doi.org/10.1016/j.taap.2017.05.025

Jacques MT, Oliveira JL, Campos EVR, Fraceto LF, Ávila DS (2017) Safety assessment of nanopesticides using the roundworm Caenorhabditis elegans. Ecotoxicol Environ Saf 139:245–253. https://doi.org/10.1016/j.ecoenv.2017.01.045

Jain A, Sharma G, Kushwah V, Ghoshal G, Jain A, Singh B, Shivhare US, Jain S, Katare OP (2018) Beta carotene-loaded zein nanoparticles to improve the biopharmaceutical attributes and to abolish the toxicity of methotrexate: a preclinical study for breast cancer. Artif Cells Nanomed Biotechnol 46(sup1):402–412. https://doi.org/10.1080/21691401.2018.1428811

Jin Y, Liu W, Li X-L, Shen S-G, Liang S-X, Liu C, Shan L (2016) Nano-hydroxyapatite immobilized lead and enhanced plant growth of ryegrass in a contaminated soil. Ecol Eng 95:25–29

Joonas E, Aruoja V, Olli K, Kahru A (2019) Environmental safety data on CuO and TiO2 nanoparticles for multiple algal species in natural water: filling the data gaps for risk assessment. Sci Total Environ 647:973–980. https://doi.org/10.1016/j.scitotenv.2018.07.446

Kah M (2015) Nanopesticides and nanofertilizers: emerging contaminants or opportunities for risk mitigation? Front Chem 3:64. https://doi.org/10.3389/fchem.2015.00064

Kah M, Beulke S, Tiede K, Hofmann T (2013) Nanopesticides: state of knowledge, environmental fate, and exposure modeling. Crit Rev Environ Sci Technol 43(16):1823–1867. https://doi.org/10.1080/10643389.2012.671750

Kah M, Kookana RS, Gogos A, Bucheli TD (2018) A critical evaluation of nanopesticides and nanofertilizers against their conventional analogues. Nat Nanotechnol 13(8):677–684. https://doi.org/10.1038/s41565-018-0131-1

Kamaluddin H (2010) Zeolites "nanoporous materials". Presented at the first scientific forum for postgraduate KAU student, King AbdulAziz University

Karlsson HL, Cronholm P, Gustafsson J, Moller L (2008) Copper oxide nanoparticles are highly toxic: a comparison between metal oxide nanoparticles and carbon nanotubes. Chem Res Toxicol 21(9):1726–1732. https://doi.org/10.1021/tx800064j

Kookana RS, Boxall AB, Reeves PT, Ashauer R, Beulke S, Chaudhry Q, Cornelis G, Fernandes TF, Gan J, Kah M, Lynch I, Ranville J, Sinclair C, Spurgeon D, Tiede K, Van den Brink PJ (2014) Nanopesticides: guiding principles for regulatory evaluation of environmental risks. J Agric Food Chem 62(19):4227–4240. https://doi.org/10.1021/jf500232f

Lahive E, O'Halloran J, Jansen M (2015) A marriage of convenience; a simple food chain comprised of Lemna minor (L.) and Gammarus pulex (L.) to study the dietary transfer of zinc. Plant Biol 17:75–81

Lee WM, An YJ, Yoon H, Kweon HS (2008) Toxicity and bioavailability of copper nanoparticles to the terrestrial plants mung bean (Phaseolus radiatus) and wheat (Triticum aestivum): plant agar test for water-insoluble nanoparticles. Environ Toxicol Chem 27(9):1915–1921

Lee CW, Mahendra S, Zodrow K, Li D, Tsai YC, Braam J, Alvarez PJ (2010) Developmental phytotoxicity of metal oxide nanoparticles to Arabidopsis thaliana. Environ Toxicol Chem 29(3):669–675

Li J, Song Y, Vogt RD, Liu Y, Luo J, Li T (2020) Bioavailability and cytotoxicity of Cerium- (IV), Copper- (II), and Zinc oxide nanoparticles to human intestinal and liver cells through food. Sci Total Environ 702:134700. https://doi.org/10.1016/j.scitotenv.2019.134700

Lima VS, Guimaraes ATB, da Costa Araujo AP, Estrela FN, da Silva IC, de Melo NFS, Fraceto LF, Malafaia G (2019) Depression, anxiety-like behavior, and memory impairment in mice exposed to chitosan-coated zein nanoparticles. Environ Sci Pollut Res Int 26(11):10641–10650. https://doi.org/10.1007/s11356-019-04536-0

Lin D, Xing B (2007) Phytotoxicity of nanoparticles: inhibition of seed germination and root growth. Environ Pollut 150(2):243–250

Linglan M, Chao L, Chunxiang Q, Sitao Y, Jie L, Fengqing G, Fashui H (2008) Rubisco activase mRNA expression in spinach: modulation by nanoanatase treatment. Biol Trace Elem Res 122(2):168–178

Liu R, Lal R (2015) Potentials of engineered nanoparticles as fertilizers for increasing agronomic productions. Sci Total Environ 514:131–139. https://doi.org/10.1016/j.scitotenv.2015.01.104

Mashock MJ, Zanon T, Kappell AD, Petrella LN, Andersen EC, Hristova KR (2016) Copper oxide nanoparticles impact several toxicological endpoints and cause neurodegeneration in Caenorhabditis elegans. PLoS One 11(12):e0167613. https://doi.org/10.1371/journal.pone.0167613

Mayo M, Seidensticker J, Hague D, Carim A (1999) Surface chemistry effects on the processing and superplastic properties of nanocrystalline oxide ceramics. Nanostruct Mater 11(2):271–282

Mohd Anuar MF, Agatz A, Hodson M, Al-Khazrajy OS, Boxall A (2018) Fate, uptake and distribution of nanoencapsulated pesticides in soil-earthworm systems, implications for environmental risk assessment: nanopesticide uptake into earthworms. Environ Toxicol Chem 37:1420–1429. https://doi.org/10.1002/etc.4094

Musante C, White JC (2012) Toxicity of silver and copper to Cucurbita pepo: differential effects of nano and bulk-size particles. Environ Toxicol 27(9):510–517

Namasivayam KR, Aruna A (2014) Evaluation of silver nanoparticles-chitosan encapsulated synthetic herbicide paraquat (AgNp-CS-PQ) preparation for the controlled release and improved herbicidal activity against Eichhornia crassipes. Res J Biotechnol 9(19):19–27

Naseri N, Valizadeh H, Zakeri-Milani P (2015) Solid lipid nanoparticles and nanostructured lipid carriers: structure, preparation and application. Adv Pharm Bull 5(3):305–313. https://doi.org/10.15171/apb.2015.043

Nekrasova G, Ushakova O, Ermakov A, Uimin M, Byzov I (2011) Effects of copper (II) ions and copper oxide nanoparticles on Elodea densa planch. Russ J Ecol 42(6):458

Nicolopoulou-Stamati P, Maipas S, Kotampasi C, Stamatis P, Hens L (2016) Chemical pesticides and human health: the urgent need for a new concept in agriculture. Front Public Health 4:148. https://doi.org/10.3389/fpubh.2016.00148

O'Donnell B, Huo L, Polli JR, Qiu L, Collier DN, Zhang B, Pan X (2017) From the cover: ZnO nanoparticles enhanced germ cell apoptosis in Caenorhabditis elegans, in comparison with ZnCl2. Toxicol Sci 156(2):336–343. https://doi.org/10.1093/toxsci/kfw258

Oliveira JL, Campos EVR, Pereira AES, Pasquoto T, Lima R, Grillo R, Andrade DJ, Santos FAD, Fraceto LF (2018) Zein nanoparticles as eco-friendly carrier systems for botanical repellents aiming sustainable agriculture. J Agric Food Chem 66(6):1330–1340. https://doi.org/10.1021/acs.jafc.7b05552

Ozel RE, Wallace KN, Andreescu S (2011) Chitosan coated carbon fiber microelectrode for selective in vivo detection of neurotransmitters in live zebrafish embryos. Anal Chim Acta 695(1–2):89–95. https://doi.org/10.1016/j.aca.2011.03.057

Parisi C, Vigani M, Rodriguez-Cerezo E (2014) Nanotechnology for the agricultural sector: from research to the field. European Commission, Seville

Pascoli M, Jacques MT, Agarrayua DA, Avila DS, Lima R, Fraceto LF (2019) Neem oil based nanopesticide as an environmentally-friendly formulation for applications in sustainable agriculture: an ecotoxicological perspective. Sci Total Environ 677:57–67. https://doi.org/10.1016/j.scitotenv.2019.04.345

Penalva R, Gonzalez-Navarro CJ, Gamazo C, Esparza I, Irache JM (2017) Zein nanoparticles for oral delivery of quercetin: pharmacokinetic studies and preventive anti-inflammatory effects in a mouse model of endotoxemia. Nanomedicine 13(1):103–110. https://doi.org/10.1016/j.nano.2016.08.033

Pereira AE, Grillo R, Mello NF, Rosa AH, Fraceto LF (2014) Application of poly(epsilon-caprolactone) nanoparticles containing atrazine herbicide as an alternative technique to control weeds and reduce damage to the environment. J Hazard Mater 268:207–215. https://doi.org/10.1016/j.jhazmat.2014.01.025

Pereira FF, Paris EC, Bresolin JD, Foschini MM, Ferreira MD, Corrêa DS (2017) Investigation of nanotoxicological effects of nanostructured hydroxyapatite to microalgae Pseudokirchneriella subcapitata. Ecotoxicol Environ Saf 144:138–147

Pradhan S, Mailapalli DR (2017) Interaction of engineered nanoparticles with the agri-environment. J Agric Food Chem 65(38):8279–8294. https://doi.org/10.1021/acs.jafc.7b02528

Raliya R, Biswas P (2015) Environmentally benign bio-inspired synthesis of Au nanoparticles, their self-assembly and agglomeration. RSC Adv 5(52):42081–42087

Raliya R, Saharan V, Dimkpa C, Biswas P (2018) Nanofertilizer for precision and sustainable agriculture: current state and future perspectives. J Agric Food Chem 66(26):6487–6503. https://doi.org/10.1021/acs.jafc.7b02178

Rao V (2014) Systems approach to biosafety and risk assessment of engineered nanomaterials. Appl Biosaf 19(1):11–19. https://doi.org/10.1177/153567601401900102

Rao CY, Sun XY, Ouyang JM (2019) Effects of physical properties of nano-sized hydroxyapatite crystals on cellular toxicity in renal epithelial cells. Mater Sci Eng C Mater Biol Appl 103:109807. https://doi.org/10.1016/j.msec.2019.109807

Salata O (2004) Applications of nanoparticles in biology and medicine. J Nanobiotechnol 2(1):3. https://doi.org/10.1186/1477-3155-2-3

Samadi N, Yahyaabadi S, Rezayatmand Z (2014) Effect of TiO2 and TiO2 nanoparticle on germination, root and shoot length and photosynthetic pigments of Mentha piperita. Int J Plant Soil Sci 3(4):408–418

Santos C, Turiel S, Sousa Gomes P, Costa E, Santos-Silva A, Quadros P, Duarte J, Battistuzzo S, Fernandes MH (2018) Vascular biosafety of commercial hydroxyapatite particles: discrepancy between blood compatibility assays and endothelial cell behavior. J Nanobiotechnol 16(1):27. https://doi.org/10.1186/s12951-018-0357-y

Savoly Z, Hracs K, Pemmer B, Streli C, Zaray G, Nagy PI (2016) Uptake and toxicity of nano-ZnO in the plant-feeding nematode, Xiphinema vuittenezi: the role of dissolved zinc and nanoparticle-specific effects. Environ Sci Pollut Res Int 23(10):9669–9678. https://doi.org/10.1007/s11356-015-5983-4

Scientific Committee of Consumer Safety SEaS-CSeee, Bernauer U (2018) Opinion of the scientific committee on consumer safety (SCCS) - revision of the opinion on hydroxyapatite (nano) in cosmetic products. Regul Toxicol Pharmacol 98:274–275. https://doi.org/10.1016/j.yrtph.2018.07.018

Sekhon BS (2014) Nanotechnology in agri-food production: an overview. Nanotechnol Sci Appl 7:31–53. https://doi.org/10.2147/NSA.S39406

Shah V, Belozerova I (2009) Influence of metal nanoparticles on the soil microbial community and germination of lettuce seeds. Water Air Soil Pollut 197(1–4):143–148

Shang Y, Hasan MK, Ahammed GJ, Li M, Yin H, Zhou J (2019) Applications of nanotechnology in plant growth and crop protection: a review. Molecules 24(14):E2558. https://doi.org/10.3390/molecules24142558

Shen Y, Zhu H, Cui J, Wang A, Zhao X, Cui B, Wang Y, Cui H (2018) Construction of lambda-cyhalothrin nano-delivery system with a high loading content and controlled-release property. Nano 8(12):1016. https://doi.org/10.3390/nano8121016

Simonin M, Colman BP, Tang W, Judy JD, Anderson SM, Bergemann CM, Rocca JD, Unrine JM, Cassar N, Bernhardt ES (2018) Plant and microbial responses to repeated Cu(OH)2 nanopesticide exposures under different fertilization levels in an agro-ecosystem. Front Microbiol 9:1769. https://doi.org/10.3389/fmicb.2018.01769

Song G, Gao Y, Wu H, Hou W, Zhang C, Ma H (2012) Physiological effect of anatase TiO2 nanoparticles on Lemna minor. Environ Toxicol Chem 31(9):2147–2152

Song L, Vijver MG, Peijnenburg WJ, Galloway TS, Tyler CR (2015) A comparative analysis on the in vivo toxicity of copper nanoparticles in three species of freshwater fish. Chemosphere 139:181–189. https://doi.org/10.1016/j.chemosphere.2015.06.021

Tang Y, Li S, Lu Y, Li Q, Yu S (2015) The influence of humic acid on the toxicity of nano-ZnO and Zn2+ to the Anabaena sp. Environ Toxicol 30(8):895–903

Tarafdar A, Raliya R, Wang W-N, Biswas P, Tarafdar J (2013) Green synthesis of TiO2 nanoparticle using Aspergillus tubingensis. Adv Sci Eng Med 5(9):943–949

Van den Brink PJ, Boxall ABA, Maltby L, Brooks BW, Rudd MA, Backhaus T, Spurgeon D, Verougstraete V, Ajao C, Ankley GT, Apitz SE, Arnold K, Brodin T, Canedo-Arguelles M, Chapman J, Corrales J, Coutellec MA, Fernandes TF, Fick J, Ford AT, Gimenez Papiol G,

Groh KJ, Hutchinson TH, Kruger H, Kukkonen JVK, Loutseti S, Marshall S, Muir D, Ortiz-Santaliestra ME, Paul KB, Rico A, Rodea-Palomares I, Rombke J, Rydberg T, Segner H, Smit M, van Gestel CAM, Vighi M, Werner I, Zimmer EI, van Wensem J (2018) Toward sustainable environmental quality: priority research questions for Europe. Environ Toxicol Chem 37(9):2281–2295. https://doi.org/10.1002/etc.4205

Wang W-N, Tarafdar JC, Biswas P (2013) Nanoparticle synthesis and delivery by an aerosol route for watermelon plant foliar uptake. J Nanopart Res 15(1):1417

Wang P, Lombi E, Zhao FJ, Kopittke PM (2016a) Nanotechnology: a new opportunity in plant sciences. Trends Plant Sci 21(8):699–712. https://doi.org/10.1016/j.tplants.2016.04.005

Wang Y, Sun C, Zhao X, Cui B, Zeng Z, Wang A, Liu G, Cui H (2016b) The application of nano-TiO 2 photo semiconductors in agriculture. Nanoscale Res Lett 11(1):529

Wu M, Hao R, Tian X, Wang X, Ma G, Tang H (2010) Effects of adding nano-carbon in slow released fertilizer on grain yield and nitrogen use efficiency of super hybrid rice. Hybrid Rice 4:034

Wu F, Falfushynska H, Dellwig O, Piontkivska H, Sokolova IM (2020) Interactive effects of salinity variation and exposure to ZnO nanoparticles on the innate immune system of a sentinel marine bivalve, Mytilus edulis. Sci Total Environ 712:136473. https://doi.org/10.1016/j.scitotenv.2019.136473

Zahra Z, Arshad M, Rafique R, Mahmood A, Habib A, Qazi IA, Khan SA (2015) Metallic nanoparticle (TiO2 and Fe3O4) application modifies rhizosphere phosphorus availability and uptake by Lactuca sativa. J Agric Food Chem 63(31):6876–6882

Zhang X, Xu Z, Wu M, Qian X, Lin D, Zhang H, Tang J, Zeng T, Yao W, Filser J, Li L, Sharma VK (2019) Potential environmental risks of nanopesticides: application of Cu(OH)2 nanopesticides to soil mitigates the degradation of neonicotinoid thiacloprid. Environ Int 129:42–50. https://doi.org/10.1016/j.envint.2019.05.022

Zhao L, Ortiz C, Adeleye AS, Hu Q, Zhou H, Huang Y, Keller AA (2016) Metabolomics to detect response of lettuce (Lactuca sativa) to Cu(OH)2 nanopesticides: oxidative stress response and detoxification mechanisms. Environ Sci Technol 50(17):9697–9707. https://doi.org/10.1021/acs.est.6b02763

Zhao L, Huang Y, Adeleye AS, Keller AA (2017) Metabolomics reveals Cu(OH)2 nanopesticide-activated anti-oxidative pathways and decreased beneficial antioxidants in spinach leaves. Environ Sci Technol 51(17):10184–10194. https://doi.org/10.1021/acs.est.7b02163

Zulfiqar F, Navarro M, Ashraf M, Akram NA, Munné-Bosch S (2019) Nanofertilizer use for sustainable agriculture: advantages and limitations. Plant Sci 289:110270

# Nanopesticides: From the Bench to the Market

**Rafael Pires-Oliveira, Marta Simão Kfouri, Bruna Mendonça, and Poliana Cardoso-Gustavson**

**Abstract** Encapsulated nanopesticides are an emerging technology with significant growing in crop protection market. These nanotechnologies are now enhancing pesticides formulations with safer, more efficient, and sustainable products, and have been adopted as an important tool in the integrated pest management. The value of this new technology has been pushed by regulatory pressure against the overuse of chemical pesticides, and the demand for healthy and environment harmless products with superior performance. The market scenario shows a continuous investment in technology and innovation to develop more effective products, in a framework of mergers, ventures, and partnerships to accelerate the development and launch of products. Encapsulated nanopesticides can be produced using appropriate combination of traditional surfactants and polymers, including natural-based and advanced polymers. This is an open field to create original materials. In this chapter, we introduce and discuss the stages from the initial laboratory development until the industrial requirements for implementation of the final product. Challenges of scale-up and strategies to overcome them are reviewed and described. The encapsulation strategies for microorganisms and dsRNA are shown here as two special cases that have emerging as disruptive technologies. An overview of the current status and trends in market and a summary of the forthcoming technologies assessed by a study of patents deposited from the pivotal companies in crop protection segment are the basis of the industry perspective introduced herein.

**Keywords** Pesticide market · Biopesticide · Controlled release · Industrial manufacturing · Encapsulation process · Scale-up · Polymerization · Formulation · dsRNA

The original version of this chapter was revised. The correction to this chapter is available at https://doi.org/10.1007/978-3-030-44873-8_12

R. Pires-Oliveira · M. S. Kfouri · B. Mendonça (✉) · P. Cardoso-Gustavson
Oxiteno S.A., Research and Development Crop Solutions, São Paulo, Brazil
e-mail: bruna.mendonca@oxiteno.com



L. F. Fraceto et al. (eds.), *Nanopesticides*,
https://doi.org/10.1007/978-3-030-44873-8_11

# 1 Introduction

Agriculture plays an essential role in food production for the exponential increasing global population that is estimated to achieve around 8.5 billion in the next decade and over 9 billion in 2050 (United Nations 2019). The protection of crops against unwanted plants, insects, and diseases is crucial to ensure productivity but the use of conventional pesticides displays several drawbacks, e.g. the activity against non-target organisms, environmental contamination, and pest resistant populations increasing. To address these issues, novel pesticide formulations are emerging using nanotechnology concepts initially developed for inks, foods, and pharmaceutics industries and they are now enhancing pesticides formulations creating safer, more efficient and sustainable products. The combination of two or three (sometimes even more) active ingredients with different modes and sites of action in a single formulation is beneficial to overcome increasing pest resistance to traditional pesticides. The replacement of chemical actives ingredients by biopesticide or the combination of them is a growing tendency to comply with the society and regulatory pressure for safer and harmless products. Hence, the emerging encapsulation technologies of active ingredients unlock the development of safer and powerful novel pesticide formulations. In this chapter we will examine a strategy and factors to be considered during the development of an encapsulated nanopesticide, focusing on chemical active ingredients, and afterwards a discussion on encapsulation details of microorganisms and RNAi.

# 2 Encapsulated Pesticides: Market and Intellectual Property Analysis

Encapsulation technology is a very important market for crop protection segment with increasing adoption by farmers aiming an integrated pest management, in addition to the regulatory pressure against chemical pesticides. The global encapsulated pesticides market is expected to reach US$800 million by 2025 at a CAGR[1] of 11.8% during the forecast period from 2019 to 2025. From the different types of pesticides (herbicides, insecticides, fungicides, and others), encapsulated insecticides segment featured as the more important with a 42.3% share of revenue in 2017. As insecticides usually contain more toxic active ingredients, the demand for safer and harmless products with superior performance pushed the encapsulated insecticides development and their commercialization growing. Following these

[1] ClearBridge All Cap Growth ETF (CACG) is an actively managed strategy to achieve long-term capital appreciation through investment in large-, mid-, and small capitalization stocks that have the potential for above average long-term earnings and/or cash flow growth. (Legg Mason Global Asset Management 2019).

increasing demands, herbicides are projected to grow at an accelerated CAGR of 13.9% over the forecast period. Due to the processes used, encapsulation has the potential to take some market share of emulsifiable concentrate (EC) formulations as well, which was around 31% in volume in 2017. In terms of region, Europe stood out in encapsulated pesticide market with a share of 44.9% in 2017. Big agrochemical companies figure as key players in the global microencapsulated pesticides market: Syngenta, BASF, Bayer, FMC, ADAMA, BotanoCap, Arysta LifeScience Corporation, GAT Microencapsulation, Reed Pacific, Belchim Crop Protection, Ecosafe Natural Products Inc, Monsanto Company, among others. These companies have shown a continuous effort in technology adoption, product research, and innovation to deliver more effective products in the industry. Mergers, ventures, amalgamations, partnerships, and promotional activities have been initiatives adopted by some companies to accelerate the launch of products.

A similar picture is observed in intellectual property literature. First patents regarding encapsulating for agricultural purposes were published around 1870 and described a seed coating process (Eckhardt 1868; Claussen 1872). Encapsulation techniques have been developed over the years for different applications—most for foods—but also for inks, pharmaceutics, and personal care uses. In the late 1990s the number of patents published regarding encapsulated active ingredients for agriculture purposes (excluding results with CPC A61K[2] to ensure the focus on this search) started to increase exponentially, as seen in Fig. 1a. From a total of almost 11,000 records, 45% has patent legal status already dead, while 44% are still alive and 11% has indeterminate legal status (most part is still on grant process, i.e. examination). China is the region with the majority of patent publications (3474), followed by the USA (1100), Europe (891), Australia (700), Japan (684), Brazil (423), Canada (375), Republic of Korea (352), India (349), and Mexico (266) (see Fig. 1b). The list of the Top 10 most patenting companies over the last 20 years (shown in Fig. 1c) exhibits the main big companies in the crop care market, in which Syngenta and Bayer are both the major contributors in this area—the former is proprietary of the Zeon® Technology for encapsulation formulation (Syngenta 2018) and the latter has been encapsulating an innovative combination of chemical with biological ingredients (Hack et al. 2012). However, few pesticide products comprising capsules are commercialized to date.

Nanopesticide formulations are classified as capsule suspensions (CS) according to the CropLife International[3] and Table 1 shows other formulations classification containing capsules. The range of active ingredients registered for an encapsulated product is reported by the Pesticide Manual Online (2019) is listed in Table 2, although only a few have been marketed as CS formulations, as mentioned earlier. Indeed, mostly are under development phase and it may take many years before they

[2] The Cooperative Patent Classification (CPC), a system used by the European Patent Office and the United States Patent and Trademark Office, A61K refers to preparations for medical, dental, or toilet purposes.

[3] CropLife International is an international trade association of agrochemical companies including BASF, Bayer, Corteva, FMC, Sumitomo Chemical, and Syngenta.

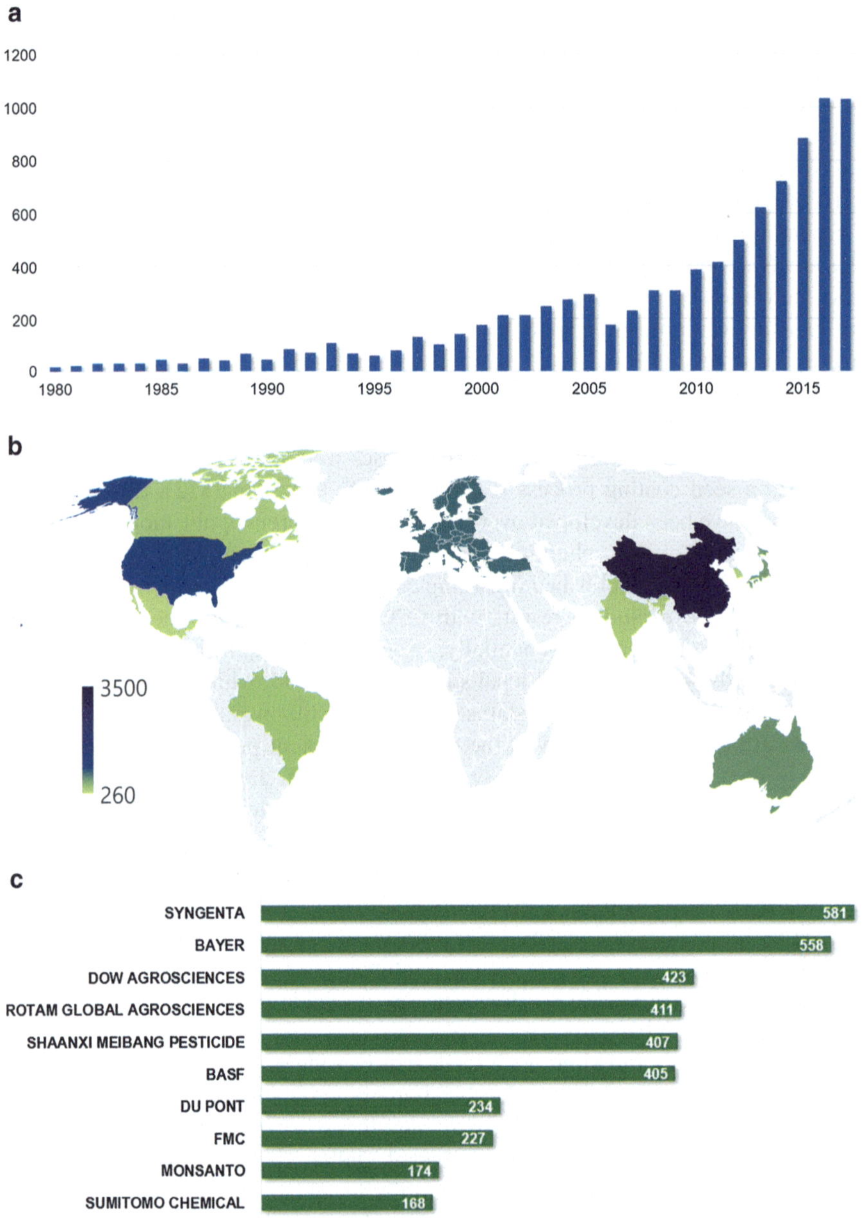


**Fig. 1** Number of patent publication about encapsulated pesticide (**a**) over the years, (**b**) sorted by country, and (**c**) classified by Top 10 companies over the last 20 years. References found by abstract, title, and claim search containing ((encapsul∗ OR ∗capsul∗ OR ∗sphere∗ OR (core near3 shell)) AND (agri∗ OR agro∗ OR (crop AND protection)) AND ((vector AND control) OR (pest AND control) OR pesticide OR insecticide OR herbicide OR fungicide OR acaricide OR nematicide) NOT IC = (A61K)). Source: Derwent™ Innovation

**Table 1** Microcapsule mixture formulations codes (CropLife International 2017)

| Code | Description | Definition |
|---|---|---|
| CS | Capsule suspension | A stable suspension of capsules in a fluid, normally intended for dilution with water before use |
| ZC | A mixed formulation of CS and SC | A stable suspension of capsules and active ingredient(s) in a fluid, normally to be diluted with water before use |
| ZE | A mixed formulation of CS and SE | A stable aqueous dispersion of active ingredients in the form of capsules, solid particles, and fine droplets, normally to be diluted with water before use |
| ZW | A mixed formulation of CS and EW | A stable aqueous dispersion of active ingredients in the form of capsules and fine oil droplets, normally to be diluted in water before use |
| CG[a] | Encapsulated granule | A granule with a protective or granule release controlling coating. Refer to GR (Granule, a free-flowing solid formulation of a defined granule size range ready for use) |

[a]Old formulation code which is no longer supported

**Table 2** Formulation type of encapsulated active ingredients in commercial products sorted by field of use and class listed by the Pesticide Manual Online (2019)

| Type of pesticide | Class | Active ingredient | Formulation type |
|---|---|---|---|
| Insecticide | Carbamate | Carbosulfan | CS |
| | | Furathiocarb | CS |
| | Juvenile hormone mimic | Methoprene | CS |
| | Organophosphate | Acephate | CG |
| | | Chlorpyrifos | CS |
| | | Fonofos | CS |
| | | Parathion-methyl | CS |
| | | Pirimiphos-ethyl | CG |
| | Pyrethroid | Beta-cypermethrin | CS |
| | | Etofenprox | CS |
| | | Gamma-cyhalothrin | CS |
| | | Lambda-cyhalothrin | CS |
| | | Tefluthrin | CS |
| Insecticide and acaricide | Chlorinated cyclodiene | Endosulfan | CS |
| | Organophosphate | Diazinon | CS |
| | | Parathion | CS |
| | | Propetamphos | CS |
| Insecticide and Nematicide | Organophosphate | Cadusafos | CS |
| Nematicide | Organophosphate | Fenamiphos | CS |
| Acaricide | Pyrethroid | Halfenprox | CS |
| Herbicide | Benzonitrile | Dichlobenil | CS |
| | Chloroacetamide | Acetochlor | CS |
| | Dinitroaniline | Pendimethalin | CS |
| | Isoxazolidinone | Clomazone | CS |
| | Thiocarbamate | Esprocarb | CS |
| | Triketone | Bicyclopyrone | ZC |
| | Other | Flurochloridone | CS |
| Herbicide safener | Dichloroacetamide | Dichlormid | CS, CG |
| Fungicide | Spiroketalamine | Spiroxamine | CS |

are commercialized. The insecticides are the larger number of active ingredients registered as a CS formulation mainly due to their high toxicity on non-target organisms. Examples of insecticide actives in the market are beta-cypermethrin, lambda-cyhalothrin, and chlorpyrifos. The herbicides encapsulated in commercial products are: acetochlor, clomazone, dichlobenil, esprocarb, flurochloridone, and pendimethalin (Calvert and Bullock 2016).

## 3 Encapsulation of Chemical Nanopesticides

Innovative encapsulated nanopesticides are formulations comprising the active ingredient surrounded by a wall material, forming a capsule (Fig. 2) in a core-shell or matrix structures. The CS product containing the nanopesticide capsule is usually diluted with water before use. After the application as a spray, the active ingredient is released from the capsule.

These capsule particles are typically in a submicron or reduced micron-size scale (around 0.05–20 μm in diameter), which increases dramatically the specific surface area (i.e., the total surface area per unit volume of particles) of these nanopesticides and provide unique characteristics compared to the same active ingredient in the conventional form. This small-scale size boosts dissolution rate of the active ingredient and enhance its distribution leading to an improved biological activity. In addition, encapsulated pesticides may exhibit several benefits which the following is worth mentioning:

- *Reduce impact on non-target organisms*, as the active ingredient may be only delivered in a target organism or released under given tailored conditions. This can limit the exposure of non-target organisms to the poisonous active ingredient (Edly and Cottle 2018), minimizing deleterious effect on other organisms including phytotoxicity to crops.

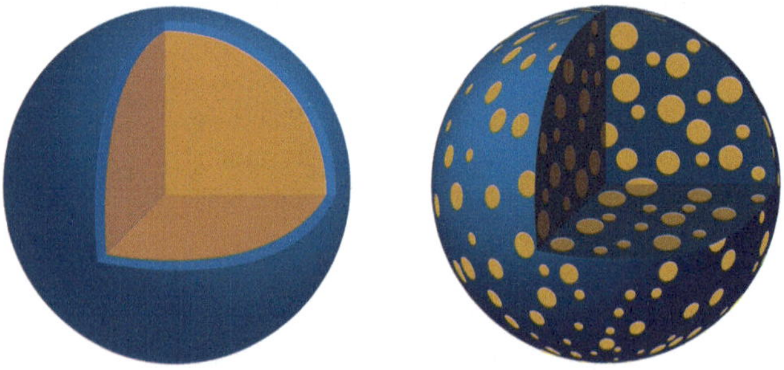

**Fig. 2** Microcapsule structure of a core-shell particle (left) and a matrix capsule (right). The capsule is represented by the blue color and the encapsulated material by the yellow color

- *Improve stability and compatibility within a formulation system and enhanced tank mix compatibility* since the active may not interact with external components of its shell. This feature allows mixed formulations with other incompatible active ingredients (Zedda et al. 2004).
- *Preserve the integrity of bio-based ingredients* (microorganisms and biomolecules) avoiding the contact with damaging components of the formulated product.
- *Protect unstable active ingredients from degradation*, e.g. hydrolysis, photolysis, enzymes, etc. (Braig et al. 2004; Scher et al. 2012).
- *Improve rainfastness of the product*, i.e. capsules can adhere to leaf surface preventing the active to be washed off with rain and watering (Phillips and Gillham 1973; Quong and Nielsen 2003; Bosch et al. 2010).
- *Enhance pesticide residual activity* extending the efficacy of the ingredient which may postpone or even eliminate a subsequent application of a pesticide (Wilson 2010; Preisler et al. 2019).
- *Reduce volatility rate* of active ingredients that have high vapor pressure, such as dicamba, clomazone, and pesticides based on essential oils derived from botanical extracts, reducing the risk to drift and ingredient loss (Markus and Linder 2015; Urch et al. 2018).

Additionally, the encapsulation of active ingredients can provide a sustained or delayed release, combining fast with slow release rates and converting liquid ingredients into solid particles. Thus, a capsule formulation technology leads to a handling and environmental safer formulation besides to boost active ingredients efficacy reducing the required quantity of the conventional analogue product applied to crop. Novel pesticide products have been launched comprising combo formulations of not only chemical ingredients but also including microorganisms, mainly for seed treatment applications. Indeed, there is a current pressure to replace chemicals with microorganisms and other bio-based molecules as active ingredients for natural and harmless components.

Encapsulation technology displays some drawbacks and difficulties that should be considered and tackled during the development of new commercial products in an industrial scale. The main obstacle is the high production cost and process complexity. Not only the manufacturing process of encapsulated product in a manner that ensures reliable quality at a reasonable cost but also the supply chain for industrialization and commercialization must be considered. One of the challenges is to find an asset with scale large enough and capability to manufacture the encapsulation technology developed in the laboratory. A way to compensate investments on asset and new equipment is using the developed encapsulation technologies in different applications or with several active ingredients, maintaining the industrial production on full capacity as much as possible. Companies often produce encapsulated product in a third party. It is not unusual when companies acquire another company that has asset and know-how on encapsulation process to formulate their actives.

Furthermore, the regulatory issue for registration of the commercial product in authority agencies should be considered, especially for nanosized capsules having dimensions of the order of 100 nm or less once the effects of nanosized material are still not fully understood (SCENIHR 2006; Cuffari 2018; IUPAC 2019). Other than these aspects, the stability of the formulated product comprising capsules may also be a challenge for long-term storage and should prevent capsule aggregation and/or phase separation. Also, capsule particle size distribution may cause nozzle clogging during spray application. These usual challenges are also faced during conventional formulation and should be evaluated to guarantee the quality of the product.

## *3.1 Strategy for Designing an Encapsulated Nanopesticide*

A simplified step-by-step strategy for a smart design of CS product is described as follows:

1. Selection of the active ingredient to be encapsulated, wall material and process.
2. Characterization of encapsulation efficiency and determination of release profile.
3. Formulation of the capsule in a CS product and its characterization.
4. Design and scale-up process for the industrial scale.

A useful approach to the selection of the active ingredient to be encapsulated is to choose a high value compound that displays enough bioefficacy at low dosage and a deficient physicochemical property which can be ameliorated by encapsulation process. The high cost of the ingredient will counterbalance the process expenses, while the low dosage will be barely affected by encapsulation efficiency and release. Besides the physicochemical properties of the active ingredient, the design of the capsule and the release system defines the most suitable shielding material and encapsulation process. The wall material option obviously must show great performance in encapsulation efficiency likewise release of the core material. Also, it must not interact or affect the bioefficacy/bioactivity of the active ingredient, display acceptable cost, and adequate supply of raw materials in the market. Furthermore, it is a convenient practice to select materials for capsule wall that are non-toxic, biodegradable, that can be stored and handled in a safe manner besides already listed by the regulatory authorities. Encapsulated nanopesticides are produced mainly by two methods (which will be described in detail in the next sections): interfacial polymerization and coacervation. The first technique is suitable for producing a core-shell structure, while the latter is adequate for microorganisms, biomolecules, and other non-toxic active ingredients.

The synthesized capsule is characterized by size, shape, and visual microscopic observation. The encapsulation efficiency is calculated by the ratio of the total amount of active ingredient in the capsule suspension and the total quantity of active ingredient added initially during preparation. The release profile must also be determined to ensure delivery of the core material. In laboratory tests, the release behavior can be determined in water, while in greenhouse test the release can be determined into the soil, evaluating bioefficacy performance as well. CIPAC[4] recommends some standard methods for determination of active ingredient release, such as MT 190 and MT 190.2 for lambda-cyhalothrin and pirimiphos-methyl, respectively (general methods are under development). The efficacy in field trials of the encapsulated pesticide should be assessed using a CS formulation, which convert the encapsulated active ingredient into an applicable plant protection product. This formulation comprises other ingredients besides the capsule itself. For instance, wetting agent ensures good dilution in spray tank and spreadability onto leaf surface; dispersant agent provides steric hindrance and or electrostatic repulsion avoiding capsule aggregation and subsequent phase separation. Rheology modifiers, such as polysaccharide gums, antifreeze agent, and biocide can also be added to improve formulation stability and shelf life. Other additives, such as uptake enhancer, sticker compounds, and UV protector agent, can also be used in the CS formulation or incorporated within the capsule wall material. It is important to note that all these additives should not impact negatively the capsule shell, e.g. solubilizing or disrupting the capsule, neither affecting the bioactivity of the ingredient. This highlights the need of conventional components and compositions to improve the CS formulation. Furthermore, a nanopesticide CS formulation must also meet some physical properties criteria, as recommended by CIPAC tests:

- Acidity and/or Alkalinity (MT 191) or pH range (MT 75.3).
- Pourability (MT 148.1).
- Spontaneity of dispersion (MT 160).
- Suspensibility (MT 184).
- Wet sieve test (MT 185).
- Persistent foam (MT 47.2).
- Particle size distribution (MT 187).
- Stability at elevated temperature (MT 46.3).

The final stage of the nanopesticide product development is the scale-up and industrial production, which is discussed in detail in a later section of this chapter.

## *3.2 Capsule Wall Material*

The capsule wall is the part surrounding the encapsulated active ingredient and it is usually a polymer. This material promotes both the isolation of the active ingredient from the environment and the warranty that it is released at the best time and rate. The characteristics of the capsule wall are affected by the monomers and other ingredients used in the process. Special features can be added to the capsule using appropriate wall material, e.g. sticking onto insect/leaf surface using an adherent material, such as acrylic, dopamine, and tannic acid-based polymers (Wege et al.

[4] CIPAC: Collaborative International Pesticides Analytical Council.

1999; Quong and Nielsen 2003; Tang et al. 2019; Yu et al. 2019). As mentioned before, capsules can also provide UV protection for unstable active ingredients. Although nanosized wall thickness by itself does not help to achieve this protection, the selection of an adequate wall material can shield UV radiation and provide photoprotection to the capsule, for instance, using dopamine- or tannic acid-based polymers (Tang et al. 2019; Yu et al. 2019) or incorporating other inorganic (e.g., $TiO_2$, ZnO) or organic photoprotector agents within the wall (Shirley et al. 2008; Shirley et al. 2015; Zhang et al. 2019). It is important to prevent loss or reduction of bioefficacy and bioactivity (in case of microorganisms) of the encapsulated material and overcome any chemical incompatibility between the active ingredient and the wall material.

### 3.2.1 Tailoring Nanopesticide Release Profile

The active ingredient is released from the interior of the capsule either by diffusion through the wall in a controlled release rate (mass/time) (described by Eq. (1)) or by mechanical fracture of the capsule in a quick/burst release fashion (Scher 1977; Tadros 2009), as depicted in Fig. 3. Examples of commercial products with slow release profile, also claimed as long-lasting, long-term or sustained release, are Stomp Aqua (Pendimethalin, BASF), Samurai II CS Insecticide (lambda-cyhalothrin, J. Oliver Products), Cyzmic CS (lambda-cyhalothrin, Control Solutions Inc), and Demand CS (lambda-cyhalothrin, Syngenta) (Hack et al. 2012).

$$\text{Release rate} = A \times K \times D \times \frac{C_{\text{inner}} - C_{\text{outer}}}{r_{\text{outer}} - r_{\text{inner}}} \tag{1}$$

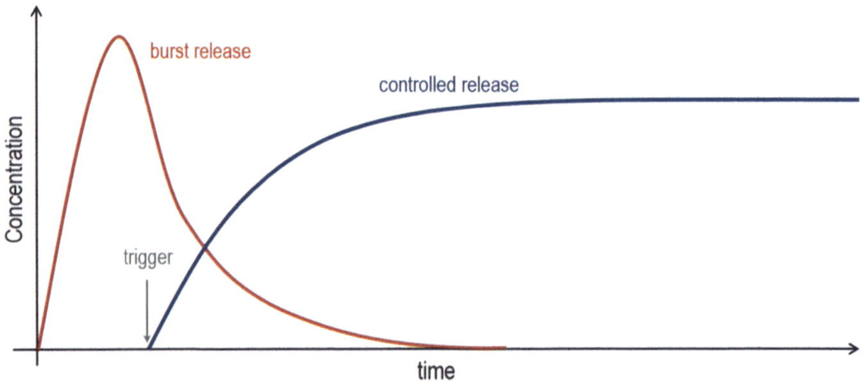


**Fig. 3** Typical encapsulated material release profile: burst (red) and controlled release rates (blue)

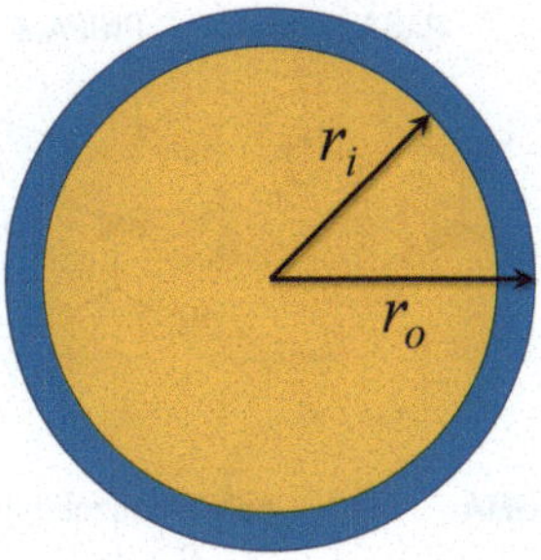


where

$A$ = capsule surface area

$K$ = partition coefficient

$D$ = apparent diffusion coefficient

$C$ = active ingredient concentration

$r_o - r_i$ = shell thickness, with outer and inner radii $r_o$ and $r_i$, respectively

The factors affecting active ingredient release rate from core-shell capsules by diffusion are capsule surface area, affected by its size, wall thickness, and porosity, that can be controlled by polymer concentration or degree of crosslinking. A general approach to adapt the desirable release profile is by changing capsule wall properties. In this way, release rate increases using smaller capsules, synthesizing thinner shells, decreasing the crosslinking degree and/or increase capsule surface area by changing the capsule morphology. On the other hand, slower release rate is achieved by producing thicker capsule shell, synthesizing a high degree of crosslinking, reducing the capsule wall permeability by incorporating objects within the capsule wall (e.g., inorganic particles, crystalline nanocellulose, nanoclay, etc.) or changing the capsule wall composition to reduce the partition coefficient of the active ingredient into the capsule wall, hence reducing the solubility of the active ingredient within the capsule wall and its release rate. Controlling release profile of the encapsulated material is a balance to avoid undesired leaching before application without compromising the active release in the desired place and time after application. Solid and rigid capsules, as in the case of encapsulated lambda-cyhalothrin from Syngenta's commercial products Karate Zeon® and Warrior II, are more prone to have burst release by mechanical rupture of the capsule wall. There is also the possibility to combine different capsule particle sizes and different capsule wall modes of active release (i.e., diffusion through shell plus burst from capsule wall rupture) to obtain both fast acting and long-lasting effects with the nanopesticide. In the case of matrix-type capsule, in which the active ingredient is dispersed throughout the matrix (usually comprising polymers such as lignin, starch, waxes, synthetic polymers, surfactants, and inorganic compounds such as silica or diatomaceous earth), the mechanism of releasing is by erosion or opening in water, as a water-soluble/dispersible granule formulation. Encapsulated microorganism formulations use the latter release mechanism.

### 3.2.2 Designing the Triggered Release System

The release of core active ingredient can be activated by variation of the external environmental conditions, using innovative stimulus-responsive polymer. These materials, also known as smart, are a sort of polymers that can self-assemble or

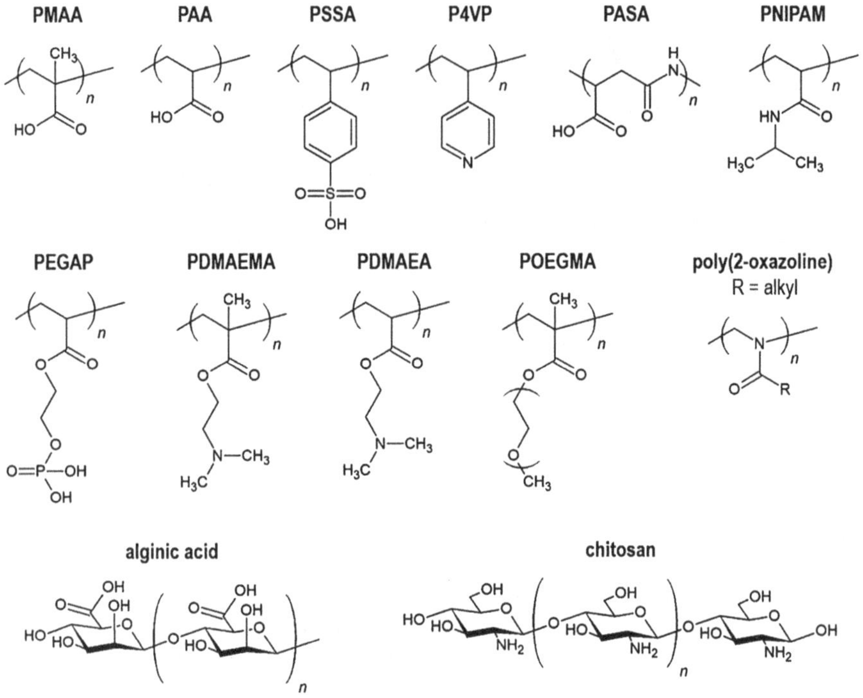


**Fig. 4** Typical pH- and/or temperature-responsive monomer units used as trigger of smart capsule wall polymers

undergo structural changes in response to small environmental changes, e.g. temperature, UV light, pH (from insect digestive tract), enzymatic activity, watering/ rain. As this kind of variations are present in the field, they can be used to trigger the release of the active ingredient in a controlled fashion. This technology has been widely explored in different areas in the last two decades and there are several review papers about this concept that can potentially benefit development of this smart system for nanopesticide formulations (Tang et al. 2015; Che and van Hest 2016; Kocak et al. 2017; Wang and Kohane 2017; Wei et al. 2017; Wertz et al. 2019). As a response to external stimuli, the polymer changes its morphology and properties, resulting in flocculation, chain collapse-extension, precipitation, self-assembly or its disruption, allowing the core material to diffuse in a faster rate or even in a burst release. The two most common and simple strategies are using pH- and temperature-sensitive monomer units in polymers, as shown in Fig. 4. pH-responsive polymers are polyelectrolytes that have in their structure groups that accept protons at low pH forming a neutrally or positively charged polymer chain, and release protons at neutral and high pH. Examples of usual pH-responsive polymer groups are poly(acrylic acid) (PAA), poly(methacrylic acid) (PMAA), poly(4-styrenesulfonic acid) (PSSA), poly(ethylene glycol acrylate phosphate) (PEGAP), poly(aspartic acid) (PASA), poly[(2-dimethylamino)ethyl acrylate] (PDMAEA), poly(4-vinylpyridine) (P4VP), alginic acid, and chitosan.

Thermoresponsive polymers change their solubility in water with temperature, leading to dehydration of the polymer and eventually phase separation, forming a self-assembly structure (like a micelle) or precipitating out from the solution depositing on the interface. Some thermoresponsive monomer units commonly used are poly[(2-dimethylamino)ethyl acrylate] (PDMAEA), poly(2-dimethylaminoethyl methacrylate) (PDMAEMA), poly(*N*-isopropylacrylamide) (PNIPAM), poly(ethylene glycol) methyl methacrylate (POEGMA), and poly(2-oxazoline).

Using the stimuli-responsive concept for agricultural applications, however, can be a more challenging task due to the unpredictable variation of environmental conditions across geographical regions (or even in a crop field) and along a year, for instance, humidity and temperature of air, sun incidence, rain frequency, and temperature and pH of soil. Nevertheless, there are some published patents describing stimuli-responsive capsules that triggers the active release (van Koppenhagen et al. 2003; Seitz and Brinker 2005) but there is no product in the market using this system to date.

## 3.3 Encapsulation Processes

Encapsulation is the technological process by which one or more active ingredients is broken down into smaller dimensions, usually nano or submicron-sized droplets/particles, and then enclosed within another material by coating or entrapping, becoming partially or totally isolated from the external environment (see Fig. 2). The nature of the coating material or matrix can be chosen or manipulated to provide manners to control the release kinetics of the encapsulated material. In the core-shell capsule, the inner part is the active ingredient normally solubilized in a solvent. In a matrix capsule, the active ingredient is dispersed in the surrounding substance. The choice of the encapsulation process must consider the physical and chemical properties of the active ingredient to encapsulate and the additional factors listed:

- Desired capsule form, i.e. core-shell or matrix structure.
- Physical characteristic of core material (solid ingredient: size, shape, and melting point; liquid ingredient: rheology).
- Hydrophilicity/lipophilicity of active ingredient.
- Solubility of active ingredient in water and organic solvents and oils.
- Sensitive of active ingredient to shear, pH, temperature, water, and chemicals (for wall generation).
- Storage conditions and shelf life of the CS product.

There are nowadays hundreds of encapsulation techniques described in the academic and patent literature but only a few are established manufacturing methods. Main encapsulation technologies used in industry include in situ and interfacial polymerizations, fluidized bed coating, spray drying, extrusion, emulsion, coacervation, and solvent evaporation, as depicted in Fig. 5. These technologies can be

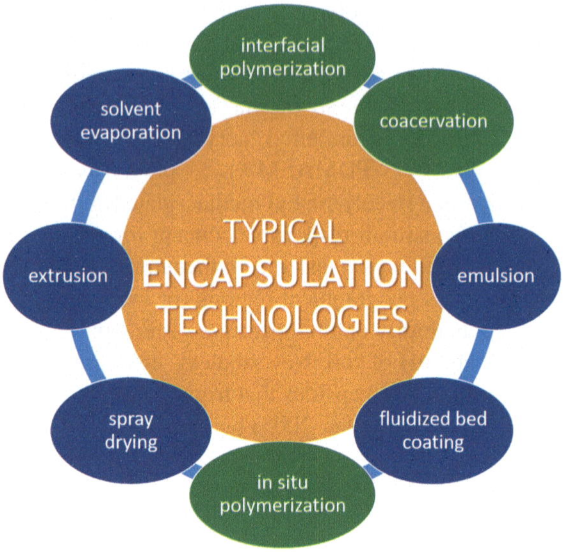


**Fig. 5** Typical industrial process technologies for encapsulation: only in situ and interfacial polymerizations and coacervation processes are suitable for nanopesticides

divided into chemical and physical methods. Among these techniques, only in situ and interfacial polymerizations and coacervation methods are suitable for nanopesticide encapsulation and will be described in the next section. In the physical methods, such as spray drying and fluidized bed coating, a liquid polymer solution is applied via an atomization process onto core particles. After coating the particle, a hard polymer wall is formed by chemical reaction, evaporation of solvent or cooling. However, these techniques are not suitable to encapsulate nanopesticides due to low control and uniformity of the particle surface coverage coupled with the limitation to produce small particles (i.e., submicron or nanosized particles) (Benita 2006; Tadros 2009; Hack et al. 2012).

### 3.3.1 Interfacial Polymerization

The interfacial polymerization is currently the most widely used technology for encapsulation in industry, being to some extent simple and cheap. For example, encapsulated chlorpyrifos (Wilson and Boucher 2014) and lambda-cyhalothrin (Neves et al. 2013) are produced by interfacial polymerization and Monsanto Chemical Company also prepare formulations with encapsulated insecticides and herbicides through this technology (Hack et al. 2012). The encapsulated product is obtained as an aqueous slurry of capsules with the range of particle size usually between 0.05 and 20 μm. The aqueous slurry form is useful for dilution in water for spray application but can also be dried to form a free-flowing powder. The interfacial polymerization method is suitable for non-polar (hydrophobic) active ingredients that can either be a liquid or be dissolved in a water-immiscible solvent, forming

**Table 3** Typical monomers and wall material of rigid capsules produced by interfacial polymerization

| Organic-phase monomer | Water-phase monomer | Polymer wall type |
|---|---|---|
| Polyfunctional isocyanate | Amine | Polyurea |
| Polyfunctional isocyanate | Water | Polyurea |
| Polyfunctional isocyanate | Alcohol | Polyurethane |
| Polyfunctional acyl halide | Amine | Polyamide |
| Polyfunctional acyl halide | Alcohol | Polyester |

an oil phase. The usual loads obtained by this method accordingly to the active ingredient physicochemical properties are around 50–70% for water-insoluble liquid, 25–30% for solid or low melting point, and 15–20% for solid dispersed in oil.

The process comprises basically two steps. First step is the emulsification, using appropriate surfactant and/or nanoparticle, of the oil phase comprising a non-polar monomer in a continuous aqueous phase containing a water-soluble monomer. Second step is the rapidly polymerization at the droplet interface forming the capsule wall at ambient temperatures (interfacial polymerization) or by heating up the system typically up to 50°C (in situ polymerization). In case of water-soluble active ingredient, the inverse emulsion can be used. The most common monomers used in this process are shown in Table 3. It is possible to emulsify the system using alkyl ethoxylate surfactants capped with polymerizable vinyl group. In a system with nanometric emulsion droplets, after the polymerization, there is a suspension of nanocapsules.

This technique allows the control of the release rate by facile designing the capsule wall thickness and porosity by adjusting the crosslinking degree and concentration of the monomers.

In addition, the capsule size can be determined by the droplet size of the oil phase comprising the active ingredient in the emulsion, i.e. adjusting the stirring rate and which emulsifying agent is used. Moreover, it is possible to add a range of additives, e.g. photoprotector agent, antioxidants, nanoparticles, etc. Indeed, nanoparticles can be added for the emulsification process, forming a stable Pickering emulsion (Auweter et al. 2009; Fowler 2010; Tang et al. 2019). It is worth mentioning that isocyanates are toxic and requires careful handling and storage with additional special safety attention. After application the capsule material must not cause any environmental contamination. Therefore, an alternative biodegradable and easy to handle polymer should be preferred.

### 3.3.2 Coacervation

Coacervation is "the separation into two liquid phases in colloidal systems" (IUPAC 2019) in which the phase more concentrated is the coacervate (or coagulate) containing the encapsulated active ingredient usually in a matrix-type capsule. This is particularly suitable for encapsulation of water-soluble nanopesticides with low toxicity and for microorganisms. It is worth mentioning that, as this encapsulation process produces a soft gel, it does not reduce the toxicity of the product like in core-shell systems. Aldehydes, such as glutaraldehyde and formaldehyde, can be used to promote crosslink and harden of the formed capsule, but some of them are toxic and their use should be avoided.

This technique is also known as phase separation or coagulation and can be described by four mechanisms: complex, simple, salting-out effect, or pH precipitation type (Benita 2006; Tadros 2009), as described below:

- Complex coacervation or phase separation occurs when two oppositely charged polyelectrolytes are mutually neutralized by one another.
- Simple coacervation takes place when a water-miscible non-solvent causes the formation of a separate polymer-rich separated phase.
- Salt coacervation, where a polymer is separated out from the aqueous solution as a result of salting-out process. This is the case of the formation of calcium alginate capsules, by adding sodium alginate into a solution of calcium chloride.
- Precipitation of a polymer by changing the pH of the aqueous solution.

Overall, coacervation consists in the eventually phase separation out from solution of a macromolecule due to change of pH or addition of a non-solvent or electrolyte, entrapping the active ingredient. These changes affect the macromolecule solubilization in water, reducing its hydration. As a result, the slightly hydrated complex colloid (coacervate) entangles the active ingredient in this agglomerate polymer phase. The polymers commonly used in this technology are the natural (alginate, carboxymethylcellulose, gelatin, gum arabic, Kraft lignin, starch, and zein) and synthetic polymers (poly(acrylic acid) (PAA), poly(acrylonitrile) (PAN), poly(methyl methacrylate) (PMMA), poly(vinyl acetate) (PVAc), and poly(vinyl pyrrolidone) (PVP)).

Solid pesticides in submicron and nanosized crystals can be easily encapsulated in a suspension through this method. Even though several patents were published along the last two decades (e.g., Medugno and Lessa 2002; Martin et al. 2008; Wagenblast 2009; Li et al. 2014; Norton et al. 2019), coacervation encapsulation technology of pesticides is currently under exploration and has not yet been applied in industrial scale. Table 4 summarizes and compares pros and cons of both encapsulation techniques for nanopesticides.

**Table 4** Pros and cons of interfacial polymerization and coacervation encapsulation techniques for nanopesticides

| Technology | Advantages | Disadvantages |
|---|---|---|
| Interfacial polymerization | Rapidly reaction<br>Low temperature process<br>High load<br>Uniform capsules | Difficulty to stabilize emulsion<br>Limited to liquid active ingredients or those readily soluble or dispersible in suitable solvents |
| Coacervation | Any active ingredient can be encapsulated (dispersible in a liquid phase)<br>Rapidly process | Hard to control coagulation stage<br>Usual crosslinkers may be toxic<br>Still not well established |

### 3.3.3 Design of Scale-up Process

The term scale-up refers to scalability, which means in chemical engineering the successful transition of a small-scale process initially designed at the laboratory to a commercial or industrial scale one. The scale-up delivers the final product manufactured at the rate that will supply the demand in volume, quality, and value to the market it was originally meant to fulfill. In the industry the scale-up process is designed by engineers to test the reproducibility and reduce risks of large investments or losses of producing a certain product at a larger scale by raising the production volume step by step. In few cases it is possible to go from the lab scale to the final manufacturing scale at once, and often high losses and quality variation happen imposing a high risk to the business and the need to go back to the development stage at the lab scale. The number of intermediate steps during the scale-up process, as well as the design of each step, will not only answer questions needed to define whether the process is robust enough to be reproduced in larger scales, but also provide data that enables the adjustment of assets, control methods, and processes to meet proper quality control, optimize operational conditions, avoid losses, and huge investments. In general terms, the scale-up design will depend on the following factors:

- *The very nature of the process*: heat, momentum, and mass transfer alone are quantifiable and easier to scale-up, while some reactions associated to the prior phenomena are less quantifiable, but more difficult to control or predict for example. The nature of the process will mostly and directly affect the design of "each experiment" planned on the scale-up and may or may not increase the number of steps.
- *Asset complexity and number of steps required*: scaling up a single-step or a batch process is simpler than a multi-step or continuous one. A higher number of steps and their duration in a process may threaten higher volumes production simply because it represents higher costs associated to the final product. In the same way, when complex or very specific operational units are required and are not commercially available, higher costs, implementation time, and risks are added and negatively impact the likelihood of a success implementation. Evaluating and reducing complexity of operational units, as well as reducing number of steps in a process, are key to a success scale-up and implementation and must start at the lab scale or in early pilot scale tests.
- *The availability of intermediate sizes of units for smaller scales* (*or capability of building them inside the industry*): one can design a scale-up with ten intermediate steps, for instance, but if the pilot scale units are not available and need to be built, cost and effort will limit the scale-up design to fewer steps and ultimately to move to an industrial scale directly.
- *The choice of process and quality control parameters to meet final quality*: choosing the right and proper parameters that will provide accurate answer whether the final product meets the desired performance or not is key to a successful scale-up design. Identifying these parameters in later early stages of

scale-up (and even at the lab scale) is the key to an effective, faster, and cheaper scale-up.

- *Precision of control parameters required to meet final quality*: very narrow ranges of control parameters can be needed but impose more risks to the success of scale-up or later production. Testing the minimum lower and higher limits in which the control parameters deliver performance in early stages of development is essential to a successful, shorter and mainly to a less risky scale-up.
- *Process factors affecting the final product*: raw material quality, order of addition, reaction time, mixing time and shear required, temperature, pressure, presence of contaminants from previous steps are examples of parameters that can affect production processes. Anticipating what level of variations can occur on a real situation (larger scale) can considerably minimize risks of failure in the final scales and help setting the correct operational conditions and restrictions to the final process.
- *Control methods or devices that can be used in a larger scale*: in a polymerization reaction, gel permeation chromatography is often used as a robust control parameter of the final product, industrially the use of this technique to cover the entire production is not always feasible as a control due to cost and time, in the same way any analytical methods that take over more than very few hours to be completed may be feasible to a lab scale development, but will probably not be in a final scale due to operation time and other factors. Control devices plugged online to control the process may also pose a large investment to a manufacturing unit if not already available and may be sensitive to installation and operation in a way that may not allow their use in a large scale of operation. Thus, scale-up must consider the feasibility of the controls to be used and apply it in the intermediate steps generating data to support a less precise control with an acceptable level of process precision if needed.
- *Safety of operation*: hazardous chemicals, unsafe process conditions or procedures in a chemical plant may pose a limitation to an industrial scale implementation, thus such factors must be considered during the scale-up to be eliminated, reduced of safely controlled.

The scale-up of well-known and widely studied products and its manufacturing processes will comprise all the factors above, but the same may not be possible for new products and processes. Innovation in chemistry and chemical processes will mean not only a challenging development, but also a challenging scale-up since knowledge is not available at the same extend (or at all). When innovation and new and complex products and processes take place, engineers often must opt in a trial and error approach rather than a structure applied science to make a product scalable. Today nanoencapsulation undoubtedly represents a promising and rising technology to many applications, and a challenge to the industry in getting it from the lab to the market since it involves many processes, a high complexity and a not well known and defined scale-up methods and parameters.

The fields of microencapsulation and nanoencapsulation are relatively new in terms of industrial implementation and literature available is mainly focused on encapsulation processes and particle preparation with the main concern of maximum loading and controlled release, in some articles process parameters such as stirring rate, addition rate of coacervation or polymerization agent, temperature influence in the process as well as others, while these studies are important for the design of encapsulation systems they do not allow a real control of the particles properties leading to trial and error approach (Benita 2006).

Encapsulated products by nature have to be manufactured in considerably smaller scales than non-encapsulated ones (considering they would go to the same end use) since the intent of the technology use is to optimize the delivery, thus reducing the amount of product used to achieve the same goal. Following on this rational smaller scale will lead to even higher costs associated to production and make investments more challenging specially to end-uses such as agriculture in which few active ingredients payoff high cost manufacturing technologies. Innovation in more efficient, scalable, and cheaper encapsulation processes and operational units (not on products) will allow a higher rate of implementation of this technology in the industry and faster expansion to segments out of pharmaceutical and food, and in addition will lead to a significant increase in available data and studies for scaling up and implementing such technology. Fortunately investments on new process technology and manufacturing units have been recently observed in the industry, Micropore has developed of a continuous, reliable, and scalable membrane emulsification process unit that allows the delivery of precision microcapsules and microparticles at any scale from lab to kiloton manufacturing in replacement for common stirred tank technology in coacervation encapsulation process (Micropore Technologies 2019).

While knowledge and new technology is been created to completely map the science of scaling up different types of encapsulation, and as the final part of the scale-up topic in this chapter, real cases of manufactured encapsulated pesticides will be review in order to exemplify the current challenge and factors involving nanopesticide production.

# 4 Special Cases: Encapsulation of Microorganisms and dsRNA

## *4.1 Market Overview of Biological Pesticides*

The global market size of the biologicals applied in agriculture (encompassing the segments biopesticides, biofertilizers, and biostimulants) was valuated at US$7.4 billion in 2018, a projection to reach US$20.6 billion by the end of 2026, and a CACG of 13.7% in the forecast period (2019–2026) (USDA, United States Department of Agriculture). The main factors driving the market are the increasing incidences of pest outbreaks in crops, a demand for supreme quality yield from organic farm produce, and the continuous rising of pest resistance to plant protection products. In addition, the increasing concern with the environmental impact of conventional agricultural practices (mainly the overuse of crop protection chemicals and fertilizers) contributed to the strong increase in the demand for biological products to be applied solely or as complement with synthetic pest resistance chemicals. Furthermore, from a regulatory point of view, biological pesticides fulfill latest stringent regulations by regulatory bodies and agencies on the use of synthetic crop protection and nutrition to safeguard the environment and health. Moreover, the lower cost of analysis of biologicals actives against the chemical ones also benefit the biologicals market (ILO 1991; Rana 2019).

Key international market players are already changing the game favoring biologicals, developing crop protection and nutrition products based on natural sources to attend the organic products, what have been and surely will benefit this market during the forthcoming years. North America and Europe highlight as the largest markets for biologicals (about 60% of the market share), mainly due to a growing demand for organic farm. Indeed, this need have been led to the development of technologies (new natural resources, equipment, digital management) that are expected to boost even more the market growth of these regions. In addition, the growing number of organic farms offer opportunities for players in this market, which had grown about 15% between 2015 and 2016 according to the USDA.

Regarding the use of biologicals in different crops, the fruits and vegetables segment is expected to retain its dominance throughout the forecast years in the market, owing to their growing demand attributed to healthy benefits, and the increase adoption of microbial by farmers to solve agricultural issues. The global use of biologicals in row crops (cereals) is projected to witness a CAGR of 13.7% during the forecast period (2020–2026). However, the use of biologicals is not restricted to these segments, but also intended (in fact, already been applied) in commodities (soybean and corn) using products designed specifically to these cultures by market players such as Marrone Bio Innovations Inc., BASF, and Bayer AG. They have launched bio-based products using a variety of microbial strains and biochemicals that are expected to further enhance the agricultural biologicals market revenue (Rana 2019).

Given the current scenario of the development of new technologies, encapsulation may contribute to the increase and widespread of the use of biopesticides, providing favorable safety profiles and stability of the microorganisms and/or other biochemicals, enhanced activity on target pests, and increased adoption by the end-users. Encapsulation techniques have been applied to protect natural-derived volatiles (such as terpenes in Mevalone®), but products specific to the category of biopesticides based on living organisms (bacteria and fungi) are still incipient in the market, although very promising in the forecast years due to the crescent patents deposited by the market players (Fig. 6).

This apparent lack in the market of products containing encapsulated microorganisms is related mainly with limitations to scale-up—equipment, method itself, and cost. It is projected that encapsulation methods will be more widespread in the upcoming years because it is a reliable strategy to allow a longer shelf life comparing to the products currently available in the market—at least similar to chemical

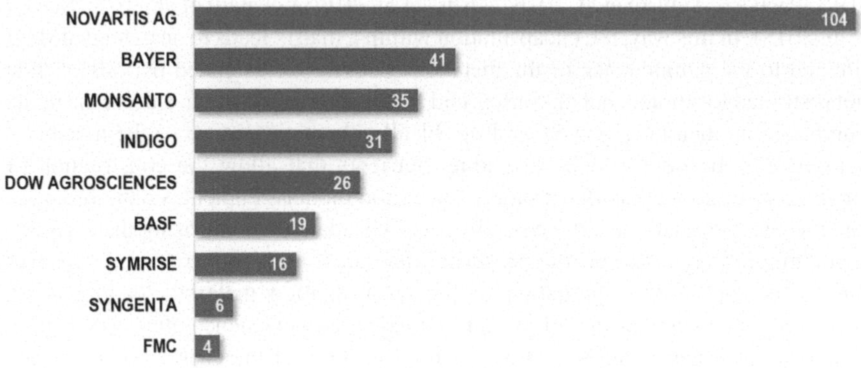


**Fig. 6** Market players and patents deposited from 2009 to 2019. References found by abstract, title, and claim search containing ((encapsul∗ OR ∗capsul∗ OR ∗sphere∗ OR (core near3 shell)) AND (fungi∗ OR bacteria) NOT IC = (A61K)). Results limited to agricultural purposes. Source: Derwent™ Innovation

actives, around 2 years (Bullock 2019). Indeed, the lower shelf life of microbial products is still faced as a hamper to the development and commercialization of this segment of agricultural biologicals.

## *4.2 Encapsulation of Microorganisms*

An effective delivery system plays an important role for successful biological control. Formulation is recognized as one of the most important priorities in biopesticide research (Castillejos et al. 2002; Damalas and Koutroubas 2018). Based on current market trends, research into new formulation methods has notably increased in the past years due to the rising demand for microbial as control agents. It is likely that encapsulating technologies will play an increasingly important role in agriculture, mainly due to the environmental impacts of nanomaterials have so far been determined to be quite limited (Zohar-Perez et al. 2005). The initial objective to encapsulate microorganisms is to ensure that the active organism is kept alive in a dormant state during manufacture and storage, but on application it must be viable to then reproduce and become effective when applied to crops. Thus, a suitable encapsulation system has the potential to improve the characteristics of a microbial-based formulation to extend shelf life, reduce the number of applications and a reduced dose, and improve handling (Bashir et al. 2016).

The method more cited in the literature for microorganisms is the encapsulation within a matrix, which protects the microorganisms from biotic and abiotic stressors (contaminations, antagonists, temperature, dryness, UV light, mechanical stress) by providing a beneficial microenvironment (Chen et al. 2013; Gašić and Tanović

2013; Pacheco-Aguirre et al. 2016; Mishra et al. 2017; Locatelli et al. 2018; Yaakov et al. 2018). In this way, the encapsulation within a matrix leads to an extended shelf life and to the maintenance of the metabolic activity for extended periods of time not restricted to storage, but also after field application, resulting in a decrease in the number of applications as well as dose. Ideally, the maintenance of the metabolic activity may be improved by providing nutrients that allow the construction of small-scale fermenters with extremely low initial biomass content, saving biomass, and turning formulations more cost-effective. The material comprising the encapsulation matrix may have specific properties that allow the slowly release by growth out of the matrix or degradation of the encapsulation material, leading to an increased establishment in soil or leaf and extended persistence after application. Finally, encapsulation allows a singular improvement of the characteristics of the microbial formulation by allowing the co-encapsulation of other microorganisms or biochemicals (semiochemicals) or chemical active ingredients, allowing a synergic interaction at reduced dose (Yaakov et al. 2018; Shang et al. 2019). Targeted delivery may be triggered by environmental conditions and material properties, thus guaranteeing the precise location and of the microorganisms or combo with chemical actives. This precision lead to a low residue and persistence in the field, in agreement with the current and upcoming tendencies in environment and sustainability. Furthermore, the improved handling by solid formulations reduce dusting and improve protection for workers (allergenic spores or co-encapsulated pesticides).

An extensive overview on current encapsulation methods tailored to living microorganisms are available in Vemmer and Patel (2013), Vinceković et al. (2016), Zhang and Wang (2016), Yaakov et al. (2018), and other chapters within this book.

## *4.3 Conventional Formulations vs. Encapsulation Strategies of Biologicals*

Besides the usual challenges faced by a formulator to guarantee the usual parameters of a conventional formulation (blooming, suspensibility, adhesiveness, spreading), one may also have to consider the influence of byproducts produced by the microorganisms that must also be formulated (e.g., extracellular matrices of bacteria or fungi), final pH of the formulation (for liquid formulations and/or tank mix in the field), the presence of electrolytes that influences the stability of the formulation, besides the use (or not) of antifoaming, antifreeze, UV protectors or preservatives that may affect the microorganisms in a deleterious way (Winder et al. 2003; Cumagun 2014; Yaakov et al. 2018). It is important to highlight, however, that the current biological fungicide by Bayer (Serenade®, *Bacillus subtilis*) has a shelf life of 2 years promoted by conventional formulation. To date, there is no encapsulated biological in the market with similar claim.

Despite the many encapsulation methods available for conventional pesticides, their adaptation to biologicals are not trivial because it usually involves reactive components which are not biocompatible with microorganisms, such as organic solvents and monomers, or even require elevated temperatures (Batista et al. 2014; Bashir et al. 2016). The encapsulation of microorganisms is often accompanied by a loss of bioactivity, and efforts have been made to overcome this challenge. A push towards is the use of natural-derived capsules. Some natural nanomaterials, such as nanocellulose, have demonstrated an outstanding performance for use with microbials, which is attributed to their recognized non-toxicity, bio-compatibility with microbials, high surface area-to-volume ratio, biodegradability, and susceptibility to be chemically modified, in addition to being commercially available from renewable sources (Plackett et al. 2014; Reid et al. 2017; Marchiol 2018). To date, nanocellulose has been applied in the literature to stabilize Pickering emulsions containing microorganisms (Fujisawa et al. 2017). Modern encapsulation science considers that the material can add function to capsules more than just being an inert matrix, allowing the tailoring of capsules to living microorganisms and reducing the still reported loss in bioactivity.

Although encapsulation may confer most of all the properties desired from a conventional formulation, some such as spreadability, retention, and rainfastness may be potentialized (or provided) by the presence of co-formulants in-can or in tank mix during application.

## *4.4 About dsRNA Encapsulation*

A promising disruptive and unique technology standing out in crop protection area the RNA interference (RNAi), a double-stranded RNA (dsRNA) able to silence a specific gene expression and act against crop pests. The RNAi mechanism works at the mRNA level by exploiting a sequence-dependent mode of action with high target specificity due to the design of complementary dsRNA molecules, allowing growers to target pests more precisely compared to conventional agrochemicals (see details of this technology, issues regarding RNAi efficiency, dsRNA degradation, environmental risk assessments, and resistance evolution on other chapters within this book). The delivery of RNAi through transgenic plants is now a reality with some products currently in the market. Conversely, it is also expected that more RNA-based products reach the market as non-transformative alternatives. A very deep and critical review of the current available strategies for non-transgenic delivery systems is available in Cagliari et al. (2019). Summarizing, there are two applications of this technology:

- *Plant-incorporated protectant*, in which the plant can produce dsRNA by itself or the introduced dsRNA into a plant deactivate a resistant gene, being considered a genetically modified organisms (GMO).
- *Non-plant-incorporated protectant*, in which a dsRNA-based product is applied in pest control. To date, this strategy is not considered a GMO (EPA 2015) have thus been widely studied as an alternative bio-based pesticide.

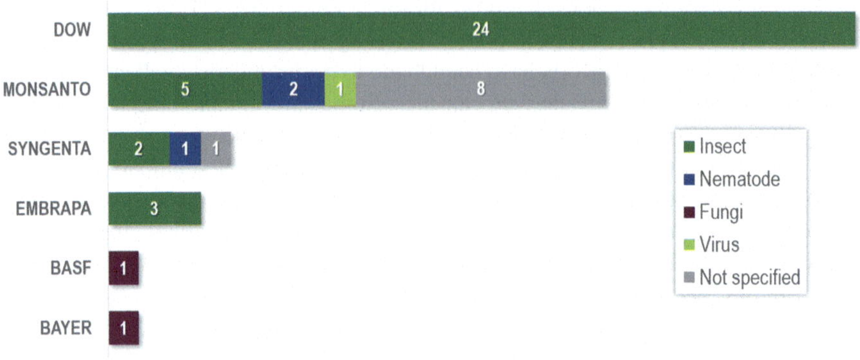


**Fig. 7** Relation of the quantity of deposited patents per target pest and assignees. References found by abstract, title, and claim search containing ((RNAi∗ OR ∗dsRNAi∗ OR ∗interference∗) AND (formulation∗ OR composition∗ OR preparation∗) NOT IC = (A61K)). Results limited to agricultural purposes. Source: Derwent™ Innovation

dsRNA as an insecticide has been the focus of the companies' most relevant R&D centers, as highlighted in Fig. 7. Probably, the main interest in insecticides can be explained by the higher toxicity of chemical ones, the increase resistance to chemical active ingredients observed in the field, and health and environmental risks. In this context, dsRNA sounds a promising friendly alternative to fight against specific insects in the field.

The delivery of dsRNA to a target may be via ingestion, soaking or microinjection mechanisms (Castellanos et al. 2019). Microinjection is highly efficient by immediately delivering dsRNA into specific insect tissues, but it is only feasible for laboratory research due to its complex requirements for both operators and experimental equipment. In the delivery by ingestion mechanism, although is more convenient, it is not feasible since it is difficult to offer an artificial diet for pests, unless it is possible to ensure that the insect feed on plant with the sprayed dsRNA formulations. Then, soaking delivery seems more suitable to field application, even though body wall barrier of pests can act as a barrier against the dsRNA penetration (Zheng et al. 2019). Other delivery methods still need to be investigated, such as seed coats or baits. The development of robust dsRNA formulations will improve efficiency and field stability. The formulation challenges are equally greater as the benefits of the use of dsRNA and will probably request uncommon strategies (and maybe creativity) of the formulator to address it.

The development and scale-up of a dsRNA formulation face the following challenges:

- *A reliable and high-yield production of dsRNA.* There are two possible methods that could be used for production of large quantities of dsRNA, a chemical synthesis of dsRNA in industrial unit, or the production of dsRNA by microbial fermentation in bioreactors (Palli 2014). The first one is easier to obtain but more

**Table 5** List of the pivotal papers (2009–2019) encompassing dsRNA encapsulation methods in both agricultural (agri) and pharmaceutics (pharma) purposes

| Reference | Area | Year | Strategy |
|---|---|---|---|
| "Guanidinium-functionalized interpolyelectrolyte complexes enable RNAi in resistant insect pests" (https://doi.org/10.1021/acs.biomac.7b01717) | Agri | 2018 | Cationic polymer: poly[*N*-(3 guanidinopropyl) methacrylamide] (pGPMA) |
| "Formulation of non-ionic surfactant vesicles (NISV) prepared by microfluidics for therapeutic delivery of siRNA into cancer cells" (https://doi.org/10.1021/acs.molpharmaceut.7b00352) | Pharma | 2017 | Non-ionic surfactant vesicles (NISV) |
| "Liposome encapsulation and EDTA formulation of dsRNA targeting essential genes increase oral RNAi caused mortality in the Neotropical stink bug *Euschistus heros*" (https://doi.org/10.1002/ps.5167) | Agri | 2018 | EDTA or liposome-encapsulated dsRNA |
| "Clay nanosheets for topical delivery of RNAi for sustained protection against plant viruses" (https://doi.org/10.1038/nplants.2016.207) | Agri | 2017 | Clay nanosheets |
| "Lipid nanoparticles for hepatic delivery of small-interfering RNA" (https://doi.org/10.1016/j.biomaterials.2012.05.002) | Pharma | 2012 | Lipid nanoparticles |
| "Systemic delivery of siRNA with cationic lipid assisted PEG-PLA nanoparticles for cancer therapy" (https://doi.org/10.1016/j.jconrel.2011.07.035) | Pharma | 2011 | Cationic lipid and polymer |
| "Viral delivery of dsRNA for control of insect agricultural pests and vectors of human disease: prospects and challenges" (https://doi.org/10.3389/fphys.2017.00399) | Agri | 2017 | Viruses |
| "Shielding of lipid nanoparticles for siRNA delivery: impact on physicochemical properties, cytokine induction, and efficacy" (https://doi.org/10.1038/mtna.2014.61) | Pharma | 2014 | Lipid nanoparticle |
| "Polymers in small-interfering RNA delivery" (https://doi.org/10.1089/nat.2011.0293) | Pharma | 2011 | Polyethylenimine, poly(lactic-*co*-glycolic acid), polypeptides, chitosan, cyclodextrin, dendrimers, and polymers-containing different nanoparticles |
| "Screening of efficient siRNA carriers in a library of surface-engineered dendrimers" (https://doi.org/10.1038/srep25069) | Pharma | 2016 | Dendrimers |
| "Amylose-based cationic star polymers for siRNA delivery" (https://doi.org/10.1155/2015/962941) | Pharma | 2015 | Cationic glyco-star polymer |
| "Cationic liposomes carrying siRNA: impact of lipid composition on physicochemical properties, cytotoxicity and endosomal escape" (https://doi.org/10.3390/nano8050270) | Pharma | 2018 | Cationic lipids (DOTAP and DC-cholesterol) |

expensive, while the latter is less expensive, though is challenging to produce concerning the yield, purity and stability.

- *A robust formulation able to stabilize, prevent dsRNA degradation, and provide the physicochemical parameters to allow its in-can commercialization.* The stabilization of the anionic double strand has been successfully achieved using cationic or amphoteric molecules. After stabilization, an encapsulation approach using materials and release systems robust enough to provide resistance against UV light, extremes pH, and RNAses along the digestive system of the target insect, are among the several claims to be addressed by the formulator. Nanoparticles (Zuckerman and Davis 2015), lipid nanoparticles (Yu et al. 2012), cationic polymers and lipids (Semple et al. 2010; Yang et al. 2011), dendrimers (Biswas and Torchilin 2013; Liu et al. 2016), and others (Setten et al. 2019) are currently applied to encapsulate dsRNA in pharmaceutical industry. The use of clays or the association of them is registered for agricultural purposes. Table 5 summarizes the main academic literature describing different strategies of dsRNA encapsulation in both agricultural and pharmaceutics purposes.
- *Proven bioefficacy* in greenhouse and further in field trials. Although the delivery of dsRNA has been conducted in greenhouse tests, there is still a lack of consistent results with formulated dsRNA. To our knowledge, there is no report of successful field trials with dsRNA formulations.

Although a market overview of the agricultural applications of RNAi technology is still somehow hazy, the investment of huge companies (Bayer, Syngenta) and universities in this disruptive technology are outstanding. The optimization of dsRNA production, delivery, and stability methods will be certainly achieved and then provide its application in the field in forthcoming future. The supply industry must be connected to this tendency and put an effort in developing solutions or components for this technology.

## 5 Challenges for Suppliers (Focus on Surfactant Industry)

The formulation of microorganisms and dsRNA consists a challenge for the surfactant industry. The lack of know-how in manipulating this active ingredient is the first one to overcome. Interdisciplinary teams (chemists, biologists, microbiologists, agronomists) have been characterized the new generation of innovation in chemical industries to face the tendency of chemical specialties for bio-based market and the difficulties in the manipulation of microorganisms or dsRNA. Indeed, the traditional chemical industry is not prepared to develop this new demand due to the need to develop in-house methods to create compositions and evaluate formulations containing such new and complex active ingredients. In this sense, the ability to cope and innovate in microbial and other bio-based formulations will be a differential among the suppliers in the future.

The development of full solutions (i.e., surfactants within compositions, or all the components of an encapsulation) is difficult due to the highly specificity of the active—the strain itself and the fermentation process (solid or liquid) which may provide different physicochemical properties to the same strain, or the precedence of the dsRNA (synthetic or via microorganisms)—in a way that only co-creation projects between the supplier and customer allow the development of solutions. Even the positioning of components for microbial formulation is not trivial. The components must exhibit a narrow range of purity to avoid harmful residues (free alcohols, hydrogen peroxide), and the percentage to be recommended must follow a previous compatibility analysis with the microbe to determine the maximum concentration limit (Cardoso-Gustavson et al. 2019).

Besides biocompatibility and function in a composition, the choice of the components or monomers/polymers have to consider the possibility to be registered by US [OMRI (Organic Materials Review Institute), EPA (Environmental Protection Agency), FIFRA (Federal Insecticide, Fungicide, and Rodenticide Act)], and European [FiBL (Research Institute of Organic Agriculture), REACH (Registration, Evaluation, Authorization and Restriction of Chemicals)] regulation agencies to be used in crop or organic markets. It is convenient (a "nice to have") for the customers (B2C[5]) of surfactant industry (B2B[6]) that components have at least one registration. Although B2C must provide registration of the final formulation, the registration of the components may facilitate the choice of some component or even direct the choice of a single specific supplier.

## 6 Final Remarks

The current trend of using pesticides with more than one mode and site of action to control pests and decrease the resistance to specific molecules employs the mixture of different active ingredients. Sometimes these molecules are not compatible or excessively toxic to non-target organisms or display undesired physical property and chemical sensibility. Encapsulation technology may improve the performance of pesticides, protecting the active ingredient from the environment constrains, as well as keeping the active ingredient enclosed in a polymeric matrix that can control its release, thus providing a long-lasting effect.

Nanopesticides using CS formulation technology are still incipient in the market, although there have been documented an increasing number of academic and industrial research, development and investment in this area. This technology has the potential to incorporate a significant number of active ingredients that are currently formulated as an EC type, notably the toxic ones. Also, it allows combo formulations of incompatible active ingredients, boosting the pest management control. Another global trend seeking to avoid the use of hazardous chemicals and applying living microorganisms leading to a biological pest control can meet the benefits granted by the encapsulation processes.

[5] Business-to-consumer, the end consumer is an individual.

[6] Business-to-business, two companies that do business as a customer and supplier.

Encapsulation technology may find first adoption in high value crops such as fruits and vegetables, and in specialty formulations. It also goes along with other innovation trends in crop protection unlocking the use of novel application and precision technologies such as drone application and disruptive mode of action promoted by dsRNA. Traditional and innovative alkyl alkoxylate surfactants and polymers chemistries enable the production of high performance encapsulated nanopesticides. Advances in process and new molecules will ensure the broad adoption and success of this revolutionary encapsulated nanopesticides formulation technologies for a better, safer, and environmental-friendly way to control pest guaranteeing and increasing sustainable food production for the world population.

## References

Auweter H, Behrens SH, Ngai T (2009) Novel agrochemical formulations. US Patent 2009/0270259, 29 Oct 2009

Bashir O, Claverie JP, Lemoyne P et al (2016) Controlled-release of *Bacillus thurigiensis* formulations encapsulated in light-resistant colloidosomal microcapsules for the management of lepidopteran pests of *Brassica* crops. PeerJ 4:e2524. https://doi.org/10.7717/peerj.2524

Batista DPC, Souza RS, Santos-Magalhães NS et al (2014) Polymer/layered silicate nanocomposite as matrix for bioinsecticide formulation. Macromol Symp 344:14–21. https://doi.org/10.1002/masy.201300208

Benita S (ed) (2006) Microencapsulation: methods and industrial applications. CRC Press, Boca Raton

Biswas S, Torchilin VP (2013) Dendrimers for siRNA delivery. Pharmaceuticals 6:161–183. https://doi.org/10.3390/ph6020161

Bosch HW, Cooper ER, McGurk SL (2010) Bioadhesive nanoparticulate composition having cationic surface stabilizers. US Patent 7,850,995, 14 Dec 2010

Braig A, Leppard DG, Wendeborn F (2004) Synergistic UV absorber combination. PCT Application WO 2004/031294 A1, 15 April 2004

Bullock J (2019) Formulating biologicals: encapsulation and stability. http://news.agropages.com/News/NewsDetail%2D%2D-31256.htm. Accessed 13 Nov 2019

Cagliari D, Dias NP, Galdeano DM et al (2019) Management of pest insects and plant diseases by non-transformative RNAi. Front Plant Sci. 10:1319. https://doi.org/10.3389/fpls.2019.01319

Calvert D, Bullock J (2016) Agrow Formulations 2016. Informa UK, London

Cardoso-Gustavson P, Pires-Oliveira R, Andrade M et al (2019) An introduction of techniques to aid in building microbial-containing formulations: from the selection of components to foliage application. In: 40th Symposium on Pesticide Formulation and Delivery Systems: Formulation, Application and Adjuvant Innovation (Houston: ASTM International) (in press)

Castellanos NL, Smagghe G, Sharma R et al (2019) Liposome encapsulation and EDTA formulation of dsRNA targeting essential genes increase oral RNAi-caused mortality in the Neotropical stink bug *Euschistus heros*. Pest Manag Sci 75:537–548. https://doi.org/10.1002/ps.5167

Castillejos V, Trujillo J, Ortega LD et al (2002) Granular phagostimulant nucleopolyhedrovirus formulations for control of *Spodoptera frugiperda* in maize. Biol Contr 24:300–310. https://doi.org/10.1016/S1049-9644(02)00030-0

Che H, van Hest JCM (2016) Stimuli-responsive polymersomes and nanoreactors. J Mater Chem B 4:4632–4647. https://doi.org/10.1039/C6TB01163B

Chen KN, Chen CY, Lin YC et al (2013) Formulation of a novel antagonistic bacterium based biopesticide for fungal disease control using microencapsulation techniques. J Agric Sci 5:153–163. https://doi.org/10.5539/jas.v5n3p153

Claussen JCH (1872) Improvement in coloring grain and seeds for sowing. US Patent 122,757, 16 Jan 1872

CropLife International (2017) Catalogue of pesticide formulation types and international coding system. Technical Monograph No. 2, 7th Edition, March 2017

Cuffari B (2018) The regulation of nanotechnology in agriculture. http://news.agropages.com/News/NewsDetail%2D%2D-27192.htm. Accessed 20 Jul 2019

Cumagun CJR (2014) Advances in formulation of *Trichoderma* for biocontrol. In: Gupta et al (eds) Biotechnology and biology of *Trichoderma*. Elsevier, Amsterdam

Damalas CA, Koutroubas SD (2018) Current status and recent developments in biopesticide use. Agriculture 8:13. https://doi.org/10.3390/agriculture8010013

Eckhardt AF (1868) Improved system of seeding and manuring. US Patent 82,503, 29 Sept 1868

Edly K, Cottle MR (2018) Microcapsules encapsulating lambda-cyhalothrin. PCT Application WO 2018/001895 A1, 4 January 2018

EPA (U.S. Environmental Protection Agency) (2015) Pesticide: regulating pesticides, plant incorporated protectants. https://www3.epa.gov/pesticides/chem_search/reg_actions/pip/index.htm. Accessed 12 Nov 2019

Fowler J (2010) Pickering emulsion formulations. US Patent 2010/0234230, 16 Sept 2010

Fujisawa S, Togawa E, Kuroda K (2017) Nanocellulose-stabilized Pickering emulsions and their applications. Sci Technol Adv Mater 18:959–971. https://doi.org/10.1080/14686996.2017.1401423

Gašić S, Tanović B (2013) Biopesticide formulations, possibility of application and future trends. Pestic Phytomed 28:97–102. https://doi.org/10.2298/PIF1302097G

Hack B, Egger H, Uhlemann J et al (2012) Advanced agrochemical formulations through encapsulation strategies? Chem Ing Tech 84:223–234

ILO (International Labour Office) (1991) Safety and health in the use of agrochemicals: a guide. International Labour Office, Geneva

IUPAC (2019) In: McNaught AD, Wilkinson A (eds) Compendium of chemical terminology, 2nd edn (the "Gold Book"). Blackwell Scientific Publications, Oxford (1997) (Online version created by S. J. Chalk). ISBN 0-9678550-9-8. https://doi.org/10.1351/goldbook

Kocak G, Tuncer C, Bütün V (2017) pH-Responsive polymers. Polym Chem 8:144–176. https://doi.org/10.1039/C6PY01872F

Legg Mason Global Asset Management (2019) ClearBridge All Cap Growth ETF. https://www.leggmason.com/content/dam/legg-mason/documents/en/product-literature/fact-sheet/fact-sheet-cbi-all-cap-growth-etf.pdf. Accessed 13 Nov 2019

Li F, Pham HH, Anderson DJ (2014) Methods to produce polymer nanoparticles and formulations of active ingredients. US Patent 8,741,808, 3 June 2014

Liu H, Chang H, Lv J et al (2016) Screening of efficient siRNA carriers in a library of surface-engineered dendrimers. Sci Rep 6:25069. https://doi.org/10.1038/srep25069

Locatelli GO, Santos GF, Botelho PS et al (2018) Development of *Trichoderma* sp. formulations in encapsulated granules (CG) and evaluation of conidia shelf-life. Biol Contr 117:21–29. https://doi.org/10.1016/j.biocontrol.2017.08.020

Marchiol L (2018) Nanotechnology in agriculture: new opportunities and perspectives. In: Çelik Ö (ed) New visions in plant science. IntechOpen, London. https://doi.org/10.5772/intechopen.74425

Markus A, Linder C (2015) Encapsulated essential oils. US Patent 9,079,152, 14 Jul 2015

Martin I, Auweter H, Koltzenburg S et al (2008) Nanoparticulate active ingredient formulations. US Patent 2008/0227646, 18 Sept 2008

Medugno CC, Lessa MM (2002) Irreversible coating particles and compositions containing these particles. PCT Application WO 02/26040, 4 April 2002

Micropore Technologies (2019). https://www.micropore.co.uk/agrochemicals.html. Accessed 21 Nov 2019

Mishra S, Keswani C, Abhilash PC et al (2017) Integrated approach of agri-nanotechnology: challenges and future trends. Front Plant Sci 8:471. https://doi.org/10.3389/fpls.2017.00471

Neves J, Esmeraldo CMPC, Bathia S et al (2013) Insecticide formulation, method for manufacturing the same and use thereof. European Patent 2,822,388, 12 Sept 2013

Norton D, Loukine N, Gong R et al (2019) Pyrethroid formulations. US Patent 10,455,830, 29 Oct 2019

Pacheco-Aguirre J, Ruiz-Sanchez E, Reyes-Ramírez A et al (2016) Polymer-based encapsulation of *Bacillus subtilis* and its effect on *Meloidogyne incognita* in tomato. Phyton-Int J Exp Bot 85:1–6

Palli SR (2014) RNA interference in Colorado potato beetle: steps toward development of dsRNA as a commercial insecticide. Curr Opin Insect Sci 6:1–8. https://doi.org/10.1016/j.cois.2014.09.011

Pesticide Manual Online (2019) BCPC (the British Crop Production Council), Bracknell. https://www.bcpc.org/bcpc_integration?application_code=PM. Accessed 8 Oct 2019

Phillips FT, Gillham EM (1973) Al comparison of sticker performance against rainwashing of microcapsules on leaf surfaces. Pestic Sci 4:51–57

Plackett DV, Letchford K, Jackson JK et al (2014) A review of nanocellulose as a novel vehicle for drug delivery. Nord Pulp Pap Res J 29:105–118. https://doi.org/10.3183/npprj-2014-29-01-p105-118

Preisler AC, Anderson ES, Pereira AES, Campos EVR et al (2019) Atrazine nanoencapsulation improves pre-emergence herbicidal activity against *Bidens pilosa* without enhancing long-term residual effect on *Glycine max*. Pest Manag Sci 76(1):141–149. https://doi.org/10.1002/ps.5482

Quong D, Nielsen KE (2003) Adherent microcapsules. PCT Application WO 03/007714 A1, 30 Jan 2003

Rana S (ed) (2019) Agrow Biologicals 2019. Informa UK, London

Reid MS, Villalobos M, Cranston ED (2017) Benchmarking cellulose nanocrystals: from the laboratory to industrial production. Langmuir 33:1583–1598. https://doi.org/10.1021/acs.langmuir.6b03765

SCENIHR (Scientific Committee on Emerging and Newly Identified Health Risks) (2006) The appropriateness of existing methodologies to assess the potential risks associated with engineered and adventitious products of nanotechnologies. Adopted by the SCENIHR during the 10th plenary meeting of 10 March 2006 after public consultation

Scher HB (1977) Microencapsulated pesticides. In: Scher HB (ed) Controlled release pesticides. ACS Symposium Series, vol 53. American Chemical Society, Washington. https://doi.org/10.1021/bk-1977-0053.ch012

Scher HB, Chen J, Mazeaud I et al (2012) Microcapsules. US Patent 8,173,159, 8 May 2012

Seitz ME, Brinker RJ (2005) Microcapsules having activated release of core material therein. PCT application WO2005/122758, 29 Dec 2005

Semple S, Akinc A, Chen J et al (2010) Rational design of cationic lipids for siRNA delivery. Nat Biotechnol 28:172–176. https://doi.org/10.1038/nbt.1602

Setten RL, Rossi JJ, Han S (2019) The current state and future directions of RNAi-based therapeutics. Nat Rev Drug Discov 18:421–446. https://doi.org/10.1038/s41573-019-0017-4

Shang Y, Hasan MK, Ahammed GJ et al (2019) Applications of nanotechnology in plant growth and crop protection: a review. Molecules 24:2558. https://doi.org/10.3390/molecules24142558

Shirley IM, Heming AM, Bon SF et al (2008) Composition for photoprotection. PCT Application WO 2008/032022, 20 Mar 2008

Shirley IM, Wright T, Perrin RM et al (2015) Capsule formulation. US Patent 9,204,630, 8 Dec 2015

Syngenta (2018) Zeon Technology enhances insecticide's performance. http://www.syngenta-us.com/thrive/production/zeon-technology-insecticide-performance.html. Accessed 8 June 2019

Tadros TF (2009) Colloids in agrochemicals. In: Tadros TF (ed) Colloids and interface science series, vol 5. Wiley, Weinheim

Tang C, Li Y, Pun J et al (2019) Polydopamine microcapsules from cellulose nanocrystal stabilized Pickering emulsions for essential oil and pesticide encapsulation. Coll Surf A 570:403–413. https://doi.org/10.1016/j.colsurfa.2019.03.049

Tang J, Quinlan PJ, Tam KC (2015) Stimuli-responsive Pickering emulsions: recent advances and potential applications. Soft Matter 11:3512–3529. https://doi.org/10.1039/C5SM00247H

United Nations, Department of Economic and Social Affairs, Population Division (2019). World Population Prospects 2019: Highlights. ST/ESA/SER.A/423

Urch H, Franz D, Nolte M et al (2018) Microcapsules comprising cinmethylin in the core and a polyurea derived from diphenylmethane diisocyanate or an oligomer thereof. PCT Application WO 2018/130589, 19 Jul 2018

Van Koppenhagen JE, Scher HB, Lee KS et al (2003) Base-triggered release microcapsules. European Patent 1,100,326, 23 Apr 2003

Vemmer M, Patel AN (2013) Review of encapsulation methods suitable for microbial biological control agents. Biol Contr 67:380–389. https://doi.org/10.1016/j.biocontrol.2013.09.003

Vinceković M, Jalšenjak N, Topolovec-Pintarić S et al (2016) Encapsulation of biological and chemical agents for plant nutrition and protection: chitosan/alginate microcapsules loaded with copper cations and *Trichoderma viride*. Agric Food Chem 64:8073–8083. https://doi.org/10.1021/acs.jafc.6b02879

Wagenblast G (2009) Method for producing polymer nanoparticles. US Patent 2009/0053272, 26 Feb 2009

Wang Y, Kohane D (2017) External triggering and triggered targeting strategies for drug delivery. Nat Rev Mater 2:17020. https://doi.org/10.1038/natrevmats.2017.20

Wege PJ, Hoppe MA, Bywater AF et al (1999) A microencapsulated formulation of lambda-cyhalothrin. In: Robinson WmH, Rettich F, Rambo GW (eds) Proceedings of the 3rd international conference on urban pests, Prague, 19–22 July 1999, pp 301–310

Wei M, Gao Y, Li X et al (2017) Stimuli-responsive polymers and their applications. Polym Chem 8:127–143. https://doi.org/10.1039/C6PY01585A

Wertz JT, Kobilka BM, Porter JT et al (2019) Light-triggered sustained release of pesticides. US Patent 2019/0350194, 21 Nov 2019

Wilson S (2010) Microencapsulated insecticide with enhanced residual activity. PCT Application WO 2010/101820, 10 Sept 2010

Wilson SL, Boucher RE (2014) Microencapsulated insecticide formulations. US Patent 8,916,520, 23 Dec 2014

Winder RS, Wheeler JJ, Conder N et al (2003) Microencapsulation: a strategy for formulation of inoculum. Biocont Sci Technol 13:155–169. https://doi.org/10.1080/0958315021000073439

Yaakov N, Mani KA, Felfbaum R et al (2018) Single cell encapsulation via Pickering emulsion for biopesticide applications. ACS Omega 3:14294–14301. https://doi.org/10.1021/acsomega.8b02225

Yang XZ, Dou S, Sun TM et al (2011) Systemic delivery of siRNA with cationic lipid assisted PEG-PLA nanoparticles for cancer therapy. J Control Release 156:203–211. https://doi.org/10.1016/j.jconrel.2011.07.035

Yu B, Hsu SH, Zhou C et al (2012) Lipid nanoparticles for hepatic delivery of small interfering RNA. Biomaterials 33:5924–5934. https://doi.org/10.1016/j.biomaterials.2012.05.002

Yu M, Sun C, Xue Y et al (2019) Tannic acid-based nanopesticides coating with highly improved foliage adhesion to enhance foliar retention. RSC Adv 9:27096–27104. https://doi.org/10.1039/C9RA05843E

Zedda A, Ferri G, Sala M (2004) Preparation of sterically hindered amine esters. US Patent 6,677,451, 13 Jan 2004

Zhang G, Wang C (2016) Pickering emulsion-based marbles for cellular capsules. Materials 9:572. https://doi.org/10.3390/ma9070572

Zhang Z, Zhang B, Grishkewich N et al (2019) Cinnamate-functionalized cellulose nanocrystals as UV-shielding nanofillers in sunscreen and transparent polymer films. Adv Sustain Syst 3:1800156–1800161. https://doi.org/10.1002/adsu.201800156

Zheng Y, Hu Y, Yan S et al (2019) A polymer/detergent formulation improves dsRNA penetration through the body wall and RNAi-induced mortality in the soybean aphid *Aphis glycines*. Pest Manag Sci 75:1993–1999. https://doi.org/10.1002/ps.5313

Zohar-Perez C, Chet I, Nussinovitch A (2005) Mutual relationships between soils and biological carrier systems. Biotechnol Bioeng 92:54–60. https://doi.org/10.1002/bit.20574

Zuckerman JE, Davis ME (2015) Clinical experiences with systemically administered siRNA-based therapeutics in cancer. Nat Rev Drug Discov 14:843–856. https://doi.org/10.1038/nrd4685

# Correction to: Nanopesticides: From the Bench to the Market

**Rafael Pires-Oliveira, Marta Simão Kfouri, Bruna Mendonça, and Poliana Cardoso-Gustavson**

## Correction to:
## Chapter 11 in: L. F. Fraceto et al. (eds.), *Nanopesticides*, https://doi.org/10.1007/978-3-030-44873-8_11

This book was inadvertently printed without the following text on page 337:

"actives, around 2 years (Bullock 2019). Indeed, the lower shelf life of microbial products is still faced as a hamper to the development and commercialization of this segment of agricultural biologicals.

4.2 Encapsulation of Microorganisms

An effective delivery system plays an important role for successful biological control. Formulation is recognized as one of the most important priorities in biopesticide research (Castillejos et al. 2002; Damalas and Koutroubas 2018). Based on current market trends, research into new formulation methods has notably increased"

This is now corrected by including the missed lines in the chapter.

The updated online version of this chapter can be found at
https://doi.org/10.1007/978-3-030-44873-8_11


L. F. Fraceto et al. (eds.), *Nanopesticides*,
https://doi.org/10.1007/978-3-030-44873-8_12

# Index




L. F. Fraceto et al. (eds.), *Nanopesticides*,
https://doi.org/10.1007/978-3-030-44873-8

**T**

The manufacturer's authorised representative in the EU is Springer Nature Customer Service Centre GmbH, Europaplatz 3, 69115 Heidelberg, Germany. If you have any concerns regarding our products, please contact ProductSafety@springernature.com

Printed and bound by CPI Group (UK) Ltd, Croydon, CR0 4YY
05/10/2026
02228126-0012